Managing Editor:	Brenda Owens
Senior Production Editor:	Bob Dreas
Cover Designer:	Tammy Norstrem
Text Designer:	Leslie Anderson
Production Specialists:	Tammy Norstrem, Ryan Hamner
Indexer:	Ina Gravitz

Care has been taken to verify the accuracy of information presented in this book. However, the authors, editors, and publisher cannot accept responsibility for Web, e-mail, newsgroup, or chat room subject matter or content, or for consequences from application of the information in this book, and make no warranty, expressed or implied, with respect to its content.

Trademarks: Some of the product names and company names included in this book have been used for identification purposes only and may be trademarks or registered trade names of their respective manufacturers and sellers. The authors, editors, and publisher disclaim any affiliation, association, or connection with, or sponsorship or endorsement by, such owners.

Acknowledgments: The author, editor, and publisher wish to thank the following individuals for their insightful feedback during the development of this text:

• Dr. Jim DeKloe, Co-Director, Biotechnician Program, Solano Community College
• Dr. Toby Horn, Co-Director, Carnegie Academy for Science Education, Carnegie Institute of Washington
• Simon Holdaway, MS, Molecular Biology & Microbiology Instructor, The Loomis Chaffee School
• Brian Robinson, PhD, Cell Biology, MD candidate, Emory University School of Medicine

We have made every effort to trace the ownership of all copyrighted material and to secure permission from copyright holders. In the event of any question arising as to the use of any material, we will be pleased to make the necessary corrections in future printings. Thanks are due to the aforementioned authors, publishers, and agents for permission to use the materials indicated.

ISBN 978-0-76386-806-2

Biotechnology
Science for the New Millennium
Second Edition

Ellyn Daugherty

EMC Publishing PARADIGM EDUCATION SOLUTIONS

ST. PAUL, MINNESOTA

To exert a lasting effect on their students and their colleagues, science educators have to be "in it for the long haul." My husband, Paul Robinson, has wielded a significant influence on science education by inspiring his own students (over 7,000 to date) and by acting as a mentor and role model for hundreds of other science teachers. This book is dedicated to him and to one of our best science students, his son Brian.

Photo by Kainaz Amaria.

About the Author

A 34-year veteran biology teacher, Ellyn Daugherty began teaching biotechnology in 1988. Ellyn founded the San Mateo Biotechnology Career Pathway (**www.SMBiotech.com**) in 1993. Her model curriculum attracts students into an intensive, multiple-year program in biotechnology that leads them to higher education and into the biotechnology workplace.

Ellyn has received several awards for her innovative teaching and curriculum development, including:

- BayBio Pantheon Award, Biotechnology Educator, San Mateo Biotechnology Career Pathway, 2010

- The National Biotechnology Teacher-Leader Award, Biotechnology Institute and Genzyme, 2004

- Presidential Award in Science Education, California State Finalist, 2000

- Intel Innovations in Teaching Award, California State Runner-Up, 2000

- Tandy Technology Prize, Outstanding Teacher Award, 1997

- LaBoskey Award, Stanford University, Master Teacher Award, 1995

- Access Excellence Award, NABT and Genentech, Inc., 1994

- National Distinguished Teacher, Commission on Presidential Scholars, 1992

Ellyn retired from the classroom in 2013 but still works with teachers and schools to improve biotechnology education. Ellyn believes strongly in teacher professional development and conducts several workshops a year in her lab and at national conferences. Her website (**www.BiotechEd.com**) contains a collection of teacher support materials and information about upcoming workshops.

An avid sports fan, Ellyn spends her time outside of the lab at baseball and basketball games, golfing with friends and family, or hiking with her husband, Paul, and their chihuahua, Rocky Balboa.

Contents

5 Introduction to Studying Proteins 148

Unit 2 Modeling the Production of a Recombinant DNA Protein Product 179

6 Identifying a Potential Biotechnology Product...................... 180

7 Spectrophotometers and Other Analytical Tools 212

8 The Production of a Recombinant Biotechnology Product...................... 238

Contents

12 Medical Biotechnologies.....................368

Biotech Careers: Genetic Nurse Practitioner 368

13 DNA Technologies............................402

Biotech Careers: Forensic Scientist/DNA Analyst................402

14 Biotechnology Research and Applications: Looking Forward 436

Biotech Careers: Research Scientist 436

Preface

Although I did not know it at the time, the work on *Biotechnology: Science for the New Millennium* began even before my biotechnology career pathway program started in 1994.

In 1983, as a first-year public school science teacher, I wanted my students to "do science," not just learn about science. In my classes, virtually every day was a lab day punctuated regularly with student sounds of delight at the sight of streaming *Amoeba*, multitudes of slimy bacteria, and the ever-lengthening intestine during fetal pig dissection. So why was it then, that every February, during the annual science fair, even my best students had trouble designing a novel, challenging, original research question and experimental plan? No matter how hard I tried, it was not uncommon each year to have dozens of "How much alcohol will inhibit a mouse's ability to go through a maze?" or "What pigments are found in roses?" Even my AP biology students rarely came up with something more sophisticated than "How does caffeine affect the heart rate of *Daphnia*?" That is, unless their parents worked in science. Then the questions were more demanding, such as, "What are the enzyme kinetics in the activation of cyclin E during rabbit pregnancy, as measured by an HPLC?"

I had mixed feelings about my success teaching biology and AP biology. My students were exposed to a wide variety of biological explorations but had few opportunities to study any subject in enough depth to ask insightful scientific questions of their own. Even though my students achieved high scores on all their standardized tests, they knew a tiny bit about a whole group of subjects and processes but were the masters of none. I often grappled with how to juggle each topic so that my students could build on their past experiences and gain some lab proficiencies.

In April of 1988, at about the same time local companies such as Chiron and Genentech were bringing the first products of genetic engineering to market, I received a flyer in the campus mail advertising a summer course, *Recombinant DNA Technology*, sponsored by Edvotek, Inc and the National Science Foundation (NSF). It caught my attention due to the news of recent advances in gene studies and manipulation. Dr. Jack Chirikjian (Georgetown University), Dr. Carol Chihara (University of San Francisco), and her assistant, Mary Connoly, taught the four-week course. I hadn't taken a "real" college course in almost 10 years. After the initial shock of Carol's expectations wore off, I learned more molecular biology than I thought possible. More important, I was given training using some easy-to-follow biotechnology lab kits supplied by Edvotek. That tiny bit of experience gave me the courage to bring these simple biotechnology labs to my students when I returned to school that fall.

Though my courage was abundant, the funds to run the biotechnology labs were not. At the time, enrollment in each biology class hovered at around 32 students with funding each year at about $3.33/student. How could all those students squeeze around one gel box? Not easily. And how was I to find the funds to purchase more kits and equipments? I approached our Assistant Superintendent Tom Mohr (the best administrator I have ever known) and proposed an extracurricular *Biotechnology* course for gifted and talented students to be taught on 22 Wednesday nights and Saturdays throughout the year. Tom somehow found $3,000 in one of his school accounts to purchase more gel boxes, power supplies, and lab kits. For five years *Biotechnology* attracted more students than the supplies could serve. These lucky students gained an appreciation for the techniques used in biotechnology research as well as some basic molecular biology laboratory skills. They had the opportunity to do some fairly sophisticated activities and were required to ask science fair questions that built on skills they learned in the class. The course was very popular and the students, to some extent, fulfilled my objectives of producing original basic research. However, the class served only our very best students, those who might end up earning PhDs regardless of what I did.

Meanwhile, the biotechnology industry was growing at a feverish pace in the San Francisco Bay Area. By 1993, I began looking for another way to bring biotechnology to a greater number and larger diversity of students. The biotech industry was beginning to shift to manufacturing, and it was obvious that in the near-future, academic and career opportunities would be opening up for all types of motivated and appropriately trained lab workers, including technicians, research associates, and scientists.

Birth of a Program

In 1994, with the support of Tom Mohr, school administrators, the San Mateo County Regional Occupational Program, and several teachers, I developed a vision for a career pathway program in biotechnology that would fuse academic and technical training with workplace experiences. We wrote grants and assembled an advisory committee composed of scientists, biotech business people, community leaders, parents, school administrators, teachers, and students. The ultimate goal was to prepare all students, including honors and previously under-motivated but able students, to pursue entry-level, technical-level, or professional-level careers, following an intensive high school and/or college focus in biotechnology.

In the Fall of 1994, the San Mateo Biotechnology Career Pathway (SMBCP) was born as an attempt to increase science literacy and preparedness for employment. It allows students to focus their studies and gain valuable skills that are directly related to advanced science courses and the biotechnology workplace. To date, more than 9,000 students have entered the pathway and completed at least one semester of biotech classes and more than 900 have completed 4–8 semesters, including a 180-hour industry laboratory internship. Until recently, approximately 25 percent of our students were adults who took courses in the evenings and summers. Now adults are directed to similar programs at local colleges. This makes room for the over 400 high school students a year in one of our 11 biotech classes taught by our four SMBCP teachers. Some 20 percent of our students have moved on to some type of part-time or full-time employment at one of our local biotechnology companies. Each student has left the program with a better understanding of how biotechnology is done because they do it. Biotech students are also better prepared to make informed decisions about their academic direction and career opportunities.

Scope and Sequence of the Curriculum

The preliminary drafts of the book and laboratory manual were developed and have evolved during the 22 years since the inception and implementation of the SMBCP. These materials are grounded in the philosophy that the concepts of science support the processes of science, and that information is accessed and learned as it is needed to conduct more experimentation and analysis.

Although the textbook and laboratory manual appear traditional in many ways, the lab manual is a "staircase" of skill development activities, each building upon previous ones, allowing students to develop and demonstrate their proficiency before moving to the next level. The text provides background to prepare students for their lab experiences as well as information to give students perspective on how and where the techniques and technologies are used in biotechnology research and manufacturing. Whether students continue in the science or business of biotechnology, or if they pursue other interests, they will be well informed citizens who can better evaluate the growing number of bioethical and bioeconomical issues.

Four Areas of Focus

Biotechnology: Science for the New Millennium is loosely grouped into three focus areas. Chapters 1–5 are what I call "SLOP," the standard lab operating procedures that every lab worker must master if he or she goes into an academic or corporate lab in pharmaceutical, agricultural, industrial, or instrumentation biotechnology. SLOP includes safety, documentation, following oral and written directions, experimental design, data analysis and reporting, volume measurement, mass measurement, solution and dilution preparation, sterile technique, cell culture, DNA isolation and analysis, and protein isolation and analysis.

Chapters 6–9 focus on the production of recombinant proteins (The rAmylase Project) created through the use of genetic engineering and recombinant DNA technologies. This unit includes the use of assays and assay development with a considerable focus on spectrophotometers and their use to quantify molecules as well as the use of ELISA and Western blotting. Students learn scale-up of transformed production organisms and the methods of product purification. Quality control, regulation, and marketing are presented in this section.

The third area of focus covers the applications of biotechnologies to fields that improve the quality of life for all humankind. Agricultural and medical biotechnologies have shown the most growth in the short history of the industry. Chapters 10–12 spotlight traditional as well as recently developed technologies for creating new and novel crops and medicines. These include plant breeding, asexual plant propagation and tissue culture, plant DNA and protein studies, plant genetic engineering, and the creation of plant-based pharmaceuticals. Chapters 13 and 14 introduce some of the most recent advances in DNA and protein studies and diagnostics as well as some of the new cutting-edge technologies applied to medicine, stem cell biology, environmental studies, and biodefense. Key topics are DNA synthesis, PCR, DNA sequencing, genomics, microarrays, proteomics, genetic testing, biofuels, and bioremediation.

Although the language of biotechnology can make reading a text an overwhelming challenge, I have attempted to write in the same student-friendly tone in which I teach, and have interspersed the text with many excellent photos and illustrations. Definitions of important vocabulary are located in the margins of the text to help the reader better understand the terminology. To expand student understanding and interests, each chapter offers *Biotech Online* Internet research activities, as well as more extensive *Biotech Live* explorations in the *Chapter Review*. One or more *Bioethics* activities at the end of each chapter help students think about the implications and applications of biotechnological advances for individuals and for society.

Using the Program Digital Resources

The digital resources packaged within the text and lab manual eBooks offer valuable resources to reinforce mastery of the concepts and skills taught in the printed products. Presented in a multimedia format, the eBooks include the following elements:

- **Lab Tutor:** a group of # tutorials that teach critical lab skills using audio, video, and photos; each tutorial concludes with 3–4 brief questions to check students' understanding of key points.
- **Glossary and Image Bank:** a database of the "Speaking Biotech" terms highlighted in bold and defined in the margins of each chapter of the text; selected terms link to full-color illustrations, displaying them along with the definition of the term when the user clicks that word in the list.
- **Quizzes:** a multiple-choice quiz for each chapter of the text and a Book Quiz that includes questions for each of the 14 chapters; two modes of quiz taking are available—in Practice mode, students can take each quiz as often as they like with the score reported to the student only, and in Test mode, the results are reported to both the instructor and the student by e-mail.
- **Flash Cards:** interactive Flash cards of chapter key terms
- **Crossword Puzzles:** a fun learning tool that checks students' understanding and recall of chapter concepts and terms
- **Internet Resource Center:** a link to comprehensive resources for instructors and students on the publisher's website

Growing the Curriculum

The *Biotechnology: Science for the New Millennium* curriculum has been used by my students and the students of other pathway teachers and has been modified annually using their feedback and the suggestions of advisory committee members from the SMBCP partner companies. Each part is reviewed in light of how well the material prepares our students for further academic, research, or manufacturing experiences in biotechnology. I welcome your comments and feedback, especially in order to improve the curriculum's quality and effectiveness. We are in the Age of Biotechnology and I foresee a time when biotechnology is taught in every high school, community college, and career or technical college. I hope this curriculum will enhance and accelerate that process.

Acknowledgments

I might never have started teaching biotech if Dr. Jack Chirikjian, and Dr. Carol Chihara hadn't let me into their Recombinant DNA workshop in 1998. When Jack gave participants 10 Edvotek, Inc lab kits, a gel box, and a power supply to take back to their classrooms, those 10 labs snowballed over 28 years into this 4-year curriculum. Even with all of their efforts, my science was rather weak (it had atrophied during my first 10 years of teaching) and I needed to relearn virtually everything from college plus all the "new" biotechnology. I gratefully thank Patricia Seawell, formerly of Gene Connection and Frank Stephenson, PhD of ThermoFisher Scientific/Applied Biosystems for the many mini-courses over the phone, online, and in the lab to help me get my science "right" and for providing support and reagents when I was testing new curricula. Frank was also the first person to brave the first draft of my manuscript. He deserves an award for that.

Maureen Munn, PhD, Project Director of the Human Genome Program, University of Washington Genome Center, and Lane Conn, former Director of the Teacher Education in Biology program at San Francisco State University, spent many hours helping me bring DNA synthesis and DNA sequencing activities to my students. Both Maureen and Lane have had an enormous impact on my curriculum and on biology education in general, in-servicing many hundreds of teachers in recombinant DNA and DNA sequencing workshops. Maureen also read and gave feedback on some sections of my manuscript. Brock Siegel, formerly of Applied Biosystems, Inc has also been a champion of biotechnology education and my program.

I am indebted to Diane Sweeney, formerly of Genencor International and now at Sacred Heart Preparatory, for the extensive teacher training and in-service she has conducted. Two of her Amylase Project labs, which she shared with me in a workshop in 1989, became the cornerstone for a few dozen amylase activities in my curriculum.

Several teachers used the early drafts of the text and lab manual and gave me valuable input. I want to thank Leslie Conaghan, Karen Watts, Josephine Yu, PhD, Tina Doss, and Dan Raffa for their constant and considerable contributions and corrections. Dan and Tina also provided technical advice regarding instructional materials and spent the better part of a summer creating the lab skills tutorials for the Encore CD that was packaged with the first edition of the text and lab manual. I am particularly appreciative of Jimmy Ikeda, my SMBCP teaching lab partner. He is the best "lab husband" a lab teacher could have. We worked very closely on all SMBCP program matters and he has helped me out of several jams including class coverage, materials acquisition and preparation, and facilities maintenance. In recent years, Jimmy's Independent Research students field tested several of the new lab activities in the second edition lab manual. I thank them all for their efforts testing and improving this curriculum.

I am immensely grateful to the science content editors of *Biotechnology: Science for the New Millennium*. Each of them accepted the daunting challenge of reading and reviewing almost 900 pages of text and labs in a short time. A huge thank you to Dr. Jim DeKloe, Co-Director, Biotechnician Program, Solano Community College; Dr. Toby Horn, Carnegie Academy for Science Education, Carnegie Institute of Washington, Dr. Brian Robinson, MD/PhD, Emory University School of Medicine, and Simon Holdaway, MS, Molecular Biology & Microbiology Instructor, The Loomis Chaffee School. I was extremely fortunate to have met Simon Holdaway, bioscience teacher/professor extraordinaire. In the 15 years I have known him, Simon has shown me how to do biotech better, faster, and cheaper. For this 2nd edition, Simon helped me identify the strengths and the weaknesses in earlier editions and helped me plan the direction to take for this one. SImon created the new pAmylase2014 plasmid used in several of the labs. Simon contributed three new labs to the lab manual and produced the new pAmylase plasmid that makes recombinant alpha-amylase, faster and better. Simon really knows his science and has helped me get that right as he served as the science reviewer.

Regarding technical assistance, the creation of the manuscript would not have been possible without the untiring efforts of my computer technician and exceptional Webmaster, Skip Wagner. At any time of the day or night, Skip can be counted on to drop everything and solve my technology crisis. In addition, a huge hug and lots of love to my mom, Lorna Kopel, who read the first editions of the text, lab manual, and instructor's guide and made grammatical corrections.

Many scientists, teachers, students, and colleagues have worked on curriculum development with me, or have provided feedback or scientific advice and support on one or more topics or techniques. I would like to thank Katy Korsmeyer, PhD, Maria Abilock, Shalini Prasad, Joey Mailman, Aylene Bao, Natasha Chen, Daniel Segal, and Luhua Zhang for their efforts in helping me complete the first draft of the manuscript. Aylene Bao was the original illustrator of the manuscript, and her extensive collection of drawings served as excellent models for the illustrators at Precision Graphics who created the beautiful drawings in the text and lab manual. Maria Abilock wrote the original test bank questions and suggested several changes as she read through the manuscript. In addition, thank you to Dr. Timothy Gregory of Genentech, Inc for letting me be the first teacher intern at Genentech, Inc and allowing me to gain real science skills for two summers with Lavon, Allison, Millie, and Dave in the Protein Process Development Department.

I am particularly appreciative of Sandra Porter, PhD, President, Digital World Biology LLC. Sandy taught me how to use and teach others to use the molecular modeling applications used in this edition. Sandy also developed a molecular model database to support this curriculum found at **digitalworldbiology.com/dwb/structure-collections**.

The activities in the lab manual are significantly challenging because of the extensive amount of reagents and equipment. Likewise, assembling a good working version of the laboratory materials list was one of the biggest challenges in this project. Many individuals have worked hundreds of hours over several years helping me determine the materials that would provide the best performance and value to my teacher users. I appreciate the hard work of all of them, but especially Amy Naum who had worked tirelessly to make sure the original materials lists were current and accurate. Amy supported the research and development of several labs in this revision and has been a champion in helping science teachers develop better lab skills.

For this edition, I am indebted to both Colin Heath, of G-Biosciences and Lindsay Kotula of Fisher Science Education. Among other projects supporting the *Biotechnology: Science for the New Millennium* lab manual, Colin and I collaborated to develop The rAmylase Project kits that support Chapters 6–9 in the curriculum. These 10 lab kits make it easier to bring a thematic lab experience in recombinant DNA protein production to any high school or college student. Colin has spent countless hours in the lab testing reagents for new or improved lab activities. I appreciate Lindsay's time and effort to create the current *Biotechnology: Science for the New Millennium* laboratory materials list that facilitates educators with the challenging task of finding the right lab materials of good quality and value.

Finding a publishing company that wanted to be a pioneer in a new area of science education was not an easy task. I was extremely fortunate to find, early on, the right publisher. I thank Dr. Elaine Johnson of Bio-Link and Kristin Hershbell Charles of City College of San Francisco for all their efforts to support and promote my program, and especially for connecting me with EMC Paradigm. I would also like to thank John Simpson for his extensive and valuable advice about getting my work published.

I am so grateful to the publishing teams at EMC School and Paradigm Education Solutions: Carley Bomstad, Mick Demakos, Brenda Owens, Deanna Quinn, Bob Dreas, and Sonja Brown. I particularly appreciate their patience as I lobbied for each item that I thought was critical for the successful implementation of this curriculum by my teacher users. "Whoa Biotech!"

In 1997, my husband, Paul Robinson, science teacher extraordinaire, casually commented, "You will not be able to personally help *everyone* start their biotechnology programs. You'd better write your stuff into a book." Thanks, honey!

Ellyn Daugherty
San Francisco, California

Biotechnology
Science for the New Millennium
Second Edition

Unit 1
Biotechnology Standard Laboratory Operating Procedures

Biotechnology is a broad term that includes the study and manipulation of cells and molecules in a laboratory setting for the purpose of producing products or services that will improve human life. Common biotechnologies are used in medical, agricultural, industrial, environmental, and research applications. A researcher or technician In a biotechnology laboratory environment must master these basic concepts and skills to be able to work in a productive manner as part of a research or manufacturing team in any setting.

The basic biotechnology standard laboratory operating procedures include being able to:

- Gather background information to understand the biology, chemistry, and engineering needed to work toward a specific lab objective.
- Understand how specific work to be done in a laboratory fits into the organization at a particular biotechnology facility.
- Find and follow oral and written instructions as well as to document and share all procedures used and information and data collected.
- Recognize safety hazards and work safely in a laboratory and business environment.

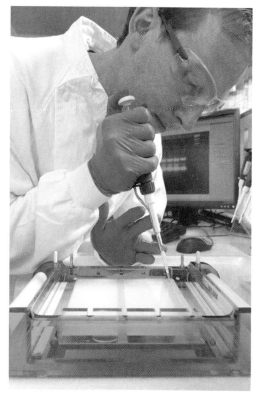

© Andrew Brookes/Getty Images.

- Make accurate measurements of samples and reagents including volume measurement, mass measurement, and sample size.
- Prepare and store reagents, solutions, media, and samples for use in DNA and protein studies.
- Conduct DNA and protein sample extractions from cells and tissues and analyze the samples using indicators and electrophoresis.

In Chapters 1–5, you will learn the basics needed to work in most entry-level biotechnology environments. These chapters serve as a platform from which you can work on specialized projects such as genetically engineering cells to produce commercially important products (Unit 2 - The rAmylase Project) or producing products important in the areas of medicine, agriculture, research, or industry (Unit 3 - Biotechnologies Feed, Heal, and Protect the World).

© Wavebreakmedia/iStock.

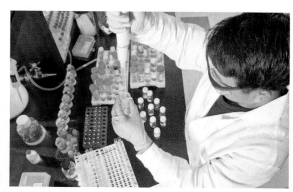

© Charlotte Raymond/Photo Researchers.

Biotech Careers

Photo courtesy of Susan Taillant.

Quality Control Analyst

Susan Taillant
Genentech, Inc.
Vacaville, CA

Susan Taillant works in one of the Genentech Inc., protein-manufacturing facilities. Genentech Inc., uses genetic engineering technology to produce new proteins for medical uses. Working in a large team under the direction of a manufacturing supervisor, Susan is responsible for testing samples at the end of production. She performs general wet chemistry assays (tests) and chromatography (separation) techniques. All testing is performed in compliance with current good manufacturing practices (cGMPs).

The instrument shown in the photo above is a type of high-performance liquid chromatography (HPLC) instrument. It separates molecules based on size, charge, and/or shape, and is used to test the purity of a sample.

Quality control analysts (QCAs) are usually hired after they have earned a Bachelor's degree in biochemistry, chemistry, or molecular biology. Some companies hire QCA candidates with a 1- or 2-year Biotechnician Certificate.

1 What Is Biotechnology?

Learning Outcomes

- Describe the science of biotechnology and identify its product domains
- Give examples of careers and job responsibilities associated with biotechnology
- Outline the steps in producing and delivering a product made through recombinant DNA technology
- Describe how scientific methodologies are used to conduct experiments and develop products
- Apply the strategy for values clarification to bioethical issues

1.1 Defining Biotechnology

Imagine going to a grocery store and having only a single choice of apple, orange, or lettuce to buy. Or picture planting a rose garden with only one type of rose bush available. Suppose you wanted to own a dog, but just three breeds existed. Life without variety could be quite dull. Humans have been manipulating living things for centuries to produce plants and animals with desired characteristics. Cows, goats, sheep, and chickens have a long history of being bred for increased milk, meat, or egg production. Wheat, rye, corn, rice, tobacco, and soybeans have been selectively bred for increased yields and healthier, disease-resistant plants. Seedless watermelons, a large variety of apples, roses of many colors and fragrances, and the huge assortment of domesticated dogs are some examples of our continuing efforts to diversify our surroundings (see Figures 1.1, 1.2, and 1.3). For the most part, we have created organisms that have both benefited people and improved our quality of life.

Scientists recently have learned to manipulate not only whole organisms, such as plants and animals, but also the molecules, cells, tissues, and organs of which they are built. In the past few decades, we have learned to manufacture large amounts of specific molecules of scientific or economic interest, such as human **insulin** made in bacteria cells to treat diabetic patients. Using cells to manufacture specific molecules is one example of the scientific field called biotechnology.

insulin (in•sul•in) a protein that facilitates the uptake of sugar into cells from blood

3

Figure 1.1. Among the more than 100 breeds of dogs, the Chihuahua and the Great Dane illustrate the variety in size that has resulted from selective inbreeding.
Photo by author.

Figure 1.2. Eight different varieties of apples are commonly available at most grocery stores. Each is a human-engineered variant of the same species.
© Royalty-Free/Corbis.

biotechnology (bi•o•tech•nol•o•gy) the application of science and engineering to manipulate cells and their components to produce products or services that may improve the quality of human life

DNA abbreviation for deoxyribonucleic acid, a double-stranded helical molecule that stores genetic information for the production of all of an organism's proteins

recombinant DNA (rDNA) technology (re•com•bi•nant DNA tech•nol•o•gy) cutting and recombining DNA molecules

polymerase chain reaction (PCR) (po•ly•mer•ase chain re•ac•tion) a technique that involves copying short pieces of DNA and then making millions of copies in a short time

cloning (clon•ing) a method of asexual reproduction that produces identical organisms

fermentation (fer•men•ta•tion) a process by which, in an oxygen-deprived environment, a cell converts sugar into lactic acid or ethanol to create energy

diabetes (di•a•be•tes) a disorder affecting the uptake of sugar by cells, due to inadequate insulin production or ineffective use of insulin

proteases (pro•te•as•es) proteins whose function is to break down other proteins

antibodies (an•ti•bod•ies) proteins developed by the immune system that recognize specific molecules (antigens)

Figure 1.3. Selective breeding, one of the oldest forms of biotechnology, has resulted in hundreds of varieties of roses with a large range of colors.
Photo by author.

Biotechnology is the study and manipulation of living things or their component molecules, cells, tissues, or organs to create products or services that benefit humankind. Biotechnology is an expansive field that includes many modern techniques that involve **deoxyribonucleic acid** (DNA) and proteins, including **recombinant DNA (rDNA)** technology (cutting and recombining DNA molecules), **polymerase chain reaction** (copying short pieces of DNA), DNA sequencing, and **cloning** (producing identical organisms). In the broadest sense, biotechnology includes many practices that have been in use for thousands of years, such as selective breeding and the **fermentation** of certain beverages and foods. However, the term "biotechnology" is relatively new; it has been used since the 1970s to reflect the application of exciting new technologies to the research and development of products from plant and animal cells. Advances, such as the ability to "cut and paste" DNA, allowed biotechnology companies to manufacture a wide variety of products that were either previously unavailable or could only be made in small quantities. Examples include human insulin for the treatment of **diabetes**, **proteases** used in many applications, including removing stains from clothing, **antibodies** for recognizing and fighting certain diseases, and enzymes for specialty apparel, such as stonewashed denim jeans (see Figure 1.4). Today, it is common to speak of the science of biotechnology as well as the biotechnology industry.

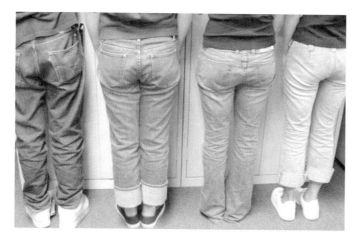

Figure 1.4. Denim jeans with a faded look were originally called "stonewashed" because they were created by washing blue jeans with rocks in the washer. Damage to the machines resulted in a high price tag for the denim product. Now, genetically engineered enzyme products, such as IndiAge® by Genencor International Inc., cause a faded look without battering the washing machine.
Photo by author.

Figure 1.5. Although not currently available to the consumer, a genetically modified tomato called FLAVR SAVR® by Calgene, Inc. was created in the 1990s to allow tomatoes to ripen on the vine longer for better flavor. Now, new versions of genetically modified tomatoes are being developed. Genetically modified tomatoes may be more uniform, stay fresh longer, and taste better.
© Tony Freeman/Photo Researchers.

Figure 1.6. Human ears are sometimes lost through accidents. Mouse cells can be "tricked" into growing the outer portion of the human ear from stem cells grown on a titanium frame. The new outer ear is then surgically transferred to the human patient. The titanium gives the new ear flexibility like the original ear.
© Associated Press Photo/Xinhua.

Researchers in biotechnology apply laboratory techniques from the fields of biology, chemistry, and physics. They use mathematics and computer skills to process and analyze data. Biotechnologists most commonly work in commercial organizations, such as companies that make **pharmaceutical**, agricultural, or industrial products, or medical or industrial instruments. The goal of most biotechnology companies is to manufacture a product that is useful to society. Some examples are a plant, such as a new variety of tomato (see Figure 1.5); an animal, such as a goat that produces a human pharmaceutical; an organ, such as a human ear grown on a mouse's back (see Figure 1.6); or a molecule, such as human growth hormone. No matter what the product, it must generate enough profit for the company to function and to fund research on additional new products.

**pharmaceutical
(phar•ma•ceu•ti•cal)** relating to drugs developed for medical use

Biotechnology Workers and the Biotechnology Workplace

Biotechnology is practiced in several different settings, including companies, universities, and government agencies. In general, the setting determines the major emphasis: scientific research or the development and manufacture of products. Universities and government agencies tend to focus on research, while private companies focus on developing and manufacturing products for sale. To develop products, however, companies also conduct extensive research.

Biotechnology Companies   Thousands of biotechnology companies around the world produce and sell a wide variety of products. An example is lovastatin, a drug isolated from a strain of the fungus *Aspergillus terreus (A. terreus)*, which is used to treat high cholesterol (see Figure 1.7). The goal of every biotechnology company is

Figure 1.7. Paul, a well known physics teacher and San Francisco Giants baseball enthusiast, suffers from high cholesterol, hypercholesterolemia. Lovastatin keeps his cholesterol below 200, lowering his risk for heart attack or stroke. To learn more about hypercholesterolemia visit www.ibiology.org/ibiomagazine/michael-brown-joseph-goldstein-hypercholesterolemia.html
Photo by author.

Figure 1.8. Many biotechnology companies produce protein products in cells grown in large fermentation tanks, such as those used in this cancer vaccine manufacturing facility.
Photo courtesy of Cell Genesys, Inc.

research and development (R&D) (re•search and de•vel•op•ment) the early stages in product development that include discovery of the structure and function of a potential product and initial small-scale production

pure science (pure sci•ence) scientific research whose main purpose is to enrich the scientific knowledge base

to produce and sell commercial "for-profit" products. In this way, a company can retain valuable employees and continue to invest in the **research and development (R&D)** of future products.

At most companies, some scientific staff (scientists, research associates, and lab technicians) conduct basic research, but usually a much greater number work toward applying science to product development or manufacturing. In addition, a large number of nonscientific staff (administrators, clerical workers, and sales and marketing representatives, etc) support research and product development to ensure the success of a product in the marketplace. A walk through a typical biotechnology company takes you through several different workspaces, including laboratories, manufacturing facilities, and offices (see Figure 1.8).

Different companies, of course, produce different products. Most biotechnology companies fall into one of the following four categories based on the products they make and sell:

1. Pharmaceutical/Medical products
2. Agricultural products
3. Industrial/Environmental products
4. Research or production instruments, reagents, or data

Some biotechnology companies sell their services rather than a specific product. For example, an Illinois company called Integrated Genomics, Inc. has customers who hire them to sequence DNA molecules, which means that they determine the genetic message on the DNA molecule. Integrated Genomics, Inc. is so efficient at sequencing that it is more cost-effective for companies to pay it to do the sequencing than to pay their own employees to do it.

University and Government Research Labs Not all biotechnologists work at "for-profit" biotechnology companies. Some biotechnologists work in university labs or at government agencies conducting what is considered **"pure science"** research. Many of the experimental techniques and scientific methodologies used in these facilities are the same as those used at biotechnology companies (see Figure 1.9). The main difference in these workplaces is that companies must provide a product or a service that results in earnings; a nonprofit research facility does not.

University and government researchers apply for grants from industry, foundations, or the government to pay for the research they do. Data are usually the products of such research, and they are shared with others through scientific journal (magazine) articles or at scientific meetings. Results reported in this way add information to the scientific community's information base and are available for "the public good." For instance, at the University of San Francisco, researchers, such as Dr. Carol Chihara (see Figure 1.10), experimented for several years to understand how organisms develop from an embryo into an adult. They used fruit flies in their experiments because fly development has some parallels with human development.

Dr. Chihara's group worked on determining which molecules were present, at what time, and in what amounts. They hoped to see the relationship between the presence of a certain chemical and a specific stage in development. When researchers like these get experimental results that represent significant advances in knowledge, they publish papers in scientific journals. By describing their results publicly, they invite other scientists to scrutinize their work and design further experiments to find out even more about development.

Research centers in several locations may have many scientists working on diverse projects. University and government labs around the world are conducting research on the human immunodeficiency virus (HIV, the virus that causes acquired immunodeficiency syndrome, or AIDS), malaria, and diabetes, as well as working to improve crop yields. For example, the Gladstone Institute at the University of California, San Francisco, is an academic research facility focused on studying **viruses** and viral therapies (see Figure 1.11). Scientists conducting pure or **applied science** can use the results to further research or to provide information for the development of new products.

Observe.
Observing a scientific phenomenon increases curiosity.

Formulate a scientific question.
The question must be testable.

Develop a hypothesis.
Predict the results of experimentation based on past research/experience.

Plan an experiment.
Design a controlled experiment with measurable data.

Conduct experiments.
Do multiple replications of the experiment.

Analyze data and report results.
Analyze data in light of expected results. Report final results in notebooks and scientific journals.

Figure 1.9. **Scientific Methods.** The "scientific method" is a collective term for the techniques that scientific researchers use to provide data and gather evidence to answer scientific questions. One approach to scientific methodology is shown here. More on scientific methodologies is presented in Section 1.4.

Figure 1.10. **Dr. Carol Chihara is Professor Emeritus of Biology at the University of San Francisco. Work in her lab focuses on the characterization of the omega gene, a recessive gene for a protein that modifies fly larva development. Scientists study this gene because it has been shown to affect the duration of larval development and the fertility of adult males. Understanding development in fruit flies is expected to shed light on development in other organisms, including humans.**
Photo courtesy of Dr. Carol Chihara.

What Is Biotechnology?

Figure 1.11. The Gladstone Institute of Virology and Immunology at the University of California, San Francisco. The focus of this research facility is the understanding of HIV and AIDS. Researchers at the Gladstone Institute are working on developing therapies for AIDS patients and vaccines to prevent HIV infection.
Photo courtesy of The Gladstone Institute.

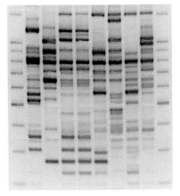

Figure 1.12. DNA fingerprinting technology can be used to study DNA from virtually any organism. Shown above is a gel that contains DNA fragments from human chromosome 22. Studies like this can identify organisms or differences in organisms.
© zmeel.

NIH abbreviation for National Institutes of Health; the federal agency that funds and conducts biomedical research

CDC abbreviation for Centers for Disease Control and Prevention; the national research center for developing and applying disease prevention and control, environmental health, and health promotion and education activities to improve public health

DNA fingerprinting (DNA fing•er•print•ing) an experimental technique that is commonly used to identify individuals by distinguishing their unique DNA code

Researchers at US government laboratories, such as the **National Institutes of Health (NIH)** and the **Centers for Disease Control and Prevention (CDC)**, along with researchers at many universities, use biotechnology research techniques when looking for treatments for major diseases, including heart disease, cancer, and Alzheimer's disease.

Scientists working in small departments of larger organizations where biotechnology may not be their main focus also do biotechnology research. Many forensic scientists, for example, work in police departments. They use biotechnology lab procedures, such as **DNA fingerprinting** (identification of a person's unique DNA code), when they analyze evidence from a crime scene. The OJ Simpson trial, in which the famous ex-football player was accused of murdering his ex-wife and her companion, demonstrated how DNA from blood cells could be used (or, some say, misused) as evidence in a criminal case.

Ecologists may use similar DNA fingerprinting techniques to identify plant or animal breeding partners to control the parentage for protected or endangered species. For example, whooping cranes are severely endangered birds. In 2000, the total North American population of whooping cranes was estimated at only 30. Due to intense breeding programs in Wisconsin and Canada, including DNA testing of all the remaining whooping cranes, the populations had increased to almost 200 birds by the spring of 2005 and up to 600 birds in 2015. Results of DNA tests help scientists determine which birds should be allowed to breed to increase genetic diversity (differences in the DNA code from organism to organism) in the population. Increasing genetic diversity through selective breeding is important because it increases the survival of the whole species.

Wildlife biologists and customs agents identify illegally transported or poached animals through biotechnology techniques. Rhinoceros horns, bear gall bladders, and exotic birds from the South Pacific, all considered "black market" items, have been identified using DNA fingerprinting studies similar to those used in human DNA studies (see Figure 1.12). In other applications, the molecules of related organisms show evolutionary biologists of common ancestry among organisms. After DNA and protein analysis, the red panda of China was shown to be more closely related to the raccoon than to the well known black-and-white panda "bear." This type of research helps explain which animals are most similar to each other and which groups may have arisen from a common ancestor.

Growth in the Biotechnology Industry Although many scientists conduct biotechnology research in universities and government agencies, most biotechnologists work in companies that produce medical instruments and diagnostic tools, drugs, industrial or environmental applications, or new agricultural crops (see Figure 1.13). The number of biotechnology companies is growing dramatically. Until recently, most were located in a few geographic locations: the San Francisco Bay Area, around Boston, Massachusetts, in Madison, Wisconsin, and in North Carolina. Now, biotechnology companies are found in most metropolitan areas. These companies need both scientific and nonscientific support staff. Some companies actually employ over one-half of their workforces in nonscientific positions, such as in marketing, legal, financial, human resources, public relations, computer technology, data analysis, and transportation (see Figure 1.14).

The *San Francisco Chronicle* predicted in March 1990 that the number of jobs in biotechnology would double by the year 2000. In reality, the biotechnology industry grew so rapidly that the growth in jobs surpassed that expectation. According to the Biotechnology Industry Organization, by the end of 2006, there were more than 180,000 biotechnology industry employees in the United States working in nearly 1500

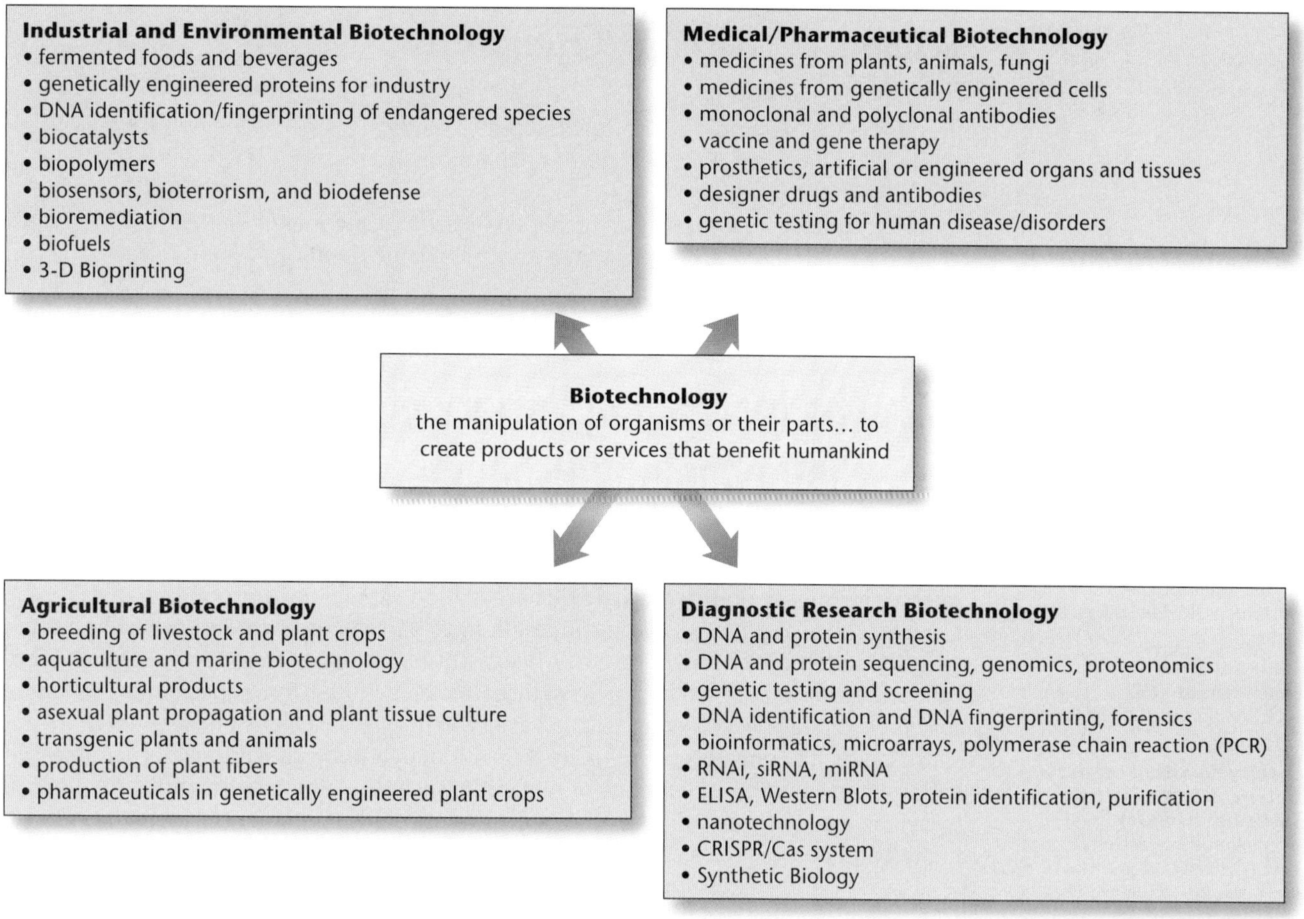

Industrial and Environmental Biotechnology
- fermented foods and beverages
- genetically engineered proteins for industry
- DNA identification/fingerprinting of endangered species
- biocatalysts
- biopolymers
- biosensors, bioterrorism, and biodefense
- bioremediation
- biofuels
- 3-D Bioprinting

Medical/Pharmaceutical Biotechnology
- medicines from plants, animals, fungi
- medicines from genetically engineered cells
- monoclonal and polyclonal antibodies
- vaccine and gene therapy
- prosthetics, artificial or engineered organs and tissues
- designer drugs and antibodies
- genetic testing for human disease/disorders

Biotechnology
the manipulation of organisms or their parts... to create products or services that benefit humankind

Agricultural Biotechnology
- breeding of livestock and plant crops
- aquaculture and marine biotechnology
- horticultural products
- asexual plant propagation and plant tissue culture
- transgenic plants and animals
- production of plant fibers
- pharmaceuticals in genetically engineered plant crops

Diagnostic Research Biotechnology
- DNA and protein synthesis
- DNA and protein sequencing, genomics, proteonomics
- genetic testing and screening
- DNA identification and DNA fingerprinting, forensics
- bioinformatics, microarrays, polymerase chain reaction (PCR)
- RNAi, siRNA, miRNA
- ELISA, Western Blots, protein identification, purification
- nanotechnology
- CRISPR/Cas system
- Synthetic Biology

Figure 1.13. Domains of Biotechnology. Biotechnology includes research and manufacturing of products and services with a focus in 1) industrial/environmental, 2) medical/pharmaceutical, 3) agricultural, and 4) diagnostic/research domains.

biotechnology companies. The biotechnology field continues to grow at an impressive rate with opportunities for all kinds of employees with all types of interests in the science and business of biotechnology and as of 2008, U.S. bioscience employment reached 1.42 million workers. More information about careers in biotechnology is presented in Section 1.5.

Looking Ahead

In upcoming chapters, you will learn about the science and business of biotechnology. You will begin with an introduction to the basic biology and chemistry concepts and laboratory techniques common to every biotechnology environment. These include cell studies and culture, solution and dilution preparation, DNA isolation and analysis, protein studies, laboratory safety, and documentation. In Chapters 6 through 9, you will learn how to produce a recombinant protein product, including how to test for its presence, purity, and activity. In Chapters 10 through 12, you will learn about the applications of biotechnology to the ever-expanding fields of agriculture and pharmaceuticals. Chapters 13 through 14 present information about some of the most recent advances in biotechnology, including new techniques for identifying proteins and DNA, as well as new scientific methods for addressing the important issues of disease, famine, and pollution.

Figure 1.14. As an Application Specialist at a biotechnology company specializing in diagnostic instruments, Philip Huang's duties include direct technical support and training of customers in the use and application of these instruments. A large majority of his time is spent answering phone calls and emails. Huang is also responsible for training the company's staff to provide better customer service.
Photo by author.

Section 1.1 Review Questions

1. What is biotechnology?
2. Name a biotechnology product that has a medical use.
3. Besides biotechnology companies, where can biotechnologists work?
4. Biotechnology companies are grouped into four categories based on the products they make and sell. Name the four categories of products.

antibiotics (an•ti•bi•ot•ics) molecular agents derived from fungi and/or bacteria that impede the growth and survival of some other microorganisms

restriction enzyme (re•stric•tion en•zyme) an enzyme that cuts DNA at a specific nucleotide sequence

DNA ligase (DNA li•gase) an enzyme that binds together disconnected strands of a DNA molecule

recombinant DNA (re•com•bi•nant DNA) DNA created by combining DNA from two or more sources

genetically modified organisms (ge•net•i•cal•ly mod•i•fied or•gan•isms) (GMOs) organisms that contain DNA from another organism and produce new proteins encoded on the acquired DNA

E. coli (E. co•li) a rod-shaped bacterium native to the intestines of mammals; commonly used in genetics and biotechnology

plasmid (plas•mid) a tiny, circular piece of DNA, usually of bacterial origin; often used in recombinant DNA technologies

1.2 The Increasing Variety of Biotechnology Products

In the past 100 years, scientists have increased the pace of searching for products that improve the quality of life. **Antibiotics** are a good example of a group of natural products whose discovery and development have had a significant impact on human longevity and quality of life. Since the 1940s, antibiotics have dramatically reduced the death and suffering caused by bacterial diseases. Penicillin, from a species of the fungus *Penicillium sp.*, has been used to treat a variety of diseases, such as pneumonia and syphilis, which at one time were likely to result in death. Modifications of the molecule have resulted in different versions of penicillin, including amoxicillin and carbenicillin. Physicians use these variations of the penicillin molecule to stop bacteria that may have mutated and become resistant to penicillin. At present, scientists are earnestly pursuing the discovery and/or development of new types of antibiotics (see Figure 1.15).

The identification and use of plant extracts has resulted in many medical and industrial products. For example, rubber extracted from rubber trees enabled the invention of the tire, which fueled industrialization all over the world. Many other plant extracts, including resins, turpentine, and maple syrup, have improved several products or processes for humans.

Bioengineered Products

As the methods of manipulating living things have become more sophisticated, the number and variety of biological products have increased at an incredible pace, characterized by the "snowball" metaphor (see Figure 1.16). By the 1970s, scientists had developed new methods that revolutionized biotechnology research and development, including the use of **restriction enzymes** (for cutting DNA) and the enzyme **DNA ligase** (for pasting pieces of DNA together) to create new combinations of DNA information. New pieces of DNA, pasted together from different sources, are called recombinant DNA (see Figure 1.17). Recombinant DNA can be inserted into cells, giving them new characteristics. These modified cells are called bioengineered or **genetically modified organisms (GMOs)**.

The GMOs contain DNA from another organism and produce new proteins encoded on the acquired DNA. The first GMOs to produce human protein were some Escherichia coli (*E. coli*) bacteria cells that were given pieces of human DNA (genes) containing the instructions to produce a human growth hormone called somatostatin. The somatostatin gene was carried into the *E. coli* cells on tiny pieces of bacterial DNA called **plasmids** (these

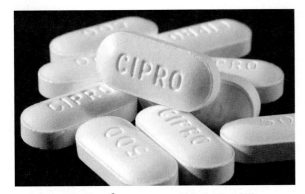

Figure 1.15. Cipro®, a strong antibiotic that kills many bacteria, including anthrax, is one of the a biotechnology antibiotics produced by Bayer Healthcare. Antibiotics can kill many dangerous bacteria, but the more antibiotics are used, the more likely it is that certain bacteria will mutate and survive antibiotic exposure. Resistant strains of bacteria are dangerous and can be lethal if no antibiotic can control them. One example is MRSA. Go to www.webmd.com and learn what MRSA is and why it is so dangerous.
© Alleruzzo Maya.

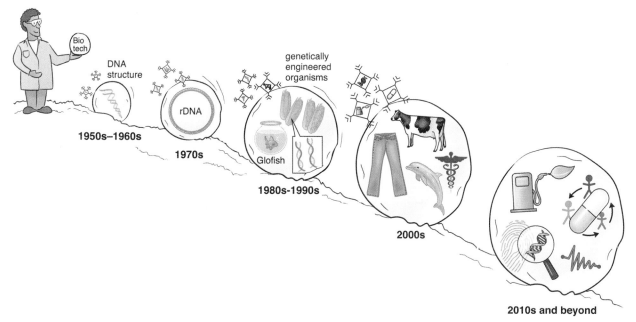

Figure 1.16. **Snowballing Biotech.** Since the 1970s, advances in laboratory techniques, research instruments, and products have "snowballed."

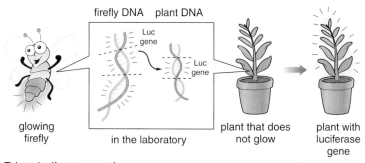

glowing firefly | in the laboratory | plant that does not glow | plant with luciferase gene

Take a luciferase gene from the firefly genome... and transfer it into the plant genome.

Figure 1.17. **Genetically Engineered Plant** Scientists have learned how to take genes that code for certain traits and transfer them from one species to another. The organism that gets the new genes will then have the potential to express the new traits coded in the newly acquired genes.

recombinant DNA plasmids contained both bacterial and human DNA). The *E. coli* cells read the human genes and produced human somatostatin, a hormone that, among other functions, inhibits the release of insulin from pancreas cells.

One of the first genetically engineered products to be sold was human **tissue plasminogen activator (t-PA)**, a blood-clot-dissolving enzyme that can be used immediately after a heart attack to clear blocked blood vessels. The body produces t-PA only in tiny amounts. To produce enough t-PA for therapeutic use, researchers genetically engineered mammalian cells using Chinese hamster ovary (CHO) cells. The ovary cells are grown in culture and given a segment of human DNA with the genetic instructions to make the human t-PA enzyme. Of course, CHO cells do not normally produce human t-PA. Under the right conditions, though, the CHO cells accept and incorporate the new DNA into their own DNA code, read the human DNA, and make the human t-PA protein. Large amounts of t-PA can then be purified from the engineered CHO cells (see Figure 1.18 and Figure 2.15).

Applications of recombinant DNA and genetic engineering technology have resulted in the launching of hundreds of biotechnology companies, specializing in all kinds

t-PA short for tissue plasminogen activator; one of the first genetically engineered products to be sold; a naturally occurring enzyme that breaks down blood clots and clears blocked blood vessels

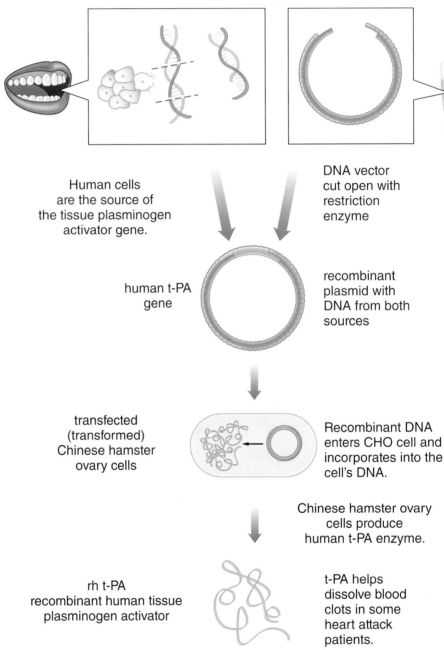

human t-PA gene

Human cells are the source of the tissue plasminogen activator gene.

DNA vector cut open with restriction enzyme

human t-PA gene

recombinant plasmid with DNA from both sources

transfected (transformed) Chinese hamster ovary cells

Recombinant DNA enters CHO cell and incorporates into the cell's DNA.

Chinese hamster ovary cells produce human t-PA enzyme.

rh t-PA recombinant human tissue plasminogen activator

t-PA helps dissolve blood clots in some heart attack patients.

Figure 1.18. Producing Genetically Engineered t-PA. Humans make only a small amount of human tissue plasminogen activator (t-PA) naturally. By genetically modifying Chinese hamster ovary (CHO) cells, scientists can make large amounts of t-PA for therapeutic purposes, such as to clear blood vessels in the event of a heart attack or stroke.

of biotechnology products, including genetically engineered or modified organisms (GMOs) and their protein products (see Table 1.1). Some of the new biotechnology products include proteins used in pregnancy tests, enzymes that increase the amount of juice that can be extracted from apples, molecules used in vaccines, and strawberry plants that can grow in freezing temperatures. According to the Biotechnology Industry Organization, more than 4000 biotechnology drug products and vaccines for more than 200 diseases were in human clinical trials in 2014. Restriction enzymes, DNA ligase, recombinant DNA, and their roles in producing genetically engineered organisms and products are discussed in more depth in Chapter 8.

Table 1.1. Biotechnology Products

Product	Application
Roundup Ready® Soybeans (Monsanto Canada, Inc)	herbicide-resistant soybeans
Alferon NIjection® (Hemispherx Biopharma, Inc)	drug used to treat genital warts
M-Pede® (Gowan Company LLC)	fungicide that prevents powdery mildew on fruits and vegetables
YERVOY® [ipilimumab] (Bristol-Myers Squibb Company)	treatment of metastatic melanoma (skin cancer)
QuantStudio™ 12K Flex Real-Time PCR System (Life Technologies Corp)	instrument used to recognize and multiply short DNA sequences
Purafect® protease (Genencor International, Inc)	protein-digesting enzyme
Posilac® bovine somatotropin (Monsanto Company)	growth hormone used in livestock
Promacta (GlaxoSmithKline)	treatment of thrombocytopenia (low platelets)
Bollgard® II cotton (Monsanto Company)	insect-resistant cotton
Xolair [Omalizumab] (Roche/Genentech)	treatment of asthma
NextSeq 500 (Illumina, Inc.)	instrument used to determine DNA nucleotide sequences
tissue plasminogen activator (t-PA); marketed as Activase® (Genentech, Inc)	enzyme that dissolves blood clots
Humulin R U-500 (Lily USA, LLC)	drug used to treat diabetes
Sunup® papaya (Papaya Administrative Committee of Hawaii)	virus-resistant papaya
Recombivax® (Merck & Co., Inc)	vaccine for hepatitis B
Premise® 75 (Bayer Corp)	termiticide (kills termites)
EPOGEN® (Amgen, Inc)	drug that produces red blood cells in anemic patients

BIOTECH ONLINE

© Reuters.

The GloFish® (Yorktown Technologies, LP)

Scientists recently produced transgenic fish, fish that contained genes from another species. Would you like to own genetically engineered fish that glow in certain types of light?

Using an Internet search engine, find one or more websites that discuss GloFish®. Summarize how the glowing fish are produced and describe what makes them "glow." Describe any controversies surrounding the development of GloFish® and explain your position on creating new "pet" organisms such as this one. List several advantages and disadvantages to creating pet organisms with the traits included in the GloFish®. List the websites you used.

The Human Genome Project

We are living in a time of great scientific advances with opportunities arising in many different areas of biotechnology research and manufacturing. With the recent completion of the **Human Genome Project** (determining the human DNA sequence), the doors have been opened wide to understanding the function of the human genetic code. Further work includes identifying all of the genes, determining their functions, and understanding how and when genes are turned on and off throughout the lifetime of an individual. Assigning functions and understanding how and when genes are translated into specific traits or actions will provide research work for many future generations of research and manufacturing scientists. Applying the new scientific knowledge will lead to the development of products for the improvement of human health.

Human Genome Project (hu•man ge•nome proj•ect) an international effort to sequence and map all the DNA on the 23 human chromosomes

Section 1.2 Review Questions

1. Name two antibiotics used as medicines.
2. The use of what kind of enzymes allows scientists to cut and paste pieces of DNA together to form recombinant DNA?
3. Explain how making human tissue plasminogen activator (t-PA) in Chinese hamster ovary (CHO) cells is an example of genetic engineering.

1.3 How Companies Select Products to Manufacture

Figure 1.19. The ABI PRISM® 310 Genetic Analyzer, one of several DNA sequencers and reagents sold by Life Technologies Corp., is an instrument that speeds DNA sequencing and analysis.
Photo by author.

Each biotechnology company usually specializes in a group of similar products. For example, Bayer Biotech produces therapeutic drugs for several diseases. Monsanto, Inc., produces plant products. Applikon, Inc., produces fermentation equipment for large-scale production of cell products. Gilead Sciences, Inc. produces viral therapies. Genomyx, Inc. manufactures DNA sequencers for research purposes. Genencor International, Inc., produces enzymes for food processing and other industrial applications.

Companies can specialize in a particular product area because the manufacturing processes and procedures are nearly the same among similar products (see Figure 1.19). The protocols for manufacturing recombinant human growth hormone are almost identical to those for producing recombinant human insulin. The majority of **reagents**, cells, and equipment are the same. It is, therefore, not surprising that a company would specialize in the manufacture of related human hormones as therapeutic drugs. Focusing on one product area is economical and saves steps in research and development as well as in manufacturing.

Developing Ideas for New Products

The ideas for deciding which biological materials should be investigated for product development and manufacturing can come from many sources. Research teams regularly discuss their work among themselves, and these discussions lead to ideas for new products (see Figure 1.20). Scientists may envision new products as they conduct their regular literature reviews or when they attend professional meetings. Sometimes an idea for a process or product comes about rather serendipitously. The technique for making billions of copies of DNA in a short time, the polymerase chain reaction (PCR), was conceived while a scientist was driving along a twisty mountain road late one night in northern California.

reagent (re•a•gent) a chemical used in an experiment

Research and Development

No matter which product(s) a biotechnology company makes, the goal is to market it as quickly as possible. However, the research and development (R&D) phase often requires several years. A drug must demonstrate "proof of concept" data in the research laboratory before the project moves into the development phase. At this stage, several aspects are assessed: for example, is it feasible to produce this new medicine in sufficient amounts to treat people? What needs to be done to ensure its safety? Which characteristics are indicative of **efficacy** (proof that it is effective)? Is it stable over time? If we produce it with a different process, will its properties change? For novel, cutting-edge biotechnology drugs, these questions are very challenging and require performing complex studies.

If the assessment is favorable, the project enters clinical development. Much testing is done, as the procedures for small-, and then **large-scale production** are determined. If the product is to become a pharmaceutical, it must undergo strict testing (**clinical trials**), under the guidance of the **Food and Drug Administration (FDA)**, before it can be marketed. Three rounds of clinical trials, over many years, test progressively larger numbers of patients to check the safety and effectiveness of the drug. If clinical trials show that a pharmaceutical product is safe and effective, then a Investigational New Drug (IND) application to market the drug is submitted to the FDA. On average, it takes about 10 to 15 years for a company to move a pharmaceutical product through all of these steps, a process called the "product pipeline" (see Figure 1.21). At any given time, a company has only a certain number of products in the pipeline. For a smaller company, only two or three products may be in the pipeline at a given time. Larger companies may have 10 to 15 products in production. Many companies consider it a success to move even a single product a year to market.

One biotechnology product to reach the marketplace recently is the enzyme, Pulmozyme®, manufactured by Genentech, Inc., a medication used to manage **cystic fibrosis (CF)**. This genetic disorder clogs the respiratory and digestive systems with mucus. It is often fatal to sufferers by age 30. Use of Pulmozyme® improves the quality of life for patients with CF by reducing the amount of mucus produced. In 2014, Pulmozyme® had sales of over $620 million. The revenue from such a product is used to fund more research and development, to defray marketing and administrative costs, and in some companies, to pay dividends to stockholders. As companies look for potential products, such as Pulmozyme®, they evaluate each for its marketability.

A Product Development Plan   Before a product makes it into the pipeline, the company's management reviews it to determine whether or not it is worth the investment of company resources (money, personnel, etc). Many companies develop a "comprehensive product development plan" for each potential product. The plan usually includes the following criteria:

- Does the product meet a critical need? Who will use the product?
- Is the market large enough to produce enough sales? How many customers are there?
- Do preliminary data support that the product will work? Will the product do what the company claims?
- Can patent protection be secured? Can the company prevent other companies from producing it?
- Can the company make a profit on the product? How much will it cost to make it? How much can it be sold for?

Each product in a company's pipeline is reviewed regularly in light of the comprehensive product development plan. During each review, if the answers to these questions are satisfactory, the company will continue development of the product. If a product does not meet the criteria, it may be pulled from the pipeline. It may be learned early in research and development that a potential product is too costly to produce. That product may be pulled from the pipeline early on, before the company has invested too many resources in its development. For a pharmaceutical product, the

Figure 1.20.   Flavia Borellini, PhD, Global Product Vice President, AstraZeneca. Dr. Borellini has worked on product development for most of her career. As a Global Product Strategist for Genentech, she led a team to develop and implement marketing plans for promising new cancer drugs. In her current position as Global Product Vice President, Dr. Borellini works on global strategies to market AstraZeneca's AZD9291, a medication for patients with advanced epidermal growth factor receptor mutation positive (EGFRm) non-small cell lung cancer.
Photo by author.

efficacy (eff•i•ca•cy)   the ability to yield a desired result or demonstrate that a product does what it claims to do

large-scale production (large-scale pro•duc•tion)   the manufacturing of large volumes of a product

clinical trials (clin•i•cal tri•als)   a strict series of tests that evaluates the effectiveness and safety of a medical treatment in humans

FDA   abbreviation for the Food and Drug Administration; the federal agency that regulates the use and production of food, feed, food additives, veterinary drugs, human drugs, and medical devices

cystic fibrosis (CF) (cys•tic fi•bro•sis)   a genetic disorder that clogs the respiratory and digestive systems with mucus

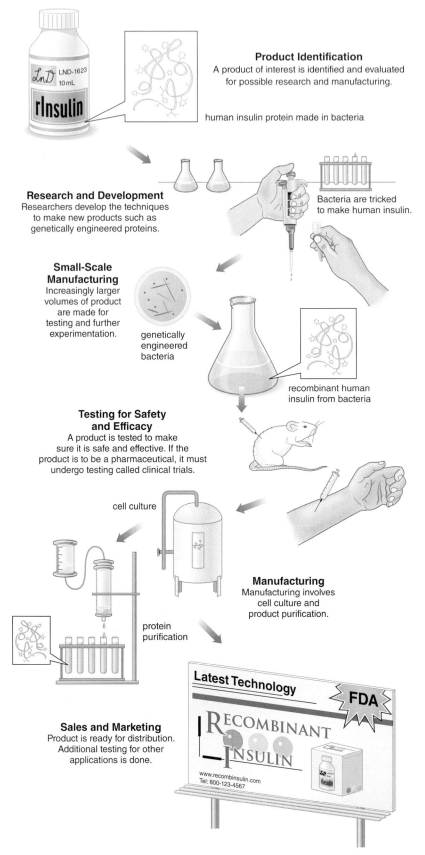

Product Identification
A product of interest is identified and evaluated for possible research and manufacturing.

human insulin protein made in bacteria

Research and Development
Researchers develop the techniques to make new products such as genetically engineered proteins.

Bacteria are tricked to make human insulin.

Small-Scale Manufacturing
Increasingly larger volumes of product are made for testing and further experimentation.

genetically engineered bacteria

recombinant human insulin from bacteria

Testing for Safety and Efficacy
A product is tested to make sure it is safe and effective. If the product is to be a pharmaceutical, it must undergo testing called clinical trials.

cell culture

Manufacturing
Manufacturing involves cell culture and product purification.

protein purification

Sales and Marketing
Product is ready for distribution. Additional testing for other applications is done.

Figure 1.21. Stages in Product Development. The stages in product development (product pipeline) are different at every company because each product has specific requirements, but most product development follows the basic outline shown here.

product must have passed through Phase III clinical trials with many thousands of patients included in the testing before an application with the FDA for permission to market the product can be filed. More on clinical trials is presented in Chapter 9.

Situations That End Product Development   Often, products are pulled from the pipeline when testing shows they are not effective (see Figure 1.22). This is what happened with Auriculin®, a **therapeutic** drug developed by Scios, Inc. for acute renal (kidney) failure. The hope was that Auriculin® would lead to a dialysis-free life for patients with diseased kidneys. Auriculin® was in the product pipeline for over five years before its development was halted. Companies like Scios, Inc. in Mountain View, California, where the original work was conducted, invested millions of dollars in Auriculin® research and development. They lost that investment because the product did not make it all the way through the pipeline to market. Even with setbacks like this, many drugs are approved for use each year (see Figure 1.23).

Regulations Governing Product Development

All biotechnology products have regulations governing their production in the pipeline (see Figure 1.24). Regulatory guidelines during the production of drugs and cosmetics, chemicals, and crops are written and overseen by such agencies as the FDA, the **Environmental Protection Agency (EPA)**, or the **US Department of Agriculture (USDA)**, respectively. Depending on the product, safety and effectiveness may have to be demonstrated. For example, a contact lens cleaning solution, such as the one shown in Figure 1.24, would undergo extensive testing. More testing, documentation, and other applications increase the time it takes a product to go through the pipeline. Some consumers think that government testing takes too long for some drugs. Such is the case for some of the drugs thought to improve the outlook for AIDS and cancer patients. A number of advocates believe that some of the risks of speeding the testing procedures are outweighed by the chance of improving the quality of life for very sick patients.

therapeutic (ther•a•peu•tic) an agent that is used to treat diseases or disorders

EPA   abbreviation for the Environmental Protection Agency; the federal agency that enforces environmental laws including the use and production of microorganisms, herbicides, pesticides, and genetically modified microorganisms

USDA   abbreviation for United States Department of Agriculture; the federal agency that regulates the use and production of plants, plant products, plant pests, veterinary supplies and medications, and genetically modified plants and animals

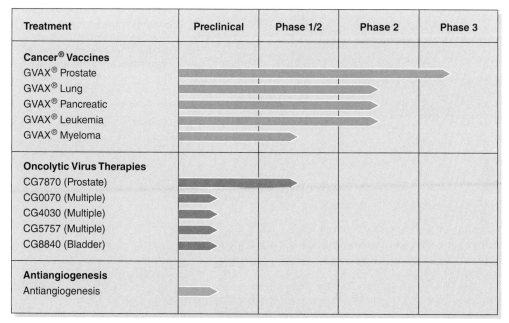

Treatment	Preclinical	Phase 1/2	Phase 2	Phase 3
Cancer® Vaccines				
GVAX® Prostate				
GVAX® Lung				
GVAX® Pancreatic				
GVAX® Leukemia				
GVAX® Myeloma				
Oncolytic Virus Therapies				
CG7870 (Prostate)				
CG0070 (Multiple)				
CG4030 (Multiple)				
CG5757 (Multiple)				
CG8840 (Bladder)				
Antiangiogenesis				
Antiangiogenesis				

Figure 1.22.   **The Cell Genesys Product Pipeline.**   The Cell Genesys product pipeline as of May 2005. The focus of Cell Genesys was on biological therapies for cancer. Unfortunately, several failed Phase 3 trials in 2008 which resulted in the closure of Cell Genesys.
Chart courtesy of Cell Genesys, Inc.

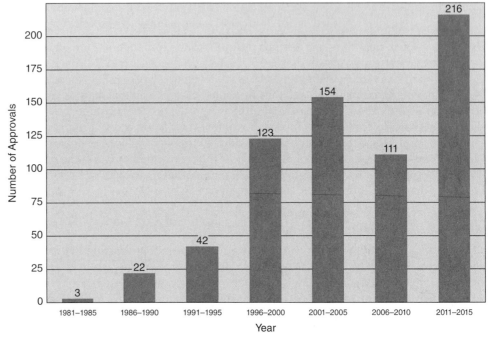

New Biotech Drug and Vaccine Approvals/New Indication Approvals by Year

Figure 1.23. **New Biotech Drug Approvals.** Even with all the government regulations, the number of new drugs approved for market increased nearly six times in the 10 years between 1990 and 2000. There were 41 new approvals by the FDA in 2014. In 2015, over 400 biotechnology pharmaceuticals were in clinical trials with 79 awaiting approval. See www.biopharmcatalyst.com/fda-calendar.

Data collected from www.fda.gov and the Biotechnology Industry Organization.

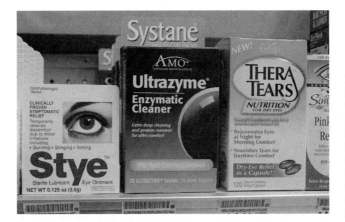

Figure 1.24. **Contact lens cleaners, both the solutions and tissues, contain bioengineered enzymes. Imagine the kind of testing done before the government regulatory agencies approve these products for use on contact lenses that will go on the human eye.**
Photo by author.

BIOTECH ONLINE

Genentech, Inc.'s Product Pipeline

Genentech, Inc. is a good example of how a company can have several different products in different stages of the production pipeline. Genentech focuses on pharmaceuticals to treat cancer, cardiovascular disease, and diseases of the immune system. In 2015, Genentech had more than 40 pharmaceutical products in research and development, clinical trials, or awaiting FDA approval.

Visit Genentech, Inc.'s pipeline status Web page at **biotech.emcp.net/genepipeline** to see where Genentech products are in the development pipeline. From the Genentech pipeline, pick a product being developed to treat a type of cancer. Record three interesting facts about the drug or its potential use.

Section 1.3 Review Questions

1. What group of potential products must be tested in clinical trials before it can be marketed?
2. A drug discovery process can take nearly 15 years. Explain why it takes so long to bring a new drug to market.
3. Which questions must be answered to the satisfaction of company officials before a product goes into research and then into development?
4. Does every product in research and development (R&D) make it to market? Yes or no? Explain.

1.4 Doing Biotechnology: Scientific Methodology in a Research Facility

Science students traditionally are taught that there is a "scientific method." In reality, there is no single scientific method that every researcher follows. Instead, there are several practices that most scientists use when conducting experimental research. These methods help ensure unbiased, reproducible **data** (scientific information). Understanding the concepts of scientific methodology is important for anyone who wants to work as a member of a scientific team in a research and development laboratory.

Usually scientific methodology is presented as a five-step process by which researchers ask and answer scientific questions (refer to Figure 1.9):

1. **State a testable scientific question or problem based on some information or observation.** Usually a question or research direction arises from previous experimental results, but sometimes a question is based on a new idea.

2. **Develop a testable hypothesis.** A testable **hypothesis** is a statement that attempts to answer the scientific question being posed. The hypothesis implies how to test and the kind of data to be collected.

3. **Plan a valid experiment.** Whenever possible and appropriate, a valid experiment contains quantitative (numerical) data, multiple replications (several trials), a single manipulated **variable** (factor being tested), and **control** groups—a **positive control** group (one that will give predictable results) and a **negative control** group (one lacking what is being tested so as to give expected negative results).

4. **Conduct the outlined experiment and collect and organize the data into tables, charts, graphs, or graphics.**

5. **Formulate a conclusion based on experimental data and error analysis.** A conclusion also suggests further experimentation and applications of the findings.

Conducting an Experiment Using Scientific Methodologies

Suppose that you wanted to create a faded look on new denim jeans. Many people have tried to fade their denim jeans by adding full-strength bleach from the grocery store (see Figure 1.25).

data (da•ta) information gathered from experimentation

observation (ob•ser•va•tion) information or data collected when subject is watched

hypothesis (hy•poth•e•sis) an educated guess to answer a scientific question; should be testable

variable (var•i•a•ble) anything that can vary in an experiment; the independent variable is tested in an experiment to see its effect on dependent variables

control (con•trol) an experimental trial added to an experiment to ensure that the experiment was run properly; see *positive control* and *negative control*

positive control (pos•i•tive con•trol) a group of data that will give predictable positive results

negative control (neg•a•tive con•trol) a group of data lacking what is being tested so as to give expected negative results

Figure 1.25. Effect of Bleach on Jeans. Bleach is a strong oxidizing agent that will fade the colored dye in fabric.

Often the results are less than desired and, sometimes, expensive jeans are destroyed in the process. Improving bleached-out jean production is a good example of how a testable scientific question can be developed, asked, and answered, as shown in the following steps:

1. **State a testable scientific question or problem.**
   A good testable scientific question based on past observation and experience might be, "What concentration of bleach is best to fade the color out of new denim material in 10 minutes without visible damage to the fabric?"

2. **Develop a testable hypothesis.**
   Based on experience, full-strength bleach will fade the color in denim, but it will also weaken the fabric and cause holes. A hypothesis could be that decreasing the **concentration** of bleach to 50% by diluting it with water will cause the desired lightening of color without visible damage to the fabric (see Figure 1.26).

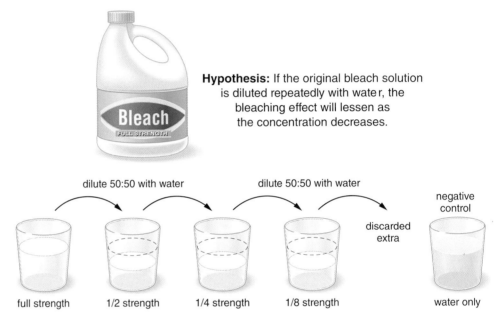

Hypothesis: If the original bleach solution is diluted repeatedly with water, the bleaching effect will lessen as the concentration decreases.

dilute 50:50 with water   dilute 50:50 with water

negative control

discarded extra

full strength   1/2 strength   1/4 strength   1/8 strength   water only

Figure 1.26.   **Diluting Bleach Hypothesis.**   Higher concentrations of bleach should cause more color fading.

3. **Plan a valid experiment.**
   To test the hypothesis that decreasing the bleach concentration will fade the color from denim samples without fabric damage, the experiment shown in Figure 1.27 is planned.
   a. Cut 15 10×10-cm squares of blue denim fabric.
   b. In 250-mL beakers, prepare dilutions (60 mL each) of five different bleach test solutions as follows: one full-strength (straight out of the bottle) and, with tap water, dilute the rest to 50%, 25%, and 12.5% strength. Measure 60 mL of tap water to use as a negative control (0% bleach).
   c. Recording the time that they are submerged, add three of the denim squares to each of five petri dishes containing 20 mL each of one of the test solutions. Submerge the squares for 1 minute.
   d. Withdraw the denim squares and lay them onto lab matting (or paper towel) for 9 minutes. Submerge the fabric pieces in 20 mL of tap water for 2 minutes to remove excess bleach and stop the bleaching process. Repeat the water wash two more times.
   e. Rank the amount of color removal from each of the denim fabric squares. Use 0 = no color change, and 10 = all the blue color is faded to white, as the ranking system.

f. Rank the amount of fabric damage on each of the denim fabric squares. Use 0 = no visible fabric damage, and 10 = holes in weakened fabric, as the ranking system. Give intermediate values for an intermediate amount of fading or fabric damage.

g. Record the data on a data table, and calculate the average color removal and the average fabric damage for each concentration treatment. Produce a bar graph to show the average results of each set of data.

4. **Conduct the outlined experiment.** Collect and organize the data into data tables, charts, graphs, or graphics.

The experiment is conducted following the procedures outlined. Data are collected. Rough drafts of observations, data tables, graphs (visualizing the results), and analyses are usually written or drawn, by hand, into a legal scientific notebook (see Figure 1.28). Final copies of data tables and graphs are produced using a spreadsheet program, such as Microsoft® Excel®. If practical, original samples are kept as evidence of the experiment and the results (see Figure 1.29). Photographs may also be made of the experimental setup and results. The original fabric squares or photographs of them are permanently affixed into notebooks.

5. **Formulate a conclusion based on experimental data and error analysis.**

Once the data are collected and organized, a researcher looks for the answer to the original research question. Does the numerical evidence support the hypothesis? Numerical data based on the averages of multiple replications give a researcher confidence in his or her findings. If it appears that there is a significant difference between one treatment and another, the researcher may recommend more testing with additional bleach dilutions. Or, once the "best condition" is determined, a recommendation might be made to use one of the treatments for large-scale faded jean production or some other application.

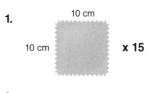

100% bleach   50% bleach   25% bleach   12.5% bleach   water only

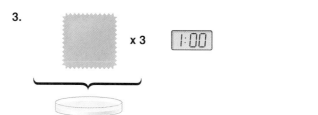

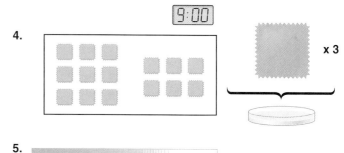

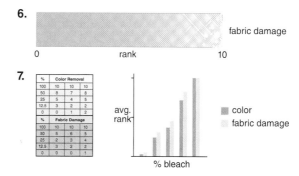

Figure 1.27. Experimental Flowchart. A valid experiment must meet certain criteria, such as having multiple replications of each experimental trial.

It is not uncommon for experimental data not to support a hypothesis or expectation. This could be due to experimental errors or to a hypothesis not being the answer to a scientific question. Experiments are repeated to make sure there are no experimental errors. Sometimes, a researcher finds that the hypothesis is just not supported even if all the procedures have been done correctly. In this case, the hypothesis is rejected and a new one might be formed.

%	Color Removal		
100	10	10	10
50	8	7	8
25	5	4	5
12.5	3	2	2
0	0	1	2

%	Fabric Damage		
100	10	10	10
50	5	6	5
25	2	3	4
12.5	3	2	2
0	0	0	1

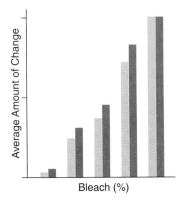

Figure 1.28. Data Table and Graph. Observations and measurements are reported in data tables. Individual trials (replications) as well as averages are shown. Numerical data are shown in picture form using graphs.

Figure 1.29. Fabric samples showing multiple trials of color removal. Each of the concentrations was tested three times. The more replications, the better the average data will represent the true value of color fading.
Photo by author.

A good approach for writing an experimental conclusion is to use the "REE, PE, PA" method. For REE (*results* with *evidence* and *explanation*) give the answer to the purpose question (results) with numerical data, if possible, as evidence. For most experiments, averaged data are the best numerical answer to a purpose question. Then, explain whether the data support or refute the hypothesis and why. Give specific examples.

For PE (*possible errors*), identify the sources of experimental design errors that would lead to fallacious (false or misleading) data, and explain the possible implications from making such errors. Two possible experimental errors in the bleaching experiment are errors in timing and solution preparation. Errors in either technique would provide data that might lead to incorrect assumptions. Once potential errors in experimentation are identified, recommendations to improve the experiment to minimize these sources of errors are given. The goal is to design experiments that have the most reproducible and reliable data.

For PA (*practical applications*), discuss the meaning or value of the experimental results in the short term and in the long term. How are the findings valuable to the scientist, the company, or the scientific community? What recommendations can be made about using the data or for planning future experimentation? Often the next experiment is only a slight modification or refinement of the previous one.

The final version of a conclusion should be a thorough analysis of the experiment and results reflecting on the uses of the new information. The final version should be proofread, or witnessed, by a colleague who understands enough about the experiment to analyze the data, but who was not involved in conducting the experiment.

Sharing Experimental Results with the Scientific Community

Once an experiment or, usually, a set of experiments is completed, the work is reported to other scientists through publications or presentations, such as at annual conferences. Scientists publish their work in scientific periodicals or magazines called **journals** (see Figure 1.30). There are many online scientific journals as well as printed scientific journals.

Formal conclusions for the notebook and reports to be published in scientific journals are written on a computer using a word processing program, such as Microsoft® Word® (see Figure 1.31). Copies of everything are permanently affixed into the notebook. It is critical to record, analyze, and reflect on all data collected, should disputes arise regarding scientific discoveries and/or intellectual property.

Designing, conducting, analyzing, and reporting a valid experiment takes practice. Throughout your biotechnology training, you will learn to practice good scientific methodology.

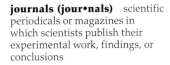

journals (jour•nals) scientific periodicals or magazines in which scientists publish their experimental work, findings, or conclusions

Figure 1.30. When scientists collect data that support a significant advancement of scientific knowledge, their findings may be published in a formal report in a scientific journal. To keep current, other scientists frequently review the articles in these journals. Most biotechnology companies and universities have extensive collections of journals in their libraries.
Photo courtesy of Sunesis, Inc.

Figure 1.31. Computers are used to produce reports of all experiments. They are also often used to record, store, and analyze data. Both personal computers (PCs) and Macintosh computers (Macs) are used in the biotechnology workplace. Therefore, it is valuable to gain experience on both types of computers.
© Cultura/Photo Researchers.

Section 1.4 Review Questions

1. Scientific methods used by scientists vary from lab to lab and situation to situation. One approach to scientific studies is to follow a five-step process in which a question is asked and answered. Outline these five steps.
2. Why do valid experiments contain many trials repeating the same version of an experiment?
3. In a conclusion, evidence for statements must be given. Describe the kind of evidence that is given in a conclusion statement.
4. Name two ways that scientists share their experimental results with other scientists.

1.5 Careers in the Biotechnology Industry

Biotechnology is one of the fastest-growing commercial industries. Career opportunities span a vast field, including bioscience (medical, agricultural, and environmental applications), applied chemistry, physics, and computer science. Due to the enormous amount of data collected in the Human Genome Project, the industry will be studying the meaning of the DNA sequence for most of the 21st century. Studying the expression and functions of the **genome** will require thousands of lab researchers as well as computer programmers and technicians.

Many biotechnology companies have focused their efforts on producing protein products. Several of these "young" biotech companies are beginning to see their products enter the marketplace and, as they do, they begin to make a profit. As venture capital (the initial investment money) is recouped and profits are generated, the companies can hire additional staff for more research and development. The need for more scientific and nonscientific staff increases, which in turn fuels the development of more products with more applications. As a company's growth spirals, it adds more employees, who generate more products (see Figure 1.32).

genome (ge•nome) one entire set of an organism's genetic material (from a single cell)

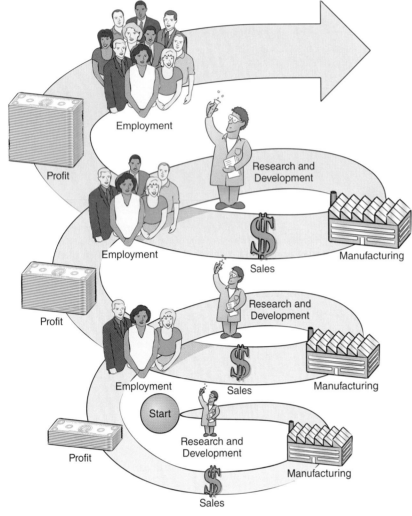

Figure 1.32. R&D, Sales, and Profit Spiral. Once a company starts marketing a new product, it earns revenue. If revenues are high enough and the company has profits, it can reinvest those profits into more research and development.

As the biotechnology industry matures, the opportunities for employees are immense. A variety of workers with a wide diversity of education, training, and experience are required. Common to all biotech employees, though, is the need for a basic understanding of the science and economics of the industry. Of course, as in any industry, job seekers with more education and experience have a better chance for employment and advancement.

Education Requirements

Most laboratory positions require a 4-year college degree, for example, a Bachelor of Science (BS) degree in **biochemistry**, **molecular biology**, **genetics**, or biology. As the industry moves beyond R&D into manufacturing, technicians with more hands-on training and experience are needed. According to the US Department of Labor's Bureau of Labor Statistics, the employment of science technicians is expected to grow 12% from 2006 to 2016, and employment of biotechnicians is expected to increase to about 91,000. Many community and career colleges have developed 2-year training programs, including internships, to address this need. Even high schools are beginning to train students in standard biotechnology laboratory techniques using state-of-the-art equipment.

Laboratory directors, senior scientists, and staff scientists usually require advanced degrees, such as a Master of Science (MS), Master of Arts (MA), or a Doctor of Philosophy (PhD) degree, and postdoctoral research experience (see Figure 1.33). Most colleges and universities have appropriate programs for undergraduate (Bachelor of Science or Bachelor of Arts) work. When continuing to advanced degrees, though, a candidate must carefully scrutinize a university's ability to provide experience, guidance, and contacts in a specific area of the industry.

Nonscientific Positions and Education Requirements

Even employees in the nonscientific areas of the biotechnology industry must have an interest in and understanding of the science of biotechnology. Sales, marketing, regulatory, legal, financial, human resources, and administrative staff must be able to communicate in the language of biotechnology (see Figure 1.34). Degrees in the scientific fields listed in Figure 1.33 are valuable even for individuals who do not intend to work in a lab. Experience in a laboratory, short- or long-term, helps employees work more productively with all members of the company.

biochemistry (bi•o•chem•is•try) the study of the chemical reactions occurring in living things

molecular biology (mo•lec•u•lar bi•ol•o•gy) the study of molecules that are found in cells

genetics (ge•net•ics) the study of genes and how they are inherited and expressed

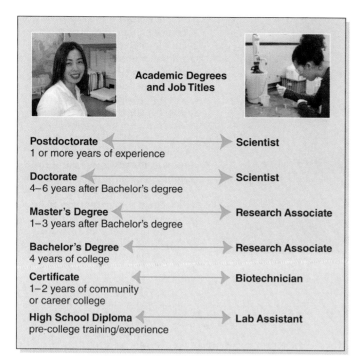

Academic Degrees
and Job Titles

Postdoctorate ⟷ Scientist
1 or more years of experience

Doctorate ⟷ Scientist
4–6 years after Bachelor's degree

Master's Degree ⟷ Research Associate
1–3 years after Bachelor's degree

Bachelor's Degree ⟷ Research Associate
4 years of college

Certificate ⟷ Biotechnician
1–2 years of community
or career college

High School Diploma ⟷ Lab Assistant
pre-college training/experience

Figure 1.33. **Academic Degrees for Jobs.** By increasing education and experience, individuals can acquire positions of more responsibility, self-directedness, and higher salary. Many companies provide incentives or reimbursement for additional schooling. Salaries differ significantly from one geographic location to another.

Figure 1.34. **Monica Poindexter (left) serves as the Associate Director of Corporate Diversity and Inclusion, a part of the Human Resources Department, at Genentech, Inc. The Human Resources Department at a biotechnology company is responsible for personnel issues, including recruiting and hiring appropriate staff, determining competitive compensation and benefits, and maintaining employee records. As part of Fluidigm Corporation, Denise Jimenez (right) served as a Human Resources Generalist with several job duties, including recruiting employees, advising employees on benefit programs, and conducting new- hire orientations. Since Fluidigm develops protein and gene chips with microscopic channels for genetic analysis and protein expression studies, employees with very specific skills are needed.**
Photos by author.

Lab techniques and experience gained in many areas are universally applicable. Many experimental and research procedures are used in a similar fashion in pharmaceutical labs, agricultural research, industrial applications, and instrument development and testing. A research associate (RA) working on breeding orchids, for example, uses many of the same techniques as an RA would use when cloning bacteria, including volume measurement, media preparation, and sterile technique. Developing skills in basic laboratory techniques increases a person's employment opportunities.

Categories of Biotechnology Jobs

Most positions in a biotechnology company are grouped into one of the eight categories listed below.

Scientific Positions:
• research and development
• manufacturing and production
• clinical research
• quality control

Nonscientific Positions:
• information systems
• marketing and sales
• regulatory affairs
• administration/legal affairs

Examples of people working in positions in these categories are present in the Career sidebar at the beginning of each chapter and in figures throughout the book, similar to those in Figure 1.35. Job descriptions and the type of education and experience needed are often included.

Figure 1.35. **As a Help Desk Technician in an Information Technology Department, Jon Ocampo troubleshoots hardware and software issues for more then 100 employees in a W2K environment. He maintains and supports printers, fax machines, video conferencing equipment, laptops, and PCs. He prepares computer, phone, and network resources for new users and provides training in commercial and custom software. His position requires an excellent understanding of PC hardware and software and the ability to troubleshoot software problems in a variety of programs.** Photo by author.

BIOTECH ONLINE

Finding "Hot Jobs"

Many websites, including biotech.emcp.net/biospace, biotech.emcp.net/biofind, biotech.emcp.net/sciencejobs, and biotech.emcp.net/lifescienceworld, post job descriptions and want ads for biotechnology employment.

Go to one of the four biotechnology job-finding websites listed above. Find a company offering a position as a quality control analyst or technician. In your notebook, record the following:

- the company offering the job
- the actual title of the job or position and a description of the job duties
- the starting salary for the position (or the salary range)
- the website URL (address) for the position

Section 1.5 Review Questions

1. For which types of biotechnology employees is there currently a large demand? What are the educational requirements for these types of employees?
2. Scientific positions in most biotechnology companies fall into one of four categories. List them.
3. Why might having laboratory experience be a benefit for a nonscientific employee at a biotechnology company?

1.6 Bioethics: Biotechnology with a Conscience

Your car is parked on a side street around the block from your friend's house. As you walk toward it, you see broken glass on the ground and realize that someone has broken into your car and stolen your built-in MP3 player out of the dashboard. What a mess! Who is going to pay to fix the damage and replace the property? Anyone in this situation would be angry and feel mistreated.

Our culture generally recognizes that all forms of stealing, including shoplifting, are wrong (see Figure 1.36). Why do some people steal even though they and we know it is wrong?

How do we learn what is right and wrong behavior? As new situations arise in your life, how do you decide what is acceptable behavior and what is unacceptable? How do you decide what is fair and just?

Moral Standards

Being able to distinguish between right and wrong and to make decisions based on that knowledge is considered "having good **morals**." Because there are some differences in people's beliefs of right and wrong, some people have different morals than others.

Many vegans, for example, believe it is immoral to eat meat of any kind or any animal products. Most vegans have decided that it is not only wrong to kill animals for food, but it is inhumane to use animals in any way to produce goods, such as leather, fur, eggs, or milk. Other vegetarians (ovolactovegetarians) may believe it is acceptable to eat dairy products and eggs, but not meat. They have different morals than vegans. Most people, though, eat and use many products from animals that are farmed (livestock). Of course, these people would never eat their pet dog or cat. Their morals about pets are different.

The study of moral standards and how they affect conduct is called **ethics**. **Bioethics** is the study of decision-making as it applies to moral decisions that need to be made because of advances in biology, medicine, and technology.

Many of the new biotechnologies are controversial because they force people to think about what they believe is right or wrong. Harvesting embryonic stem cells, genetically modifying foods, and testing for genetic diseases or conditions are just a few topics that elicit hot moral debate (see Figure 1.37). Many people have strong feelings about these and other issues, and they take personal or public positions to support or oppose these technologies.

New technologies generate ethical questions that cannot be answered using scientific methodologies. Ethical questions cannot be tested. The positions one takes on ethical issues are based on personal feelings and beliefs. A person can learn more about a technology, but determining whether it is moral to use it is not an objective decision with a clear right or wrong answer. It is a subjective decision in which a wide range of positions could be argued.

Who decides what is right and what is wrong when it comes to scientific advances and new technologies used in biotechnology? Who decides what is acceptable in research and development? Who decides which testing or products should be allowed and under what circumstances? Who makes public policy for scientific products, scientific procedures, scientific information, and new technologies? Should the decision makers be individual citizens, scientists, religious groups, government agencies, or nongovernmental organizations?

As you learned earlier, the agencies primarily responsible for regulating biotechnology in the United States are the Food and Drug Administration (FDA), the US Department of Agriculture (USDA), and the Environmental Protection Agency

moral (mor•al) a conviction or justifiable position, having to do with whether something is considered right or wrong

ethics (eth•ics) the study of moral standards and how they affect conduct

bioethics (bi•o•eth•ics) the study of decision-making as it applies to moral decisions that have to be made because of advances in biology, medicine, and technology

Figure 1.36. **Is there any justification for shoplifting? Can you name any reasons why shoplifting would ever be acceptable or, at least, justifiable? How would you feel about a homeless person, trying to get a job, who steals a toothbrush and toothpaste?**
© MachineHeadz/iStock.

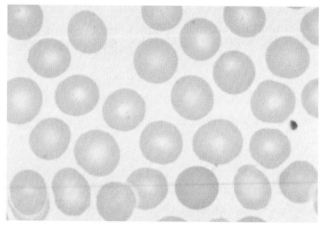

Figure 1.37. **These are human red blood cells (RBCs) derived from human embryonic stem cells by scientists at the University of Wisconsin-Madison. Can you think of any good reasons for growing human RBCs from human embryonic stem cells? Can you think of any good reasons to not use human embryos for this purpose? 500X**
© Lester V. Bergman/Getty Images.

microbial agents (mi•cro•bi•al a•gents) synonym for microorganisms; living things too small to be seen without the aid of a microscope; includes bacteria, most algae, many fungi

(EPA). Products are regulated according to their intended use, with some products being regulated under more than one agency. Often, representatives of these agencies must make decisions considering ethical issues instead of scientific ones. An example is when the USDA decides which animals it is acceptable to clone.

The USDA regulates the use and production of plants, plant products, plant pests, veterinary supplies and medications, and genetically modified plants and animals. The EPA regulates the use and production of **microbial agents**, such as bacteria and fungi, plant pesticides, new uses of existing pesticides, and genetically modified microorganisms (see Figure 1.38). The FDA regulates the use and production of food, feed, food additives, veterinary drugs, human drugs, and medical devices. Each of these agencies sets policy based on scientific facts that are viewed in conjunction with society's collective ethical positions.

A lab technician, research associate, or scientist will probably not be involved in setting public policy, but he or she must make decisions regularly that may be considered controversial either personally or in the workplace (see Figure 1.39). Data collection (accuracy), data reporting (honesty), and safety concerns (following rules) are just some of the areas where concerns about appropriate and ethical practices may arise. Taking positions and acting on bioethical issues is made easier when a technician practices the skills necessary to examine and analyze situations. For any ethical issue, strategies for examining the issue are needed so that decisions are made that are appropriate, fair, just, and based in confidence.

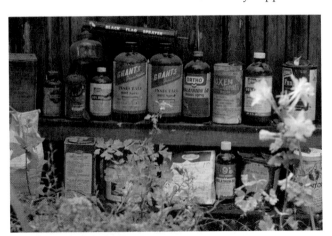

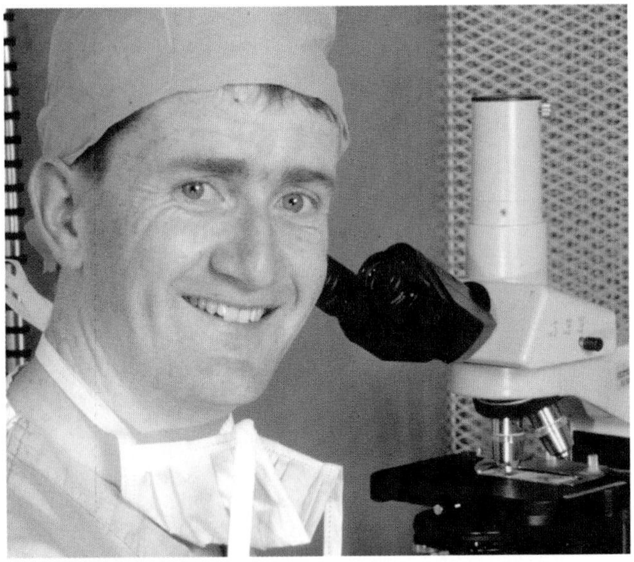

Figure 1.38. Before these plant pesticides may be sold or applied, the EPA must approve them. What do you think about approving an insecticide to kill wasps for use in and around the house and garden if it also kills the honeybees of nearby honey farmers?
© Galen Rowell/Corbis Documentary/Getty Images.

Figure 1.39. Joe Conaghan, PhD, Director of the Pacific Fertility Center in San Francisco, California, is known internationally for his work on improving embryo culture conditions. Dr. Conaghan's interests are in developing programs for the treatment of severe male factor infertility, the diagnosing of genetic disease in embryos, improving embryo culture, and improving protocols for embryo freezing. Can you think of some ethical issues that Dr. Conaghan might have to consider in his workplace?
Photo courtesy of Joe Conaghan.

Strategy for Values Clarification

One strategy for examining a bioethical issue is to clarify the values one holds (values clarification) using the following steps.

- Identify and understand the problem or issue. Learn as much as possible about the issue.
- List all possible solutions to the issue.
- Identify the pros and cons of adopting each solution. Examine the consequences of adopting one solution (or position) as opposed to another. Consider legal, financial, medical, personal, social, and environmental aspects.
- Based on the pros and cons for each solution, rank all solutions from best to worst.
- Decide if the problem is important enough to take a position. If it is, decide what your position is and be prepared to describe and defend it.

Sometimes a problem is complicated enough that it requires actually writing down all the solutions, pros and cons, and considerations. After examining these, an individual should be able to form a position on an issue and be able to justify it. In the end-of-chapter activities, ethical dilemmas will be presented. Using the Strategy for Values Clarification, you can better consider each dilemma and formulate a personal position on the issue.

Research institutions and biotechnology companies may have to develop and implement policies about controversial issues or ethical practices (see Figure 1.40). Clinical trials, animal testing,, dissection, use of cells and tissues, how to report and process data, confidentiality, hiring policies, production of genetically modified organisms, and waste management and disposal are just some of the many practices an organization might have to consider when developing a policy.

Committees of employees, including scientists, business people, and regulatory staff, work together to create policies that are fair, just, and responsible. According to U.S. Federal law, institutions that use laboratory animals for research or instructional purposes must establish an Institutional Animal Care and Use Committee (IACUC) to oversee and evaluate all aspects of the institution's animal care and use programs, facilities, and procedures. Likewise, institutions that utilize human subjects in research must establish an Institutional Review Board (IRB) to review and monitor biomedical research involving human subjects. An IRB has the authority to approve or require modifications in research plans to protect the rights and welfare of human research subjects.

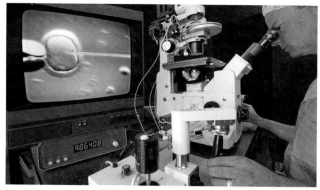

Figure 1.40. **In vitro fertilization (IVF) of a human egg cell.** *In vitro* means "in glass" and refers to processes that occur outside of a living thing. A pipet (left) holds a human egg while a scientist selects a sperm and, using a needle (right), injects the sperm into the egg. There are several pros and cons to consider when using IVF, including the following questions: What happens to the extra fertilized eggs that are not used by the clients? Who should get to pick the sperm used? What do you think of being able to pick a particular sperm and a particular egg to produce your own offspring?
© Lester Lefkowitz.

BIOTECH ONLINE

The Pros and The Cons of Biosimilars

Use the Values Clarification Model to determine if you agree or disagree with the attempt by Amgen and Genentech to restrict the ability of pharmacists to substitute generic versions of biological drugs for brand name products.

1. Use the Internet to identify and understand the problem or issue by researching the answers to these questions. Summarize what you find and list at least two websites you used to gather information.
   • What is the difference between a brand name and generic product?
   • What are some examples of brand name and generic pharmaceutical products?
   • What are the advantages and disadvantages of a generic versus a brand name pharmaceutical product?
   Learn as much as possible about the specific issue:
   • Which Amgen and Genentech products are these companies trying to protect and why?
     Read the article at www.nytimes.com/2013/01/29/business/battle-in-states-on-generic-copies-of-biotech-drugs.html?nl=todaysheadlines&emc=edit_th_20130129&_r=0
     Describe the legislative bills Amgen and Genentech are supporting.
2. List all possible solutions to the issue.
3. Identify the pros and cons of adopting each solution. Examine the consequences of adopting one solution (or position) as opposed to another. Consider legal, financial, medical, personal, social, and environmental aspects.
4. Based on the pros and cons for each solution, rank the solutions from best to worst.
5. Decide what your position is regarding legislation to restrict pharmacists' ability to fill prescriptions for biological pharmaceuticals with generic brands. Be prepared to describe and defend it.

Section 1.6   Review Questions

1. Define the term "bioethics."
2. Give an example of an event that might lead a lab employee to be faced with an ethical issue.
3. Describe how the Strategy for Values Clarification can be used to solve a problem such as the use of embryonic stem cells for basic research.

Chapter Review

Summary

Concepts

- Biotechnology includes all the technical processes that have led to improvements in products and services, and in understanding organisms and their component parts.
- Products developed through biotechnology must have a market large enough to generate the profit required to fund future research and development.
- Biotechnologists work in a variety of settings, including corporate labs, government agencies/labs, and academic (college and university) research facilities.
- Biotechnology is a broad field that includes the domains of medicine and pharmaceuticals, agriculture, industry, the environment, instrumentation, and diagnostics.
- One of the major breakthroughs in biotechnology is the ability to genetically engineer organisms by combining DNA from different sources. Human tissue plasminogen activator (t-PA), a blood-clot-dissolving enzyme used on heart attack patients, is an example of a pharmaceutical produced through genetic engineering.
- The completion of the Human Genome Project, which determined the human DNA sequence, is of major importance to biotechnologists and to society in general, since the new knowledge will lead to an increase in scientific understanding and the development of products to improve the quality of human life.
- To begin research and development on a potential product, companies must have satisfactory answers to such questions as, "Who will use the product?", "Is it economical to produce?"
- Stages in product development (product pipeline) include product identification, research and development, small-scale manufacturing (fermentation), testing for safety and efficacy (including clinical trials), manufacturing, and sales and marketing.

- Some discoveries may or may not lead to product development, but the information contributes to our scientific knowledge. This is considered "pure science."
- Although the discovery process and product pipelines are different for every product, it usually takes 10 to 15 years to bring a product to market. Some products take longer to come to market, particularly pharmaceuticals that must undergo clinical trials.
- Agencies that regulate the development and approval of biotechnology products include the FDA, the USDA, and the EPA.
- All scientists follow a set of procedures to answer their scientific questions. Most follow a scientific methodology that begins with asking a testable question. They predict the answer (hypothesis), and then design and conduct an experiment to test the question. They collect and analyze measurable data, then report their findings relative to their predictions. They report the significant findings and discoveries at scientific meetings and in scientific journals.
- Most jobs at a biotechnology company are in the following areas: research and development, manufacturing and production, clinical research, quality control, information systems, marketing and sales, regulatory affairs, and administration/legal affairs.
- Many laboratory positions require a minimum of a 4-year college degree. Manufacturing and quality control staff need either a 2- or 4-year degree. A scientific background is helpful for nonscientific employees as well.
- The study of moral standards and how they apply to biotechnology and medicine is called bioethics. Bioethical issues arise in many areas, including research, manufacturing, and product applications.
- A good method of analyzing a bioethical issue or dilemma is to use the Strategy for Values Clarification model.

Lab Practices

- For an experiment's data to be considered valid, several conditions must be met, including the following: manipulating only one variable at a time so that a cause and effect may be observed, conducting multiple replications so that averages can be determined, and controlling other variables to reduce their impact on the results.
- Data are collected and organized into data tables and presented in picture form with graphing. Final data tables and graphs are produced using software, such as Microsoft® Excel®.
- Conclusion statements should include numerical data, preferably averaged, with explanations of what the data mean. Error analysis is included to notify the reader of concerns or limitations in the experimental design. Future experiments are suggested in a conclusion, along with the value or applications of the experimental findings.
- The REE (results with evidence and explanation), PE (possible errors), PA (practical applications) approach works well for writing experiment conclusions.
- The example of the experiment designed to determine the best concentration of bleach to fade the color of new denim material in 10 minutes without visibly damaging the fabric demonstrates the importance of good experimental design. For the experimental data to be considered valid, the experiment must have only one variable tested (the concentration of bleach) and all other conditions and measurements must remain constant in each trial. Also, there must be multiple replications of each trial so that average data can be determined.

Thinking Like a Biotechnician

1. The following activities represent stages in product development and manufacturing. Rearrange them in the order in which they take place.
   Testing for safety and efficacy
   Sales and marketing
   Product identification
   Small-scale manufacturing
   Research and development
   Manufacturing

2. Match each of the descriptions or examples below with one of the five steps in the approach to scientific methodology presented in this chapter.
   a. Scientific question
   b. Hypothesis
   c. Experiment plan
   d. Experimentation
   e. Data analysis and reporting

   _____ Collect numerical data.
   _____ Create graphs of averaged data.
   _____ Ask a question that is testable.
   _____ Repeat trials.
   _____ Try to answer the question based on the experience of others.
   _____ Answer the question based on data collected during a valid experiment.
   _____ Step-by-step instructions of how to test a scientific question.

3. Biotechnologists can find job opportunities in many areas. In addition to biotechnology companies, where can a scientist with a biotechnology background find employment?

4. In the fabric-bleaching experiment in Section 1.4, all samples lost all of their coloration in the 10-minute time period. Review the proposed procedure for the experiment and suggest a change in the experimental design so that you might see a difference in the color fading due to bleach concentration.

5. When writing a conclusion, technicians use the "REE, PE, PA" method to ensure that all important data, evidence, and applications are discussed. Explain the meaning of the terms represented by the abbreviations "REE," "PE," and "PA."

6. After working for months repeating experiments, you note that your data show that purifying a particular medicinal protein works better at a certain temperature. You want to share this scientific information with the rest of the scientific world. Name two ways commonly used by scientists to share this kind of information.

7. A large amount of the human protein collagen is needed to decrease the amount of wrinkling on people's faces. Describe, in a few steps, how genetic engineering could be used to make large amounts of human collagen for the dermatology market.

8. Consider the first five products in Table 1.1, page 13. What is the market (who uses them) for these products? Consider the market size of each of these products. Which of the products do you think has the largest market (could make the most money), and which do you think has the smallest market?

9. Getting a pharmaceutical approved for sale in the United States takes much longer than getting an industrial product, such as a laundry enzyme, approved. This is because of the required clinical testing. Explain why clinical testing slows down the approval of pharmaceutical products.

10. Consider this scenario: You work for a company that has developed an AIDS drug that can prevent transmission of HIV from an infected mother to her nursing baby in 990 of 1000 cases. However, in 10 out of 1000 cases, the drug causes a severe reaction and possibly death to the mother or baby. Scientists want to conduct Phase III clinical trials in an area of Africa where the AIDS rate has doubled each year for the past 5 years. As a company employee, you are a member of a committee deciding whether or not to support and fund the trial. Use the Strategy for Values Clarification model to determine the position you will take when meeting with the committee on this issue.

Biotech Live

What is Biotechnology?

Use the Internet to find answers to the following questions about the biotechnology industry and its companies and products. Record all answers in your legal scientific notebook.

1. Locate several definitions of biotechnology from different sources on the Internet. Find one definition that you think is representative of most of the others and record it in your notebook. Make sure to record the URL (Web page address) of the definition you used as well.
2. Where are most biotechnology/biopharmaceutical companies in the United States located?
   - Go to www.genengnews.com/insight-and-intelligenceand153/top-10-u-s-biopharma-clusters/77900061 and read the article.
   - List the top 10 US biotechnology geographical clusters/regions.
   - What criteria were used to determine a metropolitan area's ranking? List 5 measurements that were used.
3. Do a search for a biotechnology company that produces a pharmaceutical (medicine).
   - What is the name and location of the company?
   - What pharmaceutical does it make?
   - For what type of patient is the medication used?
   - What is the home page (URL) of the company?
4. Do a search for a biotechnology company that produces a biotech product that is NOT a pharmaceutical.
   - What is the name and location of the company?
   - What product does it make?
   - What is the use of the product?
   - What is the home page (URL) of the company?

The Business Side of Biotechnology

How can investors and potential employees get financial and business information about biotechnology companies? A wealth of information can be found about the organization, finances, and product development and marketing of a company using company websites, investment firms, stock exchanges, and brokerage houses. Many firms make recommendations about what companies look promising and which companies do not.

A company's annual report is also a good source of information about past performance and future expectations. An annual report describes the present state of a company's scientific and business ventures. It also outlines the plans and expectations for the company's future. Many companies publish their annual reports on the Internet.

Most companies sell shares of stock (a tiny bit of the company) to the public to raise funds for research and development purposes. Before someone invests in a company by purchasing its stock, research is done to find out more about the company.

A smart investor or future employee wants to know a lot about the company including the following:
- How much a company makes (revenue)
- How much a company spends (expenses)
- Products being made or sold by the company at this time or in the future
- If the company is involved in any costly legal battles
- If the company has "performed" well in the past

Use the Internet to obtain information about a company's business and scientific interests. Pick a company to study that produces some biotechnology product and is located within 500 miles of your campus. In your notebook, create a one-page fact sheet that includes answers to the following information. It is your goal to inform potential investors so they may decide whether or not they should invest in the company.

1. Give the company's name, location, and homepage website.
2. What are the stated long-term goals of the company?
3. List the company's marketed products (common names), their trade names, and their applications or uses.
4. Using a search engine that has a finance page such as those found on www.yahoo.com or www.google.com, find the company's stock trading symbol by clicking on "Symbol Lookup." Then, using the trading symbol, find the current price per share of the company's stock. Also, print a 3-month chart of the company's stock performance.
5. Click on the "News" link and find and read a recent article that either shows that the company should be making more or less money in the future. Record the URL of the article. Explain how the article might affect stock prices.

Activity 1.3 Investing in Biotechnology

Inspired by Mark Okuda, San Jose, CA.

The "stock market" is a term that actually describes several markets such as the New York Stock Exchange and the American Stock Exchange where the stocks of companies are traded. Shares in a company are sold and the shareholders then become part owners of the company. Shareholders receive stock certificates that show the number of shares purchased.

By offering shares to the public, companies become publicly traded. Offering shares of stock raises money for continued research and development of company products or services. To determine the prices of shares, investment bankers evaluate the company's value and earnings or potential earnings. Then, stocks are offered to the public at some initial public offering (IPO) price. The public can buy shares of the stock at the given price per share.

When investing in a company, the goal is to buy shares at a low price and then sell them at a higher price. Individual stocks may go up or down independent of how "the market" is doing overall. Stock market indices such as the Dow Jones Average, the NASDAQ, and Standard and Poor's 500 report how the market is doing "on average." To check the progress of individual stocks, one can look up their price per share on one of the published indices. These are available in the business section of newspapers and on the Internet.

Many times when employees are hired at high tech companies, they are offered "stock options." Stock options allow employees the option to buy stock at a lower price than the public after a certain length of employment. Depending on the economy and the market, purchasing stocks or using stock options can be a good way of investing an employee's extra income.

How does one know which stocks to buy? No one ever knows for sure since no one knows what will happen to the economy, the market, or a company. Purchasing stocks is always a gamble, but the more you know about a company's finances and its products, the better you can decide whether a company's stock has the potential to increase in value. For example, if a new drug is just completing Phase III clinical trials and is about to go on the market, the company may expect to start making money on that product. Using annual reports and researching companies on the Internet is a good place to start.

"Purchase and track" biotechnology stocks with the goal of buying low and selling high and ending up with the highest value investment portfolio after 14 weeks.

1. Each investor begins with "$1000" and chooses two biotechnology companies to invest in. The investor must buy enough shares of each stock to spend a total of between $950 and $1000. These stocks will be held for two months.

2. Each investor must have rationale for selecting each company. Compose a short paragraph to explain the reasons for purchasing shares of each stock and exactly how much you intend to purchase based on the current price/share and total value. Consider past performance, future potential, present or future marketed products, and management or management changes.
3. Using Excel®, make an individual data table for each company's stock purchased to use as an investment record. Each data table should include:
   A. A title with the company's name, the company's trading symbol (eg DNA for Genentech), the number of shares purchased, the length of the study, and the date of the stock purchase.
   B. Columns to record for each of 14 weeks: the price per share and the total value of the shares of stock purchased.
4. Maintain the data table for 8 weeks. After 8 weeks, you may want to modify your stock portfolio. You have three investment options:
   A. You may leave your investments as is.
   B. You may redistribute all your investments to amounts of shares within the companies you already hold stock in (and start new data collection).
   C. You may sell part or all of your shares of stock and take the profit or loss and buy other companies' stocks (and start a new data collection). At no time can you own less than two companies' stocks.
   If you decide to take option "B" or "C," you must write another one-half page rationale for your new stock distribution.
5. Determine the amount of profit or loss for your portfolio of stocks after 8 and 14 weeks. In a summary data table like the one below, report the initial value of each stock, the current value of each stock at each of these times, and the total gain or loss of your portfolio.

Portfolio Value after _____ Weeks

Company	Symbol	# of Shares	Stock Value at Purchase ($)	Current Stock Value ($)	Net + or - ($)
				Total	

Final Analysis of Your Stock Portfolio

6. Using Excel®, make a summary line graph to show how each stock's price per share changed through the period you owned it. These should be different colored lines on the same graph, each line representing one of the stocks.
7. Make another graph that shows the total value of each stock for every week that you owned it. These should be different colored lines on the same graph, each line representing one of the stocks.
8. Conduct Internet research to try to determine the reasons the stocks in which you invested either went up or down in value. Citing your references, write a 10–20 sentence description of what happened to your stocks and why.
9. Prepare a PowerPoint® presentation of 5–10 minutes to be given to the other investors in the class. Include a company description and stock profile for each stock in which you held shares for the 14-week period. Include graphs of the stock's performance, the final value of your portfolio, and the percent increase or decrease from the original investment. Discuss any events (political, financial, etc.) that may have affected the stocks' performances.

Activity 1.4

How is the Biotechnology Industry Improving the Quality of Human Life?

According to the Biotechnology Industry Organization (BIO), in 2002, the total sales of biotechnology products reached approximately 24 billion dollars. These products included human healthcare products, genetically modified plants and animals, biofuels, chemicals, research instruments, and environmental products.

Go to the Biotechnology Industry Organization's website at www.bio.org/about_biotech and learn how biotechnologists are working to heal, fuel, and feed the world.

1. Read the overview. Make a 3-column chart that lists examples of how biotechnologists are working to heal, fuel, and feed the world.
2. Think about how good health, abundant energy, or abundant high-quality foods affect the quality of life for many people. Use the links on the overview page to learn more about one of these issues.

Activity 1.5

Staying Current in Biotechnology

Using the Internet to find short summary articles is an easy way for a technician to keep up-to-date. Several science news websites have searchable databases in which they catalog summary articles on biotechnology and current events. Examples of helpful websites with science news databases include biotech.emcp.net/biotechnews, www.sciencedaily.com/news/plants_ animals/biotechnology, www.genengnews.com, and www.biospace.com/news.aspx.

Your instructor may assign a domain of biotechnology for you to research, or you may choose your own biotechnology domain from the list below:

- **Agriculture**
- **Environmental/Industrial/Biodefense**
- **Medical/Pharmaceutical**
- **Diagnostic/Research/Bioinstrumentation**

Then use one of the science news searchable databases to find an article in that domain.

1. Highlight the article, copy and paste it into a Microsoft® Word® document. Reduce the document to 77% size and print a copy. Record the website address (URL) if it is not on the document.
2. "Actively" read the article highlighting the topic sentence (main sentence) in each paragraph.
3. After reading the article, place an asterisk (*) by three items in the article that you think are the most interesting and important.
4. Present the information in the article to your lab team members (if you have been assigned to a team). Include at least one reason why the article is of importance to biotechnologists. If you are working independently, write a summary of the article and its importance.

It is good scientific practice to conduct this kind of search at least once a month to gather a wide variety of articles.

Bioethics

Using Animals in Science and Industry

Humans have a long history of using animals in agriculture and industry for the following purposes:

- as sources of food (beef, pork, lamb, etc)
- as sources of raw materials (suede, leather, wool, rennin, collagen, gelatin, etc)
- as sources of medicine (insulin from a pig pancreas, growth hormone from a cow pituitary gland, etc)
- as transportation and laborers (horse, elephants, donkeys, etc)
- as laboratory test specimens (rats, mice, dogs, cats, monkeys, chimpanzees, etc)
- as educational tools (zoos, museums, etc)
- as companions/pets

In most countries, scientists are required by law to minimize to the greatest extent possible any pain and suffering they cause to animals during testing, research, and manufacturing. In the United States, government agencies, such as the National Institutes of Health (NIH), and professional organizations, such as the American Psychological Association, publish guidelines on the ethical care and use of animals.

Some people question how animals are used to improve the quality of human life. Some people question whether it is ethical to use any species of animal in any or every application. Some people are of the opinion that almost any use of animals is justified to save or improve human life. Others feel that there is almost no reason to sacrifice an animal to improve human life. Should there be regulations about how and which animals should be used for what purposes? Where do you stand on the use of animals in science and industry?

Work with the "Use of Animals in Science and Industry" chart on the next page to develop a personal position on the use of animals in science and research. If you decide that animal use is justified, then decide which animals should be approved for that purpose. For example, is it ethical to sacrifice a fish, but not a chimpanzee? For each decision, consider the Strategy for Values Clarification model, and be ready to explain your position.

1. Review the Use of Animals in Science and Industry chart. Using the code on the chart, label the animals and their uses that you think should be approved. Consider the pros and cons of using each type of animal for each type of application.
2. Compose a statement explaining how and why you have decided that certain animals should be approved for certain applications. Describe any condition that could cause you to change your position(s) on the use of these animals and the new position(s) you might take.
3. Create a group with three other classmates. In 2 or fewer minutes, present your position on animal use to the other students in this small group.
4. Each person should summarize the position of the others in the group and discuss whether or not anyone's position caused a change in their own position.

The Use of Animals in Science and Industry

Place an "X" in each box that you agree with the use of that species for that purpose. Place a "NO" in each box that you do not agree with the use of that species for that purpose. Mark "N/A" where a decision is not applicable.

Animal Use	No animal of any kind should be used for this purpose.	Any animal of any kind should be used for this purpose.	Worms (invertebrate with simple nervous system)	Octopi (invertebrate with advanced nervous system)	Fish/Zebrafish (vertebrate with relatively simple nervous system)	Frogs (vertebrate with relatively simple nervous system)	Birds/Fowl (vertebrate with relatively simple nervous system)	Rodent/Rat/Rabbit (vertebrate with more advanced nervous system)	Cow/Livestock (vertebrate with advanced nervous system)	Dog/Cat (vertebrate with advanced nervous system)	Monkey (vertebrate with advanced nervous system)	Chimpanzee (vertebrate with very advanced nervous system)
sources of food (whole animal)												
sources of food byproduct (eg, eggs, milk)												
industrial raw materials applications												
• source of fabric/clothing (eg, wool, leather)												
• source of industrial molecules (eg, rennin)												
• testing of cosmetics												
medical applications												
• source of pharmaceutical molecules (eg, insulin)												
• source of transplant organs (eg, valves, cornea)												
transportation and laborers												
laboratory test specimen												
• testing of new drugs												
• testing of environmental hazards												
• for endangered species protection												
• for broadening scientific knowledge												
educational tools/teaching purposes												
• dissection												
• surgical practice												
• behavioral observation												
• physiological observation												
companions/pets												

Extension: Developing a Policy Statement for an Institution

Often, public institutions must make policies that will apply to controversial issues. Such is the case with animal dissection conducted in university biology classrooms.

Come together as an Advisory Committee of two women and two men, and draft a half-page policy statement to guide university instructors in their design of animal dissection labs. From the list below, indicate which animals will be approved for dissection and for which courses. Everyone on the policy team must agree.

Dissection Animals
Frog
Rat
Cat
Human cadavers

College Courses with Animal Dissection
Biology labs for nonmajors
Biology labs for biology majors
Biology labs for premedical students
Anatomy and physiology labs for premedical students

Biotech Careers

Photo courtesy of Brian Robinson.

Pathologist/Research Scientist

Brian Robinson, MD/PhD
Emory University School of Medicine
Atlanta, GA

Brian began his career with a Bachelor of Science degree in Biochemistry, plus several years of laboratory experience. After graduating from the University of California at Davis, Brian worked for a year as a Research Assistant at Genentech Inc., where he screened compounds that organic chemists developed for activity against cancer cells. These studies required the use of a **fluorometer** (shown here), which measures the amount of light emitted by cancer cells as they are treated with test compounds.

Next, Brian entered an eight-year program at Emory University to become a Medical Doctor (MD)/Research Scientist (PhD). His research focused on understanding how the protein *crumbs* controls tumor development in the fruit fly, *Drosophila melanogaster*.

Currently, Brian is a 3rd-year resident training in the Department of Anatomic Pathology at Emory University. As a pathologist, Brian analyzes cells, tissues, and organs taken from patients to diagnose diseases. With his current training to become a physician-scientist, Brian hopes to continue his pursuit of investigating molecular pathways driving tumor formation and metastasis.

fluorometer (fluo•rom•e•ter)
an instrument that measures the amount or type of light emitted

organism (or•gan•ism)
a living thing

cell (cell)   the smallest unit of life that makes up all living organisms

Escherichia coli
(esch•e•rich•i•a co•li)
a bacterium that is commonly used by biotechnology companies for the development of products

multicellular
(mul•ti•cell•u•lar)   composed of more than one cell

2 The Raw Materials of Biotechnology

Learning Outcomes

- Identify the levels of biological organization and explain their relationships
- Describe cell structure and its significance in biotechnology research and product development
- Discuss the types of organisms researched and the types of cells grown and studied in biotechnology facilities plus the products with which they are associated
- Distinguish between the cellular organization of prokaryotic and eukaryotic cells
- List the four main classes of macromolecules and describe their structure and function
- Define genetic engineering and identify products created with this technology
- Explain the Central Dogma of Biology and its importance in genetic engineering

2.1 Organisms and Their Components

It is common to think of biology and biotechnology as being the same discipline since each word starts with the prefix "bio," the Greek word for "life." In many ways biology and biotechnology are so similar that it is not valuable to make the distinction. However, biology and biotechnology are different in how they are applied. Biologists seek to better understand how living things work and interact, whereas biotechnologists look for how each biological discovery can be applied to create a product or service.

To manufacture biotechnology products for medical, industrial, or agricultural applications, biotechnicians must work either directly or indirectly with **organisms** (living things) or their components. These are the "raw materials" of biotechnology, and they may be as tiny as a molecule (eg, the HER2 antibody used to treat patients with certain types of breast cancer) or a **cell**, such as *Escherichia coli* (*E. coli*). Biotechnicians also may work with large **multicellular** organisms, such as frost-resistant strawberry plants (see Figure 2.1) or endangered animals, such as the cheetah, which is being outbreed to increase its genetic diversity. Biotechnology uses biological molecules, cells, organisms and their components to create new products and services. These biological components are the raw materials of biotechnology.

Working in any area of biotechnology requires a thorough understanding of the characteristics of life and the structures that compose organisms. If, for example, one

Figure 2.1. Genetically engineered, frost-resistant strawberries contain a gene from a type of flounder, which is a cold-water ocean fish. The flounder gene codes for a protein that resists freezing temperatures. Scientists have learned how to transfer a small number of flounder genes to strawberry cells, which read the flounder deoxyribonucleic acid (DNA) and produce the "antifreeze" protein. The presence of the "antifreeze" protein allows strawberries to grow in regions that normally might be too cold.

Figure 2.3. Monsanto, Inc.'s Roundup® contains the compound glyphosate, which inhibits an enzyme, EPSP synthase, required for normal plant growth. Without EPSP synthase, plants treated with Roundup® yellow and die.
Photo by author.

Figure 2.2. Monsanto, Inc.'s Roundup® kills most weeds. It has become one of the most common herbicides used in agriculture and at home. Roundup® Ready plants contain the gene to survive Roundup® spraying. The plants on the left have not been sprayed with Roundup®. Large weeds grow within the rows, crowding the soybean plants. The plants on the right have had the Roundup®-resistant gene transferred to them and were not affected by the Roundup® spray. Weeds, however, were killed. Weed control is significantly improved compared with the unsprayed control rows on the left.
Photo courtesy of Monsanto Company.

cytology (cy•tol•o•gy)
cell biology

anatomy (a•na•to•my)
the structure and organization of living things

physiology (phys•i•ol•o•gy)
the processes and functions of living things

is working on developing a soybean plant that will be able to survive exposure to pesticides, he or she must understand normal soybean growth and the factors that influence it. Also, it is essential to be able to recognize normal stem, root, and flower development in soybeans (see Figure 2.2). It is necessary to understand the chemical structure of pesticide molecules and how these will interact with the cells and tissues on which they will be sprayed (see Figure 2.3). In other words, biotechnologists must have at least a minimal understanding of some basic biochemistry (chemistry of living things), **cytology** (cell biology), **anatomy** (structure), and **physiology** (function). As one works on the development of a particular product, he or she becomes an expert on the structure and function of the product.

It takes several years to become an expert on a specific organism, including its cells and molecules. On the following pages, though, you will be given a foundation for the further study of living things and their uses in the research and manufacture of biotechnology products. It is intended to be only a brief introduction to the cell biology and biochemistry used in biotechnology. If you hope to move up the biotechnology career ladder, you will need to take additional courses in the areas of biology, chemistry, physics, and mathematics.

The Living Condition

If you were asked to name a living thing, you would probably name some plant or animal. Most people identify animals, such as dogs, cats, fish, frogs, or worms. Some remember to name plants, such as daisies, pine trees, or grass. Very few remember to add to the list bacteria, fungi, and other microorganisms, such as protozoans that live in ponds, the ocean, or inside other organisms (see Figure 2.4). The earth is home to a tremendous diversity of living things. Biologists estimate that there are well in excess of 20 million different species, and some estimates, including those of Mark Plotkin, author of *The Shaman's Apprentice*, run as high as 150 million species. Biotechnologists may work with one or more of these varied organisms. Some popular research organisms are soybeans, cotton, fruit flies, worms, cows, chicks, mice, rats, zebrafish, yeast, and bacteria, such as *E. coli*.

An organism exhibits "the characteristics of life," including such activities as growth, reproduction, response to stimulus, the breakdown of food molecules (**respiration**), and the production of waste products. The structure of an organism may be composed of one or more cells. Organisms composed of only one cell are called **unicellular** organisms, and they include bacteria (eg, *E. coli*), protozoans (eg, amoeba), and algae. To see unicellular microorganisms, you must use a microscope. Many unicellular organisms are critical in biotechnology applications. The bacterium, *E. coli*, for example, is regularly used in genetic-engineering experiments and was the first bacterium to be used commercially in that way. Genetically engineered human insulin was first produced in *E. coli* cells.

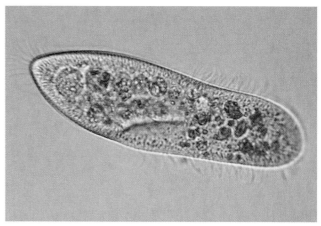

Figure 2.4. *Paramecium* **is a single-celled protozoan that lives in ponds and streams. It exhibits several traits that make it appear animal-like.** *Paramecium* **is a predator that responds quickly to prey in its environment. 100X.**
© Nancy Nehring/iStock.

respiration (res•pir•a•tion) the breaking down of food molecules with the result of generating energy for the cell

unicellular (un•i•cell•u•lar) composed of one cell

BIOTECH ONLINE

Photo by author.

Keeping the Kimchi Coming
A staple of the Korean diet, kimchi (pronounced kim-chee) is a fermented vegetable product now available in most supermarkets in the United States. Kimchi is produced during lactic acid fermentation when one or more species of bacteria utilize the sugar in the cabbage as a food source.

Use the Internet to learn about lactic acid fermentation and alcoholic fermentation.

Using two or more websites, explain what happens in:

1. The lactic acid fermentation that results in kimchi.
2. The alcoholic fermentation that results in wine.
   Record the URLs of the websites you used.
3. It is easy to find a recipe for kimchi on the Internet. Find one and record the ingredients used in the recipe. Also record the URL of the website.

Levels of Biological Organization

Plants, animals, and fungi, unlike bacteria, are composed of millions, billions, or even trillions of cells living together. The cells of multicellular organisms are usually grouped into functional units, such as **tissues** (eg, muscular or nervous tissue) and **organs** (eg, the skin, liver, and stomach). Various cells, tissues, and organs of a multicellular organism perform different jobs to ensure that the whole organism can survive. The levels of biological organization of organisms and their parts are shown in Figure 2.5.

tissue (tis•sue) a group of cells that function together (eg, muscle tissue or nervous tissue)

organ (or•gan) tissues that act together to form a specific function in an organism (eg, the stomach that breaks down food in humans)

Levels of Biological Organization

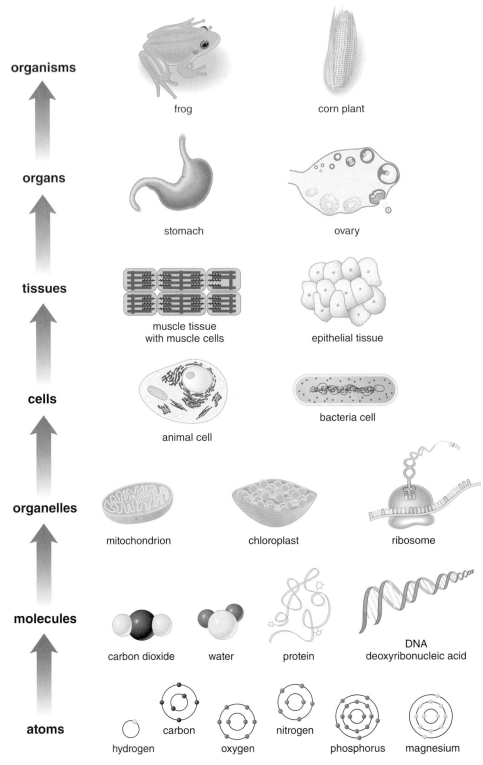

Figure 2.5. Levels of Biological Organization. Atoms are the smallest units of matter that are easily manipulated in cells and in laboratories. Atoms make up molecules, such as carbohydrates, proteins, and nucleic acids. Molecules are the building blocks of cells. Cells function in groups called tissues, and tissues function together in organs. Organs make up organ systems that work together in the multicellular organism.

As with all living things, the cells that comprise organisms function cooperatively in larger groups of tissues and organs to carry out life processes. Cells must grow and reproduce. Cells react to their environment and respond to stimuli (see Figure 2.6). Cells require energy to produce new molecules, such as **proteins**, which are the subject of research at several universities (see Figure 2.7) and often are the products that biotechnology companies want to manufacture. For example, thrombopoietin, a protein made by immune system cells, increases platelet production. Platelets act as blood-clotting agents. Several biotechnology companies are interested in producing thrombopoietin on a large scale for use in treating cancer patients undergoing chemotherapy, which often decreases platelet production.

Figure 2.6. **A mouse, like other mammals, is composed of cells, tissues, organs, and organ systems. Trillions of cells in the mouse work together in larger functional groups of tissues and organs.**
© Vasiliy Koval/Shutterstock.

Although cells are the smallest units of life, certain types of cells contain even smaller, nonliving units. Eukaryotic cells (plant, animal, fungi, and **protist** cells) are made up of smaller units, including tiny membrane-bound units called **organelles**. Organelles are not considered alive since they cannot exist outside the cell. Organelles act as specialized microscopic factories, each with specific jobs in the cell. For example, a **mitochondrion** (plural, mitochondria) is an organelle that uses **sugar** to create energy for the cell. A typical cell has many mitochondria. In fact, muscles have hundreds of mitochondria per cell because muscles require large amounts of energy.

BIOTECH ONLINE

Studying cells with a compound microscope.
© adrian beesley/iStock.

Picking the Right Tool for the Job

Different instruments are used to study different parts of an organism. Below is a list of instruments. Use the Internet to learn about the capabilities of one of these instruments. Write two or three sentences describing the instrument's uses. If directed by your instructor, share your information with your classmates. Then, identify the part of the organism you would likely study with that instrument.

Instruments	**Parts of an Organism**
atomic force microscope	organs
computed tomography (CT) instrument	tissues
magnetic resonance imaging (MRI) instrument	cells
compound microscope	organelles (larger than 1 micrometer [> 1 µm] in size)
transmission electron microscope	organelles (smaller than 1 micrometer [< 1 µm] in size)
stethoscope	
scanning electron microscope	molecules
	atoms

The organelles of cells are composed of molecules and atoms too small to be seen with a conventional microscope, however recent developments in fluorescent probes and molecular markers along with nanoparticle technology are allowing for the study of molecules and molecular interactions in living cells and tissues in real-time. In cells, a large number of different kinds of molecules, such as sugars, **starches**, proteins, **nucleic acids**, and **lipids**, are produced. Many of these are part of the structure of organelles. Other functions of these molecules include the regulation of cell activity, storage, and transport. Proteins are the most common molecules of cells. Many biotechnology products are the proteins made in certain cells. For example, amylase from soil bacteria, insulin from **pancreas** cells, and growth **hormone** from pituitary glands are currently being manufactured in genetically engineered host cells and sold by biotechnology companies for a variety of industrial and medical purposes.

Understanding how each component at each level of biological organization works is necessary for the development and manufacture of a biotechnology product. In the next few sections, you will review in more detail some of the structures and functions of the molecules and cell organelles essential to a broad understanding of biotechnology research techniques.

nucleic acids (nu•cle•ic ac•ids) a class of macromolecules that directs the synthesis of all other cellular molecules, often referred to as "information-carrying molecules"

lipids (li•pids) one of the four classes of macromolecules; includes fats, waxes, steroids, and oils

pancreas (pan•cre•as) an organ that secretes digestive fluids, as well as insulin

hormone (hor•mone) a molecule that acts to regulate cellular functions

Figure 2.7. Until 2008, Dr. Charles (Karl) L. Saxe, III, was a cell biologist at Emory University. He managed a research team of graduate students and postdoctoral fellows who studied cell behavior and how it is regulated by signaling proteins. Many of the signaling proteins may also act in several human diseases, such as cancer. Today, Dr. Saxe is a Scientific Program Director of Cancer Cell Biology and Metastasis at the American Cancer Society in Atlanta, Georgia.
Photo courtesy of Dr. Charles L. Saxe, III.

Section 2.1 Review Questions

1. Give an example of a plant that has been produced by biotechnology.
2. Knowledge of what other disciplines of science will improve the understanding of biotechnology?
3. Describe two characteristics of living things.
4. Which of the following is considered to be "alive": organs, molecules, atoms, cells, or organisms?

2.2 Cellular Organization and Processes

chlorophyll (chlor•o•phyll)
the green-pigmented molecules found in plant cells; used for photosynthesis (production of chemical energy from light energy)

photosynthesis (pho•to•syn•the•sis) a process by which plants or algae use light energy to make chemical energy

chloroplast (chlor•o•plast)
the specialized organelle in plants responsible for photosynthesis (production of chemical energy from light energy)

cytoplasm (cy•to•plasm)
a gel-like liquid of thousands of molecules suspended in water, outside the nucleus

lysosome (ly•so•some)
a membrane-bound organelle that is responsible for the breakdown of cellular waste

ribosome (ri•bo•some) the organelle in a cell where proteins are made

The cells of both unicellular and multicellular organisms are tiny microscopic factories that produce thousands of different molecules. Biotechnology companies exploit the biological manufacturing capabilities of cells and trick them into producing particular molecules in large amounts. These molecules become biotechnology products.

Depending on the type of cell, hundreds of different molecules are being manufactured at any given moment. Some of these molecules are unique to a particular kind of cell, and some of the molecules are found in every cell. For example, a human pancreas cell's main function may be to produce insulin and release it into the bloodstream. But at the same time, pancreas cells must produce all the other structural and functional molecules required to keep the cells alive. The same is true of plant cells. Plants cells produce thousands of molecules, some of which are the same molecules as those that animal cells produce. While plant cells do not produce insulin, many plant cells make **chlorophyll** molecules for **photosynthesis** (see Figure 2.8). Each type of eukaryotic cell has a unique composition of organelles needed to manufacture all of the substances and regulate all of the processes.

Some of the organelles and structural units of plant and animal cells are discussed below. This is only a partial list. The membrane-bound organelles (nuclei, **chloroplasts**, mitochondria, **cytoplasm**, and **lysosomes**) and other structures (cell walls, plasma membranes, and **ribosomes**) have been spotlighted because of their connection to biotechnology research and biomanufacturing. The diagrams that follow indicate the relative positions of these organelles in a plant cell (see Figure 2.9) or an animal cell (see Figure 2.10). The diagrams are models of what a "typical" cell might look like. No cell looks exactly like any of the cell diagrams but will have some or all of the organelles. As you read about each organelle, look over the cell diagrams to better understand their structure and function in the cell.

The Structure of Cells

Many cells require a rigid structural support. **Cell walls** provide this support to many cells around their outer boundary. In plants, cell walls are composed of rigid **cellulose** fibers. Paper is made from the cellulose fibers of plant cell walls. Cellulose is the main component of dietary fiber, which is considered important for a healthy digestive system. Bacteria and fungal cells also contain cell walls, although they are not composed of cellulose. Animal cells do not have cell walls.

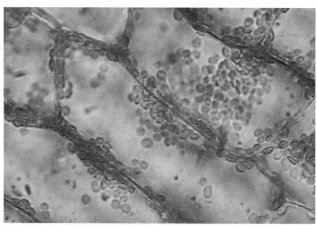

Figure 2.8. *Anacharis* is a freshwater plant. Like other plants, its cells are surrounded by a rigid cellulose-containing cell wall. Green chloroplasts containing chlorophyll surround water-filled vacuoles filling the majority of the cell. ~400X
© 2003, IABT.

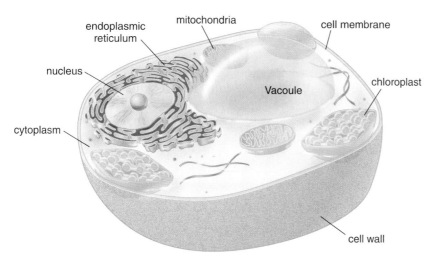

Figure 2.9. **Plant Cell.** Most plant cells contain chloroplasts and a rigid cell wall. Animal cells do not possess cell walls.

cell wall (cell wall)
a specialized organelle surrounding the cells of plants, bacteria, and some fungi; gives support around the outer boundary of the cell

cellulose (cell•u•lose)
a structural polysaccharide that is found in plant cell walls

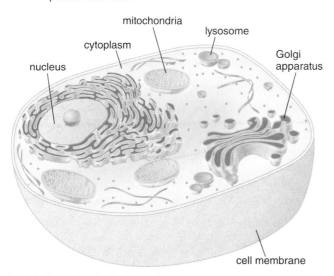

Figure 2.10. **Animal Cell.** Animal cells do not have a cell wall and, thus, do not have a rigid cell boundary. The shapes of animal cells are quite diverse due to the flexibility of the outer membrane and the responses when cells touch each other.

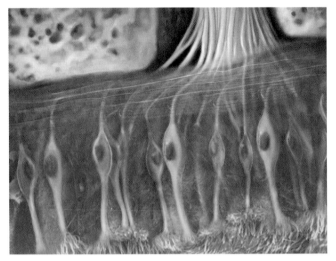

Figure 2.11a. Nerve cell extensions of the olfactory (nasal) bulb protrude through the approximately twenty openings at the top of the nasal cavities. Here, the nerve extensions receive chemical information from circulating odor molecules and convey the signals to the brain to be processed and perceived as smell.

© Anatomical Travelogue/Photo Researchers.

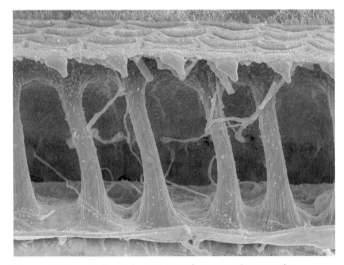

Figure 2.11b. Coloured scanning electron micrograph (SEM) of nerve cells (neurons) in the inner ear. Sound waves entering the ear, causing them to bend. The bent cells release neurotransmitter chemicals that generate nerve impulses that travel to the brain. This process transmits information about the loudness and pitch of a sound.

© Steve Gschmeissner/Getty Images.

plasma membrane (plas•ma mem•brane) a specialized organelle of the cell that regulates the movement of materials into and out of the cell

glucose (glu•cose) a 6-carbon sugar that is produced during photosynthetic reactions; usual form of carbohydrate used by animals, including humans

adenosine triphosphate (ATP) (ad•e•no•sine tri•phos•phate) a nucleotide that serves as an energy storage molecule

nucleus (nu•cle•us) a membrane-bound organelle that encloses the cell's DNA

chromosomes (chrom•o•somes) the long strands of DNA intertwined with protein molecules

Whether a cell has a wall or not, every cell is surrounded by a **plasma membrane** that regulates the movement of materials into and out of the cell. The plasma membrane is composed of lipids and proteins. Membrane proteins have several special functions. Some act as transport molecules, moving other molecules into or out of the cell. Some function as structural molecules that maintain cell shape, for example, the shape of red blood cells (RBCs) and the different shapes seen in the nerve cells of the eyes, ears, and skin (See Figures 2.11a and 2.11b).

Some membrane proteins act as identification or recognition molecules. One way that molecules can recognize specific cells is by binding with a recognition or receptor protein on the membrane surface. This is the way a molecule, such as insulin, recognizes and acts only on certain cells and not others. When an insulin molecule binds to an insulin receptor protein, a series of cellular reactions is set into motion that increase the cell's **glucose** uptake. Specifically, when insulin binds to the insulin receptor, the receptor protein transports phosphate groups from **adenosine triphosphate (ATP)** molecules to other proteins within the cell. This leads to an increase in glucose transport from outside (in the blood) to inside the muscle or adipose (fat) cells.

In the center of most eukaryotic cells is a prominent structure called a **nucleus** (see Figure 2.12). The nucleus is often called the "control center" of the cell since it contains the instructions (DNA) on **chromosomes** for constructing molecules, including proteins, which may function as **enzymes**, **pigments**, antibodies, or hormones.

Each chromosome is a long strand of DNA intertwined with protein molecules. The actual instructions for molecular construction (the genetic code of A, C, G, and T) are found within the structure of DNA. When cells make proteins, they transcribe (copy) a segment (gene) of DNA into a new version of the code, a **messenger RNA (ribonucleic acid)** molecule, mRNA for short. The mRNA transcript (A, C, G, and U) spells out how **amino acids** will be arranged into a protein. Proteins, such as insulin, hemoglobin, or amylase, are unique because of their amino acid sequence.

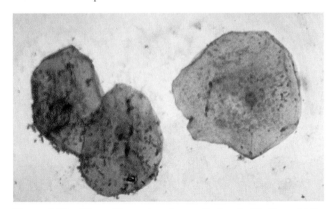

Figure 2.12. The cell's nucleus is usually located in the middle of the cell. Chromosomes within the nucleus contain DNA. The DNA strands hold the instructions for protein synthesis within the cell. 200X

© Lester V. Bergman/Getty Images.

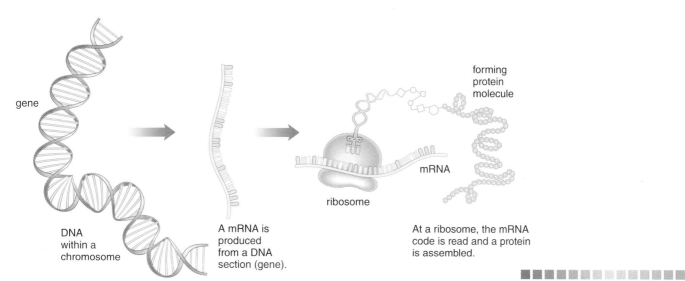

gene

forming protein molecule

DNA within a chromosome

A mRNA is produced from a DNA section (gene).

ribosome

mRNA

At a ribosome, the mRNA code is read and a protein is assembled.

Figure 2.13. **The Central Dogma of Biology.** The Central Dogma of Biology states that DNA codes for RNA and that RNA codes for proteins (DNA → mRNA → proteins). Once scientists had described the Central Dogma, they could propose and test strategies for manipulating protein production by manipulating DNA and RNA codes. Moving genes into cells to produce new proteins is the basic principle in genetic engineering.

In eukaryotic cells, the mRNA, which carries a protein code, moves from the nucleus to the surrounding cytoplasm. Cytoplasm is a gel-like liquid composed of thousands of molecules in water. In the cytoplasm, the mRNA code is translated into a protein molecule (translation). This happens at a structure called a ribosome. As the ribosome reads the mRNA (A, C, G, U) code, it assembles the 20 different amino acids into a **polypeptide** chain. The polypeptide chain folds into a protein. Since there are several hundred amino acids in a typical protein, the 20 different amino acids can be arranged in an almost infinite variety of proteins, depending on the original DNA and mRNA sequences.

The process of protein synthesis (how the DNA code is rewritten into mRNA and then decoded into a protein) is universally found in all cells. It is called the "Central Dogma of Biology" because it helps explain how virtually all molecules (proteins themselves or molecules made by proteins) in a cell are made (see Figure 2.13). Proteins are so important to cells and organisms, which are made up of cells, that their structure, function, and production are discussed in several additional sections of this text.

Chloroplasts (in plant cells) and mitochondria are involved in the production and use of cellular energy. Chloroplasts contain pigments, enzymes, and other factors that combine carbon dioxide and water. Using the sun's energy, they generate chemical energy in the form of sugar molecules (see Figure 2.14). Mitochondria use the sugar molecules made by plants to produce ATP energy molecules. Mitochondria and chloroplasts have been found to contain their own DNA, which recently has been the focus of some biotechnology research. The DNA of these organelles may help to explain the evolutionary origin of higher cells. Scientists are also interested in manipulating the DNA of these organelles to possibly increase the nutritional content of plants.

Cells in different organs and organisms may have very different structures and functions. Depending on their function, some cells have greater or fewer numbers of

enzyme (en•zyme) a protein that functions to speed up chemical reactions

pigments (pig•ments) the molecules that are colored due to the reflection of light of specific wavelengths

messenger RNA (mRNA) (mess•en•ger RNA) a class of RNA molecules responsible for transferring genetic information from the chromosomes to ribosomes where proteins are made; often abbreviated mRNA

amino acids (a•mi•no ac•ids) the subunits of proteins; each contains a central carbon atom attached to an amino group (-NH2), a carboxyl group (-COOH), and a distinctive "R" group

polypeptide (pol•y•pep•tide) a strand of amino acids connected to each other through peptide bonds

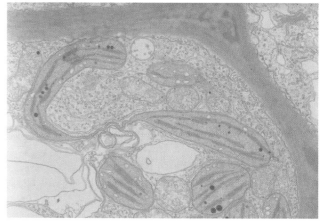

Figure 2.14. **This electron micrograph shows several chloroplasts each with the thylakoid membrane system on which the light reactions of photosynthesis occur. Light is used to energize electrons that, in turn, are used to form the bonds in sugar molecules. When cells use sugar for food, the energy in those bonds is released and used in other reactions. The dark spots are starch granules. ~8000X**
© Indigo Instruments®.

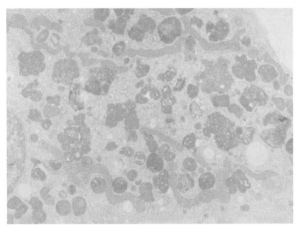

Figure 2.15. Many lysosomes are filled with enzymes (peroxidases) that break down cell waste. When this is the case, they are called perioxisomes. In this electron micrograph, dozens of lysosomes fill the cell. ~20,000X
© Indigo Instruments®.

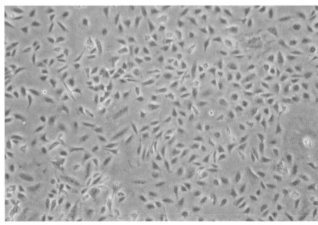

Figure 2.16. Shown in cell culture, these CHO cells are a common mammalian cell line used to manufacture recombinant protein. ~10X
© 2005, Nikon Inc.

Chinese hamster ovary (CHO) cells an animal cell line commonly used in biotechnology studies

Vero cells (ver•o cells) African green monkey kidney epithelial cells

HeLa cells (He•La) human epithelial cells

prokaryotic/prokaryote (pro•kar•y•ot•ic/ pro•kar•y•ote) a cell that lacks membrane-bound organelles

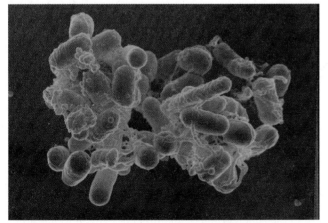

Figure 2.17. Each rod-shaped structure in this electron micrograph is an *E. coli* cell. *E. coli* cells are simple prokaryotes with no membrane-bound organelles, such as mitochondria or chloroplasts. ~20,000X
© Charles O'Rear/Corbis Documentary/Getty Images.

some structures. Muscle cells, for example, have more ribosomes and mitochondria than a "typical" cell because of their increased protein and energy production. Liver cells have more lysosomes, for waste removal, than most cells (see Figure 2.15).

The size and shape of a cell are directly related to both its structure and its function. Skin cells are flat and fit together, like jigsaw puzzle pieces, to cover and protect internal organs. In a multicellular organism, such as an orchid plant, there are many different sizes and shapes of cells to photosynthesize, conduct water, carry sugar, and store food.

In biotechnology applications, some cells are used and studied more than others. For instance, **Chinese hamster ovary (CHO) cells** are of particular interest (see Figure 2.16). Under the microscope, CHO cells resemble long, stretched-out cheek cells, all in contact with others. Many pharmaceutical biotech companies manipulate CHO cells to make them produce proteins different from those they normally create. Genes of interest from an assortment of organisms can be inserted and incorporated into the DNA of CHO cells, which read the DNA and begin producing the new proteins. Many pharmaceutical products, such as EPOGEN® by Amgen, Inc, a protein that boosts RBC production and is used to treat anemia, are produced in just this fashion for commercial purposes.

Types of Cells Used in Biotechnology

Many different kinds of plant and animal cells are grown and studied in biotechnology labs. In addition to the CHO cells mentioned above, **Vero cells** (African green monkey kidney epithelial cells) and **HeLa cells** (human epithelial cells) are often grown in large-scale cultures for biotechnology purposes.

An extensive collection of bacteria and fungal cells is used in biotech labs. *E. coli* is probably the most renowned (see Figure 2.17). The majority of biotechnology companies that grow bacteria use this well-known bacterium. Several human pharmaceuticals, including human growth hormone (hGH), were first produced commercially in *E. coli*. A variety of fungi are also utilized as production organisms, including *Aspergillus,* a type of mold, and both baker's and brewer's yeasts.

Cells differ based on the number and type of organelles present in them. Bacteria lack membrane-bound organelles. They are called **prokaryotic**. Without organelles, the

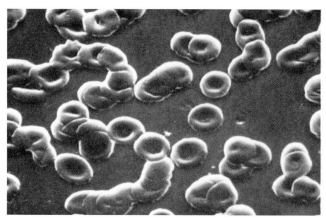

Figure 2.18. The doughnut shape of mammalian RBCs results from the loss of a nucleus during maturation. The red color is due to an enormous amount of red hemoglobin molecules filling the cell's cytoplasm. 450X
© Lester V. Bergman/Getty Images.

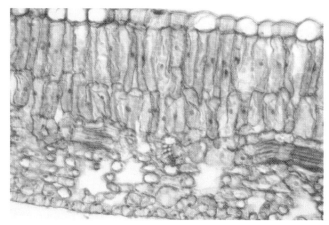

Figure 2.19. A cross section of a leaf cell shows rectangular cells lined with chloroplasts. 400X
© Clouds Hill Imaging Ltd./Corbis Documentary/Getty Images.

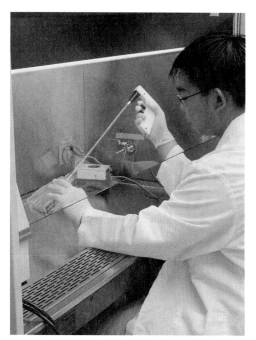

Figure 2.20. A lab technician, using good sterile technique in a laminar flow hood, transfers mammalian cell cultures from one culture flask to another. Mammalian cells need a carbon-dioxide-rich environment and a very specific broth medium to grow in.
Photo by author.

complexity of bacteria is limited compared with plant and animal cells. Most bacteria are single-celled with lengths of 1 to 10 μm. They are usually rod-shaped (bacillus), spherical (coccus), or spiral (spirillum). Prokaryotic cells also vary in how they utilize sugar. Many bacteria conduct only **aerobic respiration** (oxygen used in breaking down sugar), while others only conduct **anaerobic respiration**. Some prokaryotic cells and others can do either depending on the oxygen content in their environment.

Eukaryotic cells in plants, animals, and fungi contain membrane-bound organelles and, therefore, can be quite diverse. In higher organisms, cells specialize; that is, they have distinct jobs. Human RBCs, for example, are the shape of doughnuts; they contain no nucleus and only a few mitochondria (see Figure 2.18). Human RBCs carry oxygen through the bloodstream and do not reproduce. Plant leaf cells, specialized for photosynthesis, are packed with chloroplasts. These plant cells are usually rectangular in shape to allow sunlight to reach the greatest number of chloroplasts (see Figure 2.19).

Cell variety is largely dependent on the type and number of organelles present. The organelles are, in turn, composed of molecules ranging in size from a few to several million atoms. Ultimately, it is the molecules present in organelles, cells, organs, and organisms that determine the diversity in activity found in the living world.

Whether you become a lab technician, a research associate, or a scientist who studies molecules or cells, you will have to maintain cells and cell cultures at some point (see Figure 2.20). Developing knowledge of cell structure, cell function, and the standard procedures of cell culture is imperative.

aerobic respiration (aer•o•bic res•pi•ra•tion) utilizing oxygen to release the energy from sugar molecules

anaerobic respiration (an•aer•o•bic res•pi•ra•tion) releasing the energy from sugar molecules in the absence of oxygen

Cell Picture Show

Microscopy is critically important in finding the answers to some of the most pressing scientific questions including the research and development of products to understand and treat disease, clean the environment and fuel our economy. Some of the most fascinating micrographs (microscopic photographs) of cells and their parts have been visualized with Zeiss microscopes. ZEISS is one of the world leading manufacturers of microscopes with a diverse range of light, electron, and fluorescent microscopes.

Zeiss supports a website called the Cell Picture Show. The Cell Picture Show features an incredible array of images that researchers have posted from their trailblazing research.

1. Go to www.cell.com/pictureshow and use the labeled links to visualize four amazing micrographs.
2. For each micrograph, record a few sentences describing what the specific kind of microscopy revealed.

Reproduction   Find the light micrograph showing sperm and an egg cell during a failed fertilization attempt.

Vision   Find the image taken by a confocal microscope showing stimulated retinal neurons (color receptor nerve cells).

Cell Motility   Find the human keratinocyte (skin cell), seen using fluorescent microscopy, responding to epidermal growth factor.

Immunology   Find the scanning electron micrograph of a human dendritic immune cell interacting with a lymphocyte (white blood cell) at "The Immune Synapse".

Section 2.2   Review Questions

1. Which of the following structures are found in prokaryotic cells: a nucleus, ribosomes, mitochondria, a plasma membrane, or one or more chromosomes?
2. Which of the following structures are found in eukaryotic cells: a nucleus, ribosomes, mitochondria, a plasma membrane, or one or more chromosomes?
3. Describe the relationship between chromosomes, messenger RNA (mRNA), and proteins.
4. Explain how so many cells from the same organism can look so different from each other.

2.3   The Molecules of Cells

Engineered molecules are the basis of many biotechnology products. In subsequent chapters, we will look at how cells and molecules are engineered. This section presents a brief discussion of the **macromolecules** (large molecules) found in cells with an emphasis on the structure and function of the molecules most involved in biotechnology applications.

Cells are composed of a variety of molecules. Some molecules are rather small, made up of only a few atoms. These molecules are important for chemical reactions within the cell. The most important of these are water (H_2O), carbon dioxide (CO_2), oxygen (O_2), and salt (NaCl). Although water molecules are small, they are very significant to the living condition. Approximately 75% of the mass of a cell is water. All of the organelles and molecules of a cell are bathed in a watery solution. If the concentration of water in a cell varies significantly, it can be deadly to the cell, since it could influence the concentration of many molecules.

Many molecules found in cells are much larger than a few atoms. Most molecules involved in the structure and function of a cell range from medium size (24+ atoms) to very large molecules (billions of atoms). These **organic** molecules contain carbon and are produced only in living things. Some examples are protein molecules (see Figure 2.21), nucleic acids, lipids, and **carbohydrates**. You may be familiar with these terms from discussions of foods, since food products come from plants, animals, or fungi.

Most of the very large molecules in a cell are found in structural components, such as the cell wall, plasma membrane, or **cytoskeleton**. Many of the enzymes involved in photosynthesis, respiration, or other synthesis reactions are large and complicated. Regulatory molecules (hormones) or transport molecules (such as hemoglobin) are also made up of thousands of atoms.

Macromolecules, the large molecules of cells, are often composed of repeating units chained together. The smaller units are called **monomers** ("mono" means one). When they are linked together into larger molecules, they are called **polymers** ("poly" means many). Think of a polymer as a molecular necklace and the monomers as beads on the necklace. A necklace changes depending on the type of beads used (glass, wood, gold, or gems). Likewise, the great variety of polymers found in nature is the result of a large assortment of monomer molecules being chained together.

The following section describes the four main classes of macromolecules (carbohydrates, lipids, proteins, and nucleic acids) that give structure and function to cells.

Carbohydrates

A carbohydrate is defined as a compound with carbon, hydrogen, and oxygen atoms in a 1:2:1 ratio (or slight variations to the ratio). The term carbohydrate is often used to describe **monosaccharides** and **disaccharides**, as well as **polysaccharides**. Many people have heard of the simple carbohydrates, such as the monosaccharide (glucose and **fructose**) and disaccharide (**sucrose** and **lactose**) sugars. Polysaccharides are the largest carbohydrate molecules. They are long polymers composed of many glucose (or variations of glucose) monomers. The best known polysaccharides are plant **starches**, such as **amylose** or **amylopectin**, and **cellulose**, the long fibrous molecules of plant cell walls (see Figure 2.22).

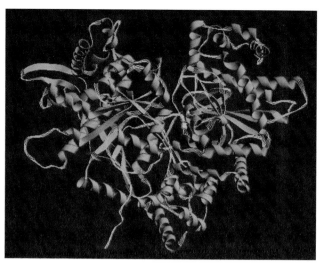

Figure 2.21. This computer-generated model shows the long, twisted nature of cathepsin K, a protein that degrades other proteins. The strand is actually a long necklace of smaller molecules (amino acids) chained together and twisted into helical shapes.
© Corbis.

macromolecule (mac•ro•mol•e•cule) a large molecule usually composed of smaller repeating units chained together

organic (or•gan•ic) molecules that contain carbon and are only produced in living things

carbohydrates (car•bo•hy•drates) one of the four classes of macromolecules; organic compounds consisting of carbon, hydrogen, and oxygen, generally in a 1:2:1 ratio

cytoskeleton (cy•to•skel•e•ton) a protein network in the cytoplasm that gives the cell structural support

monomers (mon•o•mers) the repeating units that make up polymers

polymer (pol•y•mer) a large molecule made up of many repeating subunits

monosaccharide (mon•o•sac•cha•ride) the monomer unit that cells use to build polysaccharides; also known as a "single sugar" or "simple sugar"

disaccharide (di•sac•cha•ride) a polymer that consists of two sugar molecules

polysaccharide (pol•y•sac•cha•ride) a long polymer composed of many simple sugar monomers (usually glucose or a variation of glucose)

fructose (fruc•tose) a 6-carbon sugar found in high concentration in fruits; also called fruit sugar

sucrose (su•crose) a disaccharide composed of glucose and fructose; also called table sugar

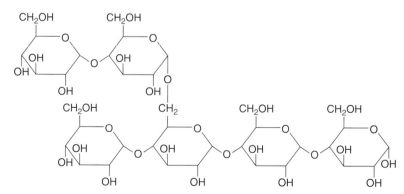

Figure 2.22. **Structural Formula of Amylopectin.** Amylopectin is one form of plant starch, and amylose is another. Plant starch, such as cornstarch, is a key ingredient in many foods.

Computer-generated Molecular Models

One of the first things a biochemist needs to know about a molecule is its three-dimensional structure, because it helps to explain the molecule's function and mode of action. Data to create a three-dimensional computer image of a protein molecule come from crystallizing the protein and using X-ray crystallography.

The National Center for Biotechnology Information (NCBI) has a web-based program called Cn3D that allows the display of numerous models of molecules important in biotechnology:

1. Go to the www.ncbi.nlm.nih.gov/home/analyze.shtml website and click on the "Cn3D" link.

2. On this page, read the summary describing what the Cn3D macromolecular structure viewer can do. If the Cn3d application is not already on your computer, click on the platform (Windows or Macintosh) and download the application.

3. Once you have the application downloaded, go to the top of the Web page and do a search for the structural model of two proteins (collagen and keratin). Collagen and keratin are structural proteins found in skin and hair, respectively.

   Make sure the database selected is "Structure" then type in 2FSE to pull up the collagen model. Click "view structure" in the right-hand box to download the model in Cn3D.

   Make sure the database selected is "Structure" then type in 3TNU to pull up the keratin model. Click "view structure" in the right-hand box to download the model in Cn3D.

4. Use the menu choices at the top of Cn3D to learn how to view and rotate molecules. Write a summary of the similarities and differences in the structures of collagen and keratin as seen in their models.

---

lactose (lac•tose) a disaccharide composed of glucose and galactose; also called milk sugar

amylose (am•y•lose) a plant starch with unbranched glucose chains

amylopectin (am•y•lo•pec•tin) a plant starch with branched glucose chains

glycogen (gly•co•gen) an animal starch with branched glucose chains

Polysaccharides Polysaccharides are excellent structural and energy-storage molecules because of their long polymer structure. Storage polysaccharides include plant starch (amylose) and animal starch (**glycogen**). Structural polysaccharides include cellulose (in plant cell walls) and chitin (found in fungal cell walls and in insect exoskeletons). Long, rigid cellulose molecules are made when glucose molecules link together in long, straight polymers. Cellulose fibers are food molecules for microorganisms that live in the gut of termites. These tiny organisms possess enzymes that can break down cellulose to glucose.

Plants store large amounts of glucose as starch molecules (amylose or amylopectin). In times of low food production (eg, winter or drought), plant enzymes split off the glucose units, making them available for respiration (energy production). The main structural difference in the polysaccharide polymers is the way in which the glucose monomers are connected, although their physical characteristics (texture and color, etc) may be quite different from one polysaccharide to another.

Carbohydrates inside and covering plant cells present problems to biotechnologists trying to isolate proteins and DNA from cells. Polysaccharides become sticky compounds, which can interfere with purification procedures.

Monosaccharides Monosaccharides ("single sugars") are the monomer units that cells use to build polysaccharides. Monosaccharides are also called simple sugars since they include several 5- and 6-carbon sugars that exist in cells as single-ringed sugar molecules (see Figure 2.23).

The most well known monosaccharide is glucose. Glucose is the sugar produced during photosynthetic reactions. Glucose is a 6-carbon sugar that has the molecular formula, $C_6H_{12}O_6$. It has 6 carbon atoms, 12 hydrogen atoms, and 6 oxygen atoms. Figure 2.23 shows the 24 atoms of a glucose molecule and how the carbons bond into a ring shape through a bond with an oxygen atom.

Glucose is an energy molecule. Biotechnologists often use glucose as the food source for cell cultures. Cells break the bonds in glucose, releasing energy in a form that cells can use. This is called **cellular respiration**. Cells also store glucose

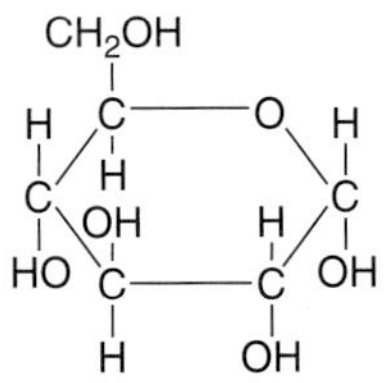

Figure 2.23. Structural Formula of Glucose. Glucose is a 6-carbon sugar ($C_6H_{12}O_6$) produced by plants during photosynthesis. Most cells use glucose as an energy source.

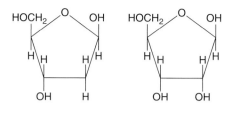

Figure 2.24. Structural Formula of 5-carbon Sugars. Deoxyribose (left) and ribose (right) are structural 5-carbon sugars found in the nucleic acids, DNA and RNA, respectively. Do you see the difference in their structure?

monomers in larger polymer molecules, including the disaccharides, maltose, sucrose, and lactose, as well as most polysaccharides. These polymers can be broken down at a later time to use the glucose. There are several other 6-carbon sugars, including fructose (the sugar that makes honey so sweet) and galactose (part of the lactose molecule found in milk). These differ from glucose in the way that their atoms are arranged. In cells, fructose and galactose are converted to glucose and used for energy.

Some other important monosaccharides are the 5-carbon sugars (see Figure 2.24). The 5-carbon sugars ($C_5H_{10}O_5$) are structural molecules that are always found as part of a larger molecule. Two important 5-carbon sugars are **deoxyribose**, found in DNA molecules, and **ribose**, found in RNA.

Disaccharides Disaccharides are produced when enzymes form a bond between two monosaccharides. Plants often manufacture disaccharides as a way to store or transport glucose for future use. The most important disaccharides are sucrose (also known as table sugar), maltose (malt sugar), and lactose (milk sugar). Figure 2.25 illustrates the structure of maltose, and how it breaks down to gluose.

Sucrose is made when fructose and glucose are chemically combined. Sugar cane and sugar beet plants produce large amounts of sucrose, which are processed and marketed. Sucrose is the molecule in which these plants store glucose. Lactose, the sugar that gives milk its slightly sweet taste, is made from glucose and galactose. Lactose is one method mammals use to store glucose energy.

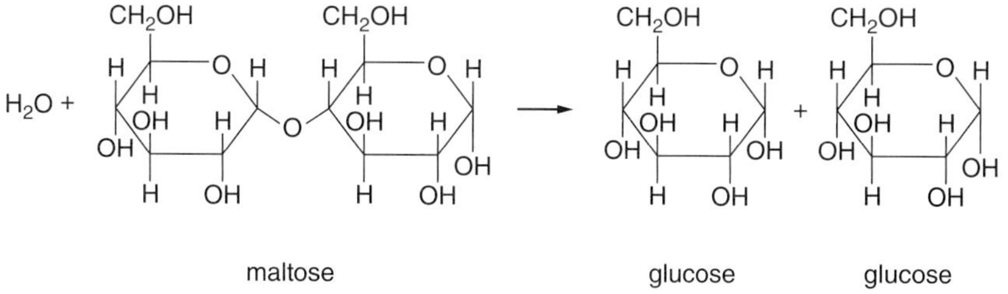

cellular respiration (cell•u•lar res•pir•a•tion) the process by which cells break down glucose to create other energy molecules

deoxyribose (de•ox•y•ri•bose) the 5-carbon sugar found in DNA molecules

ribose (ri•bose) the 5-carbon sugar found in RNA molecules

$$H_2O +$$

maltose → glucose + glucose

Figure 2.25. Structural Formula of Maltose. Maltose is a disaccharide composed of two glucose molecules bound at carbon No. 1 and carbon No. 4. When organisms digest maltose, the bond holding the glucose monomers together is broken and energy is released.

Lipids

Lipids are very different from carbohydrates in both structure and function. Lipids are often referred to as hydrocarbons because they are composed of carbon and hydrogen atoms, and they generally have only a few, if any, oxygen atoms. Their chemical nature makes them insoluble in water (**hydrophobic**).

There are three general groups of lipids, which differ chemically and functionally from each other. One group, the **triglycerides**, includes animal fats and plant oils. These molecules are nutritionally and medically important. Triglycerides are energy-storage molecules.

Phospholipids are another group of lipids. These are found primarily in a cell's membranes. Phospholipids and triglycerides are similar in structure except that phospholipids contain phosphate groups that make them slightly water-soluble on one side of the molecule.

hydrophobic (hy•dro•pho•bic) repelled by water

triglycerides (tri•glyc•er•ides) a group of lipids that includes animal fats and plant oils

phospholipids (phos•pho•lip•ids) a class of lipids that are primarily found in membranes of the cell

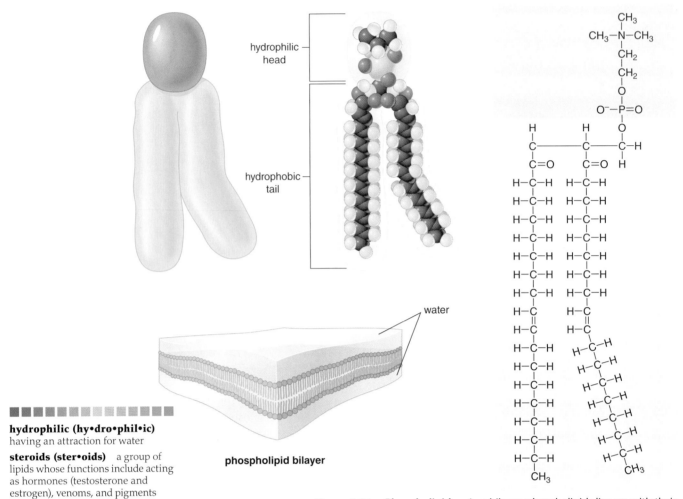

hydrophilic
head

hydrophobic
tail

water

phospholipid bilayer

Figure 2.26.   Phospholipids.   In a bilayer, phospholipids line up with their hydrophobic tails facing each other and their hydrophilic ends facing away. The bilayer creates a barrier through which only certain molecules can pass.

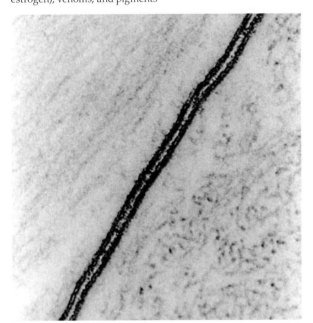

Figure 2.27.   This electron micrograph shows the boundary between two cells. Each cell has a plasma membrane composed of a phospholipid bilayer. Phospholipids must be removed from cell samples during preparations of nucleic acids and proteins. A common practice is to dissolve the plasma membranes with detergents. ~100,000X
© University of Oxford.

Phospholipids are composed of two fatty acid chains (the monomer units) attached to a glycerol molecule (see Figure 2.26). Attached to the end of the glycerol is a phosphate group. The phosphate group has a net negative charge, which makes the phosphate end polar and **hydrophilic** (does not repel water). The fatty-acid chains are hydrophobic and *do* repel water. When phospholipids are grouped together, the differences in the ends cause them to line up in a certain orientation. Fatty acids line up toward each other, and phosphate groups line up away from fatty acids and toward watery solutions within the phospholipid bilayer (see Figure 2.26). The phospholipid bilayer is a membrane through which few molecules can pass. However, protein channels embedded in the bilayer at regular intervals allow certain molecules to pass through the membrane.

Cells have an outer membrane (the plasma membrane) and many inner membranes (the endoplasmic reticulum and the membranes in and around organelles, such as mitochondria) composed of phospholipid bilayers (see Figure 2.27). To extract molecules and organelles for study, researchers must dissolve or remove the lipid membrane to release the cell contents.

The third group of lipids is the **steroids**. Steroids are composed of three overlapping 6-carbon rings bound to a single 5-carbon ring, as shown in Figure 2.28. These complex molecules have several functions, which include acting as hormones (testosterone and estrogen), venoms, and pigments. Cholesterol is an important steroid because it is found in the cell membranes of most eukaryotic cells.

Proteins

Some people might say that proteins are the most important of the cellular molecules. It is estimated that more than 75% of the dry mass of a cell is protein. In a biotechnology company, since proteins are often the manufactured product, it is typical to employ more than 50 to 75% of the scientific staff in protein research and manufacturing. The importance of proteins in the industry is reflected in an expression popular in biotechnology circles: "Where DNA is the flash of biotechnology, proteins are the cash!"

Proteins (or parts of a protein) fall into nine different categories, depending on their function (see Table 2.1). Within a group, for example, antibodies, the structures of protein molecules might be very similar. Or, the molecules within a group can be very different from one another, as in some of the protein pigments. Some proteins have different structures or functions even in the same organism. Keratin, for example, is a component of the humpback whale's hair and the baleen in its mouth (see Figure 2.29).

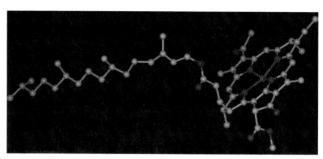

Figure 2.28. **Computer-Generated, Structural Formula of a Steroid. This molecular model shows the four hydrocarbon rings that are found in steroids. Estrogen, testosterone, and cholesterol are steroids that act as hormones.**
© New York University.

Figure 2.29. **Humpback whales, like other mammals, have hair composed of keratin protein molecules. In addition, the baleen food-filtering system in their mouths is composed of keratin protein.**
© Yann hubert/Shutterstock.

Table 2.1. **Proteins Grouped by Function**

Protein Groups by Function	Examples	Specific Function
structural	collagen	component of skin, bones, ligaments, and tendons
	dystrophin	muscle cell cytoskeleton support fibers
	keratin	component of hooves, nails, and hair
enzyme	amylase	converts starch to sugar
	alcohol dehydrogenase	breaks down alcohol
	lysozyme	breaks down bacterial cell walls
transport	aquaporin	water transport through cell membranes
	cytochrome C	moves electrons through the electron transport system
	dopamine active transporter	transports neurotransmitters back into nerve cells after a nerve impulse
contractile	myosin	involved in muscle contraction
	actin	involved in muscle contraction
	tubulin	component in spindle fibers that moves chromosomes during cell division
hormone/ growth factor	Follicle-stimulating hormone (FSH)	regulation of puberty and sex cell development
	thyroxine	modified amino acid (not a protein) that regulates cell metabolism
	Epidermal growth factor (EGF)	involved in DNA synthesis and cell division
antibody	IgA	recognizes foreign molecules in body secretions
	gamma globulin	recognizes a variety of foreign proteins
	IgE	causes allergic reactions when in too high a concentration
pigment	melanin	modified amino acid (not a protein); pigment in human cells
	rhodopsin	light-absorbing pigment in eyes
	hemoglobin	red pigment in RBCs
recognition	gp120	protein on HIV surface
	CD4	protein on T-helper cell surface
	C-reactive protein (CRP)	recognizes dead and dying cells
toxins	Botox® (botulinum)	neurotoxin (stops nerve impulses) made by Clostridium botulinum
	tetanus toxin	neurotoxin from the bacterium, Closteridium tetani
	diphtheria toxin	from Corynebacterium diphtheriae; causes heart and breathing failure

Figure 2.30. **Polypeptide Strand.**
A polypeptide strand is made of amino acids
connected to each other through peptide
bonds. A folded, functional polypeptide chain
is called a protein. Each protein has a specific
amino acid sequence and folding pattern.

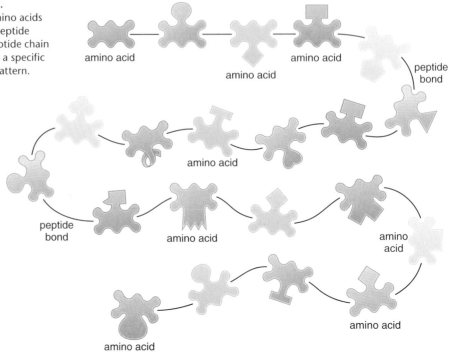

R-group the chemical side-group
on an amino acid; in nature, there
are 20 different R-groups that are
found on amino acids

glycine

serine

tryptophan

**Figure 2.31. Amino Acids—
Glycine, Serine, Tryptophan.**
Each amino acid has a core
consisting of a central carbon
atom attached to an amino group
(–NH$_2$) and a carboxyl group (–
COOH). The difference in the 20
amino acids is what is added (the
R-group) to the central carbon.
The R-group can be as simple as
an H atom, as in glycine, or an –
CH$_2$OH, as in serine, or as complex
as an indole group (2 rings), as
in tryptophan.

A typical cell produces more than 2000 different proteins, some in small quantities and
others in large quantities. For example, although insulin is only one of the proteins made
in a pancreas cell, millions of copies of insulin are produced in a single pancreas cell.

Proteins are the workhorses of the cell. Each protein conducts its particular function
because of its specific structure. The structure of a protein is determined by its amino
acid sequence (see Figure 2.30).

Amino acids are small molecules, the monomers of proteins (see Figure 2.31).
When bound together, the resulting long chains are called polypeptides. Polypeptides
are not functional until they fold into a particular three-dimensional shape. Folded
polypeptide chains are called proteins. The way a polypeptide folds into a functional
protein is determined by the amino acids in the chain and by their order. The amino
acid sequence is ultimately determined by a cell's DNA code (see Figure 2.13, which
illustrates the "Central Dogma of Biology").

There are 20 different amino acids found in proteins. The molecular formulas of the
different amino acids are shown in Table 2.2. Notice how all the amino acids have a part
that is identical and a part that is unique. The unique section is called the **R-group**,
which results in the unique characteristic of each amino acid. Also, the chemical
nature of each R-group results in attractions and repulsions of certain amino acids. For
example, at neutral pH values the R-group of glutamic acid is negatively charged and
is attracted to the positively charged R-group of arginine. The various folding patterns
for a polypeptide chain are the result of these interactions. The variety of proteins in
organisms is a result of both the sequence of amino acids and their interactions.

The R-groups of protein chains can interact between proteins as well. Many
proteins function by attracting or repelling other protein chains. Many of the
recognition proteins, antibodies, enzymes, and protein hormones work in this fashion
(see Figure 2.32). Chapter 5 presents additional information on the structure, function,
and study of proteins.

The arrangement of amino acids in a protein is determined by the genetic code in
the DNA of a cell's chromosome(s). The structure and function of DNA are discussed
briefly below and in more detail in later chapters.

Table 2.2. Molecular Structure of Amino Acids. The 20 amino acids are found in different quantities and arrangements in different proteins. "Ph" stands for phenol ring. Notice in the linear structural formula of each amino acid, the core of the amino acid is the same only the R-group is different.

Amino Acid	Three-Letter Abbreviation	One-Letter Abbreviation	Linear Structural Formula
alanine	ala	A	$CH_3 - CH(NH_2) - COOH$
arginine	arg	R	$HN = C(NH_2) - NH - (CH_2)_3 - CH(NH_2) - COOH$
asparagine	asn	N	$H_2N - CO - CH_2 - CH(NH_2) - COOH$
aspartic acid	asp	D	$HOOC - CH_2 - CH(NH_2) - COOH$
cysteine	cys	C	$HS - CH_2 - CH(NH_2) - COOH$
glutamine	gln	Q	$H_2N - CO - (CH_2)_2 - CH(NH_2) - COOH$
glutamic acid	glu	E	$HOOC - (CH_2)_2 - CH(NH_2) - COOH$
glycine	gly	G	$H - CH(NH_2) - COOH$
histidine	his	H	$NH - CH = N - CH = C - CH_2 - CH(NH_2)COOH$
isoleucine	ile	I	$CH_3 - CH_2 - CH(CH_3) - CH(NH_2) - COOH$
leucine	leu	L	$(CH_3)_2 - CH - CH_2 - CH(NH_2) - COOH$
lysine	lys	K	$H_2N - (CH_2)_4 - CH(NH_2) - COOH$
methionine	met	M	$CH_3 - S - (CH_2)_2 - CH(NH_2) - COOH$
phenylalanine	phe	F	$Ph - CH_2 - CH(NH_2) - COOH$
proline	pro	P	$NH - (CH_2)_3 - CH - COOH$
serine	ser	S	$HO - CH_2 - CH(NH_2) - COOH$
threonine	thr	T	$CH_3 - CH(OH) - CH(NH_2) - COOH$
tryptophan	trp	W	$Ph - NH - CH = C - CH_2 - CH(NH_2) - COOH$
tyrosine	tyr	Y	$HO - Ph - CH_2 - CH(NH_2) - COOH$
valine	val	V	$(CH_3)_2 - CH - CH(NH_2) - COOH$

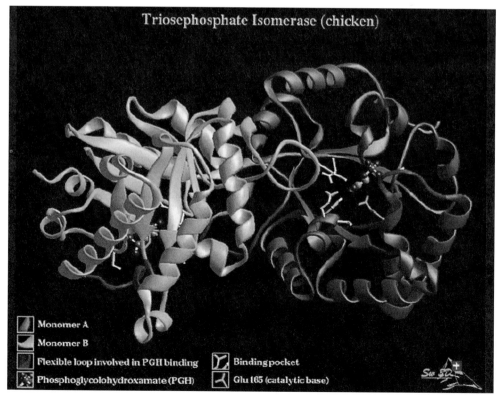

Triosephosphate Isomerase (chicken)

Monomer A
Monomer B
Flexible loop involved in PGH binding
Phosphoglycolohydroxamate (PGH)
Binding pocket
Glu 165 (catalytic base)

Figure 2.32. This is a computer image of the enzyme, triosephosphate isomerase, which converts one 3-carbon sugar to another during cellular respiration. The enzyme is composed of two subunits (shown in purple and gold). Each binds phosphoglycolohydroxamate at the active site in the center of the subunits.
© Corbis.

The Raw Materials of Biotechnology

Nucleic Acids

Nucleic acids are the fourth major group of macromolecules. These information-carrying molecules direct the synthesis of all other cellular molecules. Ultimately, each protein, carbohydrate, and lipid molecule's production can be traced back to genetic information stored in the sequence of the nucleic acid, DNA, which is packaged in the chromosomes of the cell (see Figure 2.33).

DNA is one of the two main types of nucleic acids; **ribonucleic acid (RNA)** is the other. Nucleic acids are long, complex molecules composed of four monomer units called **nucleotides**. Each nucleotide has a single- or double-ringed nitrogenous base group, a 5-carbon sugar ring, and a phosphate group (PO_4). Two of the four nucleotides found in a DNA molecule are shown in Figure 2.34.

A nucleic acid is made when a series of these nucleotides are linked together in a very long necklace. The necklace of nucleotides can be either a single chain (RNA and a few viral DNA molecules) or a double chain (most types of DNA and a few double-sided RNA molecules). Figure 2.35 illustrates the structure of DNA.

DNA and RNA molecules are similar to each other in structure, but differ from each other in how they function. DNA is located in the nucleus of eukaryotic cells and in the cytoplasm of bacteria cells. It is a very large molecule (the largest molecule in the universe), made up of two strands of nucleotides closely bound together. The entire DNA double helix may contain millions of nucleotides. The arrangement of nucleotides on the DNA

ribonucleic acid (RNA) (ri•bo•nu•cle•ic a•cid) the macromolecule that functions in the conversion of genetic instructions (DNA) into proteins

nucleotides (nu•cle•o•tides) the monomer subunits of nucleic acids

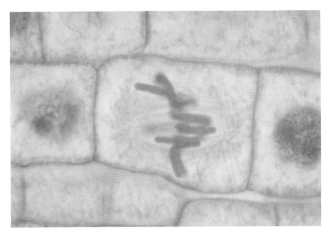

Figure 2.33. **This onion-root tip-cell shows chromosomes lining up during cell division. The chromosomes are primarily composed of long threads of the nucleic acid, DNA. 400X**
© Lester V. Bergman/Corbis/Getty Images.

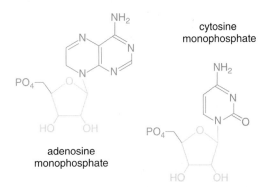

Figure 2.34. **Two Nucleotides.** A nucleotide is a molecule composed of a nitrogenous base (in pink), a 5-carbon sugar (in yellow), and a phosphate group (in blue).

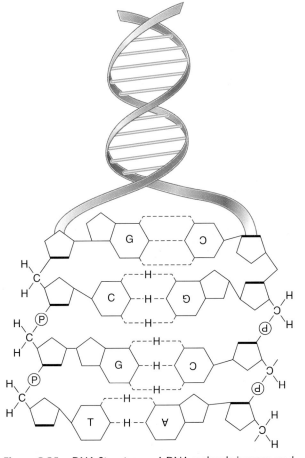

Figure 2.35. **DNA Structure.** A DNA molecule is composed of two strands of nucleotides. Each nucleotide contains a phosphate group (P), a sugar molecule (5-C ring), and a nitrogenous base, either an adenine (A), cytosine (C), guanine (G), or thymine (T). Nitrogenous bases have either a single or a double ring. Nitrogenous bases on one strand bond to a complementary nitrogenous base on the other strand. Adenine bonds with thymine, and guanine bonds with cytosine.

molecule translates into a set of instructions for the production of a cell's or an organism's proteins. This blueprint of molecular construction is what some people call "the genetic code." It is passed on from one generation of cells to another when a cell reproduces.

The RNA molecule is relatively long, but only a fraction of the size of a typical DNA molecule. There are several kinds of RNA (to be discussed later), but they are similar in that each one is composed of ribose-containing nucleotides. The RNA molecules are single-stranded. As in DNA, each nucleotide in RNA is composed of a single- or double-ringed nitrogenous base group, a 5-carbon sugar ring (ribose), and a phosphate group (PO_4).

The RNA molecule is synthesized from a DNA template molecule. Some RNA molecules (mRNA) function in the transfer of genetic information from the chromosomes (DNA) to the ribosomes where proteins are made. At the ribosome, other RNA molecules (transfer RNA [tRNA] and ribosomal RNA [rRNA]) translate the genetic code into the amino acid code of proteins.

Genetic information lies in the arrangement of nitrogenous bases on the DNA molecule. Five different nitrogenous bases are found in nucleic acids: adenine (A), cytosine (C), guanine (G), thymine (T), and uracil (U). Only A, C, G, and T are found in DNA. RNA contains A, C, G, and U (instead of T).

The sequence of bases on a DNA molecule is read three bases at a time. If the arrangement of nitrogenous bases on a DNA strand is CGGATGACCATACCCTT, then it is read as CGG/ATG/ACC/ATA/CCC/CTT and codes for the six amino acids: alanine, tyrosine, tryptophan, tyrosine, glycine, and glutamic acid. (You will learn how to decipher the code in Chapter 5.)

If the nitrogenous bases read GGGACTAGGACCT-TAAACGGC, then seven other amino acids will be in the protein. Herein lies the secret of how a DNA molecule with only four different nitrogen bases can result in the huge number and variety of proteins.

How the DNA code is read (transcribed into mRNA) and translated into the amino acid sequence is discussed at length in Chapter 5. But, for now, you can imagine how we might change, manipulate, add, or delete from the A, C, G, and T code and, thus, alter the amino acid sequence of a protein. If we could change the code, we might be able to give an organism new molecules and new characteristics, or fix genetic mistakes. This could substantially improve the quality of life for humans, including people who suffer from various illnesses caused by an error in the genetic code (see Figure 2.36).

Figure 2.36. Sickle cell disease occurs at a higher frequency in African Americans than in other groups in the United States The disease is the result of a single nucleotide substitution (error) in the hemoglobin genetic code. As a result, a single amino acid is changed in hemoglobin, and the protein folds incorrectly. Malformed (sickle-shaped) cells block blood vessels and cause organ damage.
© Associated Press Photo/Keith Srakocic.

Companies employ genetic engineers to isolate and alter the DNA codes for a particular protein or group of proteins. Sometimes the protein (eg, insulin) itself is the product of interest. The goal of the company would be to manufacture the protein in large enough amounts to sell in the marketplace. Sometimes the manipulated protein gives the organism a desired characteristic, as in the protection from the corn borer insect given to genetically engineered Syngenta Bt corn by Syngenta International AG, Switzerland. Syngenta Bt field corn is marketed in the United States, Canada, Argentina, and South Africa under the NK® brand YieldGard®, which is a registered trademark of Monsanto Company.

Section 2.3 Review Questions

1. Which of the following are monosaccharides: cellulose, sucrose, glucose, lactose, fructose, or amylopectin?
2. Which of the following molecules are proteins that function as hormones: estrogen, insulin, human growth hormone, testosterone, or cholesterol?
3. What distinguishes one amino acid from another?
4. How are the terms nucleotide, nitrogenous base, and nucleic acid related to each other?

2.4 The "New" Biotechnology

As you have learned, organisms and their products have been harvested and improved for centuries. The advances in agricultural products and medicines are numerous and well demonstrated in such examples as the breeding of Angus cattle, high-protein wheat, and the discovery and purification of antibiotics, such as penicillin (see Figure 2.37).

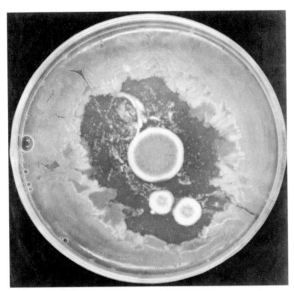

Figure 2.37. *Penicillium sp* mold inhibits the growth of bacteria around it. Scientists isolate the antibiotic penicillin from the mold, and doctors prescribe it (or modified versions of it such as ampicillin or amoxicillin) for certain bacterial infections.
© Bettmann/Getty Images.

The most significant breakthrough in the manipulation of plant and animal cells occurred when scientists learned how to move pieces of DNA within and between organisms. A key was the discovery of enzymes that cut DNA into fragments containing possibly one or more genes. These DNA pieces could be separated from each other and pasted together using other enzymes. In this way, new combinations of genetic information were formed. The resulting molecules were called recombinant DNA (rDNA).

Usually, rDNA contains fragments of DNA from different organisms. They are novel molecules, not in existence anywhere else. They have been "engineered." If the DNA fragments contain genes of interest, such as those that code for desired products, they can be pasted into vector molecules and carried back into cells. Once in cells, they are transcribed and translated into protein molecules that the recipient cells have never produced. Since these proteins code for novel characteristics in the recipient organism, a new organism is made. The organism has been "genetically engineered." Using rDNA technology and genetic engineering, scientists can synthesize new versions of organisms that have never before existed on earth.

The first genetic engineering took place in 1973 when Stanley Cohen, then of Stanford University, Herb Boyer, of the University of California, San Francisco, and Paul Berg, of Stanford University, excised a segment of amphibian DNA from the African clawed toad, Xenopus, and pasted it into a small ring of bacterial DNA called a plasmid. The new recombinant plasmid contained DNA from two species (a bacterium and an amphibian). The recombinant plasmid was inserted into a healthy *E. coli* cell. The *E. coli* cell read the toad DNA code, as if it had been there all the time, and synthesized molecules encoded for on the recombinant DNA, in this case, toad ribosomal RNA.

The first genetically engineered product to reach the marketplace was human insulin for the treatment of diabetes. People with diabetes are unable to make or respond to insulin, which is involved in the absorption of sugar into cells from the bloodstream. Using techniques similar to the toad DNA genetic engineering project, scientists transferred the human insulin gene into a bacterial plasmid. The rDNA plasmid was inserted into *E. coli* cells, which read the DNA and synthesized insulin molecules. The cells were grown in large volumes, and then the insulin protein was purified out of the cell culture. The FDA approved recombinant human insulin (rh-insulin) for marketing in 1982 (see Figure 2.38).

There are many reasons why scientists might want to make recombinant human insulin. One is that diabetic patients either do not make enough insulin or their insulin does not function properly. There is a very large market for insulin since many children born with Type I diabetes (formerly called juvenile-onset diabetes) require daily injections of insulin via needles or a newer delivery system (see Figure 2.39). Until the 1980s, diabetic patients had to use insulin derived from cow, sheep, or pig pancreases. Livestock insulin works well in many patients; however, for some patients, these forms of insulin cause allergic reactions or do not perform up to expectations.

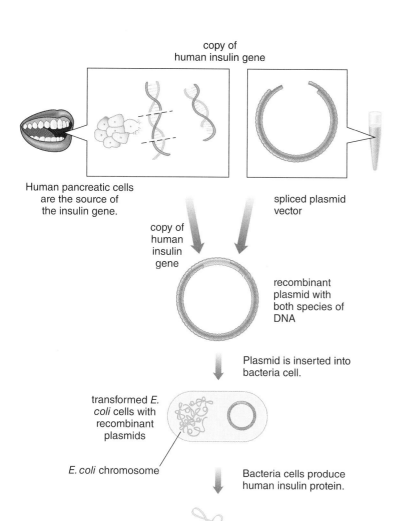

copy of
human insulin gene

Human pancreatic cells
are the source of
the insulin gene.

spliced plasmid
vector

copy of
human
insulin
gene

recombinant
plasmid with
both species of
DNA

Plasmid is inserted into
bacteria cell.

transformed *E.
coli* cells with
recombinant
plasmids

E. coli chromosome

Bacteria cells produce
human insulin protein.

rhinsulin
(recombinant human insulin)
made in bacteria

Figure 2.39. **Novo Nordisk's Innovo Insulin Delivery.**
The biotechnology industry includes companies that
produce instruments. Until recently, insulin had to be
injected using a needle attached to a syringe. Innovo® by
Novo Nordisk, Inc. was one of the first automated pen and
cartridge insulin delivery systems. Other new innovations
include inhalation insulin systems and the insulin
pump, which is programmed to automatically deliver
a predetermined amount of insulin via a tiny catheter
inserted into the abdomen.
Photo courtesy of Novo Nordisk.

Figure 2.38. **Genetic Engineering of Insulin.**
Human insulin genes were transferred into bacteria
cells. The bacteria synthesized human insulin
following the directions on the newly acquired DNA.

Another disadvantage of livestock insulin is that the price and availability fluctuate
with the price and availability of these animals.

In 1976, venture capitalist Robert Swanson and biochemist Dr. Herb Boyer, of
the University of California, San Francisco, decided there were compelling reasons
to produce human insulin commercially. With rh-Insulin as their first product, they
founded the first biotechnology company, Genentech, Inc., located in South San
Francisco, California (see Figure 2.40). Because of the success of human insulin and
other genetically engineered pharmaceutical products, Genentech, Inc. has grown into
one of the largest pharmaceutical companies in the world.

BIOTECH ONLINE

Biotech Products Make a Difference

Go to **agenda.weforum.org/2013/02/how-could-
biotechnology-improve-your-life** and summarize the 10
ways in which biotechnology could improve the quality of life
for humanity.

Figure 2.40. Genentech, Inc. started in South San Francisco, California. As Genentech grew, employees ventured out and started their own biotechnology firms. Now, "South City" is home to over 70 biotechnology companies.
Photo by author.

Genentech, Inc. currently markets, or is developing, several pharmaceuticals, including Activase®, a recombinant tissue plasminogen activator (t-PA), for the treatment of some types of heart attack and stroke; Nutropin®, a human growth hormone for the treatment of some forms of short stature; Rituxin®, an antibody for the treatment of a cancer called B-cell non-Hodgkin's lymphoma; and Pulmozyme®, an inhalation system for use by cystic fibrosis patients.

The products named above are examples of how inserting human DNA into other cells can result in the production of human proteins that can be used as pharmaceutical products. Since the 1980s, hundreds of other biotechnology companies have produced many different kinds of rDNA products. Table 2.3 gives a sample of the breadth of genetic engineering firms (from www.bio.org).

Table 2.3. **rDNA Biotech Companies** These well known rDNA biotechnology companies produce and market recombinant proteins made through genetic engineering.

Company	Location/Nearby University	Genetically Engineered Product Example
Regeneron Pharmaceuticals, Inc.	Rensselaer Polytechnic Institute, Rensselaer, New York	PRALUENT® (alirocumab) to treat familial hypercholesterolemia
Monsanto Co.	St. Louis, MO/Washington University	bovine somatotropin, a hormone to increase milk production
Promega Corp	Madison, WI/University of Wisconsin, Madison	made-to-order recombinant plasmids for R&D
Genzyme Corp	Cambridge, MA/Harvard University	Cerezyme® to treat Gaucher disease
Genencor International, Inc	Palo Alto, CA/Stanford University	Chymogen® milk-curdling enzyme
Eli Lilly and Company	Indianapolis, IN/Purdue University	Humalog® rh-Insulin
Biogen Idec.	Cambridge, MA/Harvard University	Avonex® multiple sclerosis therapy
Amgen, Inc	Thousand Oaks, CA/University of California, Los Angeles	EPOGEN® anemia therapy

Section 2.4   Review Questions

1. What term is used to describe DNA that has been produced by cutting and pasting together pieces of DNA from two different organisms?
2. What organism was the first to be genetically engineered?
3. What was the first commercial genetically engineered product?
4. Explain why South San Francisco, California calls itself "The Birthplace of Biotechology."

Chapter Review

Summary

Concepts

- Living things can be as simple as a unicellular (one cell) organism, or as complicated as a multicellular (many cells) organism, such as animals, made up of cells, tissues, organs, and organ systems.
- All living things exhibit certain characteristics of life, including growth, reproduction, respiration, and response to stimuli.
- To understand how an organism functions, one needs to know its structure of atoms, molecules, organelles, and other components.
- All cells have DNA within chromosomes, cytoplasm, ribosomes, and cell membranes.
- Eukaryotic cells contain specialized organelles that carry out complicated functions for the cell and the organism.
- Protein production is a common function in all cells. The DNA code holds instructions for protein synthesis. It is transcribed into mRNA and then translated or decoded into protein molecules at ribosomes. Differences in cells are largely due to the proteins that are produced at any given time.
- The macromolecules of the cell include carbohydrates, lipids, proteins, and nucleic acids. These large molecules are polymers made up of repeating units called monomers. The monomers

of polysaccharides are monosaccharides. The monomers of proteins are amino acids. The monomers of nucleic acids are nucleotides.

- Proteins make up over 75% of the dry weight of a cell. There are several thousand kinds of proteins, and an average cell produces more than 2000 different proteins. Differences in proteins are due to the number and sequence of amino acids in each polypeptide chain. Most proteins can be categorized into one of nine functional groups. Enzymes, antibodies, and hormones are proteins of particular importance to biotechnologists.
- In the 1970s, scientists learned how to create rDNA molecules and to transfer them into cells. This created modified organisms that could produce a new variety of proteins. Using genetic engineering advances, scientists have been able to create cells that can act as pharmaceutical factories or manufacturers of other products important to industry and consumers.

Lab Practices

- Since scientists cannot study every organism individually, model organisms are studied in detail to represent larger groups. *E. coli* is a model bacterium that has been used extensively in genetic and biotechnology studies. *Aspergillus* and yeast are also model organisms that represent eukaryotic cells.
- Each organism has a set of optimum conditions for growth, reproduction, and protein synthesis. If an organism is to be used in the laboratory or in manufacturing, the optimum conditions for growth must be determined through controlled experimentation. Growth in less-than-optimum conditions will result in less protein synthesis.
- Eukaryotic cells are much easier to observe than prokaryotic cells. Under a compound microscope they are usually 10 to 100 times larger than a typical prokaryotic cell. Many structures in eukaryotic cells are large enough to observe, including chloroplasts and other plastids, vacuoles, the nucleus, and the cell wall.
- All cells have a plasma membrane, cytoplasm, ribosomes, and one or more *chromosomes*. Sometimes it is difficult to observe these structures in the lab because compound microscopes only resolve to the size of an average prokaryotic cell, and all these structures are smaller.
- Learning how to adjust the light and to focus properly is essential for accurate microscopic observations.
- Cells are measured in micrometers (μm). A μm is equal to 0.001 mm. Most cells are between 1 and 100 μm in length. Typically, plant cells are the largest cells, and prokaryotic (bacteria) cells are the smallest.
- Small changes in molecular formula or structure can result in major differences in characteristics.
- Proteins are sensitive to slight changes in the pH (acid/base level). At some pH above or below the optimum pH, proteins begin to unwind (denature), cease to function, and will eventually come out of solution. It is important to determine the optimum pH for a protein to ensure that its function is preserved.

Thinking Like a Biotechnician

1. List three examples each of prokaryotic and eukaryotic cells.
2. Identify the four groups of macromolecules found in living things. From each group, give a specific example of a molecule and its function.
3. Describe the relationship between amino acids, polypeptides, and proteins.
4. Explain how differences in protein structure allow one protein molecule to recognize another protein molecule.
5. If more than 75% of a cell's dry weight is protein, what makes up the remaining 25% of the dry weight?
6. DNA molecules can be unzipped down the center and each side replicated. Look at Figure 2.34 (DNA structure). Propose a method by which two new strands could be produced from an existing DNA strand.
7. A colleague asks you to determine whether a cell culture is composed of prokaryotic or eukaryotic cells. You have a compound microscope at your lab station, and you can make

slides of the sample. What structures would you look for to distinguish between prokaryotic and eukaryotic cells?

8. It is often difficult to get large molecules to dissolve in a watery solution. Based on size, which of the following molecules should dissolve in water most readily: cellulose, hemoglobin, glucose, or amylose?

9. Look at Table 2.2 (Molecular Structure of Amino Acids). Based on similar R-groups, propose a scheme to divide the 20 amino acids into four smaller groups based on similarities in structure.

10. Restriction enzyme molecules cut DNA molecules into smaller pieces. DNA ligase molecules can paste cut pieces back together. Propose a method by which a scientist could create an rDNA molecule that carries genes for high levels of chlorophyll production and genes to resist frost damage. Into what kind of organism might you want to insert the new rDNA and why?

Biotech Live

Activity 2.1

Biohazards: Knowing When You Have One

In biotechnology research and manufacturing, organisms of all kinds, and the molecules in them, are grown and manipulated. Some of these materials can be dangerous when used inappropriately.

Using the Internet, find out more about biohazards and how to deal with them by finding the answers to the questions below. Answer the following questions, and list the website URL used as a reference.

1. Define the term "biological materials," and list five examples.
2. Define the term "biohazard," and describe three examples.
3. Give an example of how a specific biohazardous material should be handled and disposed of properly. Include examples of disinfectants and how they are used. Be prepared to share this information with your classmates.

Activity 2.2

Macromolecules in Your Food!

The foods we eat are all of plant, animal, or fungal origin. A potato is a plant stem. A sweet potato is a root. Hamburger is ground-up beef muscle. Cheese is fermented milk. A mushroom is a fungus.

Go to biotech.emcp.net/nutritiondata and locate information on the composition of foods.

1. Look up the nutritional information for six foods of your choice, three of plant origin and three of animal origin. Make a chart that shows the percentages of proteins, fats, and carbohydrates in each food.
2. Compare the plant and animal food nutritional data. What generalizations can you make about the molecular compositions of plants or animals?

Activity 2.3

What is the American Type Culture Collection?

Explore the American Type Culture Collection (ATCC) website at biotech.emcp.net/atcc and learn what samples and services are available from ATCC.

After studying the ATCC website, click on the links below and answer the following questions:

At "About ATTC":
1. In what year was ATCC established and for what purpose?
2. In what city and state is ATCC located?

At "Ethical Standards for Obtaining Human Materials"
3. How many cell lines are maintained by ATCC for public health research?

At the "Products" link:
4. List a few examples of the types of samples and services available through ATCC.
5. For what diseases could ATCC provide cells that could be used in research?

Use the search engine on ATCC to answer the following:
1. Name the strain(s) of *Shewanella oneidensis* bacteria available through ATCC.
2. ATCC houses samples of cells that cause several sexually-transmitted diseases. Name two.
3. How many Biosafety Level 2 viruses does ATCC carry in the virus collection?
4. If for research purposes, you need some bone marrow-derived human stem cells, does ATCC have them? If so, how much does it cost, for how much sample? Is the media to grow them available from ATCC?

Using Scientific Journals Online

Many journals publish some or all of their articles online. To find journal articles or summaries of publications, you can use a searchable database. A searchable database is a collection of Web pages or articles that have been published or posted by an interested group. There are searchable databases on virtually every topic. Well-known databases include the Smithsonian Institute's Art Collection, the National Institutes of Health (NIH) Medical Library, and Medline. Some publishers have searchable databases as well.

PubMed is a searchable database of particular interest to biotechnologists, especially those who are working on topics of medical or pharmaceutical interest.

 Access the PubMed database at the NIH website www.ncbi.nlm.nih. gov/pubmed. At the PubMed home page, do a search for HIV. The results are displayed as "hits." A hit looks like the following listing, with the journal name, *Trends in Microbiology*, displayed after the author and title:

Rinaldo CR Jr, Piazza, P. Virus infection of dendritic cells: portal for host invasion and host defense. *Trends Microbiol.* 2004:Jul;12(7):337–345.
1. How many hits (items) did you get?
2. List the first three hits (author, title, journal name, date, and page numbers).
3. Search on PubMed for the virus infection article listed above. After reading the short summary of the article, name something you would like to know more about.

How to Read a Scientific Journal Article

Scientific journal articles can be daunting to read and understand. Usually the reader knows much less than the author about the science and terminology used in the report. However, scientific articles are the best way to learn about the most current experimentation of some scientific phenomenon or process. The more you read scientific journal articles, the easier the task will be.

 Completing the steps below, read and report on a scientific journal article about telomeres, background radiation, and aging.

1. To get you "in the mood" to read a scientific journal article, view the YouTube video titled "How to read a scientific article II…."
2. Go to **biotech.emcp.net/PubMed** and do a search for *"Telomere length in human adults and high level natural background radiation."* It will bring you to a link for an article by Das B, Saini D, Seshadri M., published in PLoS One. 2009 Dec 23;4(12):e8440.PMID: 20037654.
3. Click on the **"Free PMC Article"** links that bring you to the full text article.
4. Now, do the following, and record notes for each thing you do in your notebook:
   a. Read the title and find definitions to any words you do not know.
   b. Skim the article looking at the section headings and any data tables, charts, graphs, figures, or images. Be sure to read the captions to these.
   c. Go back and read the abstract and/or summary and then, at the end of the paper, the discussion and/or conclusion.
   d. Make a list of key points you think the article is making with some data to back them up.
   e. Now, read the entire article, editing your key points for accuracy.

Conducting "Exhaustive" Research

Every scientific and nonscientific employee at a biotechnology company will have reason to conduct exhaustive research at some time. Exhaustive research means that you find, collect, and catalog virtually everything that has been written on a particular subject of interest. Exhaustive research requires the use of all the research tools available to you in libraries and on the Internet.

These could include:

- An automated card catalog (DYNIX, for example)
- InfoTrac® (by Gale Group): summaries of journal/magazine articles
- NewsBank® (by NewsBank, Inc.): summaries/abstracts of newspaper articles
- BioDigest: summaries/abstracts of journal/magazine articles on biological topics
- The Reader's Guide to Periodical Literature: citations of journal/magazine articles
- The Internet/World Wide Web (WWW) search engines and databases
- PubMed

The "Domains of Biotechnology" chart (in Chapter 1) demonstrates the wide diversity of products and applications of biotechnological research. Some of the areas in which biotechnology companies, universities, and governmental organizations focus are listed below.

 Use library resources and/or the Internet to gather information about one of these areas of biotechnology. Create a folder to contain the information you gather and a fact sheet or poster to teach about the topic you study.

fermented foods/beverages	medicines developed from molds/plants
selective breeding	DNA identification/analysis
DNA fingerprinting	genetic testing/diagnosis
genetic screening	genetically engineered bacteria and fungi
transgenic plants	transgenic animals
vaccines	human gene therapy
monoclonal antibodies	tissue culture
polymerase chain reaction (PCR)	DNA sequencing
Human Genome Project	microarray technology

1. If necessary, get instruction on how to use the library resources available to you.
2. Gather information about your topic. Conduct a thorough search. A minimum of two references from each library resource tool available is required.
3. Visit the PubMed database at the National Institutes of Health (NIH) and search the site for primary research papers. The site can be found at: **biotech.emcp.net/PubMed**.
4. In either an electronic folder or a "hard-copy" manila folder, put the information collected during your search, including bibliographical information, into a manila folder. The folder should have a table of contents listing the articles collected and enough reference information, addresses, or library call numbers to easily locate each document again. Number the articles in the folder.
5. Follow these examples to set up the table of contents:

Page	Reference Source	Bibliographical Information	Other Helpful Information
1	WWW	biotech.emcp.net/bio	pages 1–3 of 5 pages
4	DYNIX	Caldwell T., Genes, 1997:667–670.	
8	BioDigest	Sequencing Genes, Time, Sept 3, 1997:45.	

6. Create a fact sheet, PowerPoint® slide, or poster *with a minimum of text* that has the following information about your topic:
   - Title of the research topic, names of researchers, and date (centered at top)
   - Definition (a definition or explanation of the topic studied)
   - Examples (several explanations and diagrams of the products or services made or used in the topic area)
   - Companies (companies/facilities that make or use the technology)
   - Other interesting information (examples, stories, photos, diagrams, recommended websites)

Bioethics

STOP! You cannot use THOSE cells.

Stem cells have been in the news a lot lately. Many people feel strongly that we should be able to use embryonic stem cells for any medical purpose. Others believe that there is no good reason to ever use embryonic stem cells. Still others believe that embryonic stem cells are suitable for some medical applications and not others. Stem cells are so controversial that President George W. Bush created a federal policy for their use in research funded by the US government.

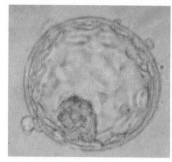

A 6-day-old human embryo (also called a blastocyst). Stem cells are obtained by growing out the inner cell mass of the blastocyst. The inner cell mass is clearly visible in this picture (it is the clump of cells at about the 6-o'clock position within the embryo). 100X
Photo courtesy of Joe Conaghan, PhD.

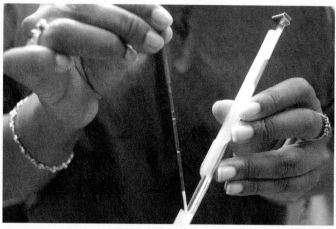

An embryologist takes up frozen embryos from a cane (white item) in which they have been stored in liquid nitrogen. Fertility clinics must store or destroy extra unused embryos after in vitro fertilization. What do you think should be done with extra embryos? ~1000X
© Carlos Avila Gonzalez/San Francisco Chronicle//Getty Images.

 Conduct research to examine the use of embryonic stem cells in research and in the development of medical therapies. Evaluate the benefits and risks of using embryonic stem cells, and present a balanced review of a controversial issue.

1. Using the Internet, find information to answer the following questions. Record all of the bibliographic information, including website addresses, for the documents you use as references.
   a. What are embryonic stem cells?
   b. How do scientists produce, harvest, and use embryonic stem cells?
   c. What is the value in using stem cells?
   d. What are the risks of and arguments against using stem cells?

2. Describe three reasons to use embryonic stem cells in research and manufacturing. Give three reasons to not use them. Consider legal, financial, medical, personal, social, and environmental aspects.

3. Create a poster that accurately explains what embryonic stem cells are and why their use is controversial. Include the pros and cons of their use from item 2. Include numbers, data, and photos to make your poster more informative and convincing. Try to avoid your personal view, and give a thorough presentation of both sides of the issue.

Biotech Careers

Photo courtesy of Wing Tung Chan.

Materials Management

Wing Tung Chan
Formerly of Cell Genesys, Inc.
Hayward, CA

In the biotechnology industry, many manufacturing facilities grow cells that produce protein pharmaceuticals. The facilities operate 24 hours a day, 7 days a week. The production staff must have the supplies and materials needed to keep the cells alive and producing protein at a maximum rate. Once the pharmaceutical protein is in high enough concentration, the production staff works to harvest the protein product from the cell cultures.

As a production planner in a cancer vaccine manufacturing facility, Wing is responsible for coordinating the ordering and availability of production and manufacturing supplies. Along with other employees who work as inventory control analysts, Wing must ensure that the raw materials, including chemical reagents, plastics, glassware, and other instruments, are available to support protein production on a timely basis. Much of her time is spent working with other employees to assess their needs and with supply vendors to schedule orders and deliveries.

3 The Basic Skills of the Biotechnology Workplace

Learning Outcomes

- Determine the most appropriate tool for measuring specific volumes or masses
- Describe how to select, set, and use a variety of micropipets within their designated ranges to accurately measure small volumes
- Convert between units of measurement using the B ⇆ S rule and appropriate conversion factors
- Recognize the different expressions for units of concentration measurements and use their corresponding equations to calculate the amount of solute needed to make a specified solution or make a dilution
- Describe what pH is and why it is important in solution preparation
- Explain what a buffer is, how one is made, and why buffers are used.

3.1 Measuring Volumes in a Biotechnology Facility

Imagine that you are working as a biotechnology technician, and you are about to begin a long series of experiments. Each experiment requires several ingredients, including solutions containing tiny amounts of DNA, enzymes, and other chemical reagents. These solutions must be prepared accurately since reactions depend on the right reagents in exactly the right amounts. To be skillful in making measurements and preparing solutions, a technician must learn how to use precision instruments with care and accuracy. In this chapter, you will learn how to make the calculations and measurements necessary to prepare solutions accurately. Measuring the **volume** of liquids is discussed below. Measuring **mass** is discussed in the next section. Solution preparation, pH and buffers are covered in later sections.

Volume is a measurement of the amount of space something occupies. In a laboratory, liquid volumes are traditionally measured in **liters** (L), **milliliters** (mL), or **microliters** (μL). A milliliter is one-thousandth of a liter or about equal to one-half teaspoon. A microliter is one-thousandth

volume (vol•ume) a measurement of the amount of space something occupies

mass (mass) the amount of matter (atoms and molecules) an object contains

liter (li•ter) abbreviated "L"; a unit of measurement for volume, approximately equal to a quart

milliliter (mill•i•li•ter) abbreviated "mL"; a unit measure for volume; one one-thousandth of a liter (0.001 L) or about equal to one-half teaspoon

microliter (mi•cro•li•ter) abbreviated "μL"; a unit measure for volume; equivalent to one-thousandth of a milliliter or about the size of the tiniest teardrop

Figure 3.1. **Quart Milk Carton and 1-Liter Carton.** A liter contains 33.81 ounces, while a quart contains 32 ounces.

of a milliliter, or about the size of the tiniest teardrop. It is helpful to try to visualize these amounts. Figure 3.1 compares the volume of liquid in a liter with the volume of liquid in a quart.

Depending on the volume to be measured, three different types of tools or instruments are used: **graduated cylinders** (Figure 3.2), **pipets** (see Figures 3.3a, b), and **micropipets** (see Figure3.4a and 3.4b). A technician must be able to select the right instrument, use it properly, and report the appropriate units of measurement for each.

Figure 3.2. **Graduated cylinders are used to measure volumes between 10 mL and 2 L.** Photo by author.

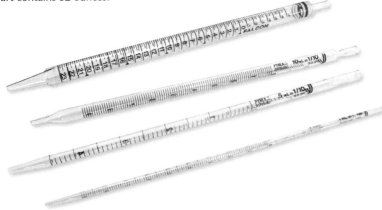

Figure 3.3a. **Pipets are available that measure volumes between 0.1 mL and 50 mL. Shown from left to right are 25-, 10-, 5-, and 1-mL pipets.** Photo by author.

Figure 3.3b. **Most biotechnology labs use 25-, 10-, 5-, 2-, and 1-mL pipets. This worker is using a 25-mL pipet and an electronic pipet pump.** Photo courtesy of Cell Genesys, Inc.

graduated cylinder (grad•u•at•ed cyl•in•der) a plastic or glass tube with marks (or graduations) equally spaced to show volumes; measurements are made at the bottom of the meniscus, the lowest part of the concave surface of the liquid in the cylinder

pipet (pi•pet) an instrument usually used to measure volumes between 0.1 mL and 50 mL

micropipet (mi•cro•pi•pet) an instrument used to measure very tiny volumes, usually less than a milliliter

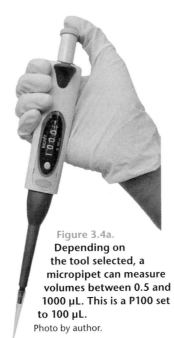

Figure 3.4a. **Depending on the tool selected, a micropipet can measure volumes between 0.5 and 1000 μL. This is a P100 set to 100 μL.** Photo by author.

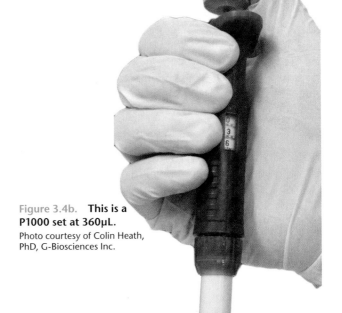

Figure 3.4b. **This is a P1000 set at 360μL.** Photo courtesy of Colin Heath, PhD, G-Biosciences Inc.

Table 3.1. Prefixes Commonly Used in Biotech Metric Measurement

Prefix	Abbreviation	Value	Meaning	Example
kilo	k	10^3 or 1000	1 thousand	kilogram
deci	d	10^{-1} or 0.1	1 tenth	decigram
centi	c	10^{-2} or 0.01	1 hundredth	centimeter
milli	m	10^{-3} or 0.001	1 thousandth	milliliter
micro	μ	10^{-6} or 0.000001	1 millionth	microliter
nano	n	10^{-9} or 0.000000001	1 billionth	nanogram
pico	p	10^{-12} or 0.000000000001	1 trillionth	picogram

Converting Units

Often, volumes are measured in one **unit of measurement** and reported in another. To do this, you must be able to convert between larger and smaller units of measurement. For example, if 0.75 mL of an enzyme is needed for a reaction in a tiny tube, a micropipet that measures in microliters may be the best instrument to use. If so, a technician must be able to quickly convert from milliliters to microliters.

It is easy to convert between metric units because they are all larger or smaller than each other by powers of 10. See Table 3.1. For example, 1 mL is 0.001 L. So, to convert between liters and milliliters, just remember that a milliliter is $1/10 \times 1/10 \times 1/10$ of a liter, which is 1/1000 (3 decimal places or 3 powers of 10) smaller than a liter. To convert from 1 L to a mL, move the decimal point to the right three places. The direction the decimal is moved depends on which way you are converting, bigger to smaller units or smaller to bigger units.

Use the B ⇆ S Rule to know which way to move the decimal. The B ⇆ S Rule shows how to move the decimal point in the value to be converted: to the right (multiplying) if converting from big units to smaller units, or to the left (dividing) if converting from small units to larger ones (see Figure 3.5).

For example, let us say a measurement of 1.25 L of solution is required, but the instrument to be used measures only in milliliters. You must convert from liters to milliliters. Since liters are bigger than milliliters, and there are 1000 mL in a liter, move the decimal to the right three places (for the 3 zeroes in 1000). Thus, 1.25 L = 1250 mL.

Mathematically, the conversion is 1.25 L × 1000 mL/1 L = 1250 mL.

unit of measurement (un•it of mea•sure•ment) the form in which something is measured (g, mg, μg, L, mL, μL, km, cm, etc)

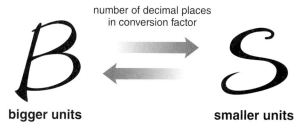

number of decimal places in conversion factor

bigger units   **smaller units**

Figure 3.5. **The B ⇆ S Rule.** To convert between metric units, move the decimals to the left or right based on the difference in the units.

BIOTECH ONLINE

Photo by Paul Robinson.

Bet You Can't Hit a 150-Meter Homer

The right field foul pole at AT&T Park, home of the San Francisco Giants, is 310 feet from home plate. How many meters is that? In biotechnology applications, often measurements made using one kind of unit needs to be converted and reported in another unit of measurement. It is easy to learn how to convert between metric units but sometimes challenging to convert between metric and other units.

A **metrics conversion table** is available at biotech.emcp.net/metric_convert. Use the conversion table to convert the following measurements.

20.0 cm = _____ in   100.0 m = _____ yd   2.0 L = _____ gal
100.0 g = _____ lb   100.0 kg = _____ lb   37°C = _____ °F

conversion factor (con•ver•sion fac•tor) a number (a fraction) where the numerator and denominator are equal to the same amount; commonly used to convert from one unit to another

metrics conversion table (met•rics con•ver•sion ta•ble) a chart that shows how one unit of measure relates to another (for example, how many milliliters are in a liter, etc)

The fraction 1000 mL/1 L is a **conversion factor**. A conversion factor is a number (a fraction) where the numerator and denominator are equal to the same amount but in different units. In this case, 1000 mL equals 1 L. Multiplying 1.25 L by 1000 mL/1 L is the same as multiplying 1.25 L by 1, except that the liter unit cancels out, converting the answer to an equivalent volume, 1250 mL.

Using the conversion factor is the "mathematical way" to do the conversion, but it is much easier to use the B ⇆ S Rule to just move the decimal point. Moving the decimal point to the right three places is the same as multiplying by the conversion factor, 1000 mL in 1 L.

For converting to a larger unit, divide by the conversion factor. How many liters is 75 mL?

$$75 \text{ mL} \times 1 \text{ L}/1000 \text{ mL} = 0.075 \text{ L}$$

Notice how the decimal point moved to the left three places.

As each of the instruments is discussed below, think about the units in which they measure (liters, milliliters, or microliters). Imagine the size of such a unit and how much space it would occupy. Also, think about how to report the volume in a different unit. The B ⇆ S Rule can be used to convert between any metric volume units. It can also be used to convert between mass units, or to convert between length units.

To measure volumes larger than 10 milliliters, technicians usually use a graduated cylinder. A graduated cylinder is a plastic or glass tube with marks (or graduations) equally spaced to show volumes (see Figure 3.6). Measurements are made at the bottom of the meniscus, the lowest part of the concave surface of the liquid in the cylinder.

Using Pipets

Measuring volumes smaller than 10 mL requires a more precise instrument called a pipet (or pipette). Similar to a straw with labeled graduations on it, a pipet is typically used for measuring volumes down to about 0.5 mL.

BIOTECH ONLINE

How Big is Big? How Small is Small?

Size is relative. A "big" ice cream scoop, which is hard to quantify (measure), means different things to different people. Scientists use units of measurements to try to make measurements more meaningful and constant. However, some of the units of measurement are hard to imagine such as a virus with a diameter of 17 nm.

An ingenious animation, called Scale of the Universe, lets you zoom through the objects in the universe comparing a full range of sizes and measurement.

1. Go to **htwins.net/scale2** and begin by sliding the horizontal scroll bar. See how objects can be compared from the entire known universe The is a range of sizes is 62 orders of magnitude—from 1027 meters down to 10⁻³⁵ meters.

2. Scroll the bar to find objects with sizes that fall into the following measurements:
   - The width of our Milky Way Galaxy
   - The distance from the Earth to the Sun
   - The diameter of The Earth
   - The length of a blue whale
   - The size of a grain of rice
   - The diameter of a white blood cell
   - The length of an *E. coli* bacterium
   - The size of the HIV virus
   - The size of a water molecule

Table 3.2. Examples of Pipet Volumes and Graduations

Pipet Volumes (mL)	Graduations (mL)
10	1/10 or 0.1
5	1/10 or 0.1
2	2/100 or 0.02
1	1/100 or 0.01

Disposable pipets are available in a variety of volumes and graduations. Commonly used pipets are listed in Table 3.2. Always pick the smallest possible pipet for the job to help decrease the amount of measuring error (see Figure 3.7).

To draw fluid into the pipet, use a pipet bulb or pump (see Figure 3.8). At no time should anyone attempt to mouth pipet! In other words, **"Never mouth pipet!"** Pipet pumps and bulbs evacuate the air in the pipet, creating a vacuum that causes the liquid to rise to a certain level. Each brand of pipet pump and bulb operates differently.

Using Micropipets

To measure very tiny volumes, less than 1 mL, a more precise instrument called a micropipet is needed. Micropipets measure in microliters. A microliter is a millionth of a liter or a thousandth of a milliliter.

$$1 \ \mu L \ (microliter) = 0.001 \ mL \ (milliliter) = 0.000001 \ L \ (liter)$$

Or another way of looking at it is:

$$1 \ L = 1000 \ mL \qquad 1 \ mL = 1000 \ \mu L$$

You can imagine that 1 μL is a very small volume since 1000 μL equals 1 mL.

Micropipets come in a variety of styles and sizes. All are basically the same in design, varying only slightly from one manufacturer to another. A micropipet is an expensive, delicate instrument that is easily damaged or mishandled. Use a micropipet with caution and handle gently.

A micropipet usually has four parts: the plunger button, the ejector button, the volume display, and the dispensing tip (see Figure 3.9). Learn how to use each part correctly to prevent damage or incorrect measurements.

Picking and Using the Appropriate Micropipet

Printed on the plunger are the maximum, and sometimes the minimum, volumes that can be dispensed with the micropipet. Look at the diagrams in Figures 3.9–3.16 to learn how to determine the volume range.

P-100 or P-200 Micropipet   The units on **P-100** or **P-200** micropipets are read in the same manner, so both are presented here. A P-100 micropipet measures accurately from 100 μL down to 10 μL (see Figure 3.10). A P-200 measures from 200 μL down to 20 μL (see Figure 3.11).

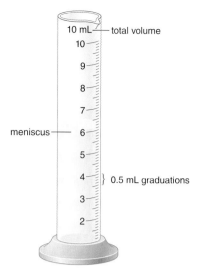
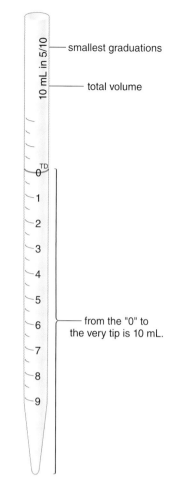

Figure 3.6.   Reading a Graduated Cylinder.   Before using a graduated cylinder, make sure that you know the total volume it will hold and the value of each of the graduations. In the lab, common graduated cylinders include 10 mL, 25 mL, 100 mL, 250 mL, 500 mL, and 1 L.

Figure 3.7.   Selecting a Pipet. Select a pipet that has the smallest volumes and graduations possible to measure the volume you need. If 7 mL is needed, you could use a 25- or 10-mL pipet, but a 10-mL pipet will give less error in measurement.

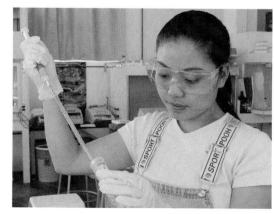

Figure 3.8.   A green pipet pump is used with 10- and 5-mL pipets. Roll up the wheel to draw fluids. Roll down the wheel to evacuate liquid. Pipetting is done near eye level to accurately judge the level of the meniscus. Other types of pumps and bulbs are also used.
Photo by author.

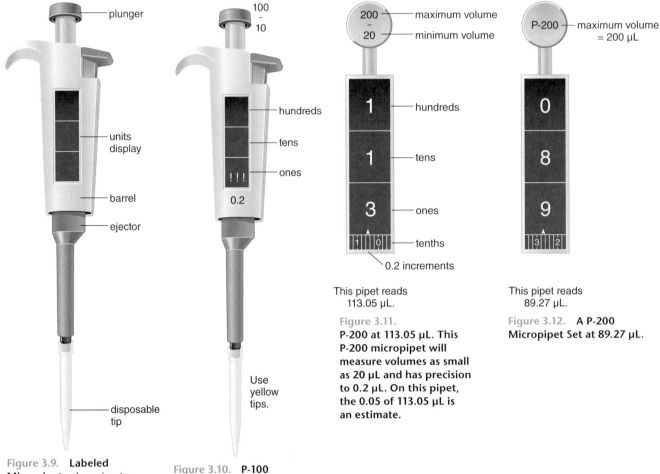

plunger

units display

barrel

ejector

disposable tip

Figure 3.9. **Labeled Micropipet.** Learning to use each part of a micropipet correctly is essential. On the micropipet shown, the plunger has two "stops." Pressing to the first stop evacuates air to the volume in the display. Pressing to the second stop evacuates that volume plus another 50% or so. To ensure accurate measurement, feel the difference between the first and second stop before using the pipet. Inaccurate measurement could waste costly reagents and cause invalid experimental results.

P-100 a micropipet that is used to pipet volumes from 10 to 100 µL

P-200 a micropipet that is used to pipet volumes from 20 to 200 µL

100 – 10

hundreds

tens

ones

0.2

Use yellow tips.

Figure 3.10. **P-100 Micropipet. This micropipet will measure volumes as small as 10 µL and has precision to 0.2 µL.**

200 — maximum volume

20 — minimum volume

hundreds

tens

ones

tenths

0.2 increments

This pipet reads 113.05 µL.

Figure 3.11. **P-200 at 113.05 µL. This P-200 micropipet will measure volumes as small as 20 µL and has precision to 0.2 µL. On this pipet, the 0.05 of 113.05 µL is an estimate.**

P-200 — maximum volume = 200 µL

This pipet reads 89.27 µL.

Figure 3.12. **A P-200 Micropipet Set at 89.27 µL.**

To set a micropipet to withdraw and dispense the proper volume, first determine how to read the volume display. The display shows three numbers. The top number is where the digit for the maximum volume is placed. In the case of a P-200, that would be "2" for 200. The P-200 displayed in Figure 3.12 shows a volume of 89 µL. Therefore, there are no hundreds, 8 tens, and 9 ones on the display. Each pipet manufacturer has a slightly different presentation of units. Look closely at each pipet you use to make sure that you are reading the units accurately.

To change the display, gently turn the adjusting knob(s). Note: Each manufacturer may place the adjustment knob in a different place. Learn how to adjust your pipet before turning any knobs. Sometimes there is an additional release button to press to turn the adjustment knob. At no time should the micropipet's display numbers be turned past their upper or lower limits.

P-10 or P-20 Micropipet Reading pipets of other volumes is not as straightforward as reading the P-200 or P-100. Remember that the maximum value the pipet can measure reveals what the first digit will be. On a **P-10**, for example, the maximum volume it can measure is 10 µL. Therefore, the top digit can go no higher than the "1" tens (for 10 µL). So, the top digit shows the tens place, the second digit shows the ones place, and the third digit shows the tenths place (see Figure 3.13).

If you are using a **P-20** and you want to display 2 µL, place a 0 in the tens place, a 2 in the ones place, and a 0 in the tenths place. To measure 12.5 µL, put a 1 in the tens place, a 2 in the ones place, and a 5 in the tenths place (see Figure 3.14). At no time should the micropipet's display numbers be turned past their upper or lower limits.

P-1000 Micropipet   For the **P-1000**, the maximum volume is 1000 µL (which equals 1 mL). Therefore, the top digit can go no higher than the 1 for 1000 µL. The top digit shows the thousands place, the second digit shows the hundreds place, and the third digit shows the tens place (see Figure 3.15). To display 200 µL, place a 0 in the thousands place, a 2 in the hundreds place, and a 0 in the tens place. The little lines on the bottom are worth 0.2 µL (see Figure 3.16). Remember, at no time should the micropipet's display numbers be turned past the upper or lower limits.

Notice how you can measure 200 µL on both a P-1000 and a P-200, but the P-200 allows you to estimate down to a hundredth of a microliter. The P-1000 does not. Picking the most appropriate measuring tool is critical to precise measurement.

Micropipets have been modified to meet several research and manufacturing needs. In many labs, there are electronic pipets that increase pipeting accuracy by controlling volume uptake and dispensing (see Figure 3.17). Another modification to a basic pipet is a **multichannel pipet** that has a plunger with 4 to 16 tips (see Figure 3.18). Pipeting can be completely automated by the use of pipeting machines or pipeting robots (see Figure 3.19).

P-10   a micropipet that is used to pipet volumes from 0.5 to 10 µL

P-20   a micropipet that is used to pipet volumes from 2 to 20 µL

P-1000   a micropipet that is used to pipet volumes from 100 to 1000 µL

multichannel pipet (mul•ti•chan•nel pi•pet)
a type of pipet that holds 4–16 tips from one plunger; allows several samples to be measured at the same time

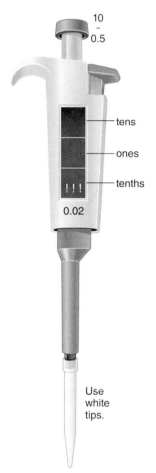

Figure 3.13.   P-10 Micropipet. P-10 micropipets are common in biotechnology labs. A P-10 micropipet will measure volumes as small as 0.5 µL and has precision to 0.02 µL. A P-10 uses tiny tips that are usually white.

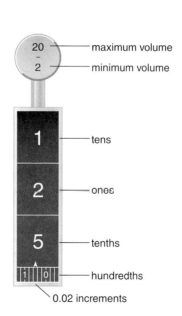

This pipet reads 112.505 µL.

Figure 3.14.   A P-20 Micropipet Set at 12.505 µL.

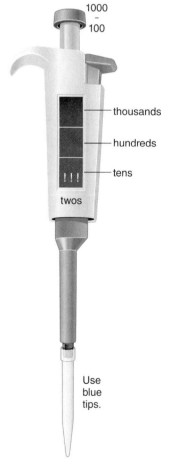

Figure 3.15.   P-1000 Micropipet. A P-1000 micropipet will measure up to 1000 µL, or 1 mL, and uses large tips that are usually blue or white in color.

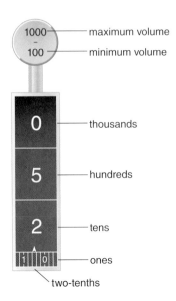

1000 — maximum volume
100 — minimum volume

0 — thousands

5 — hundreds

2 — tens

— ones

two-tenths

This pipet reads
520.5 µL.

Figure 3.16. **A P-1000 Micropipet Set at 520.5 µL.**

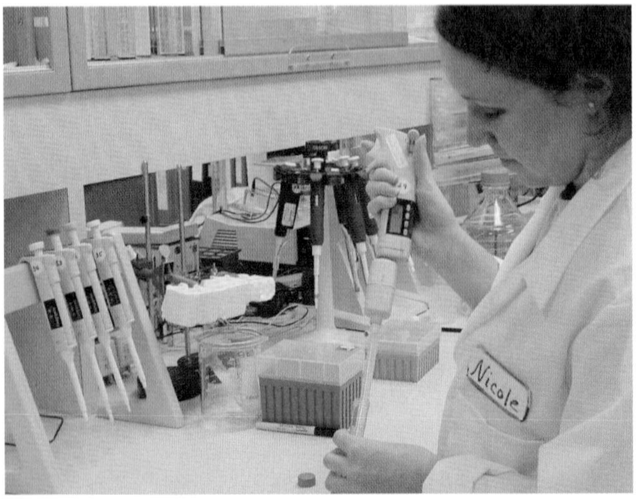

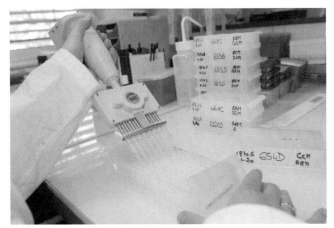

Figure 3.17. **A lab technician uses an electronic micropipet to prepare samples to load into an agarose gel. This pipet can be set to automatically measure and deliver a specific volume.** Photo by author.

Figure 3.18. **A multichannel pipet allows several samples to be measured at the same time, a feature that saves time during an experiment with multiple replications and repetitive pipeting.** © Pierre Schwartz/Corbis.

Figure 3.19. **For dispensing specific volumes to a large number of samples, a pipeting robot may be used. Here a researcher uses a pipetting robot to test enzyme activity in hundreds of samples.The pipet tips are visible at the end of long tubes on the left side of the robot.** Roy Kaltschmidt, Lawrence Berkeley National Lab, The Regents of the University of California, 2010

Positive Displacement Micropipets
A **positive displacement micropipet** is another type of micropipet commonly used in biotechnology laboratories. Find a website that describes how a positive displacement micropipet works and lists some of its uses. Print the page, with the Web address, and highlight the informative parts.

Photo by author.

Section 3.1 Review Questions

1. What instrument would you use to measure and dispense the following volumes? Pick the instrument that is likely to give the least error for each measurement.

     23.5 µL   6.5 mL   125 mL   7 µL   2.87 mL   555 µL

2. Convert the following units to the requested unit:

     1.7 L = _____ mL   235.1 µL = _____ mL   2.37 mL = _____ µL

3. What numbers should be dialed into the P-10 display if a volume of 3.7 µL is to be measured?

4. What instrument should be used if a technician wants to fill 40 sets of 16 tubes all with identical volumes?

3.2 Making Solutions for Biotechnology Applications

Solution preparation is one of the most essential skills of a biotechnology lab employee. **Solutions** are made daily in most labs. Virtually every reaction involves proteins or nucleic acids in an **aqueous** (watery) solution. How the most common types of solutions are prepared is covered in this and following sections.

Solutions are mixtures in which one or more substances are dissolved in another substance. The substance being dissolved is called the **solute**. When sugar is dissolved in water, sugar is the solute. Most often, the solute is a solid, such as sugar, salt, or some other chemical from the chemical stockroom shelves (see Figure 3.20). Sometimes, though, the solute is a liquid, as is the case when 95% ethyl alcohol (ethanol) is diluted to a 70% solution by adding water.

Solid solutes are measured on **balances** or scales. Electronic balances measure the **mass** (or **weight**) of a substance. Balances come in two forms: tabletop/portable/electronic (see Figure 3.21) or analytical balances (see Figure 3.22). The standard unit of mass is the **gram (g)**, although research and development (R&D) labs may measure small masses in milligrams (mg), whereas manufacturing facilities may measure large quantities in kilograms (kg). A gram is approximately the weight of a small paper clip. The tabletop or portable balances vary in the precision to which they measure. Some measure to within 1 g, some measure to within 0.01 g, and some measure down to 0.001 g. The last decimal place is always an approximation. Each balance has a maximum mass that may be measured.

positive displacement micropipet (pos•i•tive dis•place•ment mi•cro•pi•pet) an instrument that is generally used to pipet small volumes of viscous (thick) fluids

solution (sol•u•tion) a mixture of two or more substances where one (solute) completely dissolves in the other (solvent)

aqueous (a•que•ous) describing a solution in which the solvent is water

solute (sol•ute) the substance in a solution that is being dissolved

balance (bal•ance) an instrument that measures mass

weight the force exerted on something by gravity; at sea level, it is considered equal to the mass of an object

gram (gram) abbreviated "g"; the standard unit of mass, approximately equal to the mass of a small paper clip

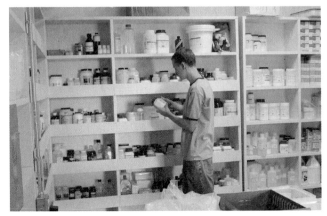

Figure 3.20. Most solutes are solid chemicals. These are stored in a cool, dark chemical stockroom, by chemical reactivity, and then, alphabetically.
Photo by author.

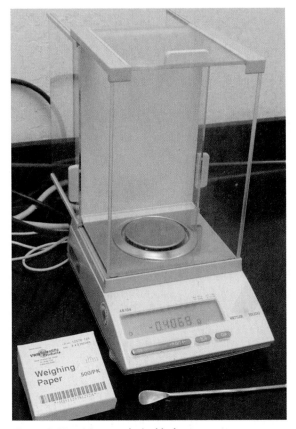

Figure 3.22. Most analytical balances measure down to milligrams, even though they usually report in grams.
Photo by author.

solvent (sol•vent) the substance that dissolves the solute

Figure 3.21. Most electronic balances measure in grams.
Photo by author.

The substance that dissolves the solute is called the **solvent**. Water is the solvent in most solutions, including the sugar in water and the alcohol examples given previously. Most molecules dissolve in water readily, or with some stirring or heating. For all laboratory solutions, either deionized or distilled water is used. Deionized and distilled water have mineral impurities removed that, if present, might interfere with reactions. Tap water is only used for glass washing.

Some molecules do not readily dissolve in water. Many organic molecules, including some proteins and most lipids, are not water soluble. These must be put into solution by dissolving them in other solvents, such as ethanol, acetone, petroleum ether, and even chloroform.

It is important to prepare solutions in clean containers to avoid contamination of solutions with chemicals that can interfere with chemical reactions (see Figure 3.23). When preparing glassware for solution preparation, wash the vessel with laboratory soap and water. Rinse with tap water until no evidence of soap remains. Then, rinse five more times with tap water, and do a final rinse with deionized water, if available (see Figure 3.24).

The amount of solute added to a solvent depends on the solution to be made. The proportion of solute to solvent is called the concentration. Most people are familiar with concentrated solutions or suspensions, such as concentrated frozen orange juice. Concentrated frozen orange juice has a higher ratio of solute molecules to solvent than is present in diluted, drinkable orange juice. To use a highly concentrated substance, such as concentrated frozen orange juice, one must add water (for OJ, usually three cans) to dilute it down to a drinkable concentration with a lower ratio of solute to solvent molecules (see Figure 3.25).

Figures 3.23. Although sterile plastic containers are used to a great extent in biotechnology, glassware is used for many things. Glassware must be very clean for solution preparation. Glass storage areas should be clean and dust free. Broken glass is only discarded in "broken-glass" cartons.
Photo by author.

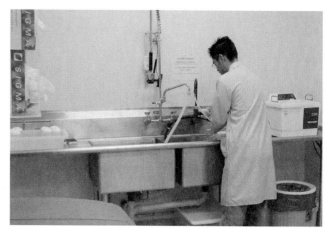

Figure 3.24. Dirt, chemicals, and soap may interfere with chemical and enzymatic reactions. Most companies have special glass-washing areas stocked with special lab detergents and purified water.
Photo by author.

Figure 3.25. Frozen, concentrated orange juice is sold at a concentration of four times stronger (4X more concentrated) than that used. It is diluted with three parts of water to one part of concentrate to make a drinkable solution (1X).
Photo by author.

Table 3.3. Common Concentration Units

Concentration	Common Unit(s) of Measurement	Examples of Solutions Measured
mass/volume	g/L mg/mL µg/mL µg/µL	2 g/L albumin solution 10 mg/mL amylase solution 2 µg/mL hemoglobin solution 4 µg/µL DNA solution
%	%	10% NaOH solution 5% $CuSO_4 \cdot 5H_2O$ 10% SDS (sodium dodecyl sulfate)
molarity	M mM µM	1 M NaCl 50 mM TRIS buffer 50 µM $CaCl_2$

Concentration is measured in several ways in biotechnology labs, including mass/volume, volume/volume, % mass/volume, **molarity**, and **normality** (for acids and bases only). The three most common expressions of concentrations are shown in Table 3.3.

Preparing solutions of mass/volume, percent mass/volume, or molarity concentrations, as well as adjusting the pH of solutions so that they are appropriate for the biomolecules they contain, is discussed in upcoming sections of this chapter.

molarity (mo•lar•i•ty) a measure of concentration that represents the number of moles of a solute in a liter of solution (or some fraction of that unit)

normality (nor•mal•i•ty) a measurement of concentration generally used for acids and bases that is expressed in gram equivalent weights of solute per liter of solution; represents the amount of ionization of an acid or base

3.3 Making Solutions from Scratch

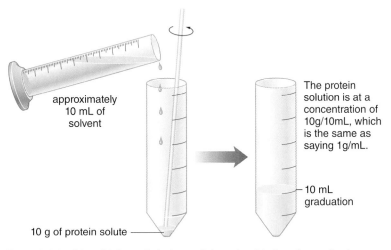

The protein solution is at a concentration of 10g/10mL, which is the same as saying 1g/mL.

approximately 10 mL of solvent

10 mL graduation

10 g of protein solute

Figure 3.26. Mass/Volume Solution. Solvent is added until a total volume of 10 mL is reached. A protein solution that has a concentration of 1 g/mL is considered fairly concentrated.

Solutions used in biotechnology facilities are prepared in one of two ways. Either,

- They are made from scratch, meaning that a certain mass of solute is measured and a specified amount of solvent is mixed into it, until the solute is completely dissolved, or
- They are made by diluting an already prepared solution down to a working concentration.

In this section, you will learn how to make solutions from scratch. These "from scratch solutions" have concentrations reported as either mass/volume, % mass/volume, or as molar amounts. No matter how the solution's concentration is expressed though, it still contains some proportion of solute (mass) in some proportion of solvent. The challenge to the lab worker is to determine what amount of solute to add to what amount of solvent.

Making Mass/Volume Solutions

For these solutions, the concentration is expressed as the amount of mass per some unit of volume (mass/volume). For example, the concentration might be reported as g/L, g/mL, mg/mL, or µg/µL (see Figure 3.26). When proteins are being purified from a solution, concentrations from 1 µg/mL to 1 mg/mL are typical. A concentration of 1 mg/mL means that every milliliter of solution contains 1 mg of protein.

Let us say that you are a lab technician in a group studying hemoglobin, the protein in red blood cells (RBCs) that binds oxygen for transport through the bloodstream. To do a particular test, you need a hemoglobin solution at a concentration of 0.05 g/mL. Calculating how much solute is used in the solution is fairly straightforward. Just multiply the volume of solution desired by the concentration desired. To make the math easy, make sure that all units have been converted so that they are as similar as possible. This equation is called the Mass/Volume Concentration Equation.

Mass/Volume Concentration Equation

$$\underset{\text{concentration desired}}{\underline{\hspace{2cm}}\ \text{g/mL}} \quad \times \quad \underset{\text{volume desired}}{\underline{\hspace{2cm}}\ \text{mL}} \quad = \quad \underset{\substack{\text{to be weighed out,}\\ \text{dissolved in the solvent}}}{\underline{\hspace{2cm}}\ \text{g of solute}}$$

Using the Mass/Volume Concentration Equation ensures that every milliliter of the solution has the same amount of solute in it. Always mix the solute and solvent until the solute completely dissolves. For example, to make 100 mL of a 0.05 g/mL hemoglobin solution, follow this process.

Using the Mass/Volume Concentration Equation,

$$0.05 \text{ g/mL} \times 100 \text{ mL} = 5 \text{ g of hemoglobin}$$

Measure out 5 g of hemoglobin and mix it with enough solvent to reach a final volume of 100 mL.

That is, a concentration of 0.05 g/mL is required, and a volume of 100 mL is desired. This means you need to measure 5 g of hemoglobin. Mix solvent with it until you reach a final volume of 100 mL.

Be careful to look at the units of measurement. It is important to use and report measurements in appropriate units (ones that do not include too many decimal places). Often, the mass/volume is given in smaller or larger units, such as µg/µL or g/L. The units must be converted to more appropriate ones that will cancel out while multiplying, using the B ⇆ S Rule.

Sometimes very dilute protein solutions are needed, such as those measured in µg/mL. A microgram is 1000 times smaller than a milligram. That is three more decimal places! For example, to make 100 mL of a 50-µg/mL hemoglobin solution, only 0.05 g of protein is required.

Using the Mass/Volume Concentration Equation,

$$50 \text{ µg/mL} \times 100 \text{ mL} = 5000 \text{ µg}$$

If your laboratory balance reports in grams, the micrograms must be converted to milligrams, and the milligrams converted to grams. Using the B ⇆ S Rule, convert micrograms to milligrams. Then convert from milligrams to grams.

$$5000 \text{ µg} = 5 \text{ mg} = 0.005 \text{ g}$$

Using an analytical balance, measure out 0.005 g of hemoglobin, and mix it with enough solvent to reach a final volume of 100 mL.

Making % Mass/Volume Solutions

The concentrations of some solutions are given as **percentages**. Remember that percent represents something that is part of 100 (or 100 parts). That is the same as saying "divide it by 100" to determine the percentage. So 50% = 50/100, which equals 0.5. Thus, 50% of 100 equals 50, or half the starting value. To determine this mathematically, let the term "of" represent a multiplication sign, and convert the percentage to its decimal value of 0.5 (by dividing by 100 or by moving the decimal point two decimal places to the left).

percentage (per•cent•age)
a proportion of something out of 100 parts, expressed as a whole number

50% of 100 equals 50, as in the equation, $0.5 \times 100 = 50$
50% of 10 equals 5, as in the equation, $0.5 \times 10 = 5$
50% of 1 = 0.5, as in the equation $0.5 \times 1 = 0.5$
20% of 40 equals what ? $0.2 \times 40 = 8$
10% of 40 equals what? $0.1 \times 40 = 4$
5% of 40 equals what? $0.05 \times 40 = 2$

The use of salt solutions, reported in percentages, is common in a lab. Let us say you need 500 mL of a 10% sodium chloride (NaCl) solution. That is, the solution would have 10 parts salt out of 100 parts of solution. To make a solution of a specific percent mass in a specific volume, convert from the percent desired to a decimal equivalent. Then, multiply the decimal value of the concentration desired by the total volume desired. This equation is called the

% Mass/Volume Concentration Equation

$$\frac{\qquad}{\text{percent value}} \% \qquad = \qquad \frac{\qquad}{\text{decimal value of the g/mL}}$$

$$\frac{\qquad}{\substack{\text{decimal value} \\ \text{(g/mL)}}} \times \frac{\qquad}{\substack{\text{total volume} \\ \text{desired (mL)}}} = \frac{\qquad \text{g of solute to be}}{\substack{\text{measured and added to the} \\ \text{volume desired of solvent}}}$$

So, to make 500 mL of 10% salt solution, convert 10% to 0.1. Then, multiply 500 (mL) by 0.1, which gives you 50. That means that 50 g of salt is needed for the solution.

A commonly used solution in the lab is 2% sodium hydroxide (NaOH). To make a 2% NaOH solution, you calculate 2% of the total volume desired. Add that mass of solute to solvent (deionized water) until the total volume desired is reached.

For example, to make 100 mL of 2% NaOH, measure 2 g of NaOH pellets and place them into an appropriate container. Add water until you reach a total of 100 mL of solution. The math looks like this:

$$\frac{2\%}{\text{percent value}} \qquad = \qquad 2/100 \qquad = \frac{0.02}{\text{decimal value}}$$

Then, using the % Mass/Volume Concentration Equation:

$$\frac{0.02}{\substack{\text{decimal value} \\ \text{(g/mL)}}} \times \frac{100 \text{ mL}}{\substack{\text{total desired} \\ \text{(mL)}}} = \substack{\text{2 g of NaOH pellets to be measured} \\ \text{and mixed with solvent until the final} \\ \text{volume desired (100 mL) is reached}}$$

Another way to think about making % mass/volume solutions is to remember that a 1% solution is 1 g of solute in a total of 100 mL of solution

Another way to look at the math is by setting up a ratio or proportion, as shown below. Remember that 15% is the same as saying 15 g per 100 mL.

$$\frac{\text{Mass}_1}{\text{Volume}_1} = \frac{\text{Mass}_2}{\text{Volume}_2} \quad \substack{\text{For this} \\ \text{example}} \quad \frac{15 \text{ g}}{100 \text{ mL}} = \frac{\text{X}}{15 \text{ mL}} \quad \substack{\text{which} \\ \text{is}} \quad \frac{15 \text{ mL} \times 15 \text{ g}}{100 \text{ mL}} = \text{X}$$

So, X = 2.25 grams dissolved into a total of 15 mL of solution.

Making Molar Solutions

Another way to report the concentration of a solution is by the number of **moles** of a solute in a liter of solution (or some fraction of that unit). This concentration measurement is called molarity, and the unit of measurement is moles/liter (mol/L). Molarity is sometimes a challenging concept to understand, but it is very important since many solutions in a biotech lab are molar solutions.

To understand how to make a molar solution, you need to know what a "mole" is. The unit "1 mole" is the number of molecules of a substance that gives a mass, in grams, equal to its **molecular weight**, also called the formula weight. The formula weight can be determined using the Periodic Table of Elements (see Figure 3.27) as described in the figure caption below.

■■■■■■■■■■■■

mole (mole) the mass, in grams, of 6×10^{23} atoms or molecules of a given substance; one mole is equivalent to the molecular weight of a given substance, reported as grams

molecular weight (mo•lec•u•lar weight) the sum of all of the atomic weights of the atoms in a given molecule

Periodic Table of Elements

1 **H** 1.01																	2 **He** 4.00
3 **Li** 6.94	4 **Be** 9.01											5 **B** 10.8	6 **C** 12.0	7 **N** 14.0	8 **O** 16.0	9 **Fl** 19.2	10 **Ne** 20.2
11 **Na** 23.0	12 **Mg** 24.3											13 **Al** 27.0	14 **Si** 28.1	15 **P** 31.0	16 **S** 32.1	17 **Cl** 35.5	18 **Ar** 40.0
19 **K** 39.1	20 **Ca** 40.1	21 **Sc** 45.0	22 **Ti** 47.9	23 **V** 50.9	24 **Cr** 52.0	25 **Mn** 54.9	26 **Fe** 55.8	27 **Co** 58.9	28 **Ni** 58.7	29 **Cu** 63.5	30 **Zn** 65.4	31 **Ga** 69.7	32 **Ge** 72.6	33 **As** 74.9	34 **Se** 79.0	35 **Br** 79.9	36 **Kr** 83.8
37 **Rb** 85.5	38 **Sr** 87.6	39 **Y** 88.9	40 **Zr** 91.2	41 **Nb** 92.9	42 **Mo** 95.9	43 **Tc** 98	44 **Ru** 101	45 **Rh** 103	46 **Pd** 106	47 **Ag** 108	48 **Cd** 112	49 **In** 115	50 **Sn** 119	51 **Sb** 122	52 **Te** 128	53 **I** 127	54 **Xe** 131
55 **Cs** 133	56 **Ba** 137	57 **La** 139	72 **Hf** 178	73 **Ta** 181	74 **W** 184	75 **Re** 186.2	76 **Os** 190	77 **Ir** 192	78 **Pt** 195	79 **Au** 197	80 **Hg** 201	81 **Tl** 204	82 **Pb** 207	83 **Vi** 209	84 **Po** 209	85 **At** 210	86 **Rn** 222
87 **Fr** 223	88 **Ra** 226	89 **Ac** 227	104 **Rf** 261	105 **Db** 262	106 **Sg** 263	107 **Bh** 262	108 **Hs** 265										

Key: 1 — atomic number; H — element symbol; 1.01 — atomic mass

Elements 58-71 and 90-103 are not shown. The atomic masses have been rounded.

Figure 3.27. Periodic Table. The Periodic Table of Elements shows the elements (atoms) found in compounds (molecules). Each element is listed along with the atomic weight (mass) of each atom in the element. A NaCl molecule has a molecular weight of about 58.5 amu (atomic mass units) because the Na atom weighs 23 amu, and the Cl atom weighs about 35.5 amu. Together, in the NaCl molecule, the atoms total approximately 58.5 amu. The mass of a hydrogen atom equals 1 amu.

An easier way to find the molecular or formula weight of a compound is to just look at the label of a chemical reagent bottle. A mole's worth of molecules is not only defined by mass, but it is also defined by evidence that a mole contains 6×10^{23} of atoms or molecules.

It sounds confusing, but just measure out the molecular weight (MW) in grams of the chemical, and you will have a mole of that compound. You will also have 6×10^{23} molecules (Avogadro's number) of the compound.

Moles are used as a way of counting molecules and/or atoms. Since molecules and atoms are too small to be counted individually, scientists can measure out a mole of molecules and know how many molecules they are getting (see Figure 3.28). Or, they can measure out 2 moles, or half a mole, and always know how many molecules they are getting.

Figure 3.28. Molecules are too small to weigh individually so scientists measure moles or parts of a mole. For a mole of NaCl, weigh out 58.5 g of it (its molecular weight in grams). How much NaCl would you weight out if you wanted 2 moles? Or 0.5 moles?

Na Cl

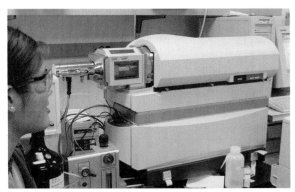

Figure 3.29. **This instrument is a mass spectrometer. Scientists use it to determine the molecular weight of a compound. A "mass spec" can also determine if a sample is contaminated with molecules of different molecular weights.**
Photo by author.

For a mole of NaCl, measure out 58.5 g since the molecular weight of a molecule of NaCl is 58.5 **atomic mass units (amu)**. From the Periodic Table, the molecular weight (or formula weight) of NaCl is 58.5 amu because one Na atom weighs 23.0 amu, and one Cl atom weighs 35.5 amu. Together, the atoms in a NaCl molecule weigh 58.5 amu. For 0.5 mole of NaCl, measure out 29.25 g, because that is one-half of 58.5 g.

Most chemical bottle labels also list the molecular weight of the compound. Note that the terms "formula weight," "molecular weight," and "molecular mass" all mean the same thing.

By definition then, every mole of NaCl weighs 58.5 g. Another way of saying it is that NaCl weighs 58.5 g per mole, or 58.5 g/mol. It is valuable to know the molecular weight of a compound for at least two reasons. One reason is that the molecular weight can be used to identify the molecule. Using an instrument called a **mass spectrometer**, one can determine the molecular weight of a compound so that it can be recognized in a mixture (see Figure 3.29).

Another good reason to know the molecular weight of a compound is that it is used to determine how to prepare molar solutions. Molar solutions have a certain number of moles per unit volume. This relationship, called molarity concentration, is usually reported as the number of moles per liter (mol/L or M), or sometimes as millimoles/liter (mmol/L or mM) or micromoles/liter (µmol/L or µM) if the concentration of solute in the solution is very small. A 1 molar solution (1 M) has 1 mole of solute for every liter of solution.

How is a liter of 1 M NaCl solution prepared? Weigh out 1 mole of NaCl (58.5 g), and place it into an appropriate container. While stirring to dissolve the salt, add deionized water up to a total volume of 1 L. The solution has 1 mole of NaCl per liter of solution and is called 1 M.

What if a 2 M NaCl solution (2 moles/liter) is needed? How much NaCl is weighed out? Multiply a mole's worth of molecules by the concentration desired. A mole's worth of NaCl is 58.5 g. Remember, it is the same as the molecular weight. The mathematical equation looks like this:

$$\underset{\substack{\text{concentration}\\\text{desired}}}{2 \text{ mol/L}} \times \underset{\text{MW}}{58.5 \text{ g/mol}} = \underset{\substack{\text{(water) to a total volume of 1 L.}}}{117 \text{ g of NaCl is mixed with solvent}}$$

Note that in the equation the "mol" units cancel out leaving g/L.

A liter of solution is too large a volume for many situations. In R&D, milliliter volumes are typically used. To determine how to mix up smaller (or larger) volumes of a molar solution, use the Molarity Concentration Equation that takes into account the volume of solution needed, making conversions as neccessary:

Molarity Concentration Equation

$$\underset{\substack{\text{wanted}\\\text{(L)}}}{\text{volume}} \times \underset{\substack{\text{desired}\\\text{(mol/L)}}}{\text{molarity}} \times \underset{\substack{\text{weight of the}\\\text{solute (g/mol)}}}{\text{molecular}} = \underset{\substack{\text{be dissolved in solvent,}\\\text{up to the total volume}\\\text{of solution desired}}}{\text{the number of grams to}}$$

Multiply the molarity desired (mol/L) by the molecular weight (g/mol) as you did above. Then, multiply by the volume desired (L). Convert smaller or larger units to L or mol/L, as necessary.

For example, suppose you want to make 20 mL of a 0.5 M CaCl$_2$ (calcium chloride) solution.

First, convert 20 mL to liters (0.02 L). Then, use the Molarity Concentration Equation.

0.02 L × 0.5 mol/L × 111 g/mol = 1.1 g of $CaCl_2$ with solvent to a total of 20 mL

Notice how the milliliter volume units are converted to liters. That way, the liters and moles can be canceled out. That leaves the answer, the mass of $CaCl_2$, in gram units.

Sometimes a very dilute molar solution is needed. Consider a solution that is 1 mM; that is, 1000 times less concentrated than a 1 M solution. To use the Molarity Concentration Equation, you must convert the molar units from 1 mM (1 mmol/L) to 1 M (1 mol/L). How would you prepare 50 mL of 10 mM $CaCl_2$ solution?

First, convert 10 mM to 0.01 M and 50 mL to 0.05 L.
Then, use the Molarity Concentration Equation.

0.05 L × 0.01 mol/L × 111 g/mol = 0.0555 g (55.5 mg) of $CaCl_2$ with solvent to a total volume of 50 mL

Sometimes the volume or concentration of a solution is so small that no balance in the lab can measure the tiny amount of solute mass that is required. In this case, you need to make a more concentrated solution of solute and dilute it to the desired concentration. Dilutions are discussed later in the chapter.

Section 3.3 Review Questions

1. Which of the following are mass/volume concentration units?
   mg/mL, g/mg, L/mg, µg/µL, or g/L?
2. What mass of salt is needed to make 150 mL of a 100 µg/mL salt solution? Describe how the solution is prepared.
3. What is the decimal equivalent of the following percentages?
   10%   15%   25%   2%   1.5%   0.5%
4. What mass of gelatin (a protein) is needed to make 0.5 L of a 3% gelatin solution? Describe how the solution is prepared.
5. What mass of sodium hydroxide (NaOH) is needed to make 750 mL of a 125 mM NaOH solution? Describe how to prepare the solution.

3.4 Introduction to pH Measurement and Adjustment

pH is a measurement of the number of hydrogen ions in a solution and it is of critical importance in biotechnology because of the effect of pH on cells and biomolecules. The pH value determines whether a solution is an **acid**, a **base**, or is **neutral**. The pH level affects molecular structure and function. Proteins maintain their structure and activity only within certain pH ranges. Solution containing proteins or nucleic acids must be prepared at a specific, constant pH. To make solutions that will maintain the structure and function of biological molecules, one must understand how pH is measured and adjusted.

Specifically, pH is a measurement of the number of **hydrogen ions** (H^+) in solution. Although it is not necessary to know how to calculate it, the pH is the negative base-10 logarithm (-log) of the H^+ concentration. So, pH=$^-$log[H^+].

acid (ac•id)   a solution that has a pH less than 7

base (base)   a solution that has a pH greater than 7

neutral (neu•tral)   uncharged

hydrogen ion (hy•dro•gen i•on) a hydrogen atom which has lost an electron (H^+)

Fortunately for a lab technician, measuring the pH (the concentration of H⁺) is easy and begins with an understanding of the characteristics of water as a solvent. Imagine a beaker of water (see Figure 3.30). Most of the water molecules in the beaker are complete molecules of H_2O. But a small fraction of the water molecules (close to 1×10^{-7} mol/L) ionize (split up) into H⁺ and OH⁻ particles (ions). In fact, one reason that water is a good solvent is because it ionizes and dissolves many molecules. The ionization of water molecules is described by the equation below:*

$$H_2O \leftrightarrows H^+ + OH^-$$

*Actually, two water molecules collide together and then ionize to $2H_2O \leftrightarrows H_3O^+ + OH^-$. To simplify, we write it as $H_2O \leftrightarrows H^+ + OH^-$

Notice that for every H⁺, an OH⁻ ion is produced. In a beaker of pure water, the number of H⁺ equals the number of OH⁻. This means that approximately 1×10^{-7} mol/L of H⁺ and 1×10^{-7} mol/L of OH⁻ are in 1 L of water. Since the + and the - charges cancel out, the solution is said to be neutral (uncharged). The pH of a neutral solution is stated as "7," the -log of the H⁺ concentration. Look at Table 3.4 to see how the pH value is the absolute value of the H⁺ concentration exponent, and to see the concentration of hydrogen and hydroxide ions at a given pH.

As the relative concentration of H⁺ ions increases, the relative concentration of OH⁻ ions decreases, and vice versa. No matter what the pH value is, the concentration of H⁺ multiplied by concentration of OH⁻ ions is 1×10^{-14}.

A solution with a pH of less than 7 has more H⁺ ions than OH⁻ ions and is called an acid. Acid solutions have certain characteristics, such as sour taste, and, depending on the strength of the acid, may cause burns. The stronger the acid is, the more H⁺ ions in solution, and the stronger the characteristics of the acid. A strong acid is dangerous to body tissues, such as eyes and skin. Digestive juices in the stomach have a pH of approximately 1.5, which would burn holes in the stomach if the inside of the stomach were not coated with a thick mucus layer (see Figure 3.31).

A solution with a pH higher than 7 has more OH⁻ ions than H⁺ ions and is called a base. Basic solutions also taste sour, feel slippery, and may cause burns. The stronger the base is, the more OH⁻ ions in the solution, and the stronger the characteristics of the base. A strong base is just as dangerous as a strong acid.

The pH of a solution, and whether it is a strong or weak acid or base, is very important in biotechnology laboratories. Virtually all proteins studied in a lab must be maintained within a certain pH range. An excess of either H⁺ ions or OH⁻ ions will change the structure and function of the protein. Each protein has an optimum pH for best structural stability or maximum activity. Amylase, the enzyme in saliva that breaks down starch into sugar, works best at a pH of around 7.5. **Pepsin**, found in gastric juice, has maximum activity at approximately pH 1.5, the pH of the stomach.

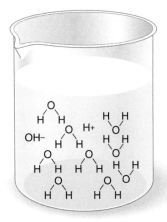

Figure 3.30. Water as a Solvent. In a sample of water, there are mostly whole water molecules. However, a tiny number of water molecules ionize into H⁺ and OH⁻ ions. In a pure sample of water, the number of H⁺ and OH⁻ ions is equal, and the water is electrically neutral. If compounds are added to water and an increase in either the H⁺ or OH⁻ occurs, the pH will change. When water molecules completely surround other molecules, the molecules are said to be dissolved in the solvent, water.

pepsin (pep•sin) an enzyme, found in gastric juice, that works to break down food (protein) in the stomach

pH paper (p•h pa•per) a piece of paper that has one or more chemical indicators on it and that changes colors depending on the amount of H⁺ ions in a solution

Table 3.4. **Concentration of Hydrogen and Hydroxide Ions (mol/L) at a Given pH**

pH	0	1	2	3	4	5	6	7
[H⁺]	1	10^{-1}	10^{-2}	10^{-3}	10^{-4}	10^{-5}	10^{-6}	10^{-7}
[OH⁻]	10^{-14}	10^{-13}	10^{-12}	10^{-11}	10^{-10}	10^{-9}	10^{-8}	10^{-7}
pH	7	8	9	10	11	12	13	14
[H⁺]	10^{-7}	10^{-8}	10^{-9}	10^{-10}	10^{-11}	10^{-12}	10^{-13}	10^{-14}
[OH⁻]	10^{-7}	10^{-6}	10^{-5}	10^{-4}	10^{-3}	10^{-2}	10^{-1}	1

Measuring the pH of a Solution

The pH of a solution is determined by measuring the number of H⁺ ions. This is done in a biotech lab in one of two ways. The easiest way to measure pH is using **pH paper** (see Figure 3.32). An indicator chemical in pH paper changes colors depending on the number of H⁺ ions in the solution. By comparing the color of the indicator strip to the "key" on the container, one can estimate a pH value.

The pH paper comes in wide-range paper and narrow-range paper. Wide-range paper shows 0 to 14 in graduations equaling 1 pH unit. Narrow-range paper can

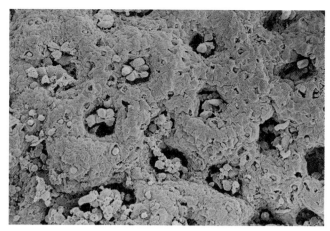

Figure 3.31. In the stomach, the cells lining these gastric folds produce hydrochloric acid (HCl), which lowers the pH to about 1.5. This is a pH at which the enzymes, pepsin and trypsin, work best. Pepsin and trypsin break down the proteins in food to amino acids. Stomach acid is so strong that special cells produce mucus to protect stomach cells from being damaged ~700X.
© Science Photo Library/Photo Researchers.

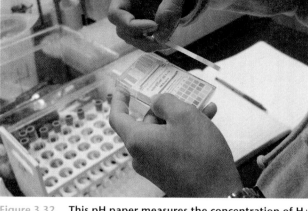

Figure 3.32. This pH paper measures the concentration of H+ ions in solution.
Photo by author.

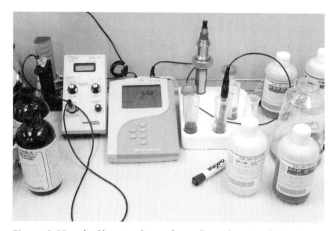

Figure 3.33. A pH meter is used to adjust the pH of solutions or to watch the pH of a solution change over time. The pH electrode is sitting in a 50-mL tube of storage solution.
Photo by author.

represent almost any pH range, at graduations of as little as 0.2 pH units. As long as the pH paper has been stored correctly, it will make a very accurate reading.

To measure a sample's pH, start with wide-range paper to determine whether the solution is an acid or a base. Once you have an idea of the approximate pH, use narrow-range paper to ascertain a more precise reading.

A **pH meter** can also be used to determine the pH of a solution (see Figure 3.33). The pH meter is calibrated using buffered solutions of known pH. An electrode then determines the pH by analyzing electrical conductivity, which varies with the number of H+ ions, and comparing the pH to the calibration buffers. A pH meter is very convenient to use when a large number of samples are to be determined or when a solution's pH is to be altered. When determining the pH of only one sample, it is usually easier to use pH paper than a pH meter.

Calibrating and Using a pH Meter

Although every brand of pH meter is different, the basic method of calibrating and using a pH meter is the same.

pH meter (p•H me•ter) an instrument that uses an electrode to detect the pH of a solution

pH Meter Use

1. Turn on the pH meter. Rinse off the electrode with distilled water. If the meter has a temperature setting knob, set it to room temperature.
2. Place the electrode in the pH 7 buffer standard. While swirling the solution, adjust the calibration knob until the display reads "7." For many pH meters, the pH 7 calibration is sufficient and once done the pH meter is ready to be used.
3. Rinse the electrode with distilled water, being very careful of the very delicate tip. Place the electrode into the solution to be tested. Swirl the solution until the pH display stops changing. Read the pH value.
4. Sometimes the solution being tested is a strong acid or base as indicated by pH paper. If that is the case, use a pH 4 buffer (for the strong acid) or a pH 10 buffer (for the strong base) and the slope or scan knob or button to give the pH meter a second calibration point before rinsing the electrode and reading the pH value.

The Beginner's Guide to pH

© Science Photo Library/
Photo Researchers.

Learn how a pH meter electrode works.

A pH meter and electrode together operate like a voltmeter. To learn how the electrode measures, go to the pH-measurement.co.uk website at: **biotech.emcp.net/pHeducation**.

Print the diagram of the electrode, and write a few paragraphs summarizing how the electrode measures the H^+ concentration in a sample.

buffer standards (buff•er stan•dards) the solutions, each of a specific pH, used to calibrate a pH meter

The solutions used to calibrate a pH meter are called **buffer standards**. They are usually purchased from supply houses, although they can be made from scratch. Buffer standards should occasionally be checked for accuracy by using narrow-range pH paper. Also, the pH electrode has an electrode-filling solution of KCl or some other salt inside. A technician must check regularly to ensure that the electrode filling solution is fresh and has the correct volume. Every brand of pH meter is slightly different, but each can be mastered easily by reading the documents that accompany the instrument.

Often, the pH of a solution needs to be adjusted. If a solution is too acidic, base (OH^-) is added. If a solution is too alkaline (basic), an acid (H^+) is added. One well known example of pH adjustment is when a swimming pool's pH needs to be adjusted with pH conditioner. Another is when stomach acid bubbles up and burns the esophagus, as in heartburn, and you take an antacid, such as Tums® (GlaxoSmithKline) to neutralize the acid. Figures 7.15a and 7.15b show examples of pH modifier products.

Section 3.4   Review Questions

1. If a sample has a pH of 7.8, is it considered an acid, a base, or neutral?
2. What does pH paper measure?
3. Before a pH meter can be used, it needs to be calibrated. To measure the pH of most solutions, the pH meter is calibrated to what pH?
4. If the pH of a hot tub is too high, say pH 8.0, then what should be added to bring it to a neutral pH?

3.5 Buffers and Their Importance in Biotechnology

Buffers are solutions that resist changes in pH. They are of critical importance in biotechnology since biological molecules (nucleic acids and proteins) must be kept within a specific pH range to maintain their structure and biological activity.

The pH of a solution affects protein structure and function because changes in pH can add or subtract charged ions from a protein. The change in charge can cause changes in a protein's shape, which may affect the protein's activity. Preparing molecules in buffered solutions, rather than in water, ensures that they will not be affected by small changes in the H^+ or OH^- concentration during reactions or storage.

A buffer solution contains molecules that resist changes in pH by interacting with hydrogen ions (H+) or hydroxide ions (OH-) that may be released or added into the solution. Most buffers are composed of a weak acid (HA) and its conjugate base (A-). When a buffering molecule is mixed in water, it ionizes to some degree, to H+ and its conjugate base.

$$HA \leftrightarrows H^+ + A^-$$

Since H+ can bind with H_2O, we can also write the equation as:

$$HA + H_2O \leftrightarrows H_3O^+ + A^-$$

The buffer works to resist slight changes in pH because, during a reaction or during storage, when small amounts of H+ or OH- are added to the solution, they will be bound by the H+ or A- in the buffer and be essentially removed from the solution (see Figure 3.34).

A good example of how a buffer works is an acetate buffer. Acetate buffers are commonly used in protein purification. You may be familiar with acetic acid (vinegar) that ionizes into hydrogen ions (H_3O^+) and acetate ions (Ac-). In an acetate buffer, the weak acid would be acetic acid (HAc), and its conjugate base would be the acetate ion (Ac-). In a solution, some of the acetic acid ionizes to

$$HAc + H_2O \leftrightarrows H_3O^+ + Ac^-$$

If a protein is dissolved in the acetate buffer, the buffer can resist changes in pH over time because the H_3O^+ can then bind to OH- ions if they are added during processing. This prevents the pH of the solution from rising. Conversely, the acetate ion (Ac-) can bind with H+ ions if they are added during a process, removing them from the solution and preventing the pH of the solution from decreasing. In this way, the acetate buffer resists changes in pH and keeps the protein at the desired pH.

Some buffers are made with buffering agents that produce a weak base and its conjugate acid (B + $H_2O \leftrightarrows BH^+ + OH^-$). Such is the case with TRIS, a very common buffer in biotechnology facilities. You will use TE buffer (TRIS, EDTA) to store DNA molecules in later chapters.. When TRIS is dissolved in deionized water at a pH between 7 and 9, it ionizes and can accept and release excess H+ and OH- ions from solution.

A buffer is prepared by dissolving one or more buffering molecules (called buffering agents) in deionized water. Some buffering agents ionize to produce a weak acid (hydroxide ion acceptor) and its conjugate base. Some buffering agents ionize to produce a weak base (hydrogen ion acceptor) and its conjugate acid. Either way, as other molecules are added to the buffer, these molecules can bind or release the excess H+ or OH-, removing them from the solution and thus preventing changes in pH. In this way, protein or nucleic acid molecules in a buffered solution are protected from slight changes in pH that might alter their structure or function.

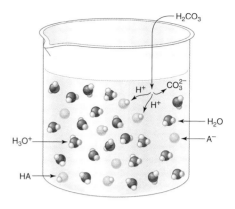

Figure 3.34. **In a buffer where HA + $H_2O \leftrightarrows H_3O^+ + A^-$, most of the water molecules exist as H_2O. When HA ionizes, some H_2O receive an extra H+ and become H_3O^+ while A- ions are released. When a small amount of acid, such as carbonic acid (H_2CO_3), is added to the buffer, the acid ionizes to release H+, which would decrease the pH if they were not bound by the A- in the buffer.**

Selecting a Buffer

Buffering agents are selected for use in a buffer based on the pH desired for the final solution. The desired final pH is determined by what protein or nucleic acid molecule will be dissolved in the buffer. pH changes add or subtract hydrogen atoms (and thus charges) from protein molecules. A change in charge will change the protein's shape or structure and may affect the protein's ability to perform its crucial biological task. At the "wrong pH," a protein might not have the desired structure, function, or biological activity. The importance of picking the right buffer for a protein cannot be overstated.

Selecting a protein buffer requires that the **pI** of the protein of interest is known. The pI is the pH where the positive and negative charges within a protein are equal and cancel out so that the protein is electrically neutral. If a protein is placed in a solution where the pH is equal to the pI, then the protein will actually precipitate (fall) out of solution. Buffers should be used that keep a protein of interest at 1 to 2 pH units

pI (p•I) the pH at which a compound has an overall neutral charge and will not move in an electric field; also called the isoelectric point

Table 3.5. Buffers Commonly Used in Biotechnology Facilities for Biological Solutions at pH 5.0–9.0

Buffering Agent	pKa	Practical Buffering Range*	Examples	Common Applications
TRIS (Trisma)	8.2	7.0 to 9.0	TE, TAE, TBE	DNA preparation, PCR, DNA electrophoresis
			TRIS/glycine/SDS buffer	protein electrophoresis (PAGE)
			TRIS-buffered saline (TBS), TRIS/CaCl$_2$	protein preparation, separation, reaction buffers
sodium phosphate	7.1	5.8 to 8.0	Na$_2$HPO$_4$/NaH$_2$PO$_4$ buffer, phosphate-buffered saline (PBS), protein purification buffers	protein preparation, separation, reaction buffers, ion-exchange chromatography, ELISA and Western blots
potassium phosphate	6.82	5.8 to 8.0	K$_2$HPO$_4$/KH$_2$PO$_4$ buffer, protein purification buffers	protein preparation, separation, reaction buffers, ion-exchange chromatography, ELISA and Western blots
borate	9.23	8.5 to 10.2	boric acid/borax lithium borate (LB) electrophoresis buffer	cytochemical reactions, DNA electrophoresis
acetate	4.74	3.7 to 5.6	acetic acid/sodium acetate buffer	protein solutions at a low pH, ion-exchange chromatography
HEPES	7.55	6.8 to 8.2	mammalian cell culture buffer	cell culture media

*Buffer ranges and pK$_a$ from biotech.emcp.net/sigmaaldrichbuffers

---

■■■■■■ ■ ■■■■ ■■ ■■ ■■

pK$_a$ (p•K•a)   the pH at which 50% of a buffering molecule in aqueous solution is ionized to a weak acid and its conjugate base; the point at which there are an equal number of neutral and ionized units.

away from their pI. Several buffers that are commonly used in a biotechnology lab are listed in Table 3.5 along with their buffering range.

Most often some buffer is selected because a researcher or technician knows from past experience that a certain buffer works for a specific application. For example, the protein amylase is known to have high catalytic activity at a pH around 7.2. Since sodium phosphate buffers are commonly used at that pH, a sodium phosphate buffer would be expected to be a good option for an amylase buffer. One of the amylase buffers used in the lab manual is 100 mM NaH$_2$PO$_4$, 1 mM Na$_2$HPO$_4$, pH 7.2.

The buffering range of a buffering compound is determined by its **pK$_a$** (see Table 3.5). pK$_a$ is the pH at which 50% of the buffering molecules dissociate or ionize. When the pH of the solution is equal to the pK$_a$ the buffer has its highest buffering capacity.

How pK$_a$ is determined is beyond the scope of this text, but it is easy to find the pK$_a$ for a buffering solute by doing a quick Internet or reference search. Once the pK$_a$ is known, then the pH range of the buffer can be estimated and the technician can confirm that a buffer will be appropriate for the application.

For an example of how pK$_a$ is important in buffer selection, consider the protein cellulase. Depending on the source, cellulase has a pI around 4.5, suggesting a buffer for cellulase might be made at around pH 6.5. Which buffer from Table 7.2 might be a good cellulase buffer?

A phosphate buffer such as 50 mM KH$_2$PO$_4$, 5 mM K$_2$HPO$_4$, pH 6.5 might be a good choice since its pK$_a$ is 6.82. The closer the pH of the buffer is to the pK$_a$ of the weak acid, the better its buffering capacity. Monopotassium phosphate (KH$_2$PO$_4$) and dipotassium phosphate (K$_2$HPO$_4$) are usually used to generate buffers of pH values around 6 to 7. The monopotassium phosphate dissolves in solution and produces a weak phosphoric acid that reaches 50% ionization at a pH of 6.8 (its pK$_a$ value). The dipotassium phosphate also ionizes to produce a weak base. Potassium phosphate buffers are commonly used for protein solutions needed at, or near, neutral pH (7.0).

Preparing a buffer is similar to preparing any molar solution, although the pH of the solution must be measured and adjusted before the final volume is reached. See the basic steps in buffer preparation given below.

■ ■ **Chapter 3**

Buffer Preparation

a. Determine the mass of each buffer solute using the molar concentration equation:

$$\text{volume (L) desired} \times \text{molarity (mol/L desired)} \times \text{formula (g/mol) weight} = \text{mass (g) of buffering solute}$$

b. Add each mass of buffer solute to a beaker, and add distilled water to approximately 75% of the final volume desired. Mix the solutes until they completely dissolved.
c. Using a pH meter, measure the pH and adjust the pH down or up are using dilute acid or base, as appropriate, to the desired pH. 1 M HCl or 1 M NaOH (or 10% NaOH) are commonly used to adjust pH.
d. Add deionized water to bring the buffer to the desired final volume.
e. Verify the pH using the pH meter and pH paper.

Let's say you need 200 mL of 50 mM NaH_2PO_4 • H_2O, 5 mM Na_2HPO_4, pH 7.0 buffer.
   a. Determine the required mass of each solute to give the desired concentration for the desired volume.

0.2L × 0.05 mol/L × 137.99 g/mol = 1.38g NaH_2PO_4 • H_2O
0.2L × 0.005 mol/L × 141.98 g/mol = 0.14g Na_2HPO_4

Measure out the mass of each solute and add to a 400-mL beaker.

b. Add deionized water to 150 mL (75% of the volume of buffer desired).

Mix the solution until the solutes are completely dissolved.

c. Using a pH meter, adjust the pH with 1 M HCl or 10% NaOH to pH 7.0 (the desired pH).
d. Add deionized water to 200 mL (100% of the volume of buffer desired).
e. Verify the pH of 7.0.

Many nucleic acid or protein buffers have additional solutes added for some specific purpose. TE buffer, commonly used for DNA storage, contains both TRIS (for buffering) and EDTA as a chelating agent to remove certain ions (such as Mg^{2+}) that could interfere with reactions from solution.

To make one liter of TE buffer (10 mM TRIS, 1 mM EDTA, pH 8.0):
   a. Determine the required mass of each solute to give the desired concentration for the desired volume.

1.0 L × 0.01 mol/L × 121.14 g/mol = 1.21g TRIS
1.0 L × 0.001 mol/L × 372.20 g/mol = 0.37g EDTA (disodium salt)

Measure out the mass of each solute and add to a 2L beaker.

b. Add deionized water to 750 mL (75% of the volume of buffer desired).

Mix the solution until the solutes are completely dissolved.

c. Using a pH meter, adjust the pH with 1 M HCl down to pH 8.0 (the desired pH).
d. Add deionized water to 1000 mL (100% of the volume of buffer desired).
e. Verify the pH of 8.0.

Buffers are used in many applications in the biotechnology lab. Buffers act in many ways, including as:

Figure 3.35. Buffers that are commonly used are commercially available in concentrated form, which are more economical to ship and store. Shown are 1 L volumes of 40X TAE buffer and 1X TRIS/glycine/SDS PAGE running buffer, and 1 mL of 10X restriction enzyme digestion buffer. What amounts of 1X buffer may be made from these concentrated stocks?
Photo by author.

- the solvent in protein and nucleic acid solutions (for pH maintenance)
- the solution in which protein and nucleic acid chemical reactions occur (for pH maintenance)
- an ingredient to resist pH changes in cell cultures (to ensure cells are maintained at proper pH)
- the liquid phase in column chromatography (for protein and nucleic acid separations)
- the liquid for conducting electricity in a gel box (separating molecules)

Several buffers used in a biotechnology facility have sodium chloride or some other salt added to them. Common saline (salt-containing) buffers include phosphate buffered saline (PBS) and Tris-buffered saline (TBS). These are used so often in protein and nucleic acid work that they can be purchased or prepared as concentrated stock solutions (10X, 5X, etc.) and diluted as needed to the 1X working concentration (See Figure 3.35). How to make diluted solutions is covered in the next section of the chapter.

Many enzymes are quite sensitive to slight changes in pH. The enzymes used in producing recombinant DNA molecules (restriction enzymes and DNA ligases) must be prepared and used in specific reaction buffers that have been optimized for each specific type of enzyme. When a restriction enzyme is purchased, it is usually shipped with a reaction buffer (usually a 10X concentrate). When all the reactants of a restriction enzyme reaction are mixed together, the resulting concentration of the reaction buffer is 1X. Restriction enzymes and their reactions are covered in Chapter 8.

Buffer recipes may appear similar but may have subtle differences for use in different applications. A good example of this is when a variety of similar buffers are used in certain kinds of protein purification column chromatography. Ion exchange column chromatography separates molecules using charged microscopic beads in a long tube or column. It requires two buffers. One buffer is used to "equilibrate the column" to the pH at which the target protein in a mixture will stick to the oppositely charged column beads. The other buffer is an elution buffer used to "exchange" or release the protein from the column beads. In amylase ion exchange chromatography, amylase has a negative charge in a pH 7.4 equilibration buffer and will bind to positively charged chromatographic beads. An elution buffer (at the same concentration and pH of the equilibration buffer) with a high NaCl concentration is used to displace the amylase molecules with highly electronegative Cl⁻ ions. The Cl⁻ ions knock off (exchange for) the amylase molecules and release them from the beads. The ion exchange chromatography buffers are used so the target protein stays at a certain pH and has the desired overall net charge so that it can bind and release from the beads as desired.

Buffers are commonly prepared and stored at room temperature or at 4°C for several weeks, depending on the buffer. To ensure that the buffer stays uncontaminated by fungi or bacteria, buffers are often filter sterilized (removing bacteria and fungi) by passing the solution through a 0.2 µm filter or by autoclaving. For some applications, a small amount of sodium azide (Na_3N) may be added to the buffer to prevent bacterial and fungal growth. If stored correctly, a buffer should maintain its pH for a long time.

Being able to make different volumes of various concentration buffers at different pH levels is of critical importance to a biotechnician. On some days, an entry-level technician or research associate may spend the majority of a workday just making the buffers needed for a lengthy experimental procedure.

The Blood Buffer System in Your Body

The pH of the blood in your circulatory system is constantly fluctuating because it is exposed to additional acids and bases routinely as part of cellular metabolism. Cells respire (breakdown glucose) and produce carbon dioxide and hydrogen ions (H+) as waste products. These diffuse out of the cells and into the blood. This makes the blood more acidic. Other things can make the blood more acidic such as over-exercising and re-breathing exhaled air. On the other hand, the blood can become more alkaline (basic) when HCO3- ions in the blood are removed by the kidneys or when the lungs remove carbon dioxide in the blood during breathing.

However for the blood proteins to function properly, the pH of blood must remain relatively constant with a pH ranging only between pH 7.35–7.45. The blood has its own buffering system that under normal circumstances does a pretty good job of keeping the pH in that range. Learn more about it below.

1. Watch the You Tube video at **www.youtube.com/watch?v=r6UAEbhRXNI** to learn how the buffering system in your blood keeps the pH of the blood at between pH 7.35–7.45.

   What are the acid and the conjugate base that buffers the blood?
   What is it called when the blood because acidic? What is it called when the blood becomes basic?
   What other body organs work to help maintain blood pH?

2. Do a search on the Internet to find a disease or disorder that causes a significant decrease in the blood's pH. What is it called? What are the symptoms? What is a treatment for the disorder? Record the website URLs you used.

Section 3.5   Review Questions

1. Why must DNA and proteins be stored in buffered solution?
2. In what kind of buffer should a DNA sample that was isolated from human cheek cells be stored?
3. The formula weight of TRIS is 121.14 g/mol. How is 100 mL of 0.02 M TRIS, pH 8.0, prepared?

3.6 Preparing Dilutions of Concentrated Solutions

Often, solutions are too concentrated to use the way they are initially prepared. When this is the case, a **dilution** (addition of more solvent) of the concentrated solution is prepared to bring the concentration down to a usable level, or "working concentration." There are several reasons why concentrated solutions are prepared, but diluted at a later date.

One reason for preparing and using concentrated **stock solutions** is that it is often easier to prepare higher-concentration solutions than lower-concentration (more diluted) ones. Sometimes, it is difficult to weigh tiny amounts of solutes, and weighing out a larger mass for a more concentrated solution may be more convenient. Also, it is easier and less expensive to ship or store small volumes of concentrated samples than larger volumes of dilute samples. The smaller, more concentrated solutions can be diluted to a working concentration as needed. Concentrated solutions are a more efficient method of preparing and storing samples that are used at a lower concentration and in large volumes.

Suppose that for an experiment you need a large volume of enzyme solution (1 L) at a certain concentration (1 mg/mL). A liter of enzyme is a large amount to ship from

dilution (di•lu•tion) the process in which solvent is added to make a solution less concentrated

stock solution (stock sol•u•tion) a concentrated form of a reagent that is often diluted to form a "working solution"

Concentrating a Sample

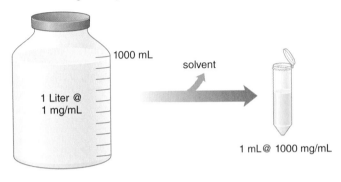

1000 mL

solvent

1 Liter @ 1 mg/mL

1 mL @ 1000 mg/mL

Diluting a Sample

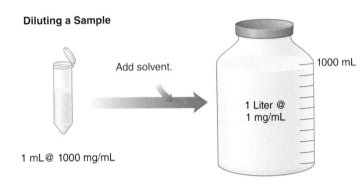

Add solvent.

1 mL @ 1000 mg/mL

1000 mL

1 Liter @ 1 mg/mL

Figure 3.36. **Concentrating a 1 L Solution.** Many chemical and biological reagents are purchased in concentrated form. Concentrated solutions can be prepared initially with a greater amount of solute to solvent, or a solution can be concentrated by removing water. A diluted solution can be prepared by adding solvent to a concentrated one.

the supplier, and takes up a lot of space in the freezer. It is much more convenient to concentrate the enzyme solution for shipping to, say, 1000 mg/mL, and send a smaller volume of 1 mL (see Figure 3.36). When you are ready to use the enzyme, it is easy to dilute the solution to the working concentration.

Another advantage of using concentrated solutions is that a large volume of a working solution may be prepared in a single dilution. The diluted solution can be used in several trials of the same experiment, which increases the consistency of sample in the trials.

To determine how to dilute a concentrated solution, use a simple ratio relating the original concentration and volume to the desired concentration and volume, a ratio called the Dilution Equation:

Dilution Equation: $C_1 V_1 = C_2 V_2$

Where C_1 = the concentration of the concentrated stock solution (the starting solution).
V_1 = The volume to use of the stock solution in the diluted sample.
C_2 = the desired concentration of the diluted sample.
V_2 = the desired volume of the diluted sample.

The $C_1 V_1 = C_2 V_2$ equation may be used with any concentration units (eg, mass/volume, %, or molar) as long as the units are the same on each side of the equation (for cancellation purposes). The equation is fairly easy to use because you usually know three of the four terms, and it is easy to solve for the fourth. When you are calculating a dilution, be sure to show each value in the equation and all units of measurement for each term. In this way, you will be certain you are representing the appropriate units and making the calculation correctly.

How would you make 1 L of 1 mg/mL protein solution from 100 mg/mL concentrated stock solution?

C_1 = 100 mg/mL
V_1 = the volume to use of the stock solution in the diluted sample (this is what is calculated)
C_2 = 1 mg/mL
V_2 = 1000 mL (1 L is converted to 1000 mL)

Set up the $C_1 V_1 = C_2 V_2$ equation and solve for V_1:

$$100 \text{ mg/mL} \times V_1 = 1 \text{ mg/mL} \times 1000 \text{ mL}$$

$$V_1 = 1 \text{ mg/mL} \times \frac{1000 \text{ mL}}{100 \text{ mg/mL}}$$

The mg/mL units cancel out and leave V_1 = 1000 mL /100 = 10 mL.

So, to dilute the concentrated protein solution to make 1 liter of 1 mg/mL solution, take 10 mL of the concentrated stock and add 990 mL of solvent, as shown in Figure 3.37.

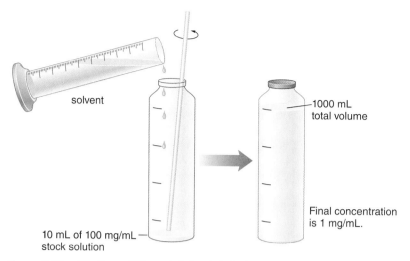

Figure 3.37. **Diluting a 100 mg/mL Stock Solution to 1 mg/mL.**

solvent

1000 mL
total volume

Final concentration
is 1 mg/mL.

10 mL of 100 mg/mL
stock solution

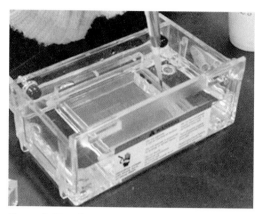

Figure 3.38. **Agarose gels are prepared using a 1X TAE buffer as the solvent. 1X TAE also serves as the buffer solution for a DNA gel box. It conducts electricity that carries molecules through the gel. The 1X TAE buffer is diluted from a 40X TAE buffer.**
Photo by author.

To make the math easy and accurate when calculating dilution, be careful that the units of measurement are the same. For example, if you are diluting an mM solution to a µM solution, one of the concentration terms must be converted to the other.

How do you make 200 mL of 75 µM CaCl$_2$ solution from 10 mM CaCl$_2$ solution?

C_1 = 10 mM CaCl$_2$ (Convert this to µM, 10 mM = 10,000 µM, so that it is
        in the same concentration units as the final solution.)
V_1 = the volume to use of the stock solution in the diluted sample
        (This is what is calculated.)
C_2 = 75 µM CaCl$_2$
V_2 = 200 mL

Set up the $C_1 V_1 = C_2 V_2$ equation and solve for V_1:

$$10{,}000 \ \mu M \times V_1 = 75 \ \mu M \times 200 \ \text{mL}$$

$$V_1 = 75 \ \mu M \times \frac{200 \ \text{mL}}{10{,}000 \ \mu M}$$

The µM units cancel out and leave V_1 = 15,000 mL /10,000 = 1.5 mL.

So, to dilute the concentrated 10 mM CaCl$_2$ and make 200 mL of 75 µM CaCl$_2$ solution, take 1.5 mL of the concentrated stock and add 198.5 mL of solvent.

As you know, proteins and nucleic acids are stored in buffered solutions. Buffers resist changes in pH, and help preserve the structure and function of molecules dissolved in the buffer. Depending on the kind of buffer needed, different salts at various concentrations are used. A common buffer salt is **TRIS**. TRIS is a complex organic molecule used to maintain the pH of a solution. TRIS is used in **TE buffer** (a buffer for storing DNA) and in **TAE buffer** (used for running DNA samples on agarose gels in horizontal gel boxes, as shown in Figure 3.38).

When TAE buffer is prepared at a concentration that is ready to use, it is called 1X TAE. 1X is the "working concentration," the recipe that is used in the gel box. Since TAE has three ingredients (TRIS, acetic acid, and EDTA), it is cumbersome to make a fresh batch of 1X TAE from scratch every time it is needed. It is much more efficient to make and store concentrated TAE and dilute it when needed. Commonly, 40X or 50X TAE buffer is prepared and stored. 40X TAE contains 40 times the concentration of TRIS, acetic acid, and EDTA as does the 1X preparation that is typically used. When 1X

buffer (buff•er) a solution that acts to resist a change in pH when the hydrogen ion concentration is changed

TRIS a complex organic molecule used to maintain the pH of a solution

TE buffer (T•E buff•er) a buffer used for storing DNA; contains TRIS and EDTA

TAE buffer (T•A•E buff•er) a buffer that is often used for running DNA samples on agarose gels in horizontal gel boxes; contains TRIS, EDTA, and acetic acid

TAE is needed, some of the 40X TAE is diluted with deionized water. Many buffers are made at a concentration that is multiple times more concentrated than the working concentration. The "X" symbol denotes the relative concentration.

How is 600 mL of 1X TAE buffer made from 40X TAE buffer stock solution?

C_1 = 40X TAE
V_1 = the volume of the 40X TAE to be used to make the diluted sample
C_2 = 1X TAE
V_2 = 600 mL

Set up the $C_1 V_1 = C_2 V_2$ equation and solve for V_1:

$$40X \ TAE \times V_1 = 1X \ TAE \times 600 \ mL$$

$$V_1 = \times \frac{(1X)\ (600\ mL)}{40X}$$

The "X" units cancel out and leave V_1 = 600 mL/40 = 15 mL.

So, to dilute the 40X TAE to a working concentration of 1X TAE, mix 15 mL of 40X TAE with 585 mL of deionized water. This makes 600 mL of 1X TAE.

Section 3.6   Review Questions

1. How do you prepare 40 mL of a 2 mg/mL protein solution from 10 mg/mL protein solution?
2. How do you prepare 200 µL of 2X enzyme buffer from 10X enzyme buffer solution?
3. How do you prepare 500 µL of 50 µM NaCl solution from 5 mM NaCl solution?
4. How do you prepare 3 L of 1X TAE buffer from 50X TAE buffer stock solution?

Chapter Review

Speaking Biotech

Page numbers indicate where terms are first cited and defined.

acid, 89
amu, 88
aqueous, 81
balance, 81
base, 89
buffer, 99
buffer standards, 92
conversion factor, 76
dilution, 97
graduated cylinder, 74
gram (g), 81
hydrogen ion, 89
liter (L), 73
mass, 73
mass spectrometer, 88
metrics conversion table, 75
microliter (µL), 73

micropipet, 74
milliliter (mL), 73
molarity, 83
mole, 87
molecular weight, 87
multichannel pipet, 79
neutral, 89
normality, 83
P-10, 78
P-100, 77
P-1000, 79
P-20, 78
P-200, 77
pepsin, 90
percentage, 85
pH meter, 91
pH paper, 90

pI, 93
pipet, 74
pKa, 94
positive displacement
   micropipet, 81
solute, 81
solution, 81
solvent, 80
stock solution, 87
TAE buffer, 99
TE buffer, 99
TRIS, 99
unit of measurement, 75
volume, 73
weight, 81

Summary

Concepts

- Liquid volumes are measured using graduated cylinders for liters and milliliters, pipets for milliliters, and micropipets for microliters.
- A µL is 0.001 mL, and 1 mL is 0.001 L.
- To convert between metric units of measure, move the decimal point left (to go to bigger units) or right (to go to smaller units) the number of zeros in the conversion factor.
- To make an accurate measurement with a graduated cylinder or pipet, make sure the bottom of the meniscus is touching the appropriate graduation.
- If you have a choice of pipets for measuring a volume, a smaller volume pipet is more likely to give an accurate measurement or reading.
- To use a micropipet correctly, you must master setting the volume and using the plunger.
- Some common micropipets and their ranges of measurement are the P-1000 (100–1000 µL), P-200 (20–200 µL), P-100 (10–100 µL), P-20 (2–20 µL), and P-10 (0.5–10 µL).
- A multichannel pipet can measure and dispense several identical volumes at the same time.
- In a solution, a solute is what is dissolved in the solvent. In most solutions, water or a water-based solution (buffer) is the solvent.
- Most solutes are measured on a balance in grams or milligrams (0.001 g).
- Concentration is the amount of solute dissolved in the solvent.
- The most common ways concentration is reported in a biotech lab are by mass/volume, % mass/volume, molarity, or by some amount "X."
- To calculate the amount of solute to use in a particular mass/volume solution, use the Mass/Volume Concentration Equation:

$$\underset{\text{concentration}}{\underline{\hspace{2cm}} \text{ g/mL}} \times \underset{\text{volume}}{\underline{\hspace{2cm}} \text{ mL}} = \underset{\text{dissolved in the solvent to the desired volume}}{\underline{\hspace{2cm}} \text{ g of solute to be measured, then concentration}}$$

- To calculate the amount of solute to use in a particular % mass/volume solution, use the % Mass/Volume Concentration Equation:

$$\underline{\hspace{2cm}}\ \% \ = \ \underline{\hspace{2cm}}\ \text{decimal value of the g/mL}$$
$$\text{percent value}$$

$$\underline{\hspace{1.5cm}} \times \underline{\hspace{1.5cm}} = \underline{\hspace{1.5cm}}\ \text{g of solute to be measured, then}$$
$$\text{decimal (g/mL)} \quad \text{volume (mL)} \quad \text{dissolved in the solvent to the desired volume}$$

- Knowing the molecular weight of a compound is important in calculating the amount of a substance to use as a solute in a molar solution. Molecular weight is reported in atomic mass units (amu). A molecule's molecular weight equals the sum of the atomic weights of the atoms that make up the molecule.
- A mole is equal to 6.02×10^{23} atoms or molecules. A mole weighs, in grams, the molecular weight of the molecule.
- To calculate the amount of solute to use in a particular molar solution, use the Molarity Concentration Equation:

$$\begin{array}{c}\text{volume} \\ \text{wanted (L)}\end{array} \times \begin{array}{c}\text{molarity} \\ \text{desired (mol/L)}\end{array} \times \begin{array}{c}\text{molecular weight of} \\ \text{the solute (g/mol)}\end{array} = \begin{array}{c}\text{grams of solute to be measured,} \\ \text{then dissolved in the solvent to} \\ \text{the desired volume}\end{array}$$

- A solution's pH is a measurement of the hydrogen ion concentration. If there is an excess of H^+ ions in a solution, the pH will be less than 7 and the solution is called an acid. If there is an excess of OH^- ions in a solution, the pH will be more than 7 and the solution is called a base.
- A solution with no excess of H^+ or OH^- ions is called a neutral solution.
- The pH value is important because excess H^+ or OH^- ions in solutions can interact with proteins, and cause a change in their shape and function.
- One can measure pH using pH paper (paper with indicators) or pH meters. The meters must be calibrated with buffer standards of known pH. These meters are appropriate instruments to measure changes in pH.
- A buffer is a solution that resists changes in pH. Biological molecules, such as proteins, DNA, and RNA, should be stored in buffers because the molecules' structure and function could be altered by changes in pH.
- Buffers work because they contain buffering compounds. Common buffering agents are TRIS and sodium phosphate. The requirements of a given experiment dictate buffer use.
- Buffers are made like other molar solutions. The amount of buffering molecule needed is calculated and mixed with solvent. The pH is adjusted to the desired level, and more solvent is added to reach the final desired volume
- A solution gets more concentrated as solvent is removed or as solute is added. A sample gets more dilute as solvent is added or solute is removed.
- To calculate how to make a specific dilution, use the Dilution Equation, $C_1 V_1 = C_2 V_2$.

Lab Practices

- Volume and mass measurements are made in metric units. You can easily convert by using the B ⇆ S Rule, moving the decimal point to the left or right depending on the units being converted.
- 10-, 5-, 2-, and 1-mL pipets are common in biotechnology labs. Before using each pipet, determine the maximum volume of the pipet and the value of its smallest graduations.
- Never mouth pipet. Use an appropriate pipet pump or pipet aid with each pipet.
- Pipet at eye-level. The bottom of the sample meniscus is where the volume is read.
- Although micropipets differ from one manufacturer to another, they are similar in that they measure in microliters (μL).
- Micropipets are named based on the maximum volume they will measure. For example, a P-1000 will measure a maximum of 1000 μL. Each pipet has a minimum and maximum volume it will measure.

- Most micropipets have a plunger apparatus for drawing up and dispensing specified volumes. The technician must learn how each plunger works, where the first and second stops are, and how to control, fill, and dispense samples. Large errors in volume can occur if the plunger is not used correctly.
- A micropipet must be used with a micropipet tip. Tips should be discarded after each use to prevent contamination.
- A microcentrifuge is used to pool samples in the bottom of a microcentrifuge tube. Tubes placed into any centrifuge must be balanced in mass and volume to other tubes in the centrifuge.
- Mass or weight is measured in the laboratory using an electronic balance. For masses above 0.1 g, a tabletop balance is usually used. For milligram masses, an analytical balance is used.
- A weigh boat or weigh paper is always used on a balance. Samples are never placed on an unprotected weighing pan.
- Level and zero the balance with nothing on it. Then add a weigh boat or weigh paper and rezero the balance before weighing a sample.
- A micropipet's accuracy can be checked by measuring water samples on a balance. For example, 1 mL of water weighs 1 g, and 1 µL of water weighs 1 mg.
- A solution is composed of one or more solutes dissolved in a solvent. When making a solution, measure the solute and then slowly mix solvent into the solute.
- The concentration of a solution is the amount of solute in a particular volume. Concentration is reported in different ways depending on the solute. Common units of concentration include mass/volume, % mass/volume, molarity, and "X" concentration.
- Indicator strips and indicator solutions can be used to measure the concentration of solute in solvent. Biuret solution (a mixture of $NaOH$ and $CuSO_4 \cdot 5H_2O$) indicates the presence of protein.
- To determine how to prepare a certain volume of a solution at a certain mass/volume concentration, use the equation:

  concentration desired (m/v) × total volume desired = mass of solute in the volume desired

- A spectrophotometer can be used to quantify the difference in colored solutions.
- To make a solution of a specific % mass in a specific volume, multiply the decimal value of the concentration desired by the total volume desired:

$$\text{decimal value of concentration desired} \times \text{total volume desired} = \text{mass of solute to add to desired volume of solvent}$$

- A one molar (1 mol/L) solution is 1 mole of solute dissolved in a volume of solvent to a total of 1 L. A one molar solution is reported as 1 M. A mole is 6.02×10^{23} molecules, and a 1 M solution contains 6×10^{23} molecules in a liter of solution.
- To determine how to prepare a molar solution, use this equation:

  volume (L) × molarity (mol/L) × molecular weight (g/mol) = grams of solute in final volume

- Use pH paper to make a single pH measurement of a solution. Use a pH meter if the pH of a solution is to be monitored or adjusted.
- Picking the best buffer to use to make a specific protein solution requires knowledge of the protein's pI and the buffers pKa.
- Dilutions of concentrated solutions are made by adding some volume of solvent to some volume of concentrated solution. Use the equation $C_1 V_1 = C_2 V_2$ to calculate the volume of concentrated solution to use in the dilution.
- To cancel units in the solution equations, you may need to convert units, for example, grams to milligrams, or milliliters to liters.

Thinking Like a Biotechnician

1. What is the best instrument to use to measure these volumes: 12.5 mL, 25 μL, 8.3 μL, 250 mL, and 571 μL?
2. What mass of gelatin is needed to make 15 mL of a 12-mg/mL gelatin solution? Describe how the solution is prepared.
3. What mass of glucose is needed to make 50 mL of 2.5% glucose solution? Describe how the solution is prepared.
4. What mass of calcium chloride ($CaCl_2$) is needed to make 125 mL of 0.55 M $CaCl_2$ solution? Describe how the solution is prepared.
5. In moles/liter, what is the concentration of H^+ in a solution that has a pH of 6.0? In moles/liter, what is the concentration of OH^- in a solution that has a pH of 6.0?
6. If a solution has a pH of 5.3, how can it be brought to a pH of 7.1?
7. Describe how to prepare 5 L of a 0.25 M TRIS buffer at pH 7.4.
8. How do you prepare 200 mL of 5 mM $CaCl_2$ solution from 2 M $CaCl_2$ stock solution?
9. Which of the following represents concentration of a solution, and which represents dilution of a solution?
   a. From 5X solution to 0.1X solution
   b. From 25 mM solution to 1 M solution
   c. From 3% solution to 0.2% solution
   d. From 0.1X solution to 0.15X solution
   e. From 2 M solution to 10 mM solution

10. A lab technician diluted 10 μL of a 50X sample with enough solvent to reach a final volume of 20 mL. What is the new concentration of the sample?

Biotech Live

Maintaining Stock Areas using Inventory Logs

Areas where equipment and chemicals are stored are called "stock areas." For safety and productivity reasons, stock areas should be kept clean and orderly, and filled with appropriate reagents, supplies, or equipment. It is the lab technician's responsibility to inventory and maintain stock areas and replace or restock missing items. Since it often takes days or weeks to reorder certain supplies, inventories are done either daily or weekly.

 Create an inventory log, and maintain and document a stock area.

1. The supervisor will assign a stock area to you and your lab partner.
2. Using Microsoft® Excel®, create an inventory log for your stock area that is similar to the example shown below. Include the following elements in the log:
   • title
   • column with each supply listed and described
   • column showing the time the inventory is expected to be checked
   • column with the date the inventory is expected to be checked.
   Set the chart so it is in "landscape" mode. This allows more dates to be included.
3. Ask your instructor or the lab supervisor to approve your inventory log. Then print it.
4. Place the log in an acetate cover. Then place or post the inventory log in a highly visible location in the assigned inventory area.
5. Monitor and maintain your assigned stock area. Document that the area is clean, properly stocked, and organized by initialing the log at the end of each lab period.

Inventory Log Lab Station No. 8
Stock Drawer No. 1

Item	Time	9/11	9/12	9/13	9/14	9/15	9/18	9/19
1 peg rack								
1 micro test tube rack								
1 box large micropipet tips								
1 box small micropipet tips								
1 P-1000 micropipet								
1 P-100 micropipet								

Finding Experimental Protocols

The procedures (protocols) for even the simplest scientific experiment can be very challenging to develop. Scientists must come up with specific recipes for reagents and buffers, and well-defined operating conditions. Sometimes researchers must develop their own procedures or instruments for a technique because no one else has conducted similar research. More often, though, researchers in a laboratory build on experimental techniques that have worked in the past for themselves and their colleagues throughout the scientific community.

For example, once scientists learn how to extract DNA from one type of cell, it is reasonable to expect that similar techniques can be used to extract DNA from other, similar cells. Relatively simple and straightforward biotechnical research techniques are shared between researchers, and can be found in easily accessible databases on the Internet.

Take, for example, recombinant DNA (rDNA) production. When technicians want to produce rDNA molecules, they need two sources of DNA: 1) the section of DNA that codes for a protein of interest, and 2) a piece of DNA to attach the DNA of interest to (called a vector) that will carry the DNA of interest into a cell. Usually the vector DNA molecule is a bacterial plasmid.

Plasmids are small pieces of DNA that must be extracted (removed) from cells to be used as vectors. Extracting plasmids from cells is called a "prep," short for preparation. When a small amount of plasmid is needed, it is called a "mini-prep." Several mini-prep protocols are available on the Internet for a technician who wants to attempt to retrieve plasmids from cells.

 Learn how to access laboratory protocols from databases on the World Wide Web. Find an experimental protocol for plasmid DNA extraction.

1. Find a protocol (set of procedures) for conducting a mini-prep by doing a "search" on one of the following databases or search engine websites:
   biotech.emcp.net/protocol_online
   openwetware.org/wiki/Protocols
   www.cellbio.com/protocols.html
   www.promega.com/resources/protocols
   www.currentprotocols.com/WileyCDA
   www.lib.ncsu.edu/guides/protocols
   www.elabprotocols.com

2. Your protocol must be different from that of anyone who sits within one seat of you in any direction. More points will be given for a protocol that includes the specific ingredients required in all solutions and reagents. Note: There are "kits" that can be purchased for mini-preps, but often the buffers in these are not detailed and are listed as only "Buffer 1" or "Lysis Buffer," etc, so you really do not know what the ingredients are.

3. Print the protocol you found (maximum of two pages, so you may have to cut and paste it into a Microsoft® Word® document). Record the bibliographical reference.

4. Compare the protocol you found to that of your lab partner. The protocols found may look too highly scientific or challenging to understand. Read through them as well as you can. Highlight those steps in the procedures that are basically the same from one mini-prep to the next. List three procedures that are similar from one mini-prep to another.

Activity 3.3 Hazardous Chemicals: Knowing When You Have One

Any chemical has the potential to be dangerous if not used properly. Even when a technician is skilled, accidents occasionally happen in the laboratory. To reduce the chance of an accident, and to minimize damage if one occurs, manufacturers of chemicals are required to provide a "Safety Data Sheet" (SDS) for every chemical distributed.

An SDS may be several pages long since it contains so much information about a chemical. Among other information, the SDS lists the molecular formula and molecular weight of the substance, whether or not it is flammable, and what to do in case of a spill, exposure to eyes or skin, or inhalation.

Most vendors have websites where a SDS for their chemicals can be found. Most lab facilities also require that a hard copy, such as a binder with SDS, is located on the premises.

 Gather important safety information about three chemicals used regularly in the biotechnology laboratory.

1. Find the source of SDS information available to you. It may be binders of SDS sheets in your lab, a CD, or on a website (such as from a vendor that sells a particular lab chemical).

2. Find a SDS for each of the following chemicals. Use the SDS links found for each on www.gbiosciences.com/ResearchProducts/BTSNMSupport_Materials.aspx

- sodium hydroxide (NaOH) pellets
- ampicillin, sodium salt
- SDS, 20% Solution

3. Make a chart in your notebook to record the following information for each chemical.
   - molecular weight of the compound
   - appearance (texture, color, phase, etc)
   - melting point
   - solubility
   - flammability
   - action to take if it gets on skin
   - action to take if it gets in eyes
   - action to take if it is ingested or inhaled
   - other safety precautions

4. Kanamycin sulfate and ampicillin are both used as medical and laboratory antibiotics to stop bacterial growth. Examine how similar they are to each other by comparing their SDS information. Record three differences in the information reported for these two chemicals.

Finding the Molecular Weights of the Solutes Used in Common Solutions

Activity 3.4

Being able to determine the molecular weight of a compound is essential when making solutions of a particular molarity. When calculating the mass of a compound to be used in a solution, multiply the molecular mass (g/mol) by the concentration (mol/L) and volume (L) desired.

Since each atom in a molecule has a certain mass, the total mass of a molecule can be calculated by adding molecular masses of all the individual atoms together. For example, the molecular mass of salt (NaCl) is 58.5 amu. This is because the sodium atom weighs 23.0 amu and the chlorine atom weighs 35.5 amu. Since atomic mass is shown on the Periodic Table, a technician can calculate the molecular mass of a compound, as long as he or she knows the formula for that compound.

Sometimes, however, the molecular mass of a compound is not obvious. For example, the compound TRIS is used in many solutions, but most people do not know the molecular formula. This is not surprising since TRIS has 19 atoms in it!

Chemical companies print the molecular mass on the bottles of most compounds, making it easy for lab technicians who use this information frequently. The molecular mass is shown as either molecular weight (MW) or formula weight (FW).

Find the molecular formula and molecular mass (formula weight) of some common solutes.

Note: Compounds may exist in several forms depending on their association with water molecules. When a molecule is bound to one or more water molecules, the compound is known as "hydrated."

Look at the bottle labels to find the molecular mass (FW) of the compounds in the chart on the next page. If the molecular formula is given, record that as well. Cupric sulfate pentahydrate, shown by the formula $CuSO_4 \cdot 5H_2O$, is one cupric sulfate molecule attached to five water molecules. This is the hydrated form of cupric sulfate. Hydrated compounds may be liquids, but usually are not, even though the molecules are associated with water molecules. They are usually powders or crystals, as is the case for $CuSO_4 \cdot 5H_2O$. On the other hand, cupric sulfate can exist in a form not bound to water molecules. This form is called "anhydrous." Cupric sulfate (anhydrous) is represented by the formula $CuSO_4$. It is important to know whether a compound is hydrated or not since hydration changes the molecular weight of the compound and, therefore, dramatically changes how many grams of a compound are added to a solution.

Create a data table similar to the one below. Fill in the "open" boxes to complete all three columns.

Compound	Molecular Formula	Molecular Mass (g/mol) Calculated from the Periodic Table	Molecular Mass (g/mol) Reported on the Stock Bottle
ammonium sulfate	$(NH_4)_2SO_4$		
cupric sulfate pentahydrate	$CuSO_4 \cdot 5H_2O$		
TRIS			
TRIS-HCl			
calcium chloride (anhydrous)	$CaCl_2$		
magnesium chloride (anhydrous)	$MgCl_2$		
magnesium chloride (hexahydrate)	$MgCl_2 \cdot 6H_2O$		
zinc chloride	$ZnCl_2$		
EDTA (disodium salt, dihydrate)			
glucose (also called "dextrose")			
sodium hydroxide			
potassium hydroxide	KOH		
sodium acetate	$NaCH_3COO$		
sodium phosphate (monobasic, monohydrate)	$NaH_2PO_4 \cdot H_2O$		
sodium phosphate (dibasic, anhydrous)	Na_2HPO_4		
sodium carbonate	Na_2CO_3		
sodium citrate (tribasic, dihydrate)	$Na_3C_6H_5O_7 \cdot 2H_2O$		
sodium bicarbonate	$NaHCO_3$		
bromophenol blue			
salicylic acid (sodium salt)			

Activity 3.5 Writing a Standard Operating Procedure (SOP)

As part of good manufacturing practice (GMP), technicians working in a biotechnology manufacturing facility must follow procedures exactly every time. To ensure that they do, companies produce detailed step-by-step instructions called a "Standard Operating Procedure," or "SOP." A SOP is a very formal set of instruction pages, and technicians are required to verify that each step is completed on a document called a "batch record." The technician initials and dates the end of each step. An example of an SOP batch record is shown at the top of the next page.

Write an SOP and batch record for using an individually wrapped, plastic, sterile, disposable 10-mL pipet to dispense exactly 7 mL of solution without contamination.

1. Using Microsoft® Excel®, write a SOP in so much detail that an untrained individual could accurately follow it to conduct the above procedure.
2. Give your SOP to a student from another biotechnology or science class. Ask the student to perform the procedure exactly as written, with you observing. Did the student follow the steps exactly? Were there any places where the student was uncertain about what to do? Was the exact desired outcome produced? Edit the SOP to achieve a functional protocol.

Step	Substep	Completion (Initial/ Date)	Colleague Verification	Procedure
1.				**Obtain stock bleach solution.**
	a.			Obtain bottle of commercial bleach from stockroom shelf.
	b.			Record Lot number and expiration date from bottle.
2.				**Prepare bleach storage bottle.**
	a.			Obtain a 250-mL amber bottle and black cap from glass stockroom.
	b.			Clean the 250-mL amber bottle and cap with prepared Alconox® solution by Alconox, Inc.
	c.			Rinse bottle and cap with tap water until all bubbles are gone.
	d.			Rinse bottle and cap three more times with tap water.
	e.			Place bottle and cap upside down on paper towel for 5 minutes to drip dry.
3.				**Prepare the bleach solution.**
	a.			Put on safety goggles and gloves.
	b.			Obtain a clean glass 25-mL graduated cylinder.
	c.			Move amber bottle and cap, stock bleach bottle, and graduated cylinder to a chemical fume hood.
	d.			Turn on the fan for the chemical fume hood.
	e.			Pour exactly 10 mL of bleach into the 25-mL graduated cylinder.
	f.			Pour the 10 mL of bleach into the clean 250-mL amber bottle.
and so on…				

pH, Buffers, and Solution Preparation

Have you ever noticed claims by shampoo manufacturers that their shampoos are "pH-balanced"? What does that mean?

Hair is composed of protein molecules called keratin. The keratin molecules are laid down in long strands that form fibers. Like all proteins, the structure of keratin is affected by environmental factors. High heat, for example, can break the bonds holding protein molecules together. If this happens at the end of a hair shaft, the result may be split ends.

Keratin protein is also affected by pH. pH is a way of reporting the acid-base level of a solution. pH measures how many hydrogen ions (H^+) are in solution. pH is measured using indicators in liquid or paper form.

pH is reported in units ranging from 0–14. In pure water, most of the water molecules exist as H_2O, but a few ionize to H^+ and OH^- ions. The numbers of H^+ and OH^- are equal so their charges cancel out. Pure water has no charge, is said to be neutral, and corresponds to a pH 7.

If solutes are dissolved in water, the ratio of H^+ to OH^- ions can change. If H^+ ions are added, the pH goes down and the solution is considered acidic. If H^+ concentration goes down, the relative number of OH^- ions goes up, the pH goes up, and the solution is considered alkaline (or basic).

pH Scale

A pH around 5 appears to be good for maintaining the bonds in keratin and keeping hair fibers strong.

In a biotech lab, proteins or DNA in cells and solutions prepared in the lab must be made and maintained at a specific pH, otherwise a change in molecular structure or function could result. Buffers are solutions that resist changes in pH. They are used as DNA and protein solvents to ensure that the pH and electrical charges around molecules stay at a constant level (pH-balanced).

To determine the pH, dip a pH strip into the solution, remove it, and compare to the color-coded key. A pH meter is shown in the background.
Photo by author.

Learn more about the pH of common solutions that are of biological importance and are used in and out of the laboratory.

Use Internet resources to find the pH of these common household and biotechnology laboratory solutions.

Item	Approximate pH (or pH range)
apple juice (apple cell cytoplasm)	
tomato juice (tomato cell cytoplasm)	
orange juice (orange cell cytoplasm)	
lemon juice (lemon cell cytoplasm)	
human blood	
stomach juices	
sea water	
liquid drain cleaner	
TE buffer—DNA storage buffer	
IX TAE buffer (TRIS/Acetic Acid/EDTA)—DNA electrophoresis buffer	
IX TBE buffer (TRIS/Boric Acid/EDTA)—DNA electrophoresis buffer	
IX PBS (phosphate buffered saline)—used for protein solutions	
IX Laemmli buffer (PAGE running buffer)—used for protein gel electrophoresis	

Activity 3.7 "pHun" at Home with pH

Many of the beverages, condiments, and cooking ingredients found in a kitchen are acids and bases. A solution with a pH of less than 7 has more H^+ ions than OH^- ions and is, therefore, an acid. Acid solutions have certain characteristics, including, depending on the strength, a sour taste, and the ability to burn. A solution with a pH higher than 7 has more OH^- ions than H^+ ions and is considered a base. Basic solutions may also have a sour taste, feel slippery, and cause burns.

Determine the pH of solutions in the kitchen.

1. Arrange to take home a box of wide-range pH paper.
2. Determine the pH of at least 10 solutions or liquid mixtures in your kitchen. Be creative. Almost anything liquid or sitting in liquid can be tested. **Caution: Read all labels before testing a solution to ensure that the solution is safe**.
3. Make a data table to record the name of each item tested, its pH, whether it is an acid, a base, or a neutral solution.
4. Which of the samples has the additional characteristics of an acid or a base?

Bioethics

Is Honesty Always the Best Policy?

In an academic research facility, such as a university laboratory, professors write grant proposals to obtain funding for research. Often, they need to acquire hundreds of thousands of dollars in grants to keep themselves and their researchers employed. They also need funding to pay for reagents, supplies, and equipment.

Grants are available from public and private sources. The majority of research funding is provided through grants from the federal government. Nonprofit organizations or foundations, such as the American Cancer Society or the American Heart Association, also provide grants and funding.

There is a lot of competition for limited funding dollars. If laboratories produce "good" quality research and large amounts of data, the grant proposal is looked upon favorably by the granting organizations. This creates pressure on scientists to get "good results."

In laboratory research, though, most hypotheses are not well supported by data. The majority of experiments give data that are unexpected or unsupported. Researchers often joke that only about 5 percent of experiments work. This is not a "bad thing." It is simply the nature of science that many unexpected factors can affect experiment results. Through setbacks and unsupported hypotheses, scientists redesign experiments and conditions to get reliable data.

The scientific environment demands honesty from researchers. Scientists must be trusted to present accurate, authentic data. Researchers who are found to have presented falsified data are barred from future funding and support. It is not difficult to imagine that scientists regularly face ethical dilemmas, given the pressure to "publish or perish," and to acquire grants and funding. It might be tempting to falsify or exaggerate data or results, but scientists must be trusted to be ethical.

 Decide what steps to take when presented by scientific dishonesty.

Read, review, and reflect on the scenario that follows. Decide what you would do if you were the technician observing potential dishonesty. Then write at least a one-half page discussion outlining your position regarding the situation, and how and why you made the decision.

Scenario: You have been a lab technician in a breast cancer research laboratory at a prestigious university for the past 8 months. Your team (professor, postdoctoral student, three research associates, and a lab technician) has been working to develop a new, modified version of a protein that will recognize breast cancer cells and cause the body to destroy them. In early trials, 80% of the breast cancer cells were recognized and destroyed after treatment with the new protein. An annual grant for another $800,000 of funding is due at the National Institutes of Health (NIH) by 6 pm on Friday. It is 4 pm on Tuesday, and the most recent data, which must be included in the grant application, show that only 20% of the breast cancer cells are destroyed after treatment. This percentage is not very impressive, especially when many of the mice tested with the protein developed sleepiness and other side effects. The professor still believes that the protein can be effective and that the second experiment was flawed. Without the grant, the lab will have to be shut down in 6 months. The postdoctoral student proposes ignoring the second set of data. He says, "Once we get the funding, we can retest. I am sure our results will support the original data." What would you say to the professor and postdoc? Do you support the postdoctoral student's position? Should the professor follow his advice? Why or why not?

Biotech Careers

Photo courtesy of Paul Kaufman.

Molecular Biologist/ Associate Professor

Paul D. Kaufman, PhD
University of Massachusetts Medical School
Worcester, MA

Dr. Kaufman heads a group that studies how eukaryotic cells assemble chromosomes and how the chromosomes function. They are especially interested in **chromatin** (DNA and protein complexes in the nucleus). The group studies several different classes of proteins used by yeast and human cells to deposit histone proteins onto DNA, as well as enzymes that chemically modify histone structure and function.

Paul's primary responsibility is scientific discovery. He conducts experiments of his own, trains graduate students, and oversees the research of postdoctoral fellows in his laboratory. His typical day involves reviewing recent experiments by members of his lab and suggesting improvements, writing papers and grants, reading scientific literature, and attending research seminars. For several weeks each year, Paul travels to scientific meetings to present research results from his laboratory.

chromatin (chro•ma•tin)
nuclear DNA and proteins

Introduction to Studying DNA

Learning Outcomes

- Describe the structure and function of DNA and explain the process by which it encodes for proteins
- Explain how molecular modeling is used to study the structure and function of DNA as well as other biomolecules.
- Differentiate between eukaryotic and prokaryotic chromosomal structure and explain how this difference impacts gene regulation in the two cell types
- Differentiate between bacterial cultures grown in liquid and solid media and explain how to prepare each media type using sterile technique
- Discuss the characteristics of viruses and their importance in genetic engineering
- Explain the fundamental process of genetic engineering and give examples of the following applications: recombinant DNA technology, site-specific mutagenesis, and gene therapy
- Describe the process of gel electrophoresis and discuss how the characteristics of molecules affect their migration through a gel

4.1 DNA Structure and Function

In the late decades of the twentieth century, a "new" biotechnology industry grew out of innovative techniques to both transfer DNA between cells and to manipulate the cells to manufacture specific proteins. These techniques revolutionized science and industry, creating new kinds of companies and new kinds of science jobs. In this chapter, some of the methods by which DNA is isolated and studied are presented. Other biotechnologies are presented in later chapters, including genetic engineering, protein studies, and biomanufacturing. Together these have led many biotechnology companies to specialize in creating cellular-protein factories and modified organisms with new and more desirable characteristics.

The manipulation of genetic information, specifically the deoxyribonucleic acid (DNA) and ribonucleic acid (RNA) codes, is at the center of most biotechnology research and development. The genetic

Figure 4.1. Pure salmon DNA isolated from sperm cells. In solution, DNA is clear, with bubbles trapped in long strands of DNA. During spooling, the solvent is squeezed out and the DNA appears white.
Photo by author.

gene (gene) a section of DNA on a chromosome that contains the genetic code of a protein

nitrogenous base (ni•trog•e•nous base) an important component of nucleic acids (DNA and RNA), composed of one or two nitrogen-containing rings; forms the critical hydrogen bonds between opposing strands of a double helix

information within cells is stored in DNA molecules (see Figure 4.1). Sections of the DNA sequence are then transcribed into RNA messages. The arrangement of nitrogen bases on a DNA strand determines the RNA sequence, and the RNA code determines the amino acids that are placed in the polypeptide chain of a protein (see Figure 4.2). Proteins do the work of cells, and they give cells and organisms their unique characteristics.

The "Central Dogma of Biology" explains, in general, how genetic information is converted into structures and functions. It describes how small sections of DNA (**genes**) are transcribed into messenger RNA (mRNA) molecules, which are, in turn, translated at ribosomes into the proteins of the cell. Many biotechnology efforts modify DNA molecules with the goal of affecting protein production.

At a biotechnology company, the proteins produced in a cell may be the actual product desired, such as an insulin molecule manufactured in a pancreas cell. Or, the protein product may be a regulatory molecule, such as an enzyme needed to make another molecule, for example, cholesterol.

Since so many proteins are necessary for the functions in an organism, the number of genes needed is enormous—about 20,000 to 25,000 genes in humans. A typical cell synthesizes more than 2000 different kinds of protein. Hundreds or thousands of copies of a particular protein may be needed at a given moment (remember that a cell's proteins may be structural or regulatory). Multiply that number of proteins by the many hundreds of different cell types a multicellular organism contains, and the number of proteins produced by a single organism may reach over a million. Each protein has one or more genes coding for its production or regulation. The result is many thousands of genes coded for by several billion DNA bases in each species.

The entire sum of DNA in a cell (the genome) is variable from organism to organism, but every cell within an organism, except sex cells, has the same genome. The sizes of the genomes from some important organisms are listed in Table 4.1.

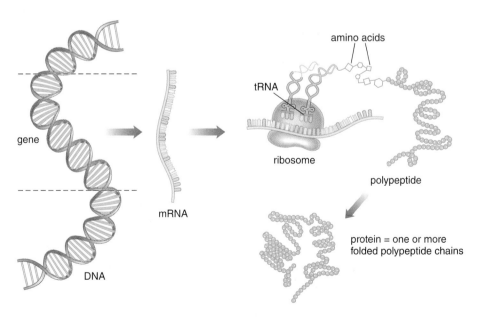

Figure 4.2. **The Central Dogma of Biology.** Proteins are produced when genes on a DNA molecule are transcribed into mRNA, and mRNA is translated into the protein code. This is called "gene expression." At any given moment, only a relatively small amount of DNA in a cell is being expressed.

Table 4.1. Sizes of the Genomes from Some Model Organisms The abbreviation bp stands for **base pair** and refers to counting both nitrogenous bases on each side of the double helix at a particular location on the strands. This is the accepted method of sizing DNA.

Organism	Number of Genes	Size of the Genome (bp)
Haemophilus influenzae: childhood ear infection	1,749	1,830,137
Mycoplasma genitalium: free-living bacterium	470	580,070
Caenorhabditis elegans: free-living roundworm	19,899	97,000,000
Homo sapiens: humans	20,000–25,000	3,000,000,000

Although there are different quantities of DNA in cells from different organisms, all DNA from all organisms share many characteristics. In fact, DNA molecules from different organisms are virtually identical, except in the order of **nitrogenous bases** (containing nitrogen). The structure of DNA was first introduced and discussed in Chapter 2. A discussion of the similarities and variations in DNA molecules among organisms is in the following pages.

Similarities in DNA Molecules Among Organisms

1. All DNA molecules are composed of four nucleotide monomers that contain four nitrogenous bases (see Figure 4.3a–d and Figure 2.34).
   adenosine deoxynucleotide (A) guanosine deoxynucleotide (G)
   cytosine deoxynucleotide (C) thymine deoxynucleotide (T)
2. Virtually all DNA molecules form a double helix (two sides) of repeating nucleotides, several million bp in length. Nucleotides connect to other nucleotides in the strand through strong **phosphodiester bonds** between sugars and phosphates of adjacent nucleotides. **Hydrogen bonds** (H-bonds) between nitrogen bases on each complementary strand hold the two sides of the DNA molecule together. Base pairing in DNA only occurs between adenine and thymine molecules (A-T), or between guanine and cytosine molecules (G-C). Figure 4.4 a-d uses different molecular models to represent the structural features of a section of DNA.

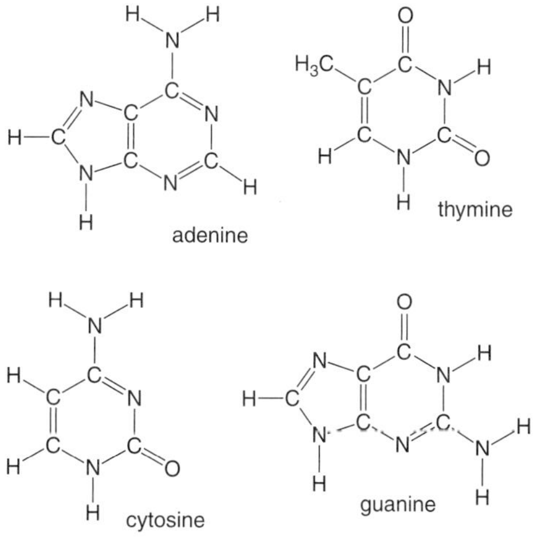

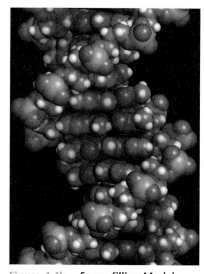

Figure 4.3a-d. Nitrogenous Bases in DNA. The four nitrogenous bases (adenine, thymine, cytosine, and guanine) are found in the four DNA nucleotides. Notice how two of them are composed of a single carbon ring (called a **pyrimidine**), and two of them are composed of a double carbon ring (called a **purine**). The entire nucleotide, including the sugar and phosphate group, is shown in Chapter 2, Figure 2.34.

Figure 4.4b. Space-filling Model of a Section of DNA In this space filling model, the phosphate groups in the backbone strands of each side of the double helix are visible (orange phosphorus with red oxygen atoms). Notice the (blue) nitrogenous bases in the center of the double helix making up the rungs in the DNA ladder.

base pair (base pair) the two nitrogenous bases that are connected by a hydrogen bond; for example, an adenosine bonded to a thymine or a guanine bonded to a cytosine

phosphodiester bond (phos•pho•di•es•ter bond) a bond that is responsible for the polymerization of nucleic acids by linking sugars and phosphates of adjacent nucleotides

hydrogen bond (hy•dro•gen bond) a type of weak bond that involves the "sandwiching" of a hydrogen atom between two fluorine, nitrogen, or oxygen atoms; especially important in the structure of nucleic acids and proteins

pyrimidine (py•rim•i•dine) a nitrogenous base composed of a single carbon ring; a component of DNA nucleotides

purine (pu•rine) a nitrogenous base composed of a double carbon ring; a component of DNA nucleotides

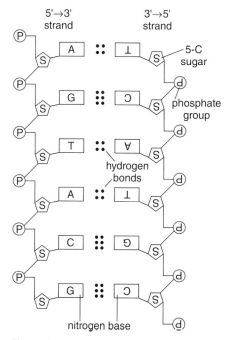

Figure 4.4a. DNA Structure. This structural diagram of a section of a DNA molecule shows how the nucleotides in one chain of the helix face one direction, while those in the other strand face the other direction. Each nucleotide contains a sugar molecule, a phosphate group, and a nitrogenous base. Nitrogenous bases from each strand bond to each other in the center through H-bonds. The H-bonds are rather weak; therefore, the two strands of DNA separate easily at high temperatures.

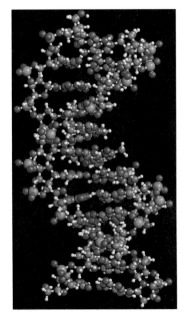

Figure 4.4c. Ball and Stick Model of a Section of DNA In this model, the same color coding is seen as in Figure 4.4b. Tracing the backbone strands down, the pentagonal deoxyribose sugar molecules can be seen between phosphate groups. Notice on one strand the sugars are facing up and one the other strand they are facing down (the strands are anti-parallel.

■■■■■■■■■■■■■■■■

antiparallel (an•ti•par•al•lel) a reference to the observation that strands on DNA double helix have their nucleotides oriented in the opposite direction to one another

semiconservative replication (sem•i•con•ser•va•tive rep•lic•a•tion) a form of replication in which each original strand of DNA acts as a template, or model, for building a new side; in this model one of each new copy goes into a newly forming daughter cell during cell division

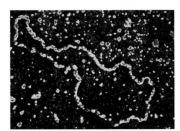

Figure 4.5. Circular DNA visualized by TEM (transmission electron microscope) using cytochrome C and metal shadowing techniques. a single, circular DNA molecule is found in bacteria cells. ~40,000X
© Corbis.

3. Because an adenosine deoxynucleotide (A) of one strand always bonds to a thymine deoxynucleotide (T) in the other strand, the amount of adenine in every double-stranded DNA molecule always equals the amount of thymine. Likewise, the amount of guanosine deoxynucleotide (G) in one strand of every double-stranded DNA molecule always equals the amount of cytosine (C) (See Figure 4.4d). Since adenine and guanine are purines, and thymine and cytosine are pyrimidines, the width of an A-T base pair is the same as the width of a G-C base pair.

4. The nucleotides in each of the two strands of DNA are oriented in the opposite direction of the other strand. On one side, the nucleotides are oriented upward, and on the other side, they are oriented downward. We say that the strands are **antiparallel**. This arrangement gives "directionality" to a DNA strand, so that the DNA strand is "read" in one direction only. All DNA molecules are antiparallel. (See Figures 4.4a–d)

5. The nitrogenous bases (A, C, G, and T) are stacked 0.34 nanometer (nm) apart with 10 nitrogen bases per complete turn of the helix. This stacking of bases ensures that the shape of the DNA molecule is constant throughout its length. The uniform shape ensures that enzymes and regulatory molecules can recognize the DNA molecule, and allows for coiling and packaging of the DNA (see Figure 4.5).

6. DNA undergoes **semiconservative replication**, making two daughter strands from a single parent strand. During semiconservative replication, a strand unzips. Each original strand acts as a template, or model, for building a new side. By the time replication is completed, two identical DNA strands have been produced. One of each copy goes into a newly forming daughter cell during cell division (see Figure 4.6).

Variations in DNA Molecules

Although DNA molecules are all remarkably similar from one organism to another, there are some variations, which may include the following:

- the number of DNA strands in the cells of an organism (ie, the number of chromosomes)
- the length in base pairs of the DNA strands
- the number and type of genes (nucleotide sequences that code for protein production) and noncoding regions
- the shape of the DNA strands (circular or linear chromosomes)

Some of the unique characteristics of DNA from different sources are discussed in later sections.

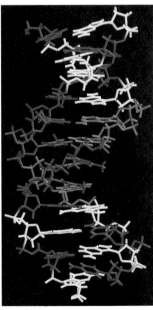

Figure 4.4d. Wire Model of a Section of DNA The same DNA strand as in Figure 4.4c represented in a wire model. The nucleotides are color-coded (yellow=G, blue=C, green=A, red=T). Do you see the G-C, A-T base pairing? The anti-parallel configuration of the strands is seen in this model.

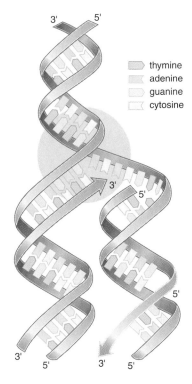

thymine
adenine
guanine
cytosine

Figure 4.6. DNA Replication. DNA replicates in a semiconservative fashion in which one strand unzips and each side is copied. It is considered semiconservative since one copy of each parent strand is conserved in the next generation of DNA molecules.

Section 4.1   Review Questions

1. Describe the relationship between genes, mRNA, and proteins.
2. Name the four nitrogen-containing bases found in DNA molecules and identify how they create a base pair.
3. The strands on a DNA molecule are said to be "antiparallel." What does antiparallel mean?
4. During cell division, DNA molecules are replicated in a semiconservative manner. What happens to the original DNA molecule during semiconservative replication?

4.2 Sources of DNA

In nature, DNA is made in cells. For sources of DNA, scientists can find cells in nature or they can grow cultures of cells in the laboratory. Scientists have learned how to grow many different cells on or in a **medium** (source of nutrients) prepared in the laboratory. Cells from cell cultures can be collected and broken open, a process called **lysis**. Lysed cells release their DNA molecules in a mixture of other cellular molecules. Through separation techniques, the DNA molecules can be isolated from the other cell molecules.

The basic DNA molecule and the genetic code are the same from organism to organism, but the packaging of the DNA and its location within the cell vary among groups of organisms. Knowing about the differences in DNA packaging in a cell will help a technician isolate DNA from different cells.

medium (me•di•um)
a suspension or gel that provides the nutrients (salts, sugars, growth factors, etc) and the environment needed for cells to survive; plural is media

lysis (ly•sis)   the breakdown or rupture of cells

R plasmid (R plas•mid)   a type of plasmid that contains a gene for antibiotic resistance

Prokaryotic DNA

Bacteria cells, such as *E. coli,* are prokaryotic, meaning they do not contain a nucleus or other membrane-bound organelles (see Figure 4.7). In a bacterium, the DNA is floating in the cytoplasm but is usually attached at one spot to the cell membrane. A bacterium typically contains only one long, circular DNA molecule (see Figure 4.5). It is usually supercoiled, folding over on itself like a twisted rubber band. Microscopically it appears as a dense area called the nucleoid. The DNA molecule is rather small and contains only several thousand genes. The well known bacterium, *E. coli,* has a single DNA strand containing approximately 4,400 genes with about 4.6 million bp. The entire *E. coli* genome codes for RNA and proteins. There is very little spacer (unused) DNA. The genes on the single DNA strand are, for the most part, necessary for survival.

Some bacteria contain extra small rings of DNA floating in the cytoplasm (see Figure 4.8). These are called plasmids. A plasmid contains only a few genes (5 to 10), and these genes usually code for proteins that offer some additional characteristic that may be needed only under certain conditions but not others. Some of the most familiar plasmids are the **R plasmids**. These contain antibiotic resistance genes. Bacteria with these resistance genes can survive exposure to antibiotics that would normally kill or stunt them.

Bacteria can transfer plasmids and, thus, genetic information between themselves. There are several ramifications of this practice. Genes, such as those for

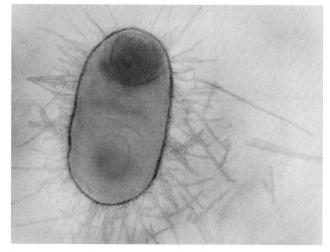

Figure 4.7.   **Prokaryotic cells, such as *E. coli*, do not contain membrane-bound organelles such as mitochondria and lysosomes. ~30,000X**
© Lester V. Bergman/Corbis/Getty Images.

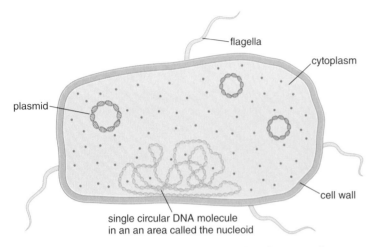

Figure 4.8. **Structure of a Bacterium.** Although prokaryotic cells are rather simple in structure, some may contain one or more plasmids. Plasmids are tiny rings of "extra" DNA. So far, scientists have only found plasmids in prokaryotes and yeast.

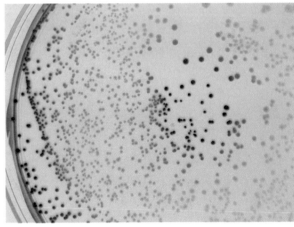

Figure 4.9. **Beige (nontransformed) and blue-black (transformed) colonies of bacteria. The cells in the dark colonies have taken in plasmids carrying a new gene that codes for the production of a blue product.**
© Ted Horowitz/Corbis/Getty Images.

transformed (trans•formed) the cells that have taken up foreign DNA and started expressing the genes on the newly acquired DNA

vector (vec•tor) a piece of DNA that carries one or more genes into a cell; usually circular as in plasmid vectors

operon (op•er•on) a section of prokaryotic DNA consisting of one or more genes and their controlling elements

antibiotic resistance, can be transferred between bacteria and lead to deadly antibiotic-resistant forms of disease-causing bacteria. Transferring plasmids may give bacteria a way of "evolving" by gaining new and better characteristics for survival. Scientists have learned how to use plasmids to transfer "genes of interest" into cells. When these cells take up foreign DNA and start expressing the genes, they are considered **transformed** (see Figure 4.9).

Different bacteria have different plasmids. Some bacteria have more than one kind of plasmid, and some bacteria contain no plasmids. Because plasmids are small and easy to extract from cells, they are often used as recombinant DNA (rDNA) **vectors** to carry new genes into cells for transformation. Foreign DNA fragments (genes) can be cut and pasted into a plasmid vector. The recombinant plasmid may then be introduced into a cell. The cell will read the DNA code on the recombinant plasmid and start synthesizing the proteins coded for on the "foreign" gene(s). This is how Genentech, Inc. first manipulated *E. coli* cells to make human insulin. The scientists tricked the *E. coli* cells into reading the human genes that had been inserted into them on recombinant vector plasmids.

Another difference between prokaryotic DNA and DNA from eukaryotic cells is that the gene expression (how genes are turned ON or OFF) in prokaryotes is rather simple with only a few controls. Figure 4.10 is a simplified model of an **operon** (two or more genes and their controlling elements) on a piece of prokaryotic geomic DNA. Remember that on a real bacterial chromosome there would be several hundred operons, one directly after another.

In the middle of the operon is one or more structural genes. A structural gene is a section that actually codes for one or more mRNA molecules, which will later

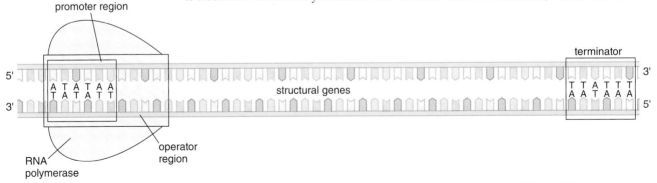

Figure 4.10. **Bacterial Operon.** An operon contains the controlling elements that turn genetic expression ON and OFF.

be translated into one or more proteins. For the structural gene(s) to be read or "expressed" as a functional protein, other accessory areas are required.

For gene expression to occur in prokaryotes, the enzyme that synthesizes a mRNA molecule, **RNA polymerase**, must attach to a segment of DNA at a **promoter** region of the operon. This essentially "turns on" the prokaryotic gene. The RNA polymerase works its way down the DNA strand to a structural gene, where it builds a mRNA molecule from free-floating nucleotides, using the DNA strand as a template. A synthesized mRNA is decoded into a peptide at a ribosome. A region called the **operator**, located just prior to the structural gene(s), can "turn off" the operon. If a regulatory molecule attaches at the operator, the operon is turned off because the RNA polymerase is blocked from continuing down the strand to the gene(s). In this case, no protein is produced. Blocking or unblocking the operator is the way bacterial cells make only certain proteins at certain times. It would be a waste, for example, for a cell to make **beta-galactosidase**, which breaks lactose down to monosaccharides, if no lactose was around (see Figure 4.11).

Genetic engineers utilize the promoter and operator regions to turn on and off the production of certain genes. When the insulin gene from humans was genetically engineered into *E. coli* cells, a bacterial promoter region had to be attached to the human insulin gene. Without it, the *E. coli* cells would not have recognized the new gene, and it would not have been transcribed and translated into protein. There would have been no gene expression.

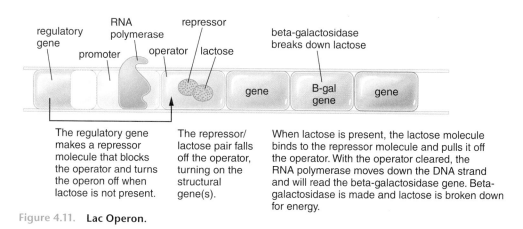

Figure 4.11. **Lac Operon.**

Cell Culture and Sterile Technique

To study or manipulate the DNA of bacteria, bacteria cells are needed in large numbers. To grow bacteria cells in the laboratory, a scientist must provide an environment, or medium, that the cells "like." Some bacteria grow well in a liquid medium (**broth**). Some bacteria prefer a solid medium, called **agar**. Some will grow well on or in either.

Agar medium is a mixture of water and protein molecules. To prepare it, researchers mix powdered agar in water and heat the mixture until the agar is completely suspended. The agar is sterilized at a high temperature (121°C or higher) at a high pressure of 15 pounds per square inch (psi) or higher for a minimum of 15 minutes (see Figure 4.12). The agar is allowed to cool to about 65°C and is poured under sterile conditions into sterile Petri dishes. The agar cools and solidifies within 15 to 20 minutes, and the poured plates may be used after about 24 hours.

Liquid or broth (water and protein molecules) cultures grow as suspensions of millions of floating cells. Starting with sterile broth, the researcher introduces a colony of cells under sterile conditions into the broth. The cells grow and divide and spread themselves throughout the liquid. Broth culture cells reproduce quickly since they have better access to nutrients than do colonies growing on the surface of solid media.

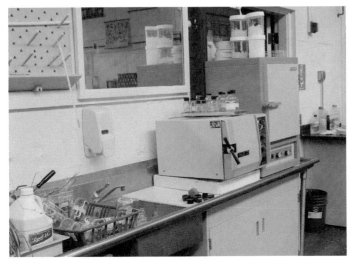

Figure 4.12. Media, glassware, tubes, and pipet tips can be sterilized in an autoclave (center). An autoclave creates high temperature and high pressure to burst all bacteria or fungus cells. Autoclaved tips and tubes that need to dry before use can be placed in a drying oven at 42°C for 24 hours (right).
Photo by author.

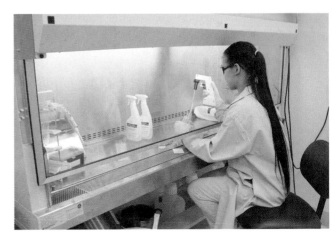

Figure 4.13a. Pouring plates and transferring media in a sterile laminar flow hood (LFH) decreases the chances of contamination. The LFH has a HEPA filter with pores so small that it removes microorganisms from the air before the air enters the working area in the hood chamber. Everything brought into the hood chamber must be free of microorganisms.
Photo by author.

■■■■■■■■■■■■■■

media preparation (me•di•a prep•ar•a•tion) the process of combining and sterilizing ingredients (salts, sugars, growth factors, pH indicators, etc) of a particular medium

autoclave (aut•o•clave) an instrument that creates high temperature and high pressure to sterilize equipment and media

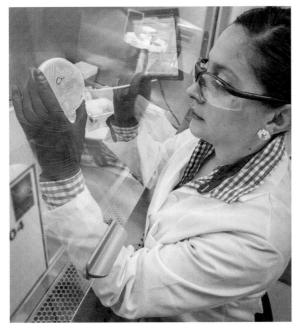

Figure 4.13b. **Cell Culture Archivist** A Research Associate uses sterile technique to collect cells from a plate culture and transfer them to long-term glycerol stock storage.
Roy Kaltschmidt, Lawerence Berkeley National Lab, The Regents of the University of California, 2010

Oxygen and food diffuse into these cells easily. Under ideal conditions, broth culture cells might replicate as often as every 20 minutes.

A technician needs to learn media preparation and sterile technique to be able to grow cells in culture. Many kinds of powdered media base can be purchased and prepared following recipes on the container. An important part of all **media preparation** is sterilizing the medium. The medium, in bottles or plates, must be free of any unwanted bacteria or fungi before it is used. An **autoclave**, or pressure cooker/sterilizer, is used to heat the medium to over 121°C for a minimum of 15 minutes to destroy any cells or spores. If the medium is to be used in other flasks, bottles, or plates, it must be transferred under sterile conditions to the new sterile vessel (see Figure 4.13a–b).

Sterile technique is the process of doing something without contamination by unwanted microorganisms or their spores. Of course, sterile technique is used in surgery. Doctors "scrub up" to get rid of "germs" before they enter the operating room. They also use disinfectants and antiseptic soaps to kill bacteria and fungi on their hands, on the patient's body, and on working surfaces. Doctors and nurses wear masks and sterile gloves to prevent transmission of microorganisms. All instruments must be sterilized by steam, heat, or radiation so that no disease-causing organisms enter a patient's body.

Technicians must practice sterile technique during cell culture. They must be certain that they are growing only the cells they want to grow (See Figure 4.13a). A single unwanted cell could ruin an experiment or a multimillion-dollar production run. Using a sterile laminar flow hood, disinfectants, sterile tools, sterile media, and attention to detail, technicians can be fairly confident they have not transferred unwanted cells into a culture.

Once cells are transferred to a sterile medium, they will start to grow and reproduce. Under ideal conditions, *E. coli* cells will divide every 20 minutes. On plate cultures, colonies can be seen within a day. A broth culture will appear cloudy as more and more cells fill it up. Bacterial cell culture concentrations may be checked using a spectrophotometer (or an instrument that contains a spectrophotometer) or by counting cells on a microscope slide (see Figure 4.14). More information on growing and monitoring cell culture and fermentation (bacterial or fungal cell culture) is presented in Chapters 8 and 9.

Eukaryotic DNA

Eukaryotic DNA from protist, fungi, plant, and animal cells is similar in many ways to DNA from prokaryotes. Eukaryotic DNA uses the same nucleotide code of A, C, G, and T. Like prokaryotic DNA, eukaryotic DNA has the same double helix structure of repeating nucleotides, with each antiparallel strand bound to the other by H-bonds.

DNA from higher organisms, however, is packaged into chromosomes, regulated, and expressed differently from that of bacteria. Most notably, eukaryotes generally have several chromosomes per cell, and each chromosome is a single, linear, very long molecule of DNA coiled around proteins (histones). Each single DNA chromosome may contain several million or more nucleotides and up to many thousands of genes (see Figure 4.15).

The genome of eukaryotes is substantially larger than that of prokaryotes (see Figure 4.16). Whereas bacterial cells have only one genomic DNA strand per cell, eukaryotic cells contain multiple numbers of chromosomes. Every cell in a multicellular organism's body has the same number of chromosomes, but depending on the species, the number of chromosomes per cell could be as few as 4 and as many as 100 or more. Humans have 46 chromosomes per cell, with about 3 billion bp making up about 20,000 genes. Fruit flies have 8 chromosomes, and some ferns have more than 1000. However, the total amount of DNA per cell is not directly related to an organism's complexity.

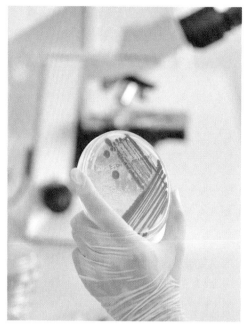

Figure 4.14. In addition to spectrophotometry, bacteria cultures can be checked using wet mount slides on high-powered light microscopes. Here, a scientist prepares to take a sample of a bacteria colony for microscopic observation.
iStock.com/tinydevil.

Figure 4.15. *Arabidopsis* plants contain 10 chromosomes in each of their cells. The genetic information stored in the genes on each of the chromosomes is similar from plant to plant. Changes in the genetic code can lead to mutant organisms, as shown in the plant on the left, which has a mutant gene for shortness.
Photo by author.

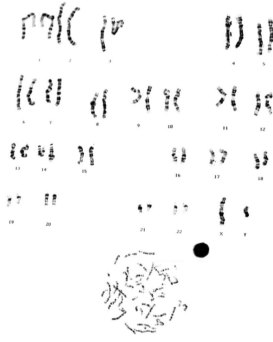

Figure 4.16. The human genome, the total DNA content of a human cell, is 46 chromosomes in 23 pairs. Chromosomes are taken from dividing cells. The chromosome pairs are lined up from large (Chromosome Pair No. 1) to small (Chromosome Pair No. 22). The sex chromosomes are the 23rd pair. The chromosome profile shown is from a male, since an X chromosome (long) and a Y chromosome (short) are shown.
iStock.com/Serpil_Borlu.

Know Your Genome

Go to biotech.emcp.net/humangenomeproject. Read the article. Write a summary of the Human Genome Project (HGP) that addresses the following topics:

- goals of the HGP
- origin of the HGP and the people involved in launching it
- When was the HGP completed and how has its completion impacted research?
- What is in store for us in the future now that we understand the sequence of the human genome?

Humans are more complex organisms than ferns, yet we have only a fraction of the amount of DNA per cell. This is because much of eukaryotic DNA is noncoding, meaning it does not transcribe directly into protein. Much of the DNA in higher organisms is spacer DNA within and between genes. There appears to be an evolutionary advantage to widely spaced genes. Genes that are far apart are often involved in recombination, which shuffles forms of a gene from one chromosome to another. This leads to new combinations of genes being sent to sex cells. The end result of variant sex cells is increased diversity in the next generation.

Determining the function of spacer DNA is an area of intense interest for researchers. What is its purpose? Why is it there? Why is there so much spacer DNA? Is it just "junk" DNA or does it have some value to the organism or the species and its function is just not yet determined?

Another difference between eukaryotic and prokaryotic DNA is the lack of operators in eukaryotes. Like prokaryotic genes, a eukaryotic gene contains a "promoter region," where an RNA polymerase molecule binds (see Figure 4.17). The RNA polymerase moves along the DNA molecule until it finds the structural gene(s). At the structural gene, the RNA polymerase builds a complementary mRNA transcript from one side of the DNA strand. The enzyme transcribes the entire gene until it reaches a termination sequence.

However, eukaryotic genes do not contain operators, so their gene expression is controlled differently than for prokaryotes. Eukaryotic genes are usually "on" and expressed at a very low level. Expression is increased or decreased when molecules interact with enhancer or silencer regions on the eukaryotic DNA strand. The enhancer or silencer regions may be within a gene or somewhere else on the chromosome. The molecules that bind at **enhancer** or **silencer** regions are called **transcription factors** (TF). The study of transcription factors in gene regulation and how to control them is a growing area in biotechnology research and development.

In prokaryotic cells, the mRNA transcript is immediately translated into a polypeptide at a ribosome. In eukaryotes, though, mRNA transcripts are often modified before translation. In eukaryotic DNA, structural genes are composed of **intron** and **exon** sections (refer to Figure 4.17). Exons are the DNA sections that actually contain the protein code. They are "*ex*pressed," which is why they are called exons. A functional mRNA molecule has the complementary code of a gene's exons.

The introns within the structural gene are sometimes spacer DNA but often are the exons for some other protein. A polymerase molecule attaches at the promoter and moves down an entire structural gene, including the intron sections, to produce a long mRNA molecule. Upon completion of the mRNA molecule, the sections on the mRNA that correspond to introns (noncoding regions) are removed so that only the coding, exon regions remain on the mRNA molecule. When reassembled, the mRNA molecule is decoded into a protein at a ribosome. Sometimes similar proteins are coded for at the same structural gene. The processing of introns differently, removing some and leaving others, can result in different forms of a protein.

---

enhancer (en•han•cer)
a section of DNA that increases the expression of a gene

silencer (si•lenc•er) a section of DNA that decreases the expression of a gene

transcription factors (tran•scrip•tion fac•tors) molecules that regulate gene expression by binding onto enhancer or silencer regions of DNA and causing an increase or decrease in transcription of RNA

intron (in•tron) the region on a gene that is transcribed into an mRNA molecule but not expressed in a protein

exon (ex•on) the region of a gene that directly codes for a protein; it is the region of the gene that is expressed

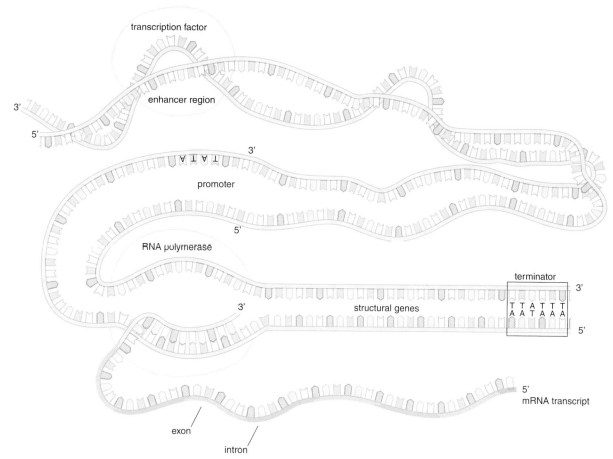

transcription factor

enhancer region

3'

5'

promoter

3'
A T A T
T A T A

RNA polymerase

5'

3'

structural genes

terminator

T T A T T T
A A T A A A

3'

5'

5'
mRNA transcript

exon

intron

Figure 4.17. **Eukaryotic Gene.** Eukaryotic genes have a promoter to which RNA polymerase binds, but they do not have an operator region. Transcription factors may bind at enhancer regions and increase gene expression.

Eukaryotic genes are also regulated by the way the chromosomes are coiled. Chromosomes in higher organisms are highly coiled around structural proteins called **histones**. The histone-DNA complex coils on itself again and again, which conceals genes (see Figure 4.18). When genes are buried this way, RNA polymerase cannot get to them to transcribe them into mRNA. The gene has, essentially, been turned off. DNA has to uncoil all the way to expose the DNA helix to be transcribed and translated to protein.

histones (his•tones) the nuclear proteins that bind to chromosomal DNA and condense it into highly packed coils

Mammalian Cell Culture

Growing mammalian cells in culture is significantly more challenging than growing bacterial cells. This is mainly because under normal circumstances, mammalian cells grow within a multicellular organism. As a part of the whole organism, mammalian cells depend on other cells for several products and stimuli. A biotechnologist growing mammalian cells in culture needs to provide an environment that is an adequate substitute for the normal environment.

On a small scale, mammalian cells are typically grown in broth culture in special tubes and bottles with a bottom surface to which the cells can stick (see Figure 4.19). In production facilities, large-scale mammalian cell cultures are grown in suspension broth cultures in fermenters. The media are specifically designed to have all the special

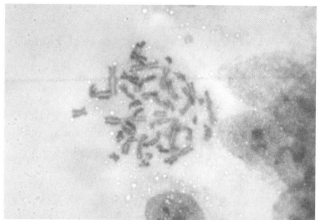

Figure 4.18. **The DNA strand is highly coiled around histone proteins. Extremely coiled, double chromosomes, such as these, are seen during cell division. 500X**
© Lester V. Bergman/Corbis/Getty Images.

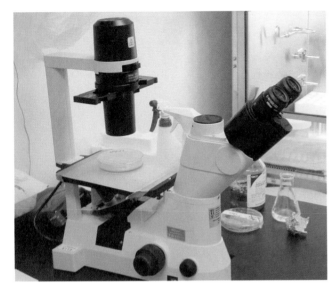

Figure 4.19. Mammalian cells are viewed on inverted light microscopes. These have the objectives mounted under the stage, closer to the mammalian cells on the bottom of the culture flask. The rate of cell growth and the health of the cells can be checked. a red indicator solution is used to monitor the pH. If the pH of the growth medium falls outside a narrow pH range, the indicator will change colors.
Photo by author.

Figure 4.20. A lab technician checks mammalian cells growing in broth culture in a carbon dioxide incubator. Most mammalian cell cultures are grown at 37°C and 5% CO_2. The cultures need to be reseeded (restarted) in fresh, sterile media every few days to give them more food and nutrients, and to prevent the buildup of toxic waste products. A phenol red indicator, added to the media, changes color as the media's pH changes.
Photo by author.

nutrients that each cell type may require. Special indicators may be added to monitor the culture. Phenol red is one such indicator. It changes from red to gold as the solution becomes acidic from cell overcrowding (see Figure 4.20).

Viral DNA

Although scientists do not consider them to be living things, viruses infect organisms and are often the target of biotechnology therapies. In addition, **nonpathogenic** viruses or virus particles are often used in biotechnology research as vectors to carry DNA between cells. For these reasons, understanding the structure of viruses and their genetic information is important.

Viruses do not have cellular structure. They are collections of protein and nucleic acid molecules that become active once they are within a suitable cell. Viruses are very tiny, measuring from 25 to 250 nm (a nanometer = 1 millionth of a millimeter). Viruses are classified into the following three main categories, based on the type of cell they attack:

- bacterial (**bacteriophages**)
- plant
- animal

Figure 4.21 shows a herpes infection, which is caused by a type of animal virus. Viruses are classified further based on the specific type of cell infected and on other characteristics, such as their genetic material and shape (see Table 4.2). No matter the type, all virus particles have a thick protein coat surrounding a nucleic acid core of either DNA or RNA.

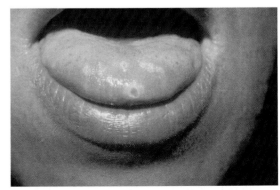

Figure 4.21. An animal virus causes the sores in oral herpes infection.
© Dr. Milton Reisch/Corbis/Getty Images.

Table 4.2. Examples of Viruses and Their Characteristics

Virus	Example of Host Cells Infected	Shape	Type of Nucleic Acid
Herpes Simplex I	human nerve and epithelial cells	spherical	double-stranded DNA
Parvovirus	human bone, blood, and cardiac cells	spherical	single-stranded DNA
Reovirus	mouse heart cells	icosahedral	double-stranded RNA
Tobacco Mosaic	tobacco and tomato, etc	rod-shaped	single-stranded RNA
HIV	human immune cells	spherical	single-stranded RNA

Within a cell, the nucleic acid of a virus is released. The viral genes are read by the host cell's enzymes, decoded into viral mRNA, and translated into viral proteins. New virus particles are assembled and released, and may infect other cells. Some viruses, for example, the lysogenic viruses, incorporate their DNA into the host chromosome, and some, for example, the lytic viruses, do not.

The structure of the virus particle is important in trying to control viruses. Many therapies utilize the fact that the human immune system can recognize virus surface proteins. Viral vaccine molecules recognize specific viral surface proteins and target them for attack. Addtionally, some biotechnology companies are developing protease inhibitors as an additional way to fight viral infection. Protease inhibitors destroy proteases made by viruses in their attempts to take over host cells. Blocking or destroying viral DNA, RNA, or protein molecules is a broadening area of disease therapy (see Figure 4.22).

Viral DNA or RNA molecules are relatively short and easy to manipulate, since viruses do not produce very many proteins compared with cells. Like plasmid DNA, viral DNA molecules are often used as vectors. They may be cut open to insert genes of interest. When sealed, they become recombinant molecules that can be inserted into new cells. If the recombinant molecules are inserted back into virus-protein coats, the viruses themselves can insert the rDNA into an appropriate host cell (see Figure 4.23).

Recombinant virus technology is one technique used in a process called **gene therapy**. Viruses can insert corrective genes (the "therapy") into cells that contain defective genes. Several companies are exploring the use of gene therapy to treat diabetes by replacing defective insulin genes in the pancreas. Gene therapy is also a possible treatment for cystic fibrosis and other genetic disorders. Although many biotechnology companies are developing gene therapies (over 400 as of 2015) to treat an assortment of disorders, to date only a few gene therapies are actually on the market. Gene therapy is discussed in more detail in later sections.

gene therapy (gene ther•a•py)
the process of treating a disease or disorder by replacing a dysfunctional gene with a functional one

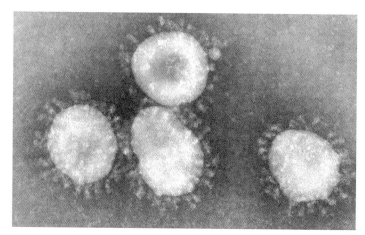

Figure 4.22. Coronaviruses are a group of viruses that have a halo or crown-like (corona) appearance due to spike-like projections. In 2005, new evidence showed that a new coronavirus is the cause of severe acute respiratory syndrome (SARS). The protein spikes could be a target for potential therapy. ~100,000X
© Mediscan/Corbis.

Figure 4.23. In France, scientists at The Institut Gustave Roussy transform mice with viral vectors carrying genes for tumor suppression or other anti-cancer proteins.
© Philippe Eranian/Corbis/Getty Images.

Using Viruses to Do Good

The word "virus" often initiates a justifiable negative response including worries of serious illness. Many viruses are responsible for diseases that cause significant sickness and even death. However, biotechnologists are learning how to manipulate some viruses to do no harm and actually to do some good. Specifically, a team has taken the poliovirus that until the 1950s, caused widespread paralysis and even death, and engineered it to fight brain cancer.

1. Go to **viralzone.expasy.org/all_by_species/678.html** and pick a human virus to learn more about. Pick one that is spread between other animals and humans. Click the link and read and record three things about the virus' characteristics or behavior.

2. Learn more about the poliovirus using information from the website, **amhistory.si.edu/polio/virusvaccine/how.htm**. Describe how the poliovirus infects cells

3. Go to **www.medicaldaily.com/polio-virus-may-cure-brain-cancer-thanks-genetic-re-engineering-327620** and read the article *Polio Virus May Cure Brain Cancer Thanks To Genetic Re-Engineering*. Describe how and why genetically-engineered poliovirus may be able to kill brain cancer cells.

Section 4.2 Review Questions

1. Plasmids are very important pieces of DNA. How do they differ from chromosomal DNA molecules?
2. Bacteria cell DNA is divided into operons. Describe an operon using the terms promoter, operator, and structural gene.
3. Describe the human genome by discussing the number and types of chromosomes, genes, and nucleotides.
4. What is gene therapy? Cite an example of how it can be used.

4.3 Isolating and Manipulating DNA

Modifications of DNA can be as simple as changing a single nitrogen base (A, C, G, or T) in a gene sequence, or as complicated as cutting out entire genes or gene sections and inserting new ones (genetic engineering). Changing DNA sequences may alter the production of proteins in a cell or an organism. New proteins may be created, or the production of some proteins may be arrested. The characteristics of cells or whole organisms may be affected by the change in the protein population.

The term "genetic engineering" was first used to describe the production of rDNA molecules (pieces of DNA cut and pasted together) and their insertion into cells. The cells were considered genetically engineered because their genetic code and, thus, their protein production, had been manipulated. Now, the phrase "genetic engineering" is used to describe virtually all directed modifications of the DNA code of an organism. Keep in mind that when scientists attempt to alter the genetic code the goal is to alter protein production.

The process of genetic engineering, by any method, requires the following steps:

1. Identification of the molecule(s), produced by living things, which could be produced more easily or economically through genetic engineering; for example, insulin for diabetic patients.
2. Finding and isolating the DNA and gene sequence for the target product molecule; for example, finding the human insulin gene in DNA isolated from human cells.

3. Manipulation of the DNA instructions by either changing them inside the cell/organism or by putting those instructions into another organism/cell that can produce the molecule more easily, in larger amounts, or less expensively. For example, the human insulin gene is pasted into a plasmid and inserted into *E. coil* cells.

4. Harvesting of the molecule or product, testing it, and marketing it to the public (for commercial products). For example, recombinant human insulin is recovered from the fermentation tanks growing huge volumes of transformed *E. coli* cells. The insulin product is tested and formulated for distribution as a therapeutic drug, such as Humalog® by Eli Lilly and Company.

Recombinant DNA Technology

Recombinant DNA technologies are the methods used to create new DNA molecules by piecing together different DNA molecules. When cells accept rDNA and start expressing the new genes by making new proteins, they are considered genetically engineered. When written by scientists, the names of proteins or DNA produced in this way are written with a "r" in front of them. For example, rInsulin is recombinant insulin, and rhInsulin stands for recombinant human insulin.

Many items currently on the market are produced through rDNA technology. These products are all versions of naturally occurring molecules or organisms improved through genetic engineering. Some well known examples of items produced through rDNA technology include the following: rhInsulin (Humalog®, marketed by Eli Lilly and Company), recombinant human growth hormone (Nutropin®, marketed by Genentech, Inc.), recombinant gamma interferon (marketed by Genentech, Inc.), recombinant HER2 antibody (produced by Genentech, Inc.), and recombinant rennin (chymosin or ChyMax®, manufactured by Pfizer, Inc.).

Each of these recombinant DNA products is either synthesized in too small a quantity in the organism, synthesized at the "wrong" time, or lacks an important characteristic. For example, tissue plasminogen activator (t-PA) is a naturally occurring protein that decreases the time it takes to dissolve blood clots. The human body makes t-PA only in very small amounts. Scientists quickly realized that if t-PA could be made on a large enough scale, it could be used to treat blood clots that occur during some heart attacks and strokes. The product is now marketed under the trade name Activase®, by Genentech, Inc. (see Figure 4.24).

Constructing rDNA requires the isolation of DNA molecules. For the production of t-PA, two kinds of DNA molecules had to be isolated: human genomic DNA containing the t-PA gene and a bacterial plasmid DNA for use as a vector. First, the human t-PA gene is identified. The t-PA gene has several introns. The mRNA exon sequence is used to produce a piece of complementary DNA (cDNA). Next, the human t-PA cDNA and the vector plasmid are each cut with a restriction enzyme and then pasted together using DNA ligase. The bacterial plasmid carrying the human t-PA genetic information is inserted into appropriate cells, in this case, Chinese hamster ovary (CHO) cells growing in broth culture (see Figure 4.25). The CHO cells then transcribe and translate the human t-PA gene and produce the human protein in the CHO cells. The t-PA molecules are harvested from the CHO broth culture and formulated for sale.

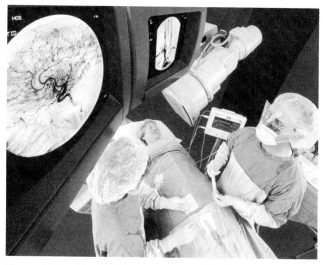

Figure 4.24. **Blood clots can occur and block blood vessels during some heart attacks and strokes. To make the blockages visible, doctors inject dye into the blood vessels in a process called angiography. If a heart attack occurs, the enzyme, Activase® (recombinant t-PA), may be injected to help "dissolve" the blockage.**
© Lester Lefkowitz/Corbis.

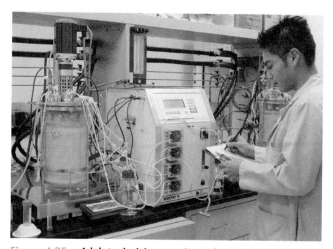

Figure 4.25. **A lab technician monitors the cells growing in broth culture. The mammalian cells in the fermentation flask on the left are making human proteins for therapeutic purposes. When enough protein is produced, the protein is separated from the cells and all other contaminants.**
Photo by author.

Recombinant Pharmaceuticals – Designed to Take Your Breath Away?

Do you have asthma or do you know someone who does? Chances are the answer is yes. According to the American College of Chest Physicians, 10% of North Americans have asthma. Several biotechnology companies are developing products to interrupt the cascade of events that result in asthmatic symptoms. Some of these asthma therapies use recombinant protein products produced through genetic engineering.

TO DO

Learn more about asthma and a genetically engineered product used to treat it. Use the Internet to answer the following questions. In your notebook, record your answers and the websites you used.

1. Define/describe the causes and symptoms of asthma.
2. Learn about Xolair® (omalizumab) by Genentech USA, Inc. and Novartis Pharmaceuticals Inc. a genetically engineered human antibody that is used as an asthma medication. Naturally produced by humans, the omalizumab protein is manufactured by Chinese hamster ovary cells grown in culture. Do a search on the Internet to find our how it works and some of the pros and cons of its use.

Site-Specific Mutagenesis

site-specific mutagenesis (site-spe•ci•fic mu•ta•gen•e•sis)
a technique that involves changing the genetic code of an organism (mutagenesis) in certain sections (site-specific) of the genome

The phrase **"site-specific mutagenesis"** or "site-directed mutagenesis" refers to the process of inducing changes (mutagenesis) in certain sections (site-specific) of a particular DNA code. The changes in the DNA code are usually accomplished through the use of chemicals, radiation, or viruses. Bacteria, fungi, plant, or cell cultures may be treated with one or more mutagens. Exposure of bacteria colonies to different amounts of ultraviolet (UV) light radiation, for example, can cause increased cell growth, new pigment production, or even cell death, to name just a few outcomes.

Mutagenic agents have varied effects, depending on the type and amount of exposure. They may cause substitutions of one base for another, for example, an "A" for a "G." A substitution may cause a positive or negative change in a protein's structure or function. Some mutagens cause additions, or deletions, to large sections of the genetic code, which may alter or prevent a protein's production.

Sometimes site-specific mutagenesis is "directed," meaning a scientist is trying to make certain changes in a protein's structure that will translate into an improved function. Such is the case with the protein, subtilisin, marketed by Genencor International. Subtilisin is an enzyme that degrades other proteins (it is a type of protease). It is added to laundry detergent to remove proteinaceous stains, such as blood or gravy, from soiled clothing (see Figure 4.26). To improve the activity of subtilisin, fungi were treated with chemicals that caused changes in their subtilisin DNA code. A random change in DNA code led to a change in protein structure such that the protein would work more effectively in the alkaline (high pH) detergent solution.

Since mutagenic agents act rather randomly, the outcome often is unexpected. The mutagenesis may stunt or inhibit product production. Sometimes genetic engineers get new or improved products. At Genencor International, Inc. and other companies with a similar focus, scientists screen large libraries of mutated fungi and bacteria while looking for new or improved ways to produce a product.

Figure 4.26. Tide® laundry detergent by Procter & Gamble contains rSubtilisin, a recombinant protease that helps remove blood, gravy, and other protein-based stains from clothing.
Photo by author.

Gene Therapy

Gene therapy is the process of correcting or modifying DNA codes that cause genetic diseases and disorders. There are two common ways to introduce new genes into defective cells. The first is to use a virus to carry a new or normal gene into target cells. A second way is to package a "good" gene in a synthetic lipid envelope (liposome) and use the envelope to bring the gene into a cell. By either method, the hope is that the gene inserts into a chromosome, is transcribed into mRNA and produces the needed functional proteins.

One of the first attempts at human gene therapy was with cystic fibrosis (CF). CF causes a buildup of thick mucus that clogs the respiratory and digestive systems and predisposes a patient to lung infections and breathing problems. One of every 3,000 babies is born with CF, and the average lifespan of patients is only about 35 years. In 2002, a modified cold virus was used to transfer a normal copy of the gene cystic fibrosis transmembrane conductance regulator, or CFTR, to cells lining the nose. The CFTR gene regulates the flow of chloride ions into cells and is defective in CF patients. The transferred gene corrected this defect. This was one successful attempt of many that are still ongoing to treat CF with a variety of gene therapies.

Companies that specialize in developing gene therapies are working on correcting other genes, such as those responsible for Parkinson's disease, diabetes, and several cancers. Researchers at the National Cancer Institute made significant progress in developing viable gene therapies when they demonstrated in 2006 that gene therapy could be used in the treatment of advanced melanoma (skin cancer). The team of researchers successfully added genes to patients' own white blood cells. When the modified white blood cells were reintroduced into the patient, they better recognized and attacked melanoma cells. The researchers hope that the gene therapy techniques they used can be applied to other cancers.

BIOTECH ONLINE

Two Therapies Are Better Than One

Gene therapy and stem cell research are two hot topics in the news. Recently, scientists in Pittsburgh, PA, have learned how to use these two therapies together to develop a better treatment for CF patients.

Go to biotech.emcp.net/CFcombined. Read the article and write a summary that describes the combined treatment and how it could help CF patients. Explain how the scientists knew the treatment was working.

Section 4.3   Review Questions

1. Genetic engineering by any method requires certain steps. Put the following steps in the correct order:
   - isolation of the instructions (DNA sequence/gene)
   - harvest of the molecule or product; then marketing
   - manipulation of the DNA instructions
   - identification of the molecule to be produced
2. What "naming" designation is used with recombinant products made through genetic engineering?
3. What is the smallest change in a DNA molecule that can occur after site-specific mutagenesis? What effect can this change have?
4. What gene has been the target of CF gene therapy? What does this gene normally do?

gel electrophoresis (gel e•lec•tro•phor•e•sis) a process that uses electricity to separate charged molecules, such as DNA fragments, RNA, and proteins, on a gel slab

agarose (a•gar•ose) a carbohydrate from seaweed that is widely used as a medium for horizontal gel electrophoresis

Gel electrophoresis uses electricity to separate molecules in a gel slab. By using electrophoresis, researchers can easily separate and visualize charged molecules, such as DNA fragments, RNA, and proteins. These molecules separate based on their size, shape, and charge.

Components of a Gel Electrophoresis

Gel material for DNA separation behaves a lot like gelatin, or Jello® by Kraft Foods, Inc. To make the gel, powdered **agarose**, a carbohydrate derived from seaweed, is dissolved in a boiling buffer solution. The solution is poured into gel trays that act as a rectangular mold (see Figure 4.27). A plastic comb is placed in the hot liquid gel so that tiny rectangular sample wells are made as the gel cools in the mold.

The solidified gel is placed in a gel box and then covered with a buffer solution. The gel box has electrodes at each end. When the power is turned on, an electric current runs into the gel box, and an electric field is established between the positive and negative electrodes (see Figure 4.28).

A sample of charged molecules is loaded into the sample wells of the gel (see Figure 4.29). When the power is turned on, and the electric field is established, the charged molecules move into the gel from the wells. If the molecules have a net negative charge, they move toward the positive end of the gel. If the molecules have a net positive charge, they move toward the negative end of the gel.

The gel material acts as a molecular strainer, separating longer molecules from shorter ones, fatter ones from thinner ones, and positively charged molecules from negatively charged molecules (see Figure 4.30). Table 4.3 lists the behavior of various molecules during electrophoresis.

Since DNA and RNA are negatively charged molecules (due to their phosphate groups), they move toward the positive pole. The agarose strands in the gel act as a strainer to separate the molecules. Small fragments move faster and farther from the well compared with larger fragments that have a more difficult time getting through the strainer. RNA molecules are smaller than most DNA molecules and,

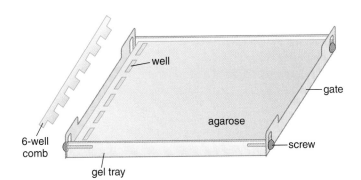

Figure 4.27. Agarose Gel Tray. Gel trays differ depending on the manufacturer. Each has some method of sealing the ends so that liquid agarose can mold into a gel. Some gel trays, such as those made by Owl Separation Systems, make a seal with the box, so casting a gel is simple. Other trays require masking tape on the ends to make a mold. Still others, like the one shown here, have gates that screw into position: up for pouring the gel and down for running the gel.

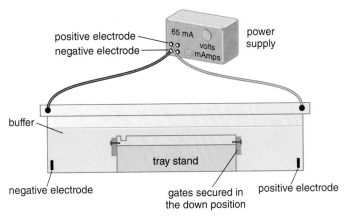

Figure 4.28. Gel Box with Buffer. For the gel box to conduct electricity and establish an electric field with a positive end (red wire) and a negative end (black wire), the solution in the gel box must contain ions. Sodium chloride (NaCl) solution can be used, but other salts, such as TRIS or lithium borate, dissolved in water (called a "running buffer"), are better for conducting electricity.

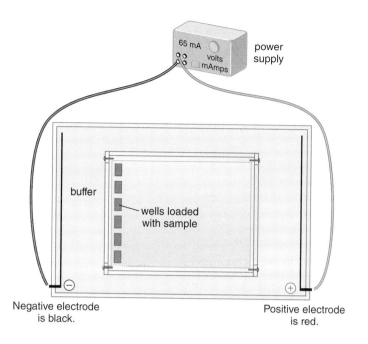

Figure 4.29. **Gel with Loaded Wells.** Since DNA and most other molecules are colorless, samples are mixed with loading dye to make the sample easy to see. The dyes also contain glycerol or sugar to make the sample dense, so that it settles into the wells and does not float away.

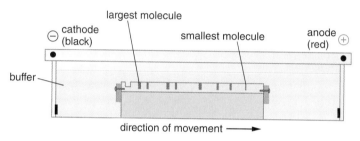

Figure 4.30. **Molecules in a Gel Box.** If negatively charged molecules are loaded into the wells and run on the gel, the smaller ones run faster and farther than the larger ones toward the positive electrode. This is because smaller molecules pass more easily through the tiny spaces of the gel network.

Table 4.3. **Behavior of Molecules During Gel Electrophoresis**

Molecule	Charge	Size	Behavior
DNA	negative	500 –25,000 bp	moves to the positive pole; smaller molecules move faster
RNA	negative	less than 1000 bp	moves to the positive pole; smaller molecules move faster
proteins	positive	1000 –350,000 Da	move to the negative pole; smaller molecules move faster
	negative	1000 –350,000 Da	move to the positive pole; smaller molecules move faster
	neutral	1000 –350,000 Da	do not have net movement to either pole
carbohydrates	Most are neutral.	variable	do not have net movement to either pole
lipids	Most are neutral.	variable	do not have net movement to either pole

because they are single-stranded, they move quickly through the gel. Since DNA molecules are usually much larger than RNA molecules, it is more difficult for them to actually penetrate the gel and move through it.

Although scientists use several different substances as gels, the two most common are agarose and **polyacrylamide**. Agarose gels are used in horizontal gel boxes (see Figure 4.31) to study medium to large pieces of DNA, such as those produced during restriction digestion as well as some RNA molecules. Polyacrylamide gels (PAGE), on the other hand, are used to separate smaller molecules, such as proteins and very small pieces of DNA or RNA. The use of PAGE gels is discussed in later chapters.

polyacrylamide
(pol•y•a•cryl•a•mide)
a polymer used as a gel material in vertical electrophoresis; used to separate smaller molecules, like proteins and very small pieces of DNA or RNA

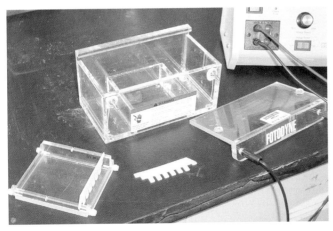

Figure 4.31. This is a horizontal gel box setup. Molten agarose is poured into the gel tray (gates up, comb in position). The agarose solidifies and the gel on the gel tray is placed in the gel box. Buffer is poured over the gel, and the comb is removed, leaving sample wells. Samples mixed with loading dye are loaded into the wells. The size, shape, and charge of the molecules determine the rate at which they move across the gel. When using the gel box, never pull the red and black electrodes.
Photo by author.

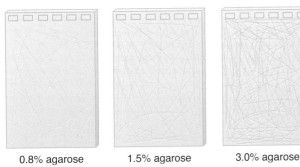

| 0.8% agarose | 1.5% agarose | 3.0% agarose |

Figure 4.32. **Agarose Concentrations.** Long, colorless agarose molecules create a network of molecules in the gel. The higher the concentration, the more molecules are crammed into the same space, and the smaller the spaces are for other molecules to filter through.

Agarose Gel Concentrations

The gel box in Figure 4.31 is designed for agarose gels. Agarose gels are most commonly used when separating pieces of DNA no smaller than 500 bp and no larger than 25,000 bp. These gels are made at a specified concentration by mixing some mass of powdered agarose with some volume of electrophoresis buffer. Electrophoresis buffers maintain pH while conducting electricity efficiently, without out creating too much heat. This allows molecules to maintain shape but still move in the electric field in the gel box. The most common electrophoresis buffers are made with the buffering molecules TRIS or lithium borate. Buffers and their characteristics are discussed in Chapter 3.

Agarose gels are commonly made with concentrations ranging from 0.6% to 3% agarose in buffer (see Figure 4.32). The concentration of a gel is of critical importance. The more agarose molecules in solution, the more strands intertwine to make the "strainer." It is difficult for large molecules to get through the long, woven agarose molecules.

The most common gel used for DNA fragment separation is 0.8% agarose. At this concentration, most plasmid and restriction digestion fragments separate well. "Tighter" gels (2% or 3%) are used to separate smaller molecules. Agarose gels with these higher concentrations are more difficult to prepare. As the need for high concentration gels increases, the advantages of using acrylamide gels likewise increase.

Compared with 0.8% agarose gels, 0.6% agarose gels barely hold together and will separate only very large DNA molecules. Experimentation is the most effective way to determine which gel material and concentration to use for a particular separation.

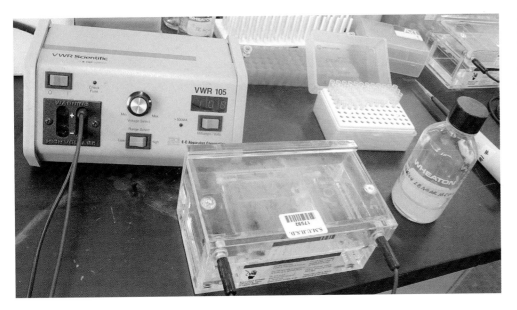

Figure 4.33. **When running an agarose gel, you can set the voltage or the current. Make sure current (mA) is flowing (displayed here in volts) and seen as bubbling at each electrode wire. The kind of buffer used determines the maximum current that may be applied.**
Photo by author.

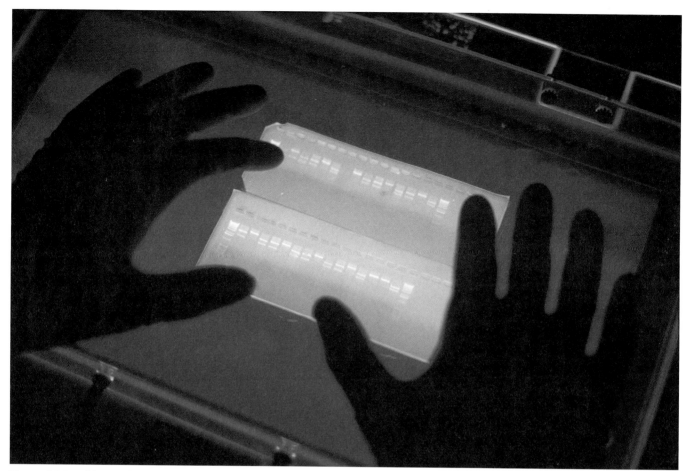

Figure 4.34. **This is a photo of an EtBr-stained gel. Each white band is composed of thousands of DNA molecules of similar sizes. These bands contain molecules of approximately 500 to 1000 bp in length.**
© Robert Essel NYC/Corbis/Getty Images.

Running and Visualizing Gels

Once a gel is prepared and loaded with sample, it is ready to "run." The power to the gel box is set at some voltage (V) or current (mA) depending on the gel electrophoresis buffer being used. Tris-buffered gels are commonly run at 110V (approximately 65 mA). Lithium borate gels are commonly run at 300V (at about 100mA). Since lithium borate gels are run at a higher current, the molecules move faster through the gel and the electrophoresis is completely more quickly (see Figure 4.33). As current moves through the gel, the molecules in a sample move into the gel and travel through it at different rates according to their size and shape. The gel is run until molecules of different sizes are thought to have completely separated.

Since nucleic acids are colorless, the technician must "stain" the gels to see the bands of separated molecules. There are a few DNA stains to choose from. **Ethidium bromide (EtBr)** is the most common DNA gel stain and considered the most sensitive to low concentrations of DNA. EtBr glows orange when it is mixed with DNA and exposed to UV light (see Figure 4.34). Since EtBr is a suspected mutagen, other stains such as LabSafe, SYBR Safe, GelRed, GelGreen, or methylene blue have become popular. Each of these stains interacts with the nucleic acid molecules in a way that they can be visualized by either UV or white light. Each has some pros and some cons in its use. Depending on the indicator/stain, it may more or less sensitive to detecting low concentrations of DNA.

ethidium bromide (eth•i•di•um bro•mide) a DNA stain (indicator); glows pinkish-orange when it is mixed with DNA and exposed to UV light; abbreviated EtBr

Figure 4.35 shows a gel diagram with sample lanes stained to reveal a variety of nucleic acid samples. Each dark band represents a large number of molecules of similar size. Only molecules of negative charge would run on this gel. If molecules are too large, they sit in the wells, unable to load into the gel. If molecules are too small, they load in and move through the gel too fast, and thus run off the end.

Lane 1 shows what a common DNA sizing standard or standard marker would look like. This sizing standard, Lambda/HindIII, is one of the most common sizing standards used in biotechnology labs. Lambda is a virus that infects some bacteria. Its DNA is isolated and cut into known pieces using a restriction enzyme called HindIII. By running the known cut pieces of lambda DNA on a gel, other samples of unknown sizes that run similar distances can be sized. The pieces of lambda DNA that are visible on a typical agarose gel range in size from 23,130 (near the top) to 564 base pairs (bp) near the bottom.

Lane 2 contains only one type of DNA sample. The molecules in the band have an approximate size of about 7000 bp (comparing them with the standards in Lane 1). This is the approximate size of a large plasmid. Remember, to be able to actually see the band, there must be thousands of plasmid molecules in it.

Lane 3 shows the probable results of a plasmid restriction digestion. In this case, the plasmid contains three recognition sites for the restriction enzyme. How is this known from the gel? What are the approximate sizes of the restriction digestion fragments? What is the approximate size of the entire plasmid?

Lane 4 shows the appearance of a sample of bacterial DNA. Even the smallest genomic DNA molecules are at least 1,000,000 bp in size.

Lane 5 shows how a sample of mRNA would run on a gel compared with DNA fragments.

Lanes 6 and 7 show samples with "smears." Smears contain thousands of different sizes of molecules in small concentration.

Lane 8 contains no nucleic acid.

Lane 9 contains DNA strands so large that they will not even "load" into the gel. Eukaryotic genomic DNA behaves in this way because it contains so many very long molecules.

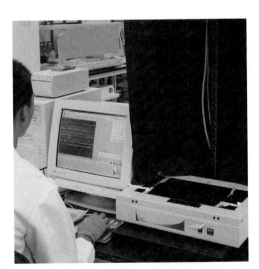

Figure 4.35. **How DNA samples may appear on a gel.** This gel represents samples from eukaryotic and prokaryotic sources.

Figure 4.36. **Technicians run large-sized gels with many samples at one time, usually looking for a common piece of data. This gel (right) has approximately 200 samples on it. The technician (left) is looking for evidence of a single-sized DNA fragment that is the result of the polymerase chain reaction (PCR), which is discussed in later chapters.**
Photos by author.

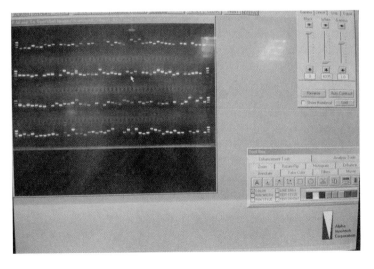

In many labs, hundreds or thousands of samples are run on agarose gels at the same time. This is called **high through-put screening**. Commonly, in genetic or evolutionary studies, large gels are poured with several rows of wells (see Figure 4.36) so that many samples may be simultaneously compared.

BIOTECH ONLINE

Chop and Go Electrophoresis

Long DNA molecules can be cut into manageable pieces using restriction enzymes. With agarose electrophoresis, the DNA fragments can be separated on a gel, based on their lengths. To visualize the fragments, they are stained with ethidium bromide and exposed to ultraviolet (UV) light. The size of each fragment can be determined by comparing each one to DNA molecular weight markers of known size.

Go to biotech.emcp.net/gelelectrophoresis to learn how to analyze a DNA gel electrophoresis. Follow the directions below to conduct the simulation at the "Virtual Lab: Agarose Gel Electrophoresis of Restriction Fragments" website. Write answers to the questions.

1. Read through the background information on how to set up and run the gel simulation.
2. Then, in the simulation, for "Choose a DNA to cut" choose "pBR322 plasmid."
   a. What is the size of pBR322 in base pairs (bp)?
   b. How many times is the plasmid cut by each of the following restriction enzymes?

   EcoRI, *Hinc*II, Ple1, BglI

3. Choose the following for the restriction digestions:

   Lane 1—EcoRI digest
   Lane 2—*Hinc*II digest
   Lane 3—Ple1 digest
   Lane 4—BglI digest

4. Click the buttons to load each lane.
5. Click "Turn ON Power." Let the gel run until the darker blue loading dye gets almost to the bottom of the gel.
6. Click "Turn OFF Power" and click "Turn ON UV."
7. What are the sizes, in bp, of the molecular weight standards?
8. What are the sizes of the restriction enzyme-digested pieces of pBR322?
9. Add up the sizes of the fragments in each lane. What value does each lane's DNA come close to? Does this make sense? Why or why not?
10. Click Reset anytime you want to start the simulation over.

Section 4.4 Review Questions

1. Agarose gels can be used to study what size of DNA fragments?
2. If agarose gel material is labeled 1%, what does the 1% refer to?
3. What causes molecules to be separated on an agarose gel?
4. Name two common DNA stains that are used to visualize DNA on agarose gels.

Chapter Review

Summary

Concepts

- The development of technology for modifying DNA molecules and manipulating protein molecules has been a key factor in the biotechnology revolution.
- In cells, DNA holds the code for proteins. To synthesize a protein, sections of the DNA code (genes) are transcribed into mRNA that is translated at ribosomes into the amino acid sequence of the protein.
- A typical cell makes over 2000 different proteins. Not all of the proteins are made at the same time. Their production is regulated through gene expression.
- All DNA molecules are composed of four nucleotides containing the four nitrogen bases: adenine, thymine, guanine, and cytosine. The order of nucleotides determines which amino acids are found in a protein.
- In DNA, the nitrogen bases from one strand bind to the nitrogen bases of the other strand through weak H-bonds. The nitrogen bonds occur between adenine and thymine molecules, and between guanine and cytosine.
- The DNA strands run antiparallel to each other. This gives DNA strand directionality, which is important to gene storage and expression.
- Advances in computer molecular modeling provide better opportunities to study the structure, function, and interactions of important biomolecules.
- DNA replicates through semiconservative replication. The resulting replicate DNA molecules move to daughter cells during cell division.
- The DNA of a prokaryote is different from a eukaryote in that a prokaryote has a single, circular DNA molecule sectioned functionally into operons. The DNA is significantly shorter than in a eukaryote and holds fewer genes. Prokaryotic cells may also contain plasmid DNA.
- Operons are the main way that prokaryotes regulate gene expression (protein production). RNA polymerase attaches to the promoter region of the operon and moves toward the structural gene to start transcribing mRNA. If a regulatory molecule is attached at the operator region of the operon, the RNA polymerase is blocked from reaching the structural gene, and no mRNA is made and, thus, no protein is produced.
- Plasmid DNA is small, circular DNA containing a few nonessential genes. These genes code

for extra traits that help bacteria survive some extraordinary circumstances, such as antibiotics or extreme temperatures.

- Plasmids can be used as vectors to carry foreign DNA into cells. Plasmid vectors are important tools for genetic engineers.
- Eukaryotic cells have several chromosomes that are long linear DNA strands, coiled around histone proteins, and interrupted by noncoding regions. Eukaryotic DNA does not have operators, but does have promoters and structural genes. Enhancers, silencers, and transcription factors often affect RNA polymerase interaction at the promoter to modify gene expression.
- Virus particles contain either small DNA or RNA molecules as their genetic material. Scientists have been able to use viral DNA as a vector for transferring genes used for genetic engineering and gene therapy.
- The genetic engineering process includes the following: identification of a target molecule(s) for production, isolation of the DNA instructions for that molecule's production, manipulation of the DNA instructions into cells that use the DNA to produce the target molecule, and harvest of the target product.
- Site-specific mutagenesis, which may cause changes in specific sections of DNA, may be used to try to achieve specific modifications in genetic codes. Chemicals and radiation are common methods of site-specific mutagenesis.

Lab Practices

- Alcohol can be used to precipitate DNA out of an aqueous solution. The technique can be used in many applications to separate DNA from other cell components.
- DNA strands that are long enough can be spooled onto a glass rod during alcohol precipitation.
- Pure DNA in aqueous solution is clear. As water is squeezed out, the DNA becomes white.
- TE buffer (TRIS, EDTA) is a commonly used DNA storage solution.
- The presence of DNA in a sample can be detected by using an indicator or stain such as the following: diphenylamine (DPA) turns blue in the presence of DNA, ethidium bromide (EtBr) which glows orange with DNA and UV light, or using one of the "safer" stains such as LabSafe, SYBR green, GelRed, or methylene blue.
- An EtBr Dot Test can quickly show the presence of DNA in a solution.
- *E. coli* cells grow well on Luria Bertani (LB) agar and in LB broth. Media can be made from dehydrated media base available from supply houses. Using the recipe on the bottle and $Mass_1/Volume_1 = Mass_2/Volume_2$, the recipe for any volume of media can be determined.
- Media base is mixed with dH_2O, and then sterilized in an autoclave until the temperature and pressure are high enough to kill all microorganisms contaminating the media, usually 15 to 20 minutes at 15 to 20 psi.
- Agar plates are used to grow isolated colonies of bacteria. Using the "triple-Z streaking" technique, isolated colonies of cells can be grown. Colonies are made up of cells that all are descendants from a single parent cell. Broth cultures are started by introducing a single colony into sterile broth media. The cells of the colony grow to fill the space and use the nutrients.
- Pouring plates and cell culture (plate and broth) require good sterile technique, including use of sterile, laminar flow hoods or biosafety cabinets, disinfectant, sterile utensils and vessels, and good aseptic technique.
- Gel electrophoresis separates molecules based on their size, charge, and/or shape.
- Agarose gel electrophoresis is conducted in a horizontal gel box containing electrophoresis buffer that creates an electric field with a positive side and a negative side. Depending on the charge of the molecules in the sample, bands of molecules migrate toward one side or the other. Since DNA molecules have a net negative charge, they move toward the positive, red electrode.
- Smaller molecules move through an agarose gel at the fastest pace. Agarose concentration affects the migration of molecules; the higher the concentration, the more difficult it is for molecules to migrate and the more slowly they move through the gel. Agarose gel electrophoresis works best with molecules of 500–25,000 bp.
- On agarose gels, DNA is visualized using EtBr stain and UV light. Alternatively, one of the "safer" stains such as LabSafe, SYBR green, GelRed, or methylene blue may be used although sensitivity may be compromised.

- For medium-sized DNA fragments, lambda DNA cut with the restriction enzyme *Hind*III is commonly used as a sizing standard on agarose gels. It shows seven pieces of known sizes between 500 and 23,000 bp.
- On a gel, eukaryotic genomic DNA molecules are usually too long to load, but if sheared, they will load and appear as a large glowing smear. Prokaryotic genomic DNA loads into the gel, but it becomes hung up at above the 25,000-bp standard band because it is too long to be resolved on the gel. Uncut plasmid DNA appears as two to four discreet bands representing circular plasmid, linear plasmid, or multiple rings of plasmid.

Thinking Like a Biotechnician

1. Name the four nitrogenous bases found in the nucleotides of the DNA molecule. List two ways that these bases are similar to or different from each other.
2. If a piece of one strand of a DNA molecule has the following sequence on it, what would be the nitrogen base sequence on the opposite DNA strand?

   ATG CCC GTG TTA AAA TGT GGG ATC CCC GGT GTG CCC TTA

3. A sample contains three DNA fragments with sizes of 3000, 20,000, and 80,000 bp. The sample is loaded onto a gel and run for the same amount of time as the gel diagramed in Figure 4.35. After staining, what will the samples look like on the gel?
4. A DNA molecule has a constant width. This is due to the nitrogenous base pairing. Explain how the nitrogen base pairing maintains the double helix's constant width.
5. The adenovirus has been genetically engineered to act as a vector to bring genes into cells for human gene therapy. Suggest a virus that could be used as a vector to carry genes into plants for plant gene therapy.
6. When working in the lab, a sample is thought to contain DNA. What method could be used to test for DNA in the sample?
7. Suppose that the DNA code in question 2 is part of the DNA sequence for an enzyme involved in milk digestion. What effect could a change (mutation) in this sequence have in the cell or in the organism?
8. Positive feedback occurs when the presence of something causes an increase in some other molecule or process. Negative feedback occurs when the presence of something causes a decrease in some other molecule or process. Lactose molecules turn on the Lac Operon in bacteria that have it. Is this an example of positive or negative feedback? Explain why.
9. A sample has a mixture of 10 pieces of DNA that are very close in size (between 700 and 1000 bp). You want to separate them on an agarose gel. What can you do to increase the chances for good separation?
10. A biotechnologist attempts to genetically engineer a cell to make a certain protein. It appears that the transformed cell is making the protein, but only in small amounts. Suggest a method to increase protein production in the transformed cell.

Biotech Live

Build Your Own DNA!

DNA is the largest molecule, but because it is composed of repeating units, it is not too complicated to build a model to represent a section of it.

Build a model of the DNA molecule showing the double helix, nucleotides, and base pairing for a strand that is at least 15 base pairs in length. Be creative in the building materials (the Internet has lots of ideas and examples), but include a coded key to identify the parts of the model.

E. coli: A Model Organism for Geneticists and Biotechnologists

Since *E. coli* has been studied for so long, it is better understood than any other bacteria or living thing. Geneticists and genetic engineers prefer to use *E. coli* for experiments and production because growing it is easy, fast, and inexpensive. Based on experience, conditions can be created to ensure the best results.

Using the Internet, find the answers to the following questions about *E. coli*. Record the information and the website URL you accessed.

1. What is the full scientific name for the *E. coli* bacteria?
2. In nature, where do *E. coli* cells grow?
3. Give an example of an *E. coli* strain that causes disease.
4. At what temperature does *E. coli* prefer to grow?
5. *E. coli* prefers to grow on LB agar or LB broth. What does LB stand for?
6. What are the ingredients in LB agar?
7. How many base pairs long is in the genomic DNA of *E. coli*?
8. How many genes does the genomic DNA of *E. coli's* chromosome contain?
9. Find an example of a product made in genetically engineered *E. coli*.
10. List two websites that give more information on how to grow or experiment with *E. coli*.

Bacteria Cell Growth Curve

Developed with the assistance of Tina Doss, Biotechnology Instructor, Belmont, CA

Under optimum conditions, *E. coli* cells double every 20 minutes. The doubling of cell numbers in the colony is called exponential growth. Optimum conditions include the "right" temperature, food, pH, amount of oxygen, and amount of light. If a single cell is dropped on an agar plate, there will be two cells in 20 minutes. After another 20 minutes, each of the two cells will divide into four. Then in 20 more minutes, the four cells will divide into eight, and so on. Since *E. coli* cells are so tiny, the growing colony is not visible without a microscope until there are several billion cells present. A colony will continue to grow exponentially until its food, space, or other nutrients begin to run low. At this point, the colony is said to be in an environment at carrying capacity or in stationary phase.

Determine how fast a colony grows under optimum conditions.

1. Assume that a plate has been streaked with *E. coli* using the triple-Z method. The goal is to produce an isolated colony started by depositing a single, isolated cell.
2. Assume that the plate is incubated at optimum conditions overnight for use the next day (approximately 17 hours).
3. On a chart similar to the one on the following page, determine the number of cells that would be present in the colony after doubling every 20 minutes. Several entries have been made. Notice that numbers with more than 3 digits are reported in scientific notation.

Table 4.4. The Number of *E. coli* Cells in a Colony over Time

Time (min.)	No. of Cells	Time (min.)	No. of Cells	Time (min.)	No. of Cells
0	1	340		680	
20	2	360		700	
40	4	380		720	6.87×10^{10}
60	8	400		740	6.12×10^{10}
80	16	420		760	6.77×10^{10}
100	32	440		780	5.99×10^{10}
120		460		800	6.02×10^{10}
140		480		820	6.89×10^{10}
160		500		840	6.80×10^{10}
180		520		860	6.81×10^{10}
200	1.024×10^{3}	540		880	6.77×10^{10}
220		560		900	6.65×10^{10}
240		580		920	6.80×10^{10}
260		600		940	6.90×10^{10}
280		620		960	5.90×10^{10}
300		640		980	6.90×10^{10}
320		660		1000	6.81×10^{10}

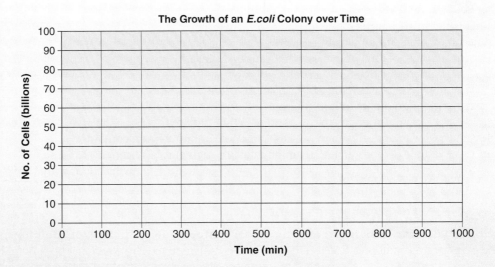

The Growth of an *E.coli* Colony over Time

y-axis: No. of Cells (billions) — 0, 10, 20, 30, 40, 50, 60, 70, 80, 90, 100
x-axis: Time (min) — 0, 100, 200, 300, 400, 500, 600, 700, 800, 900, 1000

4. Communities of organisms (like bacteria colonies) will grow as long as their resources are not limited. A community will start growing slowly since there are only a few organisms to reproduce. This time period is called the "lag phase." Create a graph similar to the one shown on the following page to chart the growth of the *E. coli* colony. Label the lag phase on your graph.

5. Once substantial numbers of organisms are present in a community/colony, doubling causes large changes in community size. On a graph, this is shown by a fast rise in the number of organisms correlating to exponential growth. Label the exponential phase on your graph.

6. All communities of organisms eventually slow their reproductive rate as the individuals start to be crowded. The line on the growth curve starts to level out. This is called the "stationary phase," and the culture is said to be at "carrying capacity." Label the stationary phase on the graph and place an asterisk (*) by the point at which carrying capacity is reached.

7. Explain how a colony on a Petri plate could reach its carrying capacity. Suggest why biotechnologists would want to keep colonies of cells in the exponential phase. Suggest methods of how biotechnologists might keep cells in exponential growth.

NCBI and Bioinformatics

Since James Watson and Francis Crick first described the structure of the DNA molecules (see the original article at **biotech.emcp.net/dnaarticle**), scientists have been racing to learn the entire nucleotide sequence of the human genome and what the sequence means. Some authorities believe it will take a century for all of the research needed to understand how the entire genetic code is expressed, with certain genes expressed at some times and not others.

The number of scientists working on DNA studies is enormous, and each one can produce vast amounts of data that need to be stored and analyzed. To handle and analyze all the data, a new discipline called **bioinformatics** has emerged. Bioinformatics uses computers and databases to analyze and relate large amounts of biological data. A number of agencies have led the way in creating databases where biological data can be archived and accessed when needed. Several databases that are commonly used to study DNA and protein sequence data and other biotechnology experimental findings are found at the National Center for Biotechnology Information (NCBI) website at **www.ncbi.nlm.nih.gov**.

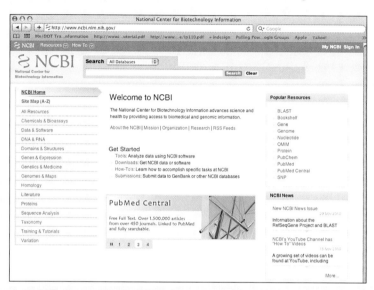

Learn more about what information is available at the NCBI site and how to use the available databases. Gather the information asked for below from the NCBI website and its databases to create a poster that explains the bioinformatics tools that are available through www. ncbi.nlm.nih.gov. Use text and graphics to provide information on your poster.

1. Go to NCBI's home page and click on "About the NBCI" and then click on "Our Mission." Summarize the mission of NBCI and the programs and activities it offers.

2. Go back to the NCBI home page and click on the link "Analyze." Look at the list of "All Tools." For the following tools, describe what they can do.
   • Genetic Codes
   • Amino Acid Explorer
   • Basic Local Alignment Search Tool (BLAST)
   • Cn3D

3. Go back to the NCBI home page and examine the "Popular Resources" list on the right-hand column. Click on each of the following links and in a few words, describe what each of these three databases does:
   • PubMed
   • Nucleotide
   • Protein

4. Click on "PubMed" and use it to find an article on colon carcinoma or colon cancer. In a sentence, tell what information is found in the article.
5. Click on "Nucleotide" and use it to find an entry for colon carcinoma. What data was found in the first result?
6. Click on "Protein" and use it to find an entry for "colon carcinoma." What data was found in the first result?
7. Pick one other "Popular Resources" database and describe what that database can do. Use it to find an entry for colon carcinoma. What data was found in the first result?
8. All the NCBI databases are linked by a search and retrieval system called Entrez. To see the power of Entrez, Pick the "Protein" database next to the search bar. Type in "alpha-amylase Geobacillus" into the search bar and hit return. Click on the first hit. Look to see what information is available for this protein and from what databases (look along the right side of the page in "Related Information").

Activity 4.5   Viewing the 3-Dimensional Structure of a DNA Molecular Model

Development of this activity was done with the assistance of Sandra Porter, Ph.D., President, Digital World Biology LLC, www.DigitalWorldBiology.com

DNA, RNA, and protein molecules are very complex in their structure and function. When studying one of these biomolecules it is important to understand its structure to understand its function. Scientists conduct structural studies using sequencing, x-ray crystallography, nuclear magnetic resonance and other methods, to collect data to add to databases that allow the drawing of structural images.

The Molecular Modeling DataBase (MMDB) is an example of a database of experimentally determined three-dimensional biomolecular structures. MMDB is also referred to as the Entrez Structure database and it is actually just one of the databases of three-dimensional structures from the RCSB Protein Data Bank (PDB). The PDB assigns a code of 4 letters and numbers to identify the data and images for a specific biomolecule. If a researcher finds a molecule that has been added to the MMDB and has a PDB code, its structure can be viewed by one of the molecular modeling applications such as Cn3D or Molecule World.

In this activity, you will learn to how to view and study a section of DNA that has had data entered into PDB. You will have the option of using the Molecule World application, available through iTunes for the iPad or iPhone. Or, you will be able to use Cn3D application, which is a program to view 3-dimensional structure using your computer's web browser.

 Using the Molecule World application, available through iTunes for the iPad or iPhone, to view a DNA structural model

1. Make sure The Molecule World application has been installed on your iPad or iPhone by you or your supervisor. Click on the application icon to open Molecule World.
2. There may be structures already downloaded onto your device. In the upper left corner, find the pull down menu "My Structures" and see what is already available for you to view.
3. Look for the structure named, "High Resolution NMR Structure of DNA Dodecamer Determined in Aqueous Dilute Liquid Crystalline Phase." Its PDB code is 1NAJ. It is a 12-nucleotide piece of a DNA that has been constructed in a lab. If this structure is not in your "My Structures" list, then click on the "+", click on PDB, type in "1NAJ, and hit return." This will bring the structure into your list and when you click on the My Structures entry you will see the molecular model (in "Element" view) on the screen. The source of the data that Molecule World is using for this structure can be found at www.ncbi.nlm.nih.gov/Structure/mmdb/mmdbsrv.cgi?uid=50438.
4. Molecule World gives lots of options for studying the structure of this (and other) molecule. Explore what can be done following the steps below:

A. Changing the size of the image. Touch your thumb and index (pointer) finger to the screen and then spreading them apart or bring them together.

B. Rotate the model.
   - To start the whole molecule rotating, tap on the rotation icon at the top right. Tap it again to stop it.
   - To control the rotation, place your index finger on the screen and move it, this will cause the molecule to rotate making some features of the structure more visible.
   - Look for the 2 sides of the double helix. Look at the repeating sugar and phosphate molecules in these two backbone strands. What color are the phosphate groups (phosphorus and oxygen atoms)?
   - Click on the palette on the top right for a color code of the atoms in the molecule.
   - Continue to rotate the DNA double helix. Look closely at the outside of the spiral as it turns. Do you notice the space between turns of the spirals 2 different sizes, one twice as large as the other? These are called the major (larger one) and minor grooves of the backbone. These are important because many enzymes that interact with double stranded DNA recognize either the major or the minor groove but not both. The major and minor grooves are even more obvious if you look at the structure in the "Spacefill" drawing style.

C. Visualize the sequence.
   - Click on the "Sequence" button on the bottom left to see the sequence of nucleotides in each of the two strands. Click on the first strand sequence "1NAJ-A." This will highlight that strand.
   - Go to the "Show/Hide" button on the bottom right and click on "Hide unselected." The second strand disappears and you can then view/rotate the single strand to see the repeated nucleotide structure better. Click the second strand sequence to reset the appearance.

D. View Different Model Styles.
   - Look at the top right for the ball and stick icon and click it to see options for viewing the atoms and subunits of the molecule. Click residues to see the color-coded nucleotides (with the nitrogenous bases A, C, G, T). Rotate the molecule to see each nucleotide's nitrogenous base. Look in both ball and stick style and spacefill style. Can you confirm the sequence shown in the key?

E. Study More Structure.
   - Reset the appearance and click on "complete backbone" in Show/Hide menu on the bottom. The Sugar-phosphate groups are shown. Change to "Spacefill" in the Drawing styles menu. Look closely at the phosphate groups. Can you confirm the each phosphorus atom has four oxygen atoms attached?
   - Reset the appearance (in Drawing Styles menu) and then rotate the structure so you can see the top of each sugar-phosphate backbone strand. Find the five-carbon deoxyribose (sugar) in each strand. Do you see how the 5-carbon ring point up (oxygen atom) on one strand and down on the other strand? This shows the anti-parallel nature of each side of the double helix. The oxygen up ribose shows the strand running from the 5' phosphate group down to the sugar and nitrogenous base. This is the 5'>3' strand. The other strand is 3'>5' since it is running in the opposite direction. The antiparallel strands are important because most DNA recognition molecules recognize the strand usually in one direction and not the other. For example, RNA polymerase recognizes a DNA strand only in the 3'>5' direction.
   - Take the time to explore the molecular structure choosing other menu options. When you have an image of the structure that best represents explains to you how the molecule functions, use the camera icon to save or email yourself an image. Print that image and place it in your notebook. On the printed image, describe 3 features of the structural model that are important for its function.

Using the Cn3D application, for use with a computer's Web browser, to view a DNA structural model

1. Make sure Cn3D application has been installed on computer by you or your supervisor. Click on the application icon to open Cn3D.

2. Structures that have a Protein Database Identification Code (PDB) will be imported by Cn3D and displayed. If your supervisor has not already done so, use the NCBI Entrez Structure database to find the structure file to download in Cn3D.

   - Go to www.ncbi.nlm.nih.gov. Change from "All Databases" search to "Structure."
   - Type in the PDB Id code "1NAJ" and hit "Search." The MMDB Structure Summary page for the structure named, "High Resolution NMR Structure of DNA Dodecamer Determined in Aqueous Dilute Liquid Crystalline Phase" will appear. This structure is for a 12-nucleotide piece of a DNA that has been constructed in a lab. Record the URL for this website in your notebook.
   - Look on the right side of the page for the "View or Save 3D Structure" box. Click on "View structure." This will download the structural image to your computer and when it is opened, it will appear in Cn3D. You can confirm it is the correct structure by checking that "1NAJ" is in the title of the image file. When the file is open, it is in the most basic style showing only the two-strand backbone of the double helix and a Sequence/Alignment Viewer box below it. If the Sequence/Alignment Viewer is not visible, show it using the "Show" menu at the top.

3. Cn3D gives options for studying the structure of this (and other) molecule. Explore what can be done following the steps below. In the "View" menu, at any time, you can choose to restore the image to the original structure.
   A. Changing the size of the image. Using the "View" pull down menu, you can zoom in or zoom out.

   B. Rotate the model.
   - To start the whole molecule rotating, pull down the View" menu and use "Animation" to select the "Spin" and "Stop" choices. Notice how there are shortcut keystrokes for these options.

   C) View the structure.
   - Go to the "Style" Menu. Choose "Rendering Shortcuts" and then choose "Ball and Stick."
   - Next go back to the "Style" Menu. Choose "Coloring Shortcuts" and then choose "Residue." You should now be able to see the color-coded nucleotides (with the nitrogenous bases A, C, G, T). What are the colors of the four nucleotides? Rotate the molecule to see each nucleotides nitrogenous base. Look in both ball and stick style and spacefill style (in Rendering Shortcuts). Can you confirm the sequence shown in the sequence viewer below the image? You can click on one or more nucleotides in the sequence viewer to highlight it in the structural model image. If you don't see the Sequence/Alignment Viewer under the image, click on its link in the "Show" menu.

- Go to "Wire" in the "Rendering Shortcuts" menu. Look for the 2 sides of the double helix. Look at the repeating sugar and phosphate molecules in these two backbone strands. Choose "Coloring Shortcuts" and then choose "Element." What color are the phosphate groups (phosphorus and oxygen atoms)?
- Change the style to "Space Fill" using the rendering shortcuts and to "Residue" using the coloring short cut. Rotate the DNA double helix stopping it as necessary. Look closely at the outside of the spiral as it turns. Do you notice the space between turns of the spirals 2 different sizes, one twice as large as the other? These are called the major (larger one) and minor grooves of the backbone. These are important because many enzymes that interact with double-stranded DNA recognize either the major or the minor groove but not both.
- Change the style to "Ball and Stick" using the rendering shortcuts. Go to the "Select" menu on the top and click on "Pick Everything." Then, select "1NAJ_A" to highlight just the single strand. The second strand disappears and you can then view/rotate the single strand to see the repeated nucleotide structure better. Go back to "Pick Everything" to bring back the second strand.
- Change to "Spacefill" in the rendering shortcuts menu and change the coloring shortcut to "Element". Look for the phosphate groups (phosphorus is the green atoms. Can you confirm the each phosphorus atom has four oxygen atoms attached?
- Reset the appearance to "Wire" (in rendering shortcuts menu) and then rotate the structure so you can see the top of each sugar-phosphate backbone strand. Find the five-carbon deoxyribose (sugar) in each strand. Do you see how the 5-carbon ring point up (the pentagon) on one strand and down on the other strand? This shows the anti-parallel nature of each side of the double helix. The ribose molecules pointing up shows the strand running from the 5' phosphate group down to the sugar and nitrogenous base. This is the 5'>3' strand. The other strand is 3'>5' since it is running in the opposite direction. The antiparallel strands are important because most DNA recognition molecules recognize the strand usually in one direction and not the other. For example, RNA polymerase recognizes a DNA strand only in the 3'>5' direction.
- Take the time to explore the molecular structure choosing other menu options. When you have an image of the structure that best represents explains to you how the molecule functions then, save or email yourself an image. Print that image and place it in your notebook. On the printed image, describe 3 features of the structural model that are important for its function.

Human Cells are Fussy Eaters

Activity **4.6**

Growing mammalian, fungal, and bacterial cells in or on sterile, prepared media is critical for their study. In the photo, mammalian cells are growing in broth culture in a carbon dioxide incubator. The broth cultures need to be reseeded (restarted) in fresh sterile media every few days to give them more food and nutrients and to prevent the buildup of toxic waste products.

Photo by author.

Compare and contrast ingredients required by bacteria cells and human cells in culture.

Go to the website **biotech.emcp.net/BiologyPages** and review the information about the ingredients required by *E. coli* bacteria cells versus human cells in culture. Explain why mammalian cells, such as human cells, have so many more required ingredients in their growth media.

Activity 4.7   Transcription Factors and Protection from Alzheimer's Disease

According to the Alzheimer's Association, over 5.3 million Americans are living with Alzheimer's disease. For the sufferers of the disease and their families, the mental anguish of Alzheimer's is truly a horrible burden to bear. Eventually Alzheimer's patients cannot take care of themselves, but they can live a long time with advance stages of the disease. This creates an enormous economic burden on families and society. Many companies and governmental agencies are funding research for better diagnostic tools to recognize Alzheimer's as early as possible and for better therapies to treat it once it has been diagnosed.

Learn more about Alzheimer's disease and how gene expression transcription factors may be a strategy for combating it.

1. Use the Alzheimer's Association's website at **biotech.emcp.net/alzheimers** to learn about Alzheimer's disease, who it impacts, the suspected causes and symptoms, and the current treatments.
2. Use the Internet to find definitions and examples of what transcription factors are and how they work.
3. Then, go to **biotech.emcp.net/alzsummary** and search for and read the summary article about Alzheimer's disease and two transcription factors (FOXO and HSF-1) that are thought to be involved in protecting proteins important in nerve cells. Think about how this new research could lead to future treatments for Alzheimer's.
4. Use the new information you have gathered to create an 11 × 17 inch mini-poster that teaches others about Alzheimer's and about this new research on the role of the transcription factors, FOXO and HSF-1. Make sure your mini-poster is easy to understand and has lots of images, descriptions, and labeled diagrams. You can create these yourself or use ones that you have found (as long as you cite the reference URLs.)

Bioethics

The Promise of Gene Therapy

Gene therapy was first proposed in the early 1960s as a way to correct a multitude of genetic disorders. To what extent has that possibility developed into reality?

Part I   Learn about the successes and setbacks of gene therapy attempts.

Study the Gene Therapy Timeline at:
www.reuters.com/article/2015/04/27/health-genetherapy-timeline-idUSL5N0XK41J20150427
Summarize the history of gene therapy for the following periods of time:

- 1970 to 1979
- 1980 to 1989
- 1990 to 1999
- 2000 to 2015

Give your opinion as to how well gene therapy attempts have met past expectations and whether or not these attempts hold promise for the future.

Part II   Take a position as to whether or not you support continued research and funding to accomplish gene therapy for a condition called Tourette's syndrome.

1. Go to the Tourette's syndrome information website at: biotech.emcp.net/tsa-usa.
2. In a few sentences, summarize the symptoms and the suspected causes of Tourette's.
3. Suppose you are serving on a funding committee that has only a certain amount of money available to award to gene therapy research. A research group is looking for a genetic fix for Tourette's syndrome, and they think they have identified at least one faulty gene that could be a target for gene therapy. They have asked for $1.5 million dollars to fund 3 years of research and development of the new Tourette's syndrome gene therapy. With a total of $10 million to award, your committee is considering this proposal as well as others from groups supporting research for cancer, multiple sclerosis, sickle cell disease, diabetes, and cystic fibrosis. Each of these groups is asking for $2 to $5 million. Decide if you believe the investment in Tourette's syndrome gene therapy is worthwhile to fund. Write a one-page summary of your position. You may consider the ethical issues presented at the US Department of Energy's Human Genome Project website at: biotech.emcp.net/genetherapy. The site lists several questions to consider for using gene therapy, including the following:
- What is normal and what is a disability or disorder, and who decides?
- Are disabilities diseases? Do they need to be cured or prevented?
- Does searching for a cure demean the lives of individuals presently affected by disabilities?
- Somatic gene therapy is performed on adult cells of persons known to have a specific disease. Germline gene therapy is performed on egg and sperm cells. In which type of gene therapy could a trait be passed on to further generations? Is one type of gene therapy more ethical than the other?
- Preliminary attempts at gene therapy are exorbitantly expensive. Who will have access to these therapies? Who will pay for their use?

Think about these questions before, during, and after you have formed your opinion on the funding of Tourette's syndrome research and gene therapy.

Biotech
Careers

Photo by author.

Staff Research Associate

Jasmin Wright
Sunesis Pharmaceuticals, Inc.
South San Francisco, CA

Jasmin Wright worked in the Cell Biology Department at Sunesis, Inc. Sunesis is a biopharmaceutical company focused on the development and commercialization of new oncology therapeutics for the treatment of solid (tumor) and hematologic (having to do with blood) cancers. The company discovers, develops, and commercializes small molecule therapeutics that act as inhibitors of DNA replication and cell division in cancer cells.

Currently, Sunesis has four products in the product pipeline. In Phase III trials for treatment of a type of leukemia, is a replication-dependent DNA-damaging agent, named Vosaroxin that stops mitosis and causes apoptosis (cell death). Another product, MLN2480 is an inhibitor of an enzyme need for cell division. MLN2480 inhibits tumor growth and is being tested for treatment of several kinds of tumors. Learn more about Sunesis' product pipeline at www.sunesis.com /products-in-development/product/MLN2480.php.

Jasmin worked on developing and running cell-based and biochemical assays directed toward cancer drug development. In the photo above, she is shown using a multichannel pipet to transfer a reagent that measures cell viability (the ability to survive) to plates of cancer cells exposed to different concentrations of anticancer agents.

5 Introduction to Studying Proteins

Learning Outcomes

- Describe the structure of proteins, including the significance of amino acid R-groups and their impact on the three-dimensional structure of proteins
- Explain the steps of transcription and translation in protein synthesis
- Discuss the role of naturally occurring proteins and recombinant proteins in biotechnology
- Differentiate proteins that function as part of structure, as antibodies, and as enzymes
- Describe the structure of antibodies and explain the relationship between antibodies and antigens
- Discriminate among the classes of enzymes and discuss the effect of reaction conditions on enzyme activity
- Summarize polyacrylamide gel electrophoresis and identify its usefulness for studying proteins

5.1 The Structure and Function of Proteins

Virtually all of the many different kinds of biotechnology products have something to do with proteins. Many biotechnology products, such as recombinant insulin (rhInsulin), are actually whole protein molecules. Other products contain protein molecules as a key ingredient, such as the enzymes found in contact lens cleaner. Some products contain parts of protein molecules. For example, the artificial sweetener aspartame is a peptide of two linked amino acids. Many biotechnology products are whole organisms characterized by making a new or novel protein. One example is Roundup Ready® soybeans, by Monsanto, Inc., that contain an added protein for herbicide resistance. Some products are instruments used to study or synthesize proteins, for example, CS Bio's automated peptide synthesizer. Protein production is so important in biotechnology that many biotech companies may employ more than half of their scientific staff in protein chemistry or protein process development.

To produce a protein product, researchers must learn about the structure and function of the protein, as well as the amino acid sequence. Several

Figure 5.1. Mark Cancilla, formerly a protein scientist at Sunesis, Inc. and currently at Merck, Inc., uses a mass spectrometer to shoot samples through an ionizer. The sample travels at a rate proportional to its mass and charge. This allows the user to determine the molecular mass of the molecule(s) in a sample, which is important for determining the protein composition and purity of a sample.
Photo by author.

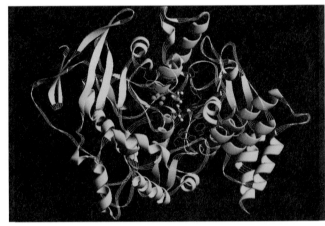

Figure 5.2. This is a computer-generated model of the structure of acetylcholinesterase, an enzyme that breaks down molecules (acetylcholine into acetate and choline) in the junction betweeen nerve cells. This process is important for the regulation of nerve impulses. The computer generated model is constructed from data collected during x-ray crystallography.
© Corbis.

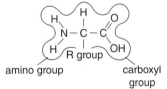

x-ray crystallography (x-ray crys•tal•log•ra•phy) a technique used to determine the three-dimensional structure of a protein

polar (po•lar) the chemical characteristic of containing both a positive and negative charge on opposite sides of a molecule

Every amino acid has the same basic structure, differing only in "R group."

Figure 5.3a. **Structure of an Amino Acid.**

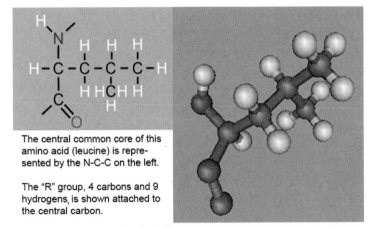

The central common core of this amino acid (leucine) is represented by the N-C-C on the left.

The "R" group, 4 carbons and 9 hydrogens, is shown attached to the central carbon.

Figure 5.3b. **Molecular Models of the Amino Acid, Leucine** Linear and computer- generated space-filling model of the amino acid, leucine.
Courtesy of *Molecule World*, Digital World Biology, LLC.

instruments and techniques are used. One important determination is the molecular mass of a protein molecule, which is achieved using an instrument called a mass spectrometer (see Figure 5.1).

It is also important to know the three-dimensional structure of a protein. This is accomplished through **x-ray crystallography** and computer analysis of the x-ray diffraction data. An x-ray beam is shined on a very pure crystal of the protein of interest. As the beam hits the atoms of a protein molecule in the crystal, the x-ray light is diffracted off the atoms. A detector records the pattern of x-ray diffracted light. A trained technician with the aid of a computer can interpret the x-ray diffraction data and generate a three-dimensional image of the protein molecule (see Figure 5.2). Several computer-generated images, determined through x-ray crystallography, are shown in this chapter.

In an effort to research and develop new products, scientists also study the chemical behavior of a protein, such as its activity, solubility, and electrical charge. Once the structure and function of a protein are ascertained, researchers develop and improve methods of isolating, purifying, and analyzing the protein. They also study the interactions of a protein's structure with other molecules. It takes a great deal of lab work and understanding of the protein to develop a reliable process for producing the protein on a commercial scale.

Protein Molecule Structure

Protein molecules are polymers composed of amino acids. Amino acids are relatively small molecules (see Figure 5.3a). Each has a central carbon atom with a carboxyl (COOH) group on one side and an amino group (NH_2) on the other side. Each amino acid has an R group that distinguishes it from other amino acids. The R group is attached at the central carbon and varies in length and shape. "R" is used in molecular formulas to indicate a nonspecified side chain (see Figure 5.3b). It is the R group that primarily determines an amino acid's interaction with other amino acids in a protein chain.

Twenty different amino acids are found in proteins. They are categorized based on the chemical nature of their R groups. The R groups may be charged (+ or -), **polar** (water soluble), or uncharged (not water soluble) at a neutral pH (see Table 5.1).

Table 5.1.   **Amino Acids Found in Proteins**

Amino Acid	R-Group*	Chemical Nature
alanine	CH$_3$-	uncharged, nonpolar
valine	(CH$_3$)$_2$-CH-	uncharged, nonpolar
isoleucine	CH$_3$-CH$_2$-CH(CH$_3$)-	uncharged, nonpolar
leucine	(CH$_3$)$_2$-CH-CH$_2$-	uncharged, nonpolar
proline	NH-(CH$_2$)$_3$-	uncharged, nonpolar
methionine	CH$_3$-S-(CH$_2$)$_2$-	uncharged, nonpolar
phenylalanine	Ph-CH$_2$-	uncharged, nonpolar
tryptophan	Ph-NH-CH=C-CH$_2$-	uncharged, nonpolar
glycine	H-	polar
cysteine	HS-CH$_2$-	polar, -SH bonds with other -SH groups
serine	HO-CH$_2$-	polar
threonine	CH$_3$-CH(OH)-	polar
tyrosine	HO-Ph-CH$_2$-	polar
asparagine	H$_2$N-C=O-CH$_2$-	polar
glutamine	H$_2$N-C=O-(CH$_2$)$_2$-	polar
arginine	HN=C(NH$_2$)-NH-	basic (positively charged)
histidine	NH-CH=N-CH=C-CH$_2$-	basic (positively charged)
lysine	H$_2$N-(CH$_2$)$_4$-	basic (positively charged)
aspartic acid	HOOC-CH$_2$-	acidic (negatively charged)
glutamic acid	HOOC-(CH$_2$)$_2$-	acidic (negatively charged)

* - means that the "R" group is attached to the central carbon of the amino acid core

Most proteins contain tens or hundreds of amino acids chained together by peptide bonds. A peptide bond is formed between the carboxyl group of one amino acid and the amino group of an adjacent one. The bonding of amino acids, through peptide bonds, into long polypeptide molecules occurs at a cell's ribosomes. A polypeptide chain is referred to as a protein's **primary structure** (Figure 5.4). It is the messenger ribonucleic acid (mRNA) instructions, from one or more genes (DNA) on the cell's chromosomes, that detail which amino acids are to be placed into the polypeptide chain and in what order. Protein synthesis is discussed in more detail in a later section.

As the polypeptide chain is assembled, it begins to fold into a protein. The three-dimensional folding of a protein, which is so vital to its function, depends completely on how the different amino acids in the chain interact with each other. In the polypeptide chain, hydrogen bonding between hydrogen, oxygen, and nitrogen atoms results in helices (each called an alpha helix) and folds (beta sheets) as shown in Figures 5.5a/b and 5.6a/b. These folds and helices make up what is called the **secondary structure**.

Additional folding in proteins is called **tertiary structure**. Tertiary folding is due mainly to the presence of charged or uncharged R groups. For example, amino acids with charges are attracted by amino acids of an opposite charge and repelled by those of the same charge. Thus, positively charged arginine molecules are attracted to negatively

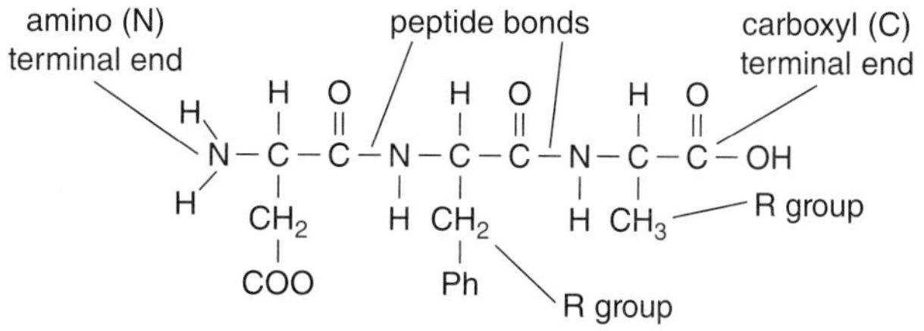

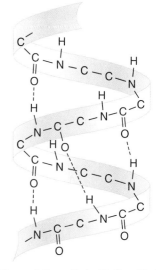

amino (N) terminal end   peptide bonds   carboxyl (C) terminal end

Figure 5.4.   **Peptide Chain.**   A peptide bond forms when the carbon of one amino acid's carboxyl group bonds to the nitrogen of another amino acid.

Figure 5.5a.   Alpha Helix.   Tight coils due to hydrogen bonding can be found in several proteins, resulting in helices as shown here. The diagram represents H-bonds as tiny dashes between carbon, oxygen, or nitrogen atoms and a hydrogen atom in amino acids along the peptide chain.

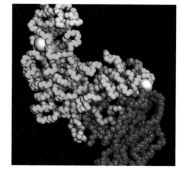

Figure 5.5b.   Alpha Helices of a Bacterial Amylase Enzyme   In this computer-generated space-filling model of the enzyme, a bacterial amylase, several alpha helices are visible in the green section of the peptide chain.
Courtesy of *Molecule World*, Digital World Biology, LLC.

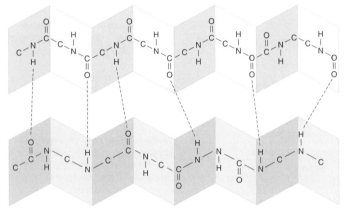

Figure 5.6a. **Beta-pleated Sheets.** Hydrogen bonding may also result in folds, or beta (beta-pleated) sheets.

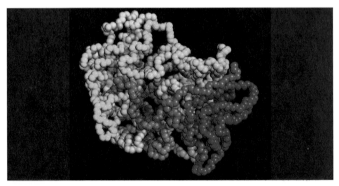

Figure 5.6b. **Beta-pleated Sheets of Human Salivary Amylase** In this computer-generated space-filling model of the enzyme, human salivary amylase, beta-pleated sheets are clearly visible in the blue-colored section of the peptide chain.
Courtesy of *Molecule World*, Digital World Biology, LLC.

charged aspartic acid molecules and repelled by histidine molecules. Sections of a strand are pulled or pushed as these charged amino acids try to get closer to or farther from each other.

Equally important to tertiary folding are the interactions between polar and nonpolar amino acids. Folding occurs when nonpolar amino acids (which are hydrophobic and repelled from water) crowd together. In contrast, polar amino acids (which are hydrophilic and attracted to water) move away from polar or hydrophobic regions and toward the outside of the molecule. Polar amino acids attract other polar amino acids and repel nonpolar ones. Thus, glycine, serine, and tyrosine molecules will try to move close to each other, while leucine, proline, and tryptophan will move away from the polar molecules and try to clump up with other nonpolar molecules.

BIOTECH ONLINE

What Sound Does a Protein Make?
Proteins are too small to be seen with instruments currently available to researchers. The diameter of a typical globular protein may be just a few nanometers (a thousandth of a micrometer). Since proteins can't be seen, to learn about the structure and function of a protein, scientist use methods such as amino acid sequencing, mass spectrometry, and x-ray crystallography to collect data that allows them to them make inferences about the molecules and make molecular models.

Nanotechnology is a relatively new field of science that studies objects less than 100 nanometers. The first nanotechnology products were carbon nanotubes and have been used as very tiny electrical and thermal conductors and as transistors (similar to silicon chips). Nanotechnology is now being used to study of proteins.

Learn how researchers are studying protein structure and activity using nanotechnology.

1. Go to www.popsci.com/science/article/2012-01/tiny-transistor-listens-proteins-human -tears and read the article about using a nanoscale transistor to study the protein, lysozyme. Summarize in 4–5 sentences how and why nanotechnologists "listened" to lysozyme.

2. Proteins are not measured in length units but instead in mass units. Lysozyme, the protein studied in the article, is a small protein, as proteins go. It measures only about 14 kilodaltons (kDa). Use the Internet to find the answers to:

   • What is a dalton?

   • How much is 14 kDa?

   • How many amino acids are found in lysozyme?

Figure 5.7. The Quaternary Structure of Hemoglobin This computer-generated model of hemoglobin shows 4 polypeptide chains (4 different colors) making up the quaternary structure of the protein. The four polypeptide chains are bound together by ionic and hydrogen bonds. In the middle of each polypeptide chain is a heme group, an organic, ringed structure containing the Fe2+ ion that binds oxygen when hemoglobin carries oxygen in the bloodstream. If the Molecule World or Cn3D app is available, you can study this structure by loading up the hemoglobin molecule designated by the PDB ID "2DN2."
Courtesy of *Molecule World*, Digital World Biology, LLC.

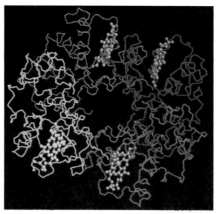

Figure 5.8a. Scientists use computer modeling to study and understand the precise shapes and interactions between molecules. Molecular recognition is important in viral infection and therapies.
© Lester Lefkowitz.

Nonpolar amino acids clump together and try to get away from water molecules surrounding the protein.

Finally, disulfide bonds, which occur between cysteine molecules, also produce and stabilize tertiary folding in and between polypeptide chains. Within a polypeptide chain, cysteine-cysteine disulfide bonds can make large loops. In proteins with more than one polypeptide chain, such as hemoglobin, a variety of ionic bonds, hydrogen bonds, and hydrophobic interactions hold the chains to each other (see Figure 5.7). This is called **quaternary structure**.

Most of the folding pattern characteristic of a specific protein results from the attraction and repulsion of amino acids within and between polypeptide chains (tertiary and quaternary structures). Therefore, the amino acid order coded for on the DNA is critical to determining the ultimate structure and function of a protein.

The Impact of Structure on Protein Function

Chapter 2 introduced protein structure and function relative to several important protein groups, including enzymes and hormones. This and later sections present additional information on the structure and function of some important protein groups.

Several proteins demonstrate well the relationship between structure and function. A good example is a viral recognition protein, glycoprotein 120 (gp120). A **glycoprotein** is a protein on which sugar groups have been added. Glycoprotein 120 exists on the surface of the human immunodeficiency virus (HIV), the virus that causes acquired immunodeficiency syndrome (AIDS). For an HIV particle to recognize, attach, and infect a T-helper cell, the gp120 structure must be a precise shape and must exactly match its human cell membrane receptors (see Figure 5.8a).

Glycoprotein 120 is a single polypeptide chain of hundreds of amino acids folded into five looped domains. The loops are formed because of several disulfide bonds that stabilize the shape of the functional protein. The chains are highly **glycosylated** (bound with sugar groups) projecting out from the amino acids at regular intervals. The loops, which jut out from the center, act as recognition sites. These regions match protein receptors on the **CD4 cells** that HIV infects. Antibodies also recognize these HIV looped domains (see Figure 5.8b).

One of the looped domains has a shape that is an exact match to the CD4 molecule, which is a recognition protein on the surface of human white blood cells (WBCs). When the HIV's gp120 surface protein bumps into a CD4 molecule, it triggers a set of reactions that results in the HIV particle being taken up by the cell. In this way, HIV infects cells.

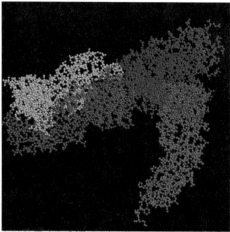

Figure 5.8b. HIV Viral Recognition Protein In this model, the gp120 protein (pink) from the surface of a HIV virus recognizes the CD4 protein (blue) found on the surface of a T-cell. This recognition is how the HIV virus targets a cell. The green and brown molecules are antibodies that recognize the gp120-CD4 complex and may be possible molecules for preventing HIV.
Courtesy of *Molecule World*, Digital World Biology, LLC.

quaternary structure (qua•ter•na•ry struc•ture) the structure of a protein resulting from the association of two or more polypeptide chains

glycoprotein (gly•co•pro•tein) a protein which has had sugar groups added to it

glycosylated (gly•co•sy•lat•ed) descriptive of molecules to which sugar groups have been added

CD4 cells (CD4 cells) the human white blood cells which contain the cell surface recognition protein CD4

Gluten-free? No Bread for You.

Do you know someone who has a gluten-free diet? Some people have discovered that gluten doesn't agree with their digestive tract and for some people gluten in the diet can cause a life-threatening allergic response. Gluten intolerance and gluten allergies have spurred a $6 billion gluten-free food market.

TO DO

Find out more about gluten disorders and what biotechnology is doing to address them. Record answers to the following questions and the websites you used to get the answers.

1. What is gluten and what about it makes it hard to digest?

2. Celiac disease is aggravated by ingestion of wheat gluten by Celiac patients. What are the causes and symptoms of celiac disease?

3. Several biotechnology companies are currently developing therapeutics for celiac disease. Visit celiac.org/celiac-disease/future-therapies-celiac-disease and find three therapies in development. Describe each and how they work.

---

antigens (an•ti•gens) the foreign proteins or molecules that are the target of binding by antibodies

epitope (ep•i•tope) the specific region on a molecule that an antibody binds to

Antibody Structure and Function

Another group of proteins, the antibodies, is structurally interesting and functionally very important. The function of an antibody is to recognize and bind foreign proteins or other molecules (called **antigens**), ultimately for removal from the body. Since there are potentially thousands of different foreign invaders, the body must be able to make thousands of different antibodies to recognize them.

Each type of antibody has the same basic shape. Antibodies are composed of four polypeptide chains (quaternary structure) attached through disulfide bonds (see Figure 5.9a). The chains are arranged into a shape resembling the letter "Y." There are four polypeptide chains: two heavy chains and two light chains.

The base of each antibody has an identical primary sequence of amino acids. This area is called "the constant region." The variability seen in antibodies, which allows them to recognize different antigenic molecules, is found at the top of the "Y," in the variable region. The DNA code for the primary sequence of this region is shuffled (different combinations of exons) to produce an infinite variety of A, C, G, and T codes. Thus, there are an infinite number of amino acid sequences that produce thousands of different antigen recognition sites for the ends of antibodies.

Most antibodies are very specific, binding only to distinct molecules (see Figure 5.9b) or specific regions, called antigenic **epitopes**, on a specific molecule. In the lab, antibody specificity is used to recognize and bind certain molecules. Antibody specificity is particularly useful in the purifying of proteins from cell cultures. Under the right conditions, a single protein can be isolated from a mixture of hundreds of proteins using an antibody chromatography column. In a column, beads with antibodies are used as a means of separating a target molecule from a mixture. The antibodies attach only to a recognized molecule, allowing all other undesirable molecules to pass through the column. Chapter 9 discusses column chromatography in detail.

Often, antibodies are used as commercial testing reagents or as test kits during research and manufacturing (see

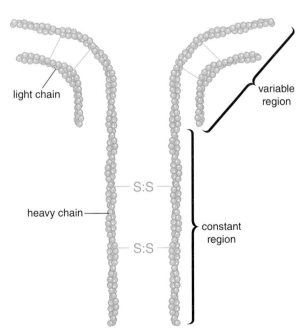

Figure 5.9a. Antibody Structure. Thousand of antibodies are produced in the body by using the same genetic code as a starting sequence. Then, by shuffling DNA sections, new variable regions are created to produce thousands of different kinds of antibodies.

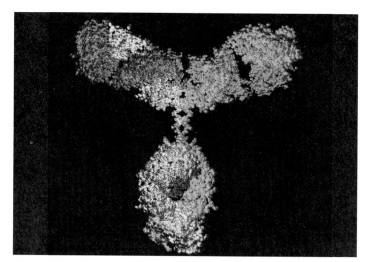

Figure 5.9b. Immunoglobulin E (IgE) is one class of the antibodies that circulate in mammals. IgEs are strongly attracted to and bind to the mast cells in blood, skin, and connective tissues. When pollen grains (gold) attach to the IgEs at the tips of the top of the "Y," the allergen (pollen)-IgE combination triggers the mast cells to release granules (tiny blue dots), causing an inflammatory response.
© Science Photo Library/Photo Researchers.

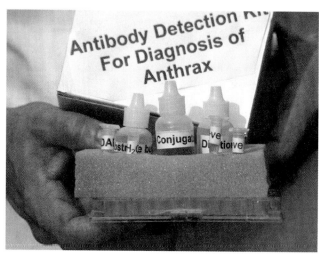

Figure 5.10. Rapid antibody-antigen assays for the detection of several viruses, bacteria, or other disease-causing agents, including HIV, hepatitis, and influenza, are now commercially available. In this anthrax antibody test kit, test solutions contain an antibody to a protein found on the surface of the anthrax bacterium.
© Pallava Bagla/Corbis Historical/Getty Images.

Figure 5.10). Pregnancy tests are an example of a commercial testing kit that uses antibodies to detect proteins, in this case, proteins associated with pregnancy.

A common test used to determine the presence and concentration of a protein in solution is an enzyme-linked, immunosorbent assay (**ELISA**) (see Figure 5.11). In an ELISA, a sample that is suspected to contain a particular protein is tested with a special antibody-enzyme complex that recognizes the protein of interest. When the antibody recognizes and binds to the protein in solution, the attached enzyme causes a color change in a reagent. The amount of color change depends on the amount of antibody that bound to the protein in the sample. ELISAs are very important in protein studies and manufacturing and are discussed in more detail in Chapter 6.

A special group of antibodies is the **monoclonal antibodies**. Monoclonal antibodies are produced in cells made by fusing immortal tumor cells with specific antibody-producing WBCs (B-cells). The resulting cells, called **hybridomas**, grow and divide at a very rapid rate, making large amounts of the specific antibodies that were coded for in the original B-cells. The advantage of monoclonal antibody technology is that many

ELISA (E•LI•SA) short for enzyme-linked immunosorbent assay, a technique that measures the amount of protein or antibody in a solution

monoclonal antibody (mon•o•clon•al an•ti•bod•y) a type of antibody that is directed against a single epitope

hybridoma (hy•brid•om•a) a hybrid cell used to generate monoclonal antibodies that results from the fusion of immortal tumor cells with specific antibody-producing white blood cells (B-cells)

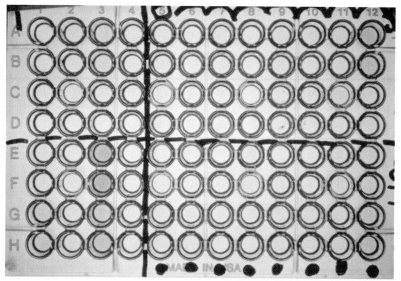

Figure 5.11. An ELISA recognizes the presence and concentration of a particular molecule in a sample. A sample is put into each well. An antibody that recognizes a protein in the sample is added to each well. The antibody has an enzyme bound to it that causes a color change when a certain other substrate is added. The more target protein in a sample, the more the antibody binds to the protein, and the darker the color change.
© Lester V. Bergman/Corbis/Getty Images.

BIOTECH ONLINE

Figure 5.12. **Researchers have developed monoclonal antibodies that detect two disease-causing organisms, *Bagesia equi* and *Babesia caballi*. These pathogens cause the disease piroplasmosis, or equine babesiosis, which is deadly to horses. Fortunately, at this time, these diseases do not exist in the United States.**
Photo courtesy of Scott Bauer, ARS/USDA.

identical antibodies to specific antigenic epitopes are produced in large quantities. These can be used in genetic testing, research, and manufacturing (see Figure 5.12).

At several biotechnology companies, many interesting antibodies are currently being produced through genetic engineering. One, the anti-HER2 antibody (Herceptin® by Genentech, Inc.), recognizes an overproduction of the HER2 protein, which is a growth factor protein found in abnormally large amounts on the surface of cells responsible for an aggressive form of breast cancer. About 25% to 30% of patients with breast cancer have this aggressive form.

The anti-HER2 antibody variable region matches and binds only to the HER2 protein epitope. Since the match is so precise, the antibody can be used to diagnose and treat the forms of breast cancer expressing these proteins. In afflicted patients, the HER2 protein genes are present in multiple copies. Since HER2 proteins are involved in regulating growth, numerous copies of the gene cause an excess of HER2 protein production. That, in turn, speeds tumor growth and the progression of the cancer.

Through genetic engineering technology, the human HER2 antibody has been produced in, and purified from, Chinese hamster ovary (CHO) cells. The antibody attaches only to cells expressing the HER2 protein and blocks or inactivates them. These antibodies only target certain cells, resulting in treatment that is more effective, with fewer side effects than traditional chemotherapies. The anti-HER2 antibody dramatically illustrates how the tertiary structure of an antibody is critical to its function.

Section 5.1   Review Questions

1. How many different kinds of amino acids are found in proteins? What distinguishes one amino acid from another?
2. What causes polypeptide chains to fold into functional proteins?
3. How many polypeptide chains are found in an antibody, and how are they held together in the protein?
4. What is the value of monoclonal antibody technology?

5.2 The Production of Proteins

Until fairly recently, proteins could be made only in cells. Now, with new technologies, small polypeptide chains can be synthesized in the laboratory (see Figure 5.13). These have been used mostly in research to understand how cells and tissues work, and how molecules interact. Recently, though, some small peptides are also being used in medicinal applications and therapies.

Proteins are very long and complex molecules that require cellular apparatus for their synthesis. Biotechnologists exploit cells' protein-making abilities to produce high yields of native proteins or novel, different proteins. The proteins are harvested from large volumes of cell cultures and are used for a variety of applications.

protein synthesis (pro•tein syn•the•sis) the generation of new proteins from amino acid subunits; in the cell, it includes transcription and translation

Overview of Protein Synthesis

Cells are protein-producing powerhouses. Proteins are so vital to every cellular activity that **protein synthesis** occurs continuously throughout a cell's life. At any given moment, thousands of genes are being decoded into millions of protein molecules. Over a typical cell's lifetime, it will produce more than 2000 different kinds of proteins.

Although prokaryotes and eukaryotes are different in many ways, the basic process of protein synthesis is strikingly similar in all cells. DNA molecules code for protein production, mRNA is decoded off DNA sections, and mRNA is processed at ribosomes to make polypeptide chains (see Figure 5.14).

In cells, the instructions for building a protein are stored within one or more structural genes on a DNA molecule of a chromosome. A typical chromosome has hundreds or

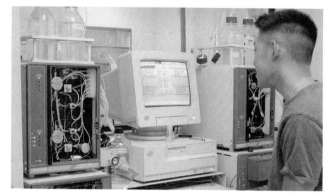

Figure 5.13. **Hanson Chan, a lab technician at CS Bio Company, Inc., Menlo Park, CA, operates a high-performance liquid chromatography instrument. CS Bio makes "to order" peptides for their clients. They also make and sell peptide synthesizers. The HPLC is used to check the purity of peptides before they are shipped.**
Photo by author.

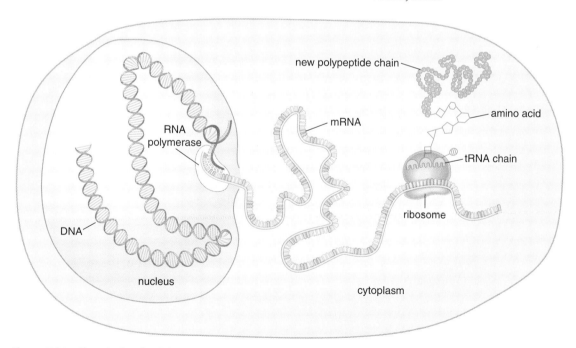

Figure 5.14. **Protein Synthesis in a Eukaryotic Cell.** In a eukaryotic cell, DNA is located within chromosomes in the nucleus. The mRNA transcripts carry the DNA code out to the ribosomes, which translate the code into a strand of amino acids. Commonly, the mRNA transcript is modified on its way to the ribosome. Post-transcriptional modification may include removing introns, shuffling exons, adding or removing nucleotides.

transcription (tran•scrip•tion)
the process of deciphering a DNA nucleotide code and converting it into an RNA nucleotide code; the RNA carries the genetic message to a ribosome for translation into a protein code

codon (co•don)   a set of three nucleotides on a strand of mRNA that codes for a particular amino acid in a protein chain

translation (trans•la•tion)
the process of reading a mRNA nucleotide code and converting it into a sequence of amino acids

tRNA (tRNA)   a type of ribonucleic acid (RNA) that shuttles amino acids into the ribosome for protein synthesis

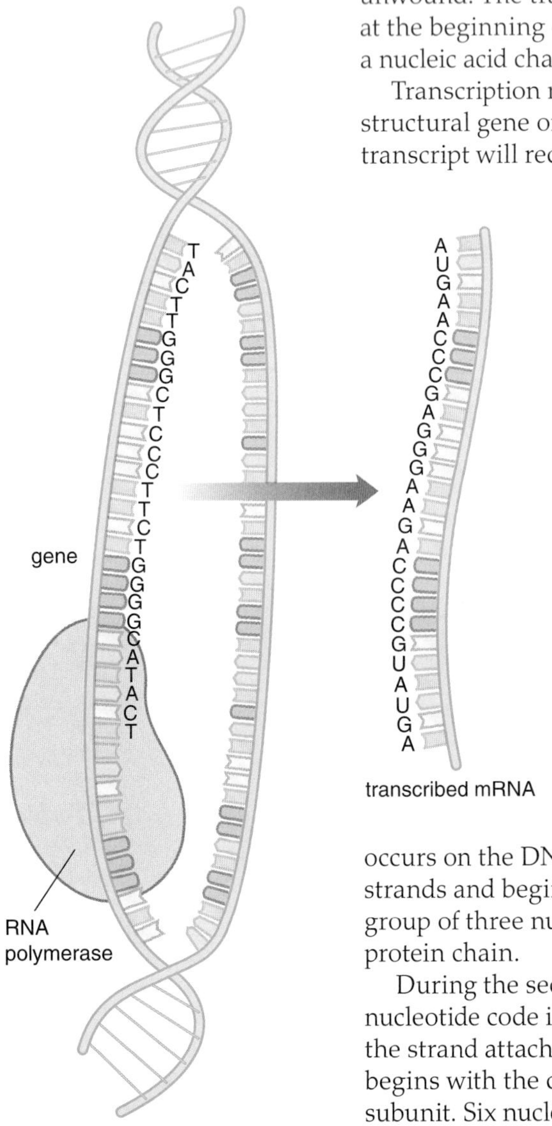

Figure 5.15. Transcription Process in Protein Synthesis. The mRNA is a complementary code to the DNA at the structural gene; T transcribes to A, A to U, C to G, and G to C.

thousands of genes. During gene expression, a structural gene is rewritten in the form of a short messenger molecule, mRNA. Depending on the protein to be synthesized, one or many mRNA transcripts may be made. The mRNA transcripts may be used "as is," or may have introns removed to create functional mRNA strands. The functional mRNA strand floats to a ribosome, where the nucleotide code is translated, and a polypeptide strand of amino acids is compiled. The polypeptide chain then folds into a protein due to the attraction and repulsion of the amino acid's R-groups. The protein may remain and function inside the cell, or it may be transported outside the cell to work on or near the membrane. If the cell is a part of a multicellular organism, the protein may function in a distinct part of the body.

Transcription and Translation

Protein synthesis occurs in a two-step process. First, the genetic code must be rewritten onto a messenger molecule (mRNA). This step is a process called **transcription** (see Figure 5.15). During transcription, sections of DNA (genes) are unwound. The transcription enzyme, RNA polymerase, attaches to a promoter region at the beginning of a gene. The RNA polymerase reads the structural gene and builds a nucleic acid chain (mRNA), which is a complement to the strand being transcribed.

Transcription results in a mRNA molecule with a complementary code of a structural gene on the DNA strand. If the DNA contains a G (guanine), the mRNA transcript will receive a C (cytosine). If the DNA contains a C, the mRNA transcript will receive a G. If the DNA contains a T (thymine), the mRNA transcript will receive an A (adenine). If the DNA contains an A, the mRNA transcript will receive a U (uracil). The RNA molecule is very similar to the DNA strand, except that it is single-stranded, has ribose molecules instead of deoxyribose molecules, and instead of thymine, it uses uracil as a complementary base for adenine.

If the gene to be transcribed had the following sequence (remember, only one side of the DNA is read):

TAC TTG GGC TCC CTT CTG GGG CAT ACT DNA strand,

the mRNA molecule produced would have the complementary code replacing thymine nucleotides with uracil nucleotides as follows:

AUG AAC CCG AGG GAA GAC CCC GUA UGA mRNA strand.

In eukaryotic cells, transcription takes place in the nucleus. Post-transcriptional excision of introns may occur before the mRNA transcript moves out of the nucleus on its way to a ribosome. Post-transcriptional modification is one way that a single gene can code for many gene products.

If the cell is a prokaryote, where there is no nucleus, transcription occurs on the DNA floating in the cytoplasm. Ribosomes move to the mRNA strands and begin translating them immediately, three nucleotides at a time. Each group of three nucleotides, called a **codon**, codes for an amino acid in the eventual protein chain.

During the second step of protein synthesis, called **translation**, the mRNA nucleotide code is rendered into a sequence of amino acids. Translation begins when the strand attaches to the bottom unit of a ribosome. A mRNA molecule usually begins with the code AUG, the "start" codon, which attaches to the bottom ribosomal subunit. Six nucleotides at a time fit in the ribosome, but the code is read three nucleotides (a codon) at a time (see Figure 5.16). Each codon corresponds to one of the 20 amino acids.

Transfer RNA (**tRNA**) molecules pick up amino acids in the cytoplasm and shuttle them into the ribosome. When a tRNA molecule brings the correct amino acid into place, a resident enzyme in the ribosome, **peptidyl transferase**, bonds the amino

**peptidyl transferase
(pep•tid•yl trans•fer•ase)** an
enzyme found in the ribosome
that builds polypeptide chains by
connecting amino acids into long
chains through peptide bonds

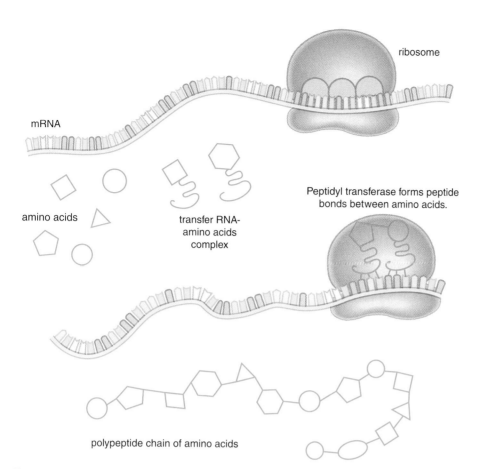

ribosome

mRNA

amino acids

transfer RNA-
amino acids
complex

Peptidyl transferase forms peptide
bonds between amino acids.

polypeptide chain of amino acids

Figure 5.16.   Translation Process in Protein Synthesis.   The genetic language of nucleotides
used in DNA and RNA molecules is translated into a new language of amino acids in a protein.

acids together with a peptide bond. The ribosome shifts to the next triplet codon and
allows the next tRNA amino acid complex to attach. Another bond is made by the
peptidyl transferase, and so on, until the mRNA transcript has been completely read.

It is the code on the mRNA that determines which tRNA amino acid complex will
bond, in which order, in the ribosome. The mRNA codon chart (see Table 5.2) shows the

Table 5.2.   **Codons in mRNA**

1st Base	U		C		A		G		3rd Base
					←	2nd Base	→		
U	UUU UUC	phenylalanine	UCU UCC	serine	UAU UAC	tyrosine	UGU UGC	cysteine	U C
	UUA UUG	leucine	UCA UCG		UAA UAG	stop	UGA	stop	A
							UGG	tryptophan	G
C	CUU CUC CUA CUG	leucine	CCU CCC CCA CCG	proline	CAU CAC	histidine	CGU CGC CGA CGG	arginine	U C A G
					CAA CAG	glutamine			
A	AUU AUC AUA	isoleucine	ACU ACC ACA	threonine	AAU AAC	asparagine	AGU AGC	serine	U C A
	AUG	methionine	ACG		AAA AAG	lysine	AGA AGG	arginine	G
G	GUU GUC GUA GUG	valine	GCU GCC GCA GCG	alanine	GAU GAC	aspartic acid	GGU GGC GGA GGG	glycine	U C A G
					GAA GAG	glutamic acid			

phosphorylation
(phos•phor•y•la•tion)   adding
phosphate groups

cleavage (cleav•age)   process of
splitting a polypeptide into two or
more strands

Taq polymerase (Taq
po•ly•mer•ase)   a DNA
synthesis enzyme that can
withstand the high temperatures
used in PCR

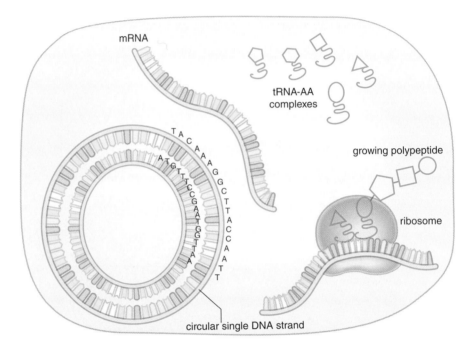

Figure 5.17.   Bacteria Protein Synthesis.   Protein synthesis occurs in the cytoplasm. As the mRNA rolls off the DNA, ribosomes attach and start to assemble the protein. Determine what mRNA strand and what polypeptide would be produced from the outer DNA strand in this diagram.

amino acid that will be added to the polypeptide strand for a given codon on the mRNA strand. To read the chart, look down the left column, then across the top, and then down the right column. For example, if the mRNA code is AAG, find A in the left column, A along the top, and G in the right column. Where they intersect is the amino acid, lysine.

Every time another amino acid is added to a growing polypeptide chain, the ribosome shifts one codon down the mRNA strand. Then another amino acid is added, and so on, until the ribosome reaches a "stop" codon. When the ribosome reaches a UAA, UAG, or UGA, the polypeptide chain disconnects. After translation, the polypeptide folds into a final three-dimensional configuration. Some proteins may need posttranslational modifications, such as the addition of sugar groups (glycosylation). Once folded and modified correctly, so that it is functional, the polypeptide is considered a protein. Figure 5.17 illustrates protein synthesis in a bacteria cell.

In cells, virtually every protein is made in the fashion described above. Once a polypeptide folds, it may be further modified by additional groups attaching. Such is the case with gp120 from the surface of the HIV virus. Recall that the "gp" in gp120 stands for glycoprotein. After the polypeptide is constructed, enzymes add sugar groups (glyco-) to some of the amino acids. Other posttranslational modifications include **phosphorylation** (adding phosphate groups) or **cleavage** (splitting the polypeptide into one or more strands). Insulin is an example of a protein that is cleaved from proinsulin (a larger, inactive form) to insulin (a shorter, active form).

The Importance of Proteins in Biotech R&D

The ability to synthesize and modify peptides or proteins is crucial to the production of virtually every biotechnology product (see Figure 5.18). Sometimes a protein is the desired marketable product of a company, as in the case with the proteases (protein-digesting enzymes) manufactured by Genencor International, Inc. or protein hormones by Genentech, Inc. But even for companies whose products are not proteins, these important molecules have many uses in research and product development.

One very important group of proteins is the enzymes. Just as enzymes catalyze (speed up) nearly every chemical reaction in cells, enzymes also control most steps in the production or breakdown of a biotechnology product. Enzymes are added to reactions performed in test tubes to mimic reactions that could normally only occur in cells. In this way, scientists can make cellular products in test tubes. Such is the case with **Taq polymerase**, used for polymerase chain reaction (PCR). During PCR, a fragment of DNA is copied and recopied to produce millions of identical DNA fragments. *Taq* polymerase is the enzyme that actually assembles the new DNA strands during PCR. In nature, DNA is only synthesized in dividing cells using several enzymes including DNA polymerase, but by using the *Taq* polymerase enzyme, millions of identical DNA pieces can be made in test tubes, within a few hours, for research or diagnostic purposes.

A large segment of the biotechnology industry is currently making enzymes for commercial purposes. Enzymes are so important that they are the focus of the next section of this chapter.

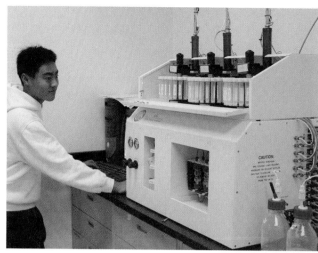

Figure 5.18. Peptide synthesizers can synthesize small peptide chains. These machines use samples of amino acids and other reagents to make peptides that are 2 to 25 amino acids long. The peptides may be used in research and development, or as therapeutic molecules.
Photo by author.

BIOTECH ONLINE

Photo by Paul Robinson.

Couch Potatoes, Relax

It is well known that physical exercise offers many benefits, but some people are not able to exercise for medical reasons and, thus, cannot enjoy those benefits. Scientists recently have discovered a biochemical pathway (a series of enzymatic reactions) that could bestow the benefits of exercise without actually exercising.

Learn about an enzyme that boosts metabolism during exercise and genetic engineers' efforts to regulate its production to mimic the beneficial effects of exercise.

Read the article at: www.nhs.uk/news/2008/08August/Pages/Exercisepill.aspx. Then explain what GW1516 is and how it might be used to give sedentary patients the same benefits as vigorous exercise. Describe how GW1516 has been studied and tested.

Section 5.2 Review Questions

1. Distinguish between transcription and translation.
2. If a structural gene's code is "TAC GGC ATG CCC TTA CGC ATC," what will the mRNA transcript be?
3. If the mRNA transcript from question No. 2 were translated into a peptide, what would the amino acid sequence of the peptide be?
4. What is the name of the machine that can make short amino acid chains?

5.3 Enzymes: Protein Catalysts

Enzymes are proteins that act as catalysts. They speed up biochemical reactions, building up or breaking down other molecules. DNA polymerase is an example of an enzyme that speeds a synthesis reaction. DNA polymerase puts nucleotides together into a growing DNA strand. On the other hand, DNase is an enzyme that speeds the breakdown of DNA into short chains of nucleotides or all the way down to individual nucleotides.

Enzymes are involved in virtually every reaction in a cell. Without them, cells would die waiting for reactions to happen. When scientists conduct reactions in test tubes, enzymes are usually involved. Usually, only tiny volumes of enzymes are needed since they are not used up in the reaction. Some enzymes are sensitive to environmental conditions such as high temperatures. Technicians must learn how handle enzymes to ensure maximum enzyme activity.

Since enzymes are made only in living cells, if an enzyme is needed in significant amounts for research or manufacturing, it must be extracted from existing or engineered cells. Manufactured enzyme products are produced in large fermentation tanks, which ensures that concentrations and volumes are high enough to generate revenue.

Many companies have focused on producing enzymes for sale. For example, Genzyme Corporation produces an enzyme called Cerezyme®. Cerezyme® is used to replace an enzyme (glucocerebrosidase) that is lacking in patients with Gaucher disease, an inherited disorder that causes fatty buildup in cells. Depending on the type of Gaucher's disease, patients can have almost no symptoms, or, in some severe cases, life expectancy may not extend beyond childhood. Glucocerebrosidase is a lipase that breaks down fats and makes fat removal possible from the afflicted cells (see Figure 5.19). Cerezyme® is a recombinant form of glucocerebrosidase, which can be given to patients who do not produce their own glucocerebrosidase. Producing replacement enzyme therapy, such as Cerezyme®, is a strategy of several biotechnology companies.

Figure 5.19. Computer Generated Model of the Enzyme Glucocerebrosidase The structure of glucocerebrosidase reveals 2 identical polypeptide chains. The enzyme breaks down lipids in a cell's lysosomes. If the glucocerebrosidase mutates, lipid can buildup and cause a life-threatening disease. In each polypeptide, two amino acids (glutamic acid), in red, show the enzymes active site.
Courtesy of *Molecule World*, Digital World Biology, LLC.

Enzymes and Their Substrates

substrate (sub•strate) the molecule that an enzyme acts on

The molecules upon which enzymes act are called **substrates**. Enzymes are usually highly specific in that each enzyme catalyzes only one type of chemical reaction and has one or only a few substrates. Some enzymes, such as proteases, break down large protein molecules into smaller ones, as shown in the following equation:

$$\text{keratin} \xrightarrow{\text{protease}} \text{amino acid}_1 + \text{amino acid}_2 + \text{etc.}$$

Some enzymes build larger molecules. These biosynthetic enzymes produce the molecules needed in organisms for structural purposes or other chemical reactions. In ribosomes, peptidyl transferase builds polypeptide chains by bonding amino acids into long chains. DNA polymerase bonds nucleotides into DNA strands during DNA replication.

Enzymes are usually named for their substrates or for the function they perform, with the suffix "-ase" added to the end of the name (see Table 5.3). For example, DNase decomposes DNA. Another enzyme, lipase, acts on lipids, degrading them to glycerol and fatty acids. Some of the first enzymes discovered were not named in this way. Among them is the enzyme pepsin (a protease), which breaks down proteins in the stomachs of mammals.

Many enzymes require additional units, called **cofactors**, to operate (see Figure 5.20a). Cofactors can be as simple as an ion, such as calcium, or a large organic compound, called a coenzyme. The coenzyme is often a vitamin or part of a vitamin. Modified thiamin and niacin vitamin molecules are common coenzymes. Some enzymes require ions, such as Ca^{2+} or Mg^{2+}, as cofactors. For example, DNA polymerase and many of the DNA degrading enzymes require Mg^{2+} ions to be active. Amylase, which breaks down starch into sugar units, requires Ca^{2+} ions.

Most enzymes are huge molecules consisting of hundreds of amino acids and sometimes containing multiple polypeptide chains. Enzymes are produced in the same manner as other proteins. The genes of a species control the kinds of enzymes its members make. Enzyme genes are expressed when mRNA is transcribed off the enzyme gene and the enzyme is assembled at a ribosome.

cofactors (co•fac•tors) an atom or molecule that an enzyme requires to function

Table 5.3. **Enzyme Groups and Functions**   Enzymes are divided into six categories based on their function.

Enzyme Group	General Function	Example
hydrolases	split their substrates with the aid of water	amylase, pepsin, lipase, sucrase, maltase, and DNase
lyases	split their substrates without aid	pectase lyase
transferases	transfer chemical groups between different molecules	hexokinase
isomerases	rearrange the molecules of their substrates	fructose isomerase
oxidoreductases	transfer hydrogen ions and electrons	peroxidase
synthetases	bring molecules together to create larger ones	ATP synthetase, DNA polymerases, and RNA polymerases

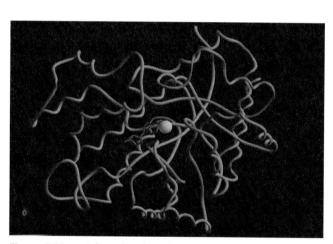

Figure 5.20a.   **Adenosine deaminase (green molecule) binds one of its substrates, 6-hydroxy-1, 6-dihydro purine nucleoside (red). Zinc ions (purple) are a required cofactor for adenosine deaminase activity. Adenosine deaminase is of critical importance in humans. If it is not present or does not function correctly, the immune system cannot fight infections.** © Corbis.

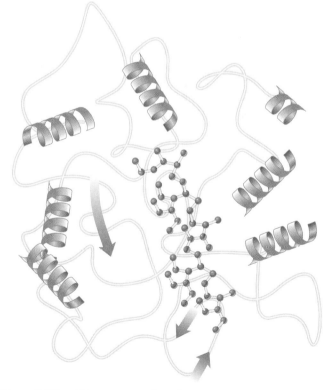

Figure 5.20b.   **This representation of a computer-generated model of cellulase shows the secondary and tertiary structure that gives cellulase its functional shape. Do you see the seven alpha helices and the beta (pleated) sheet? Part of a cellulose molecule (ball and stick model) is embedded in the active site in the center of the molecule.**

Introduction to Studying Proteins

lock and key model (lock and key mod•el) a model used to describe how enzymes function, in which the enzyme and substrate make an exact molecular fit at the active site, triggering catalysis

induced fit model (in•duced fit mod•el) a model used to describe how enzymes function, in which a substrate squeezes into an active site and induces the enzyme's activity

optimum temperature (op•ti•mum tem•per•a•ture) the temperature at which an enzyme achieves maximum activity

denaturation (de•na•tur•a•tion) the process in which proteins lose their conformation or three-dimensional shape

optimum pH (op•ti•mum pH) the pH at which an enzyme achieves maximum activity

The catalytic action of enzymes occurs in a small region on the enzyme called the active site, where the substrate and the enzyme fit together (see Figures 5.19 and 5.20b). This accounts for an enzyme's specificity for a particular substrate. When attached at the active site, a substrate either combines with another or is broken down, depending on the enzyme's action on the substrate's bonds.

There are two theories that describe enzyme catalysis, or how enzymes function. One, the **lock and key model**, suggests that the enzyme and substrate make an exact molecular fit at the active site, which triggers catalysis. More recently, an **induced fit model** was proposed. In this model, a substrate squeezes into an active site, inducing the enzyme's activity.

Factors That Affect Enzyme Activity

Several factors affect enzyme activity, which are important considerations for technicians working with enzymes in a research lab. For example, the amount of substrate in a solution affects how quickly enzymes work. Up to a point, the higher the substrate concentration, the more likely it will be for a substrate and enzyme to meet.

The temperature of a reaction has a significant impact on enzyme activity. Each enzyme has an **optimum temperature** for maximum activity (see Figure 5.21). At this temperature, the enzyme is acting on a maximum number of substrate molecules. For salivary amylase, the maximum number of starch molecules is broken down into maltose units at about 35°C. The optimum temperature for the enzyme, pepsin, is approximately 37°C. As the temperature decreases, fewer substrate molecules bump into enzyme molecules. At high temperatures, too much stress is placed on the H-bonds holding the enzyme's shape, and the enzyme may begin to unravel. The process of proteins losing their structure is called **denaturation**. In the lab, most enzymes are kept cold to decrease the chance that they may denature (see Figure 5.22 a–b).

Another factor affecting enzyme activity is pH, the degree of acidity or alkalinity. In solutions of high or low pH, charged ions may interfere with an enzyme's activity or cause it to denature. Pepsin is an enzyme that works best at a low pH, such as a pH of 1.5. Most other enzymes prefer a more neutral pH. Most amylases works best at a pH of 7.3. The pH at which an enzyme shows the most activity is called the **optimum pH** (see Figure 5.23). Enzymes are stored in buffers that maintain an enzyme at a desired pH. Chapter 7 discusses pH and buffers in more detail.

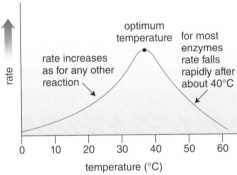

Figure 5.21. Effect of Temperature on Enzyme Activity. Each enzyme has a temperature at which it works best. Enzyme activity decreases substantially at a temperature much higher or lower than the optimum one. At colder temperatures, enzymes do not come into as much contact with their substrate, so enzyme activity is low. Keep samples on ice to decrease the decomposition of valuable DNA and protein samples by enzymes.

Figure 5.22a–b. Most proteins are kept at cold temperatures to prevent degradation. They are stored in refrigerators, cold rooms, or freezers. This cold room is maintained at 4°C. The door is kept closed (left), and electronic monitors keep track of temperature fluctuations. Inside the cold room (right), samples are stored and some experiments are conducted.
Photos by author.

Protein chemists and research associates spend months or years determining the optimum conditions for the enzymes they are producing. They try to optimize the activity of the enzyme at certain concentrations, temperatures, and pH levels. A good example is the protein subtilisin. Subtilisin is a protease that degrades proteins in clothing stains, such as gravy or blood. Companies manufacture subtilisin and sell it to laundry detergent manufacturers as an additive to make the detergent more effective. When Genencor International, Inc. scientists learned how to clone and purify the enzyme subtilisin, they first had to determine how subtilisin degrades proteins under normal conditions. Next, they determined the optimum conditions for subtilisin activity. Finally, they manipulated the enzyme's genetic code to change its structure so that it would still work in the less-than-optimal conditions of detergents and washing machines (high temperature and high pH). You can imagine how it takes years of experimentation to produce enzymes for sale that work just right.

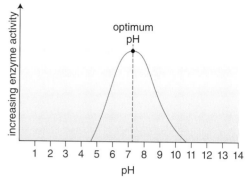

Figure 5.23. **The Effect of pH on Enzyme Activity.** Each enzyme has a pH at which it works best. This is called the optimum pH. In this example, the optimum pH is close to 7. At pH values above and below the optimum pH, enzyme activity can decrease dramatically.

BIOTECH ONLINE

Enzymes: Catalysts for Better Health
Some diseases or disorders are due to missing or faulty enzymes. Many medicinal therapies use enzymes from natural or genetically engineered sources.

Use Internet resources to fill in the remainder of the chart that identifies the enzymes involved in some diseases or disorders.

Disease/Disorder	Description of Disease/Disorder	Enzyme Involved	Treatments/ Therapies	website
phenylketoneuria (PKU)				
galactosemia				
	protein kinase C (PKC)			
glutaric acidemia				
alkaptonuria	black urine			

Section 5.3 Review Questions

1. Name three examples of enzymes and their substrates.
2. What happens if an enzyme is at a temperature significantly above its optimum temperature?
3. What happens if an enzyme is at a pH significantly above or below its optimum level?
4. What would an enzyme be called if it moved methyl groups (-CH$_3$) between molecules?

5.4 Studying Proteins

PAGE (PAGE) short for polyacrylamide gel electrophoresis, a process in which proteins and small DNA molecules are separated by electrophoresis on vertical gels made of the synthetic polymer, polyacrylamide

Proteins are usually colorless molecules that are always submicroscopic. To be studied, researchers must separate them from other molecules and determine their specific characteristics. There are thousands of different kinds of proteins, but all of them have certain characteristics in common. Each protein has a complicated three-dimensional structure. A protein's polypeptide chain is composed of 20 different amino acids in some specific length and order. The number and arrangement of amino acids in the polypeptide determines a specific folding pattern for each polypeptide. Since some amino acids are charged, proteins have a characteristic overall "net charge," or total charge, on the whole molecule, positive or negative, depending on which amino acids make up the protein. Knowing the net charge and other characteristics, technicians can separate the protein during research and manufacturing.

To develop a process for purifying a protein, a researcher must learn the protein's size, shape, amino acid composition, overall charge, and solubility. There are several sophisticated techniques for studying protein characteristics, including amino acid sequencing, mass spectrometry, and column chromatography. A good first step is to "run" a sample on a gel (see Figure 5.24).

Using gel electrophoresis, a technician can easily separate charged molecules by size and shape. Medium-sized DNA molecules are separated and studied using horizontal agarose gel electrophoresis. Proteins are not run on agarose gels because protein molecules are much smaller than most DNA fragments. On an agarose gel, most proteins would quickly pass through the spaces in the gel with little or no separation.

Figure 5.24. **A technician loads protein samples on a vertical gel. Vertical gel boxes operate in a fashion similar to horizontal gel boxes and are used to identify proteins by size and the number of polypeptide chains.**
Photo by author.

Proteins and small DNA molecules are most commonly separated on vertical gels made of the polymer, polyacrylamide. This process is called polyacrylamide gel electrophoresis (**PAGE**). Polyacrylamide gels have a high concentration of molecules ranging from 4% to 18%, so they have a greater ability to separate or resolve small

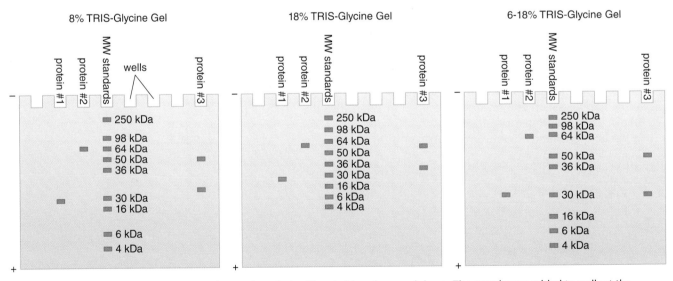

Figure 5.25. **Gels of Different Concentrations.** A vertical gel box holds gels up and down. The samples are added to wells at the top of the gel. When electric current is applied, the samples run down the gel. The molecules move toward the positive electrode, with the smallest ones moving the fastest. The concentration of the gel affects the speed at which the molecules separate. The same three protein samples were added to all three gels, along with the same MW standards. Notice the different migration patterns in each gel.

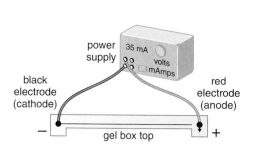

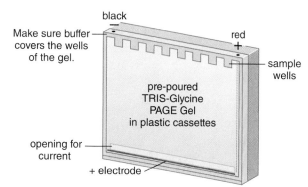

Figure 5.26. **Vertical Gel Electrophoresis.** Although vertical gel boxes vary from one manufacturer to another, all are basically of the same design. The gel cassettes are snapped or screwed in place (right). Running buffer is added behind the gel, covering the wells. Buffer is poured in front of the gel cassette to cover the front opening. When the top is placed on the box (left) and the power is turned on, electricity flows from top (negative charge) to bottom (positive charge). Negatively charged samples move down the gel toward the positive electrode.

molecules. Although PAGE gels can be made from scratch and poured in the lab, commercially prepared, prepoured gels are now more routinely purchased and used. This is because liquid polyacrylamide is highly toxic and time consuming to prepare. PAGE gels can be purchased in concentrations from 4% to 20%, depending on the size of the molecules to be resolved (see Figure 5.25). Smaller molecules are run on higher-concentration gels; larger ones are run on lower-concentration gels. Gels can also be purchased with gradients of polyacrylamide. A 4–16% gel is more concentrated on the bottom of the gel than on the top. Depending on the mixture of proteins to be separated, a gradient gel often works best, especially if you are not sure of the size of the protein(s) of interest.

Vertical gels are run in vertical gel boxes. Like horizontal gel boxes, these have electrode wires at opposite ends of the box (see Figure 5.26). When an appropriate buffer is placed in the box and a power supply is attached, an electric field is established in the box. The samples in vertical gel electrophoresis must have a net negative charge to be able to move into the gel and travel through it toward the positive side. As in horizontal electrophoresis, the longer the molecule, the more difficulty it has moving through the gel matrix. Thus, given the same amount of time, smaller molecules travel farther in a gel than do larger ones.

Samples are most commonly prepared for PAGE with a special sample buffer containing a denaturing agent, such as sodium dodecyl sulfate (SDS). SDS linearizes, or denatures, the protein to polypeptide chains so that each protein's size is based on the number of amino acids it contains, not on its shape.

SDS also coats the polypeptide with a negative charge so that all the proteins have the same charge. Thus, the rate at which the polypeptide chains move to the positive electrode is determined solely by their size. The loading dye, buffer, and gels used also contain SDS and may contain additional denaturing agents such as dithiothreitol (DTT) or beta-mercaptoethanol (BME). These gels are called "denaturing gels" since the proteins are unwound to their primary structure when run. When these samples are loaded into the wells and the power is turned on, the polypeptides travel to the positive electrode at a rate proportional to their molecular weight. The smaller the molecule, the faster it moves through the gel (see Figure 5.27).

Since most proteins are colorless, loading dye must be added to monitor the loading and running of samples. Protein sizing standards of known molecular weight are usually run in other lanes. These help the technician to monitor the progress of the gel

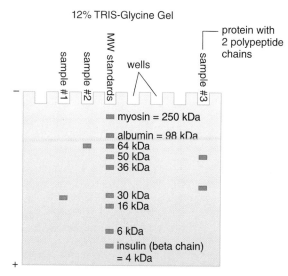

Figure 5.27. **PAGE with Standards.** The smaller the peptide chain, the faster it moves through the gel. Protein-sizing standards can be used to determine the size of unknown samples. Proteins sizes are reported in kilodaltons (kDa).

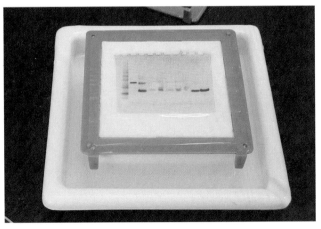

Figure 5.28. The process of staining and destaining protein gels (Coomassie® Blue stain is shown here) takes 30 minutes to a few hours.
Photo by author.

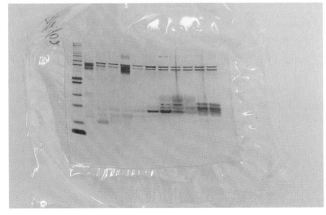

Figure 5.29. Although more expensive, silver stain is much more sensitive than Coomassie® Blue. When samples have low concentrations of protein or DNA, silver-staining is the method of choice.
Photo by author.

Coomassie® Blue (coo•mas•sie blue) a dye that stains proteins blue and allows them to be visualized

silver stain (sil•ver stain) a stain used for visualizing proteins

run. After staining, the standards are used in molecular weight determination of the unknown protein samples.

Standard PAGE gels are usually run at around 35 mAmp. Depending on the size of the gel, its composition, and its concentration, a gel may run for 1 to 3 hours. Recently manufacturers have developed new formulations of PAGE gel matrix and buffers that conduct electricity more efficiently. These allow for faster runs at higher currents. Many of these new PAGE gels run in under 30 minutes. The proteins in the gel are usually colorless and must be visualized. After the gel is run, it is stained.

Several stains exist for visualizing proteins. The most popular ones are **Coomassie® Blue** by QIAGEN (see Figure 5.28) and **silver stain** (see Figure 5.29). Silver stain is more sensitive (down to microgram amounts) than Coomassie® Blue (down to milligrams), but it is more expensive to use, and it requires much more time and labor. For these reasons, Coomassie® Blue stain is more often used. Recent advances in pre-poured gels include some that have visualization dyes already in the gel, saving the technician several hours of staining/destaining time.

After staining, the technician observes the protein-banding pattern for each sample to determine how many peptide chains or proteins are present in a sample and the differences in the proteins' sizes. The molecular weight of the unknown bands can be determined by comparison to the protein molecular weight standards or markers.

Section 5.4   Review Questions

1. What does "PAGE" stand for, and what samples are studied using PAGE?
2. What separates molecules on a PAGE gel?
3. PAGE gels are usually run at what amount of current?
4. A technician has a stock protein solution with a concentration of 1 µg/mL. He prepares a 1:4 serial dilution of the stock and runs the samples on a PAGE gel. What is the preferred method of staining and why?

5.5 Applications of Protein Analysis

Throughout the chapter, you have learned how the study of proteins in biotechnology labs has led to the development of such products as contact-lens cleaners, detergent boosters, and herbicide-resistant soybeans. Tests for pregnancy and influenza are examples of the results of protein research in the human health area, which is perhaps the area of greatest significance to the average citizen. Biotechnology researchers worldwide are experimenting with proteins and gathering information to develop therapies to combat a range of serious diseases.

One area of studies focuses on the protein profile of cells and tissues. The goal of a protein profile is to identify and quantify all of the proteins present in a sample. Comparing the protein profile of one type of cell to another may explain any observed differences in the structure or function of the tissue or cells. This approach could help researchers understand a cell-structure-related disease, such as sickle cell disease. People with this disease have abnormally shaped red blood cells (RBCs) that do not function normally (see Figure 5.30). Analyzing how the protein composition or structure of normal RBCs varies from that of abnormal, "sickled" cells might lead to a corrective therapy.

Sometimes a scientist may be interested in a particular protein's structure because it helps explain the protein's function. In muscle cells, for example, there is an abundance of the protein, myosin. To understand how a muscle does its job of contracting and relaxing, a researcher must understand the structure of myosin. With data gathered from protein studies, researchers can create computer-generated models of the protein's structure and, therefore, better understand the function of myosin in the muscle cell.

Many scientists study proteins to understand the chemical processes in cells. Thousands of metabolic reactions occur in cells while millions of molecules are combining and breaking down. Enzymes and other regulatory proteins control these reactions. Ascertaining which protein is made, and when, helps explain the growth, development, and aging of cells, tissues, organs, and organisms. For example, in the last decade, several biotechnology companies began to study the proteins involved in angiogenesis, or blood vessel growth (see Figure 5.31). Certain proteins trigger angiogenesis in tumors, and certain tumors are known to produce proteins that encourage angiogenesis. By blocking these proteins, tumors may be starved of their blood supply and die. Understanding which proteins are present, at what concentration, and when during angiogenesis is the first step in developing a therapeutic product to inhibit blood vessel growth in tumors.

It requires sophisticated equipment and techniques to learn the characteristics of a protein, but this knowledge provides a wealth of information that can be applied to developing products. To understand a protein's function and mode of action, its amino acid sequence (protein sequencing), three-dimensional structure (x-ray crystallography and computer images), charge, and size (PAGE) must be known. For example, it has been known for a long time that insulin is involved in sugar metabolism. But it was not until the structure of insulin was determined that scientists could fully explain its mode of action and develop therapies to treat diabetes.

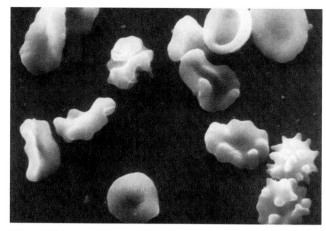

Figure 5.30. Scientists are working on new gene therapies to correct the cause of the abnormal shape seen in red blood cells (left) of sickle cell disease patients. One DNA nucleotide mistake causes a single amino acid substitution, which, in turn, causes an incorrect polypeptide folding and protein shape. This was discovered because the single amino acid substitution in the B-chain of "sickle cell" hemoglobin caused a different banding pattern than is normal during electrophoresis. ~1000X
© Bettmann/Getty Images.

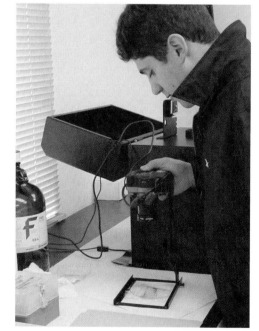

Figure 5.31. A lab technician studies proteins involved in angiogenesis a process important in tumor growth. Here, he photographs proteins from cell cultures that have been run on a PAGE gel and silver-stained.
Photo by author.

Photo by author.

Protein Sequencers

Knowing the amino acid sequence helps scientists understand the three-dimensional structure of a protein and how it works. The first proteins sequenced took months to decipher. The use of automated protein sequencers makes it possible to determine the amino acid sequence of a protein in a few days.

As described in this section, some diseases such as cystic fibrosis and sickle cell anemia are due to slight changes in a proteins amino acid sequence.

Find at least two websites that describe how the change in the amino acid sequence of some protein causes some other human disease. Describe the disease and the protein sequence responsible. List the websites that were used as references.

---

taxonomic relationships (tax•o•nom•ic re•la•tion•ships) how species are related to one another in terms of evolution

The activity, or lack of activity, of various proteins is the cause of virtually all genetic disorders. For example, an abnormal cell membrane transport protein is responsible for cystic fibrosis. Sickle cell disease results from a single mistake in the amino acid composition of the protein, hemoglobin. Determining the differences between normal and defective proteins is an expansive area of biomedical research.

Protein studies are often conducted to understand evolution and **taxonomic relationships**. Since proteins are coded for on DNA, studying the similarities and differences in proteins gives clues to the similarities and differences in DNA molecules. A difference in proteins implies that there are differences in the DNA sequence. Thus, the more different the proteins of two species are, the more likely it is that the DNA will be different. Changes in DNA indicate speciation and evolutionary change. Determining the degree of similarity in proteins allows scientists to make inferences about the evolutionary history of different species.

An example of how protein structure indicates evolutionary relationships is seen in the hemoglobin molecule. The sequence of amino acids in human hemoglobin molecules is 98% the same as in chimpanzee hemoglobin. When compared to gorilla hemoglobin, the two are 96% similar. Horse hemoglobin has only a 76% similarity to human hemoglobin. These protein studies reveal a closer ancestry between humans and chimps than between humans and gorillas or horses. Anatomical and DNA studies have confirmed these conclusions. Scientific information on protein sequences, as well as DNA sequences, is available on the website of the National Center for Biotechnology Information (NCBI) at: **www.ncbi.nlm.nih.gov** (see Figure 5.32).

Hundreds of biotechnology companies focus on the production of one or more proteins for commercial purposes, including drugs, industrial products, agricultural crops, and research and manufacturing instruments and reagents. Many studies must be conducted on proteins to understand how to store and use them to maintain proper structure and function (see Figure 5.33). Identifying the chemical behavior of a protein in different environments is critical for designing purification, assay, and manufacturing protocols.

In cells, proteins never exist in isolation. They are always found in a mixture of other molecules, including carbohydrates, lipids, and other proteins. In manufacturing, the protein of interest must be isolated from all others at a high enough concentration to result in a marketable product. If a protein's molecular weight, charge, or shape is known, it may be isolated on columns, through a process called column chromatography (discussed in more depth in Chapter 9). Chromatography results are confirmed by visualizing purification fractions on PAGE gels.

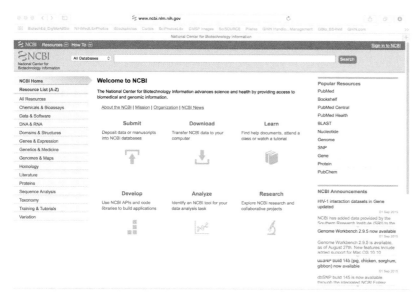

Figure 5.32. The National Center for Biotechnology (NCBI) is a collection of databases and analysis tools used for biomedical and biotechnology research. Protein and DNA sequences are stored on databases and shared among scientists at the NCBI site.
Courtesy of NBCI.

Figure 5.33. Molly He, a staff scientist at Sunesis Pharmaceuticals Inc., concentrates proteins for experiments to discover their structure. She will be running PAGE gels and conducting other characterizations. The proteins she studies could become cancer therapeutics.
Photo by author.

Since proteins are the molecules that actually do the work in cells and organisms, to understand all metabolic processes and disease, basic protein research must be done. At every university, many government agencies, and virtually all biotechnology companies, protein researchers and technicians make up a large portion of the scientific staff. Their basic protein research is funded either by shareholders or by grants from government agencies or nonprofit foundations. Funding keeps them working and allows them to employ other protein scientists, graduate students, research associates, and lab technicians. The National Science Foundation (NSF), an agency of the US government, funds thousands of protein research labs and researchers. The American Cancer Society is an example of a nonprofit foundation that distributes funds for basic protein research on the cellular mechanisms of cancer.

Protein research is critical to all fields of biotechnology. At a typical biotechnology company, a majority of the scientific staff is involved in some aspect of protein science. Thousands of jobs in protein analysis, protein engineering, protein manufacturing and applications (using marketed proteins in new ways) exist in these companies. As a relatively young biotechnology industry gets older and better established, it is moving rapidly from a research and development focus toward a product-manufacturing focus, or **biomanufacturing**. This trend is also leading to a growing demand for protein-manufacturing technicians in private industry as well as in research labs at universities and government agencies.

biomanufacturing (bi•o•man•u•fac•tur•ing) the industry focusing on the production of proteins and other products created by biotechnology

Section 5.5 Review Questions

1. What causes the difference between normal and sickled cells in sickle cell disease?
2. Give an example of proteins studied to understand evolutionary relationships.
3. What is NCBI, how can you access it, and what important information is found there?
4. Do all protein scientists work at biotechnology companies? Explain.

Chapter Review

Speaking Biotech

Page numbers indicate where terms are first cited and defined.

antigens, 154
biomanufacturing, 171
CD4 cells, 153
cleavage, 160
codon, 158
cofactors, 163
Coomassie® Blue, 168
denaturation, 164
ELISA, 155
epitope, 154
glycoprotein, 153
glycosylated, 153

hybridoma, 155
induced fit model, 164
lock and key model, 164
monoclonal antibody, 155
optimum pH, 164
optimum temperature, 164
PAGE, 166
peptidyl transferase, 159
phosphorylation, 160
polar, 150
primary structure, 151
protein synthesis, 157

quaternary structure, 153
secondary structure, 151
silver stain, 168
substrate, 162
Taq polymerase, 160
taxonomic relationships, 170
tertiary structure, 151
transcription, 158
translation, 158
tRNA, 158
x-ray crystallography, 150

Summary

Concepts

- Most biotechnology products are proteins or protein-related products.
- Protein structure is determined by several techniques, including x-ray crystallography, protein sequencing, and PAGE. Based on this data, molecular modeling programs create three-dimensional images of protein structure that help explain protein function.
- Proteins are composed of some assortment of the 20 amino acids, held together by peptide bonds. The DNA code on the structural gene determines both the number and arrangement of amino acids in a protein.
- The 20 different amino acids vary by the type of R group, which can be charged, uncharged, or polar.
- R groups interact with other R groups to cause the folding pattern characteristic of a protein. Interactions include H-bonding, disulfide bonds, and nonpolar interactions.
- The HIV coat protein, gp120, has a 3-D structure that is complementary to the 3-D structure found on CD4 cells of the human immune system. The gp120 structure mutates so quickly that it is difficult to develop an antibody vaccine to fight it.
- Antibodies are complex proteins composed of four chains. Antibodies recognize and bind specific antigen molecules. The tips of the antibody are variable in the amino acid sequence and recognize specific, unique antigens.
- Antibodies are used in research and diagnostic testing, including tests for pregnancy, contamination, or disease. Monoclonal antibody technology can produce many identical antibodies for these purposes.
- An ELISA is a test that uses antibodies to recognize and quantify the amount of a specific protein in a sample.
- Protein synthesis is similar in all cells and occurs in two steps: transcription and translation. During transcription, a mRNA molecule is made at a section of DNA. The mRNA moves to a ribosome, where it is read, and a peptide chain of amino acids is produced. The ribosome reads the mRNA three nucleotides at a time (codon). The ribosome facilitates the correct tRNA-AA (amino acid) complex to bring in the next amino acid. Peptidyl transferase binds adjacent amino acids. Due to the secondary and tertiary interactions, the lengthening polypeptide chain folds.

- Enzymes are proteins that speed the synthesis or decomposition of substrate molecules. Enzymes are named by their substrate or a function they perform, and with an "-ase" ending.
- Enzymes and their substrates have to get very close for catalysis to occur. Two models of enzyme-substrate action are the lock and key model and the induced fit model. Certain enzymes require cofactor ions or molecules. Because enzymes are sensitive to temperature and pH, technicians need to know the optimum temperature and pH for a molecule they are studying or using in a reaction.
- Protein size, the number of polypeptide chains in a protein, and the approximate concentration of proteins in a solution can be determined by running a PAGE gel.
- Researchers study proteins to understand the structure and function of cells, tissues, and organisms, as well as their behavior and processes. Protein studies may be used to explain evolutionary relationships, as well as identification for some species. Protein studies lead to understanding of diseases and how to treat them.

Lab Practices

- Enzymes speed reactions and are often the result of biotechnology product development. Several commercial enzymes are available that make paper softer, remove stains from clothes, make meat tender, and clarify juices. Cellulase and pectinase are two enzymes that increase the amount of juice released from apple cells. Increasing the enzyme concentration in a juicing sample will increase juice yield, up to a point.
- An enzyme assay can be designed to indicate the presence and activity of an enzyme. A valid assay results in measurable data.
- An indicator, Biuret reagent, turns violet-blue in the presence of protein. The higher the protein concentration, the darker violet-blue the Biuret reaction is. The lower the concentration, the lighter violet-blue the Biuret result is. Eventually the concentration is so low that a Biuret cannot detect it. Biuret is not a protein indicator of choice for many applications because the protein precipitates in the reaction. Bradford Reagent is another protein indicator that changes from green-brown to deep blue with increasing amounts of protein in solution. Bradford Reagent testing is commonly used in protein research and development labs to measure the presence and concentration of protein samples.
- Syringe sterilization filters bacteria and fungi out of protein samples for long-term storage. Filter sterilizing removes unwanted or contaminating microorganisms without increasing the temperature of a sample.
- PAGE can be used to understand the function or behavior of a sample by determining the protein composition of cells or tissues.
- PAGE denaturing gels are used to estimate the size and number of polypeptide chains in a pure protein sample. From the size data, the number of amino acids in a protein can be estimated. Similar tissues should have similar protein content.
- PAGE is performed in vertical gel boxes. Prepoured commercial TRIS-glycine gels are most commonly used to analyze protein samples. Samples are loaded at the top of the gel and move to the bottom of the gel (positive electrode) at a rate proportional to their sizes. PAGE gels are commonly run at about 35 mAmp. Loading dye is added to track the progress of the gel. Advances in PAGE technology have brought faster gel running and visualization.
- SDS and other denaturing agents are added to samples in denaturing gels to linearize the polypeptide chains. The gel is stained with either Coomassie® Blue or silver stain to visualize colorless proteins.
- Laemmli buffer is a common running buffer for PAGE. It contains SDS to denature the protein samples.
- As the concentration of a protein sample decreases below about 1 mg/mL, it is difficult to visualize using Coomassie® Blue and silver staining may be required. Too high a concentration of protein on a gel causes large blobs or smears.
- Molecular weight sizing standards are used to estimate the molecular weight of peptide chains.

Thinking Like a Biotechnician

1. How do the 20 amino acids differ from each other?
2. What bonds or forces hold a protein together in a functional three-dimensional shape?
3. Enzyme solutions are always prepared using a buffer at a specific pH as the solvent. Why is a buffered solvent important for enzymes and other protein solutions?
4. Describe the relationship between an antibody and an antigen. Explain how the human body can make so many antibodies.
5. A technician needs to determine the size and shape of a protein. Which of these methods could be used to gain the appropriate data for protein size and shape determinations?
   a. Mass spectrophotometry
   b. PAGE
   c. Protein indicator testing
   d. X-ray crystallography
   e. Protein sequencing
   f. Protein synthesis
   g. Visible spectrophotometry

6. If a structural gene has the code TAC CCC ATG GGG TAA GGC GTC, what mRNA transcript will be made, and what peptide will be produced?
7. If a mutation occurs, substituting the "A" with a "G" at the seventh nucleotide in the structural gene, what are the consequences of the mutation?
8. A technician prepares a 1-mg/mL hemoglobin solution and leaves it on the lab bench over the weekend. When the concentration of the sample is checked on Monday, its value is significantly less than expected. What might have caused the difference in concentration between Friday and Monday? How should the solution have been stored?
9. DNase is an enzyme that chops DNA into tiny pieces. It is evident that DNase is working when a thick mucus-like, DNA-containing solution becomes watery and runny. Design an experiment that would determine the optimum temperature for DNase activity.
10. Protein is an important food nutrient. A technician working at a food company is interested in the nutritional content of seeds/nuts. She runs a PAGE with samples from two different seed extracts. She runs multiple lanes of the samples. The gel is stained with Coomassie® Blue. Seed Extract No. 1 has three faint bands at 25, 30, and 35 kDa. Extract No. 2 has two bands at 25 and 35 kDa, and the bands are very dark. What might she conclude from her results about the nutritional value of the seed extracts?

Biotech Live

Activity 5.1

Gathering Information on the Structure and Function of Proteins

When designing experiments to identify and characterize a protein, scientists need to find out what is already known about a protein or other related proteins. They conduct background literature searches to find articles published in scientific journals, such as *The Journal of Cell Biology*, or on the Internet in scientific databases, such as Medline. Scientists report the results of structural and functional analysis of many proteins. From these reports, further studies can be planned. One site that contains several databases that allows scientists to report and find information about protein structure and function is the NCBI at: www.ncbi.nih.gov. At this site, you can query for a protein's structural information, including amino acid composition and protein size.

Find existing information on the size, structure, and function of a protein. Create an informational poster about the protein's structure and function.

1. The instructor will assign a protein of interest from the list in Table 5.4.
2. Use search engines on the Internet and the protein database at NCBI to collect information about the protein of interest. Include the following information in your literature search:
   a. Where the protein can be found in nature, including photographs or diagrams, if possible.
   b. Specific functions of how the protein works or what it does, including any diagrams or photos, if possible.
   c. Use information from the Protein Data Bank (PDB) at www.rcsb.org/pdb/home/home.do to learn about the structure of the protein molecule, including the number of polypeptide chains, number of amino acid residues, molecular weight (in kilodaltons), and its three-dimensional structure.
   d. Two or more additional interesting facts about the protein structure or function.
3. Cut and paste photos, diagrams, and other information from websites into a document you create in Microsoft® Wordf®. From this document, cut, print, and paste the information onto a poster board. Be sure to record the reference/source of each piece of information.

Table 5.4.   **Protein Groups and Their Functions**

Type of Protein (Function)	Examples of Specific Proteins with These Functions
structural	collagen Dystrophin keratin
enzyme	cellulase alcohol dehydrogenase lysozyme
transport	Aquaporin cytochrome C dopamine active transporter
contractile	myosin actin tubulin
hormone/growth factor	Follicle-stimulating hormone (FSH) human growth hormone Epidermal growth factor (EGF)
antibody	IgA gamma globulin (IgG) immunoglobulin E (IgE)
pigment	melanin (modified amino acid) rhodopsin hemoglobin
recognition	gp120 CD4 C-reactive protein (CRP)
toxins	Botox® (botulinum) tetanus toxin diphtheria toxin

Determining the Amino Acid Sequence of Insulin

Activity **5.2**

Inspired by an activity by Charles Zaremba in Activities-to-Go, Access Excellence. © The National Health Museum, biotech.emcp.net/accessexcellence.
In 1953, Fredrick Sanger developed a method to determine the amino acid sequence of a polypeptide chain. The Sanger Method breaks the disulfide bonds holding the tertiary structure of the protein. Then, some peptide bonds between amino acids are hydrolyzed (water is added at the bond that breaks). Next, short fragments of the polypeptide are sequenced by enzymatic or chemical hydrolysis. The number and types of amino acids are then determined. The peptide sequences are compared for overlaps, and the entire polypeptide sequence is determined. The Sanger Method is a laborious process that has recently been automated using a protein sequencer.

Insulin was the first protein to be sequenced using Sanger's method, and in 1958 he received the Nobel Prize for developing the process. Insulin is a hormone involved in sugar metabolism. It enhances the transport of glucose from blood, across the cell membrane, into cells. Produced only in mammals, it is a small protein with a molecular weight of about 5800 daltons (Da). It is made when a precursor molecule, proinsulin, has a 35-amino acid section removed. The final form of insulin has two polypeptide chains—an α chain of 21 amino acids and a β chain of 30 amino acids. Three disulfide bonds hold the chains together in the functional protein.

The sequence of proinsulin can be determined by transcribing the gene sequence on the next page into a mRNA molecule and translating it into the amino acid sequence:

TACAAACATTTAGTTGTAAACACACCCTCAGTGGACCAACTCCGCAACATAAA-
CCAAACACCGCTCGCGCCGAAAAAGATATGGGGGTTTTGGTCTTCCCTCGCG-
CTCCTAAACGTTCAACCGGTTCAACTTAATCCGCCGCCAGGGCCCCGCCCCT-
CAGAAGTTGGTGATGCGAATCTCCCATCAGACGTTTTTGCCCCGTAACAACTT-
GTTACAACATGGTCATAAACGTCAGAGATGGTCAATCTCTTAATGACGTTAACT

By excising amino acids No. 32 to 66, the insulin sequence is revealed.

Determine the amino acid sequence of insulin.
Determine the number of amino acids in the final protein.
Determine where disulfide bonds may be found in the final protein.
Propose a three-dimensional structure for the insulin molecule.

1. Using the sequence of nucleotides (above) that represents the proinsulin DNA code, write down the mRNA code that would be transcribed from this gene. Draw a line after every third nucleotide in the mRNA sequence so that the codons are easy to read.
2. Cut a piece of adding machine tape/paper approximately 60 cm long and 1 cm wide. The long, narrow paper represents the proinsulin polypeptide chain. Using the mRNA codon/amino-acid chart (Table 5.2), determine the name of the amino acid for which each RNA codon "codes." Write the three-letter abbreviation (Table 2.2) of each amino acid in the polypeptide chain in their proper order on the tape (about 1 cm/amino acid).
3. The proinsulin model represents the inactive, precursor form of insulin. To function, a segment of the proinsulin, amino acids No. 32 to 66, must be removed. The two remaining polypeptide sections (chains α and β) are reconnected to form insulin.
4. Circle the cysteine molecules on the α and β chains. Cysteine molecules form disulfide bonds with other cysteine molecules.
5. Using Table 5.1, place a "+" by each of the positively charged amino acids (at pH 7). Place a "−" by each of the negatively charged amino acids (at pH 7). At pH 7, what is the overall charge on an insulin molecule? The charge on a protein is interesting because of how it affects folding and for purification purposes.
6. How many amino acids long is each chain in the insulin molecule? If a typical amino acid has a molecular weight of about 137 daltons (137 Da), then what is the expected molecular weight of the entire insulin molecule sequenced above?
7. Now fold/bend the insulin chains so that three disulfide bonds can be made between cysteine molecules on the α and β chains. Be careful to consider the positively charged and the negatively charged amino acids and how they would interact with each other within and between chains. Tape the disulfide bonds together since they represent fairly strong bonds. The resulting three-dimensional model represents one way the insulin molecule could fold.
8. Knowing the sequence, size, charge, and shape of a protein, such as insulin, devise some method by which you could isolate insulin molecules from other proteins found in insulin cells.

Activity **5.3**

Insulin: Complicated Choices for a Simple Protein

Insulin is a small protein composed of two polypeptide chains. Its structure was determined and it was available for diabetic therapies in the 1920s. The first therapeutic insulin for humans was purified from pig and cow pancreas. However, even though insulin from a different species is similar to human insulin, "foreign" insulin may cause allergic reactions in some patients. In the 1980s, biotechnology research and development produced human insulin (rhInsulin) using engineered bacteria. Now, recombinant human insulin, synthetic human, or human insulin analogs are available as insulin therapies.

Learn about the characteristics of human insulin and options that biotechnology has provided for diabetic patients.

1. Using an Internet search engine, find a description of the structure and function of human insulin. In 4–5 sentences describe the molecule.
2. Find a molecular model for human insulin in the NCBI's Structure database (www.ncbi.nlm.nih.gov). One model of human insulin has the PDB ID, 5CNY. Download the structure file and open it in either the Molecule World application or the Cn3D application. When you are able to view it, work through the tasks below:
   - Look at the model, there are actually six identical proteins (insulin), each with 2 polypeptide chains. The six insulin molecules surrounding a Zn^{2+} ion.
   - Make sure you can view the amino acid sequence (in the Cn3D menu, choose "show sequence viewer"). Find the "Pick Structures" choice and highlight a single 5CNY-A and 5CNY-B to see an entire insulin molecule made up of its A and B chains.
   - Record the one letter amino acid sequence for each chain. Then, use Table 2.2 in Chapter 2 to record the names of the amino acids in each chain.
   - View the insulin with "Wire" configuration. Highlight the all the cysteine amino acids in each chain. These bond to each other with disulfide bonds and hold the A and B chains together. How many cysteine-cysteine bonds do you see? Rotating the molecule will make these more visible.
3. Biotechnology companies are now producing synthetic analogs of insulin that are slightly modified in structure from naturally occurring human insulin. Often these have advantages over the native protein. Insulins are categorized by differences in how they are produced, how they are administered, how quickly and how long they work. Learn about different types of insulin on the Diabetes Education Online website at **dtc.ucsf.edu/types-of-diabetes/type2/treatment-of-type-2-diabetes/medications-and-therapies/type-2-insulin-rx/types-of-insulin**.
   - Using the Diabetes Education Online websites, for the insulin analog, Aspart, record how it is different than regular human insulin.
   - Using either the Molecule World application or the Cn3D application as with the human insulin above, determine the amino acid sequence of this insulin analog. Its PDB ID is 4GBN. How is it structurally similar to and different from the regular insulin you studied?
4. Advances in delivery of insulin into the bloodstream have given patients many options. In the past, therapeutic insulin could only be delivered through injection. Obviously, a less painful method would be attractive. Now, there are inhalable forms of insulin. Use www.webMD.com to learn about the pros and cons of inhalable insulin. Record several benefits and concerns about inhalable insulin.

Prions: Enough to Drive You Mad

Activity 5.4

Some very serious diseases are caused by a group of unusual proteins that are thought to replicate themselves without using DNA for protein synthesis! These self-replicating proteins are called **prions** and scientists believe they function, like viruses, by taking over a cell.

One type of prion appears to cause the fatal nervous system condition Mad Cow disease, more appropriately called **bovine spongiform encephalopathy** (BSE). Scientists have developed tests using antibodies that will recognize the BSE prion protein in brain tissue of suspected BSE subjects.

Find more information on bovine spongiform encephalopathy (BSE) and learn about the efforts to diagnosis and treat it.

Use Internet resources to create a one-page fact sheet describing and illustrating the agent that causes BSE, the symptoms and treatments for BSE, and other interesting information about the disease. Find at least one company that has or is developing a test or treatment for BSE. Include graphics and reference URLs on the fact sheet.

Bioethics

Who Owns the Patent on the Genetic Code for Your Proteins?

With the recent decoding of the human genetic sequence (the Human Genome Project), it is feasible that in the future everyone could have their own DNA sequenced. This would give a "DNA fingerprint" of all the genes and, therefore, all the proteins a person synthesizes. Who will decide who should have access to this genetic information? Is genetic information private or is it important for the public good?

For each issue listed below, provide a three-part answer that includes (a) a supporting argument for when, if ever, the genetic code should be available; (b) a supporting argument for when, if ever, the genetic code should not be available; and (c) an explanation of any conditions that would be an exception to your position.

Issue: Should medical authorities get your genetic fingerprint?

Issue: Should insurance agencies get your genetic fingerprint?

Issue: Should the military get your genetic fingerprint?

Issue: Should law enforcement agencies get your genetic fingerprint?

Issue: Should prospective spouses get your genetic fingerprint?

Issue: Should employers get your genetic fingerprint? Should some employers have the right to the genetic fingerprint and others not?

Issue: Should scientists conducting gene therapy/corrective therapy get your genetic fingerprint? Should you get royalties (get paid) on your genetic information if it is used to correct faulty or inferior DNA?

Unit 2
Modeling the Production of a Recombinant DNA Protein Product

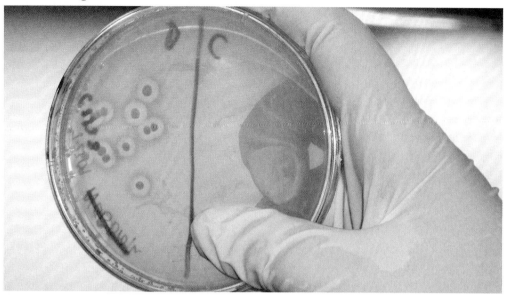

Photo by author.

When asked what biotechnology is, most people would say it has to do with genetic engineering of cells. This is probably because the first biotechnology companies tricked certain target cells into taking up foreign pieces of DNA which then were transcribed and translated into proteins that could be marketed as a commercial product. By the 1980s, several biotechnology companies were using genetically engineered bacteria, fungus and mammalian cells to produce human proteins as medical therapies, industrial products, and as research molecules.

In Chapters 6–9 of *Biotechnology: Science for the New Millennium*, you will learn more about how biotechnology products are selected for research and manufacturing. Specifically, what questions must be asked and answered before a recombinant DNA protein is targeted as a potential product? What laboratory tests or assays must be developed to monitor the progress of a targeted protein product as it goes through development? What molecular biology techniques are needed to find and use the DNA code for the production of the target protein? What methods will be used to produce, purify, and formulate the protein in sufficient amounts to provide it to customers?

The production protocols are unique to each protein product but there are some basic tools and techniques that are common to all recombinant protein production including:
- Develop methods to recognize and measure DNA and proteins in solution.
- Learn how to create and study recombinant DNA molecules
- Produce and monitor genetically engineered cells and grow them in increasingly large volumes
- Use purification technologies to separate a protein of interest from contaminant molecules

You will be using several proteins to learn recombinant protein methodologies. However, since the protein amylase is of considerable commercial importance and is rather easy to use in academic labs, it will be the model protein for our study (*The rAmylase Project*). In *The rAmylase Project*, you will learn:
- About the economic value of amylase and how it is important in biofuel production, the beverage industry and in textile manufacturing
- How to measure the presence, activity, and concentration of amylase in solution using several tests (assays) and instruments
- How to insert an amylase gene into targeted bacteria cells (genetic engineering) and grow the cells in sufficient amount to harvest (purification) the amylase from them.

Biotech Careers

Photo courtesy of Colin Heath, PhD.

Director of Research & Development and Marketing

Colin M Heath, PhD
Geno Technology Inc., G-Biosciences
St. Louis, MO

As Director of R&D and Marketing for G-Biosciences, Colin's primary role is to oversee the development of new products from initial concept through to product launch. G-Biosciences provides reagents, experimental kits, and other research tools used for protein isolation and analysis in academic research and the biotechnology industry. As part of Geno Technology Inc., G-Biosciences' primary goal is to target the key techniques of protein research and find affordable methods to simplify and improve on these techniques.

Colin oversees the lab managers and technicians involved in research and development at G-Biosciences. A large part of his day is involved in coordinating the staff and their multiple projects, as well as troubleshooting the unexpected, albeit common, surprise issues of science research.

Colin began his scientific career at The University of Bath, UK, where he completed a BSc in Biochemistry. His degree allowed him to work for 6 months at both St Jude's Children's Hospital and Glaxo SmithKline. Colin undertook a PhD at the Institute for Animal Health, Pirbright, UK, where he successfully completed a thesis on "The role of proteolytic processing of the 220 kDa polyprotein of African swine fever virus and aggresomes during virion assembly". Colin moved to St Louis, MO, USA to undertake a research associate position in the Department of Cell Biology, Washington University School of Medicine researching the role of lipid kinases in membrane trafficking. In 2005, Colin moved into the biotech industry when he joined Geno Technology, Inc.

6 Identifying a Potential Biotechnology Product

Learning Outcomes

- Give examples of biotechnology products derived from plant and animal sources and discuss the challenges of identifying potential product sources
- Identify the steps in a Comprehensive Product Development Plan and use it to determine whether a potential biotechnology product is worth manufacturing
- Use the enzyme, amylase, as an example of how a commercially-important protein product is targeted by a biotechnology company for research and manufacturing (The rAmylase Project).
- Discuss the types of assays done as potential products move through process development and identify the additional assays required for pharmaceutical development
- Describe how an ELISA or a Western blot is conducted and what the results of each assay can reveal
- Describe the role of CHO cells in protein product development
- Describe the typical recombinant DNA protein product pipeline, additional steps required by the FDA for pharmaceutical proteins, and possible formulations of the final product

6.1 Sources of Potential Products

Biotechnology products come from many sources. For thousands of years, people have collected plant and animal organs and used them either in their entirety or in part for an assortment of purposes. A good example of this practice is the chewing of willow branches to lessen toothache pain. The willow bark contains salicin, a precursor to aspirin. For hundreds of years, people have used and engineered animals and plants into new breeds or varieties, essentially creating new products.

More recently, scientists have learned how to use plant and animal parts as sources of specific macromolecular products. For example, for decades the pancreas of many kinds of livestock was ground up and used as a source of the protein, insulin (see Figure 6.1). Insulin is used to treat diabetes. Another example is the use of the foxglove plant as a source of digitalis, a chemical used to regulate an irregular heart rate. The majority of pharmaceutical, agricultural, and industrial products still come from nature.

Figure 6.1. **Cows are important to biotechnology for several reasons. First, cows are an essential agricultural product. In addition, the cow is an important model organism for biomedical and veterinary research. Cows and other livestock have been used as sources of pharmaceuticals for a long time. Until recently, the pancreas of cows was the main source of insulin for human diabetic patients.**
© levers2007/iStock.

Figure 6.2. **Drug Production Factory.**   Large metal fermentation tanks hold cell cultures producing recombinant proteins. The top of each tank is seen above the floor level.
© Michael Rosenfeld/Science Faction/Getty Images.

Harnessing the Potential of Materials Produced in Nature

Scientists can significantly improve access to naturally occurring products. Sometimes, a potential product is made in very small quantities in nature, so it is impractical to extract it from a natural source. In other words, there is not enough source material available. Such is the case with the protein tissue plasminogen activator (t-PA). As discussed in earlier chapters, t-PA is an enzyme that dissolves blood clots. It is now administered to many patients after a heart attack or stroke to clear blocked blood vessels. Humans make t-PA naturally but in such small quantities that it cannot be harvested from the blood for therapeutic purposes. In the mid-1980s, scientists at Genentech, Inc. cloned the human t-PA gene in Chinese hamster ovary (CHO) cells. The CHO cells took up the human DNA and transcribed it into human t-PA protein. The cloned cells are grown in large (several thousand liters) fermentation tanks in broth culture. (see Figure 6.2) As the cells grow, they produce t-PA in large quantities. The t-PA is harvested and purified from the broth. It is formulated for the market to be used by doctors and emergency medical technicians.

Another example is the production of large quantities of antibiotics. Antibiotics are molecules produced in bacteria and fungi to inhibit the growth of other bacteria. Before the widespread use of antibiotics in the 1940s, many people died from common diseases that we now treat fairly easily, such as strep throat, bronchitis, and pneumonia. In underdeveloped countries, the death rate from these, and many other diseases, is still high because they do not have sufficient access to antibiotics. You can imagine the volumes of antibiotics that are needed for patients in and out of hospitals. Antibiotic production is a large business.

Antibody products, including monoclonal antibodies, are also a growing segment of the biotechnology industry. Antibodies are a focus of product development because antibodies may be used to recognize very specific molecules. Antibody recognition is important in several medical therapies and in research. For example, Cosentyx™ (secukinumab, Novartis Pharmaceutical Corporation) is a human IgG1 monoclonal antibody that selectively binds to the interleukin-17A (IL-17A) diminishing the symptoms of plaque psoriasis. Antibodies of medicinal interest are naturally made in small amounts in humans. For therapeutic purposes, scientists have engineered production cell lines to make large amounts of specific medically important antibodies. According to centerwatch.com, by 2015 the FDA had approved 192 immunological therapies, the majority of which are antibodies or antibody conjugates. More on antibody therapeutics is found in Chapter 12.

Many biotechnology products have medical applications, however many biotechnology products for use in industry, agriculture, biodefense, environmental and research applications are on the market or currently in the product pipeline. Table 6.1 lists a range of biotechnology products that come from a variety of sources. Some are still harvested directly from existing organisms in nature. Some are synthesized in the laboratory, while others are produced in genetically engineered organisms.

Table 6.1. Sources of Selected Biotechnology Products

Product	Source/Description
Roundup Ready® soybeans (Monsanto Canada, Inc)	herbicide-resistant soybeans with a resistance gene from bacteria
IndiAge® cellulose (Genencor International, Inc)	cellulose-digesting enzyme made in *Bacillus subitilis* bacteria
nerve growth factor	stimulates nerve cell growth in humans; made in *E. coli* bacteria
Premise® 75 (Bayer Corporation)	termiticide (kills termites); synthetic, organic compound
Posilac® bovine somatotropin (Monsanto Co)	growth hormone for livestock; made in *E. coli* bacteria
thrombopoietin	human blood-clotting agent made in CHO cells
Videx® (Bristol-Myers Squibb Co)	nucleoside analog for HIV treatment, synthesized in a laboratory
BollgardII® cotton (Monsanto Co)	insect-resistant cotton with resistance genes from a bacterium
XOLAIR® (omalizumab) (Genentech, Inc, Novartis Pharma AG, and Tanox, Inc)	anti-IgE monoclonal antibody that binds and removes immunoglobulin E (IgE), which is involved in allergic responses; made in CHO cells
aspartame	artificial sweetener produced in the laboratory by combining two amino acids
BXN™ cotton seed (Monsanto Co)	herbicide-resistant cotton with a resistance gene from a bacterium
Vectibix® (panitumumab, ABX-EGF) (Amgen, Inc.)	monoclonal antibody that targets the epidermal growth factor receptor (EGFr), which is overexpressed in a variety of cancers including metastatic colorectal cancer; made in XenoMouse™ technology (by Abgenix, Inc)
EPOGEN® (Amgen, Inc)	a human protein that stimulates red blood cell (RBC) production
Rocephin® (Roche Pharmaceuticals, Inc)	cephalosporin antibiotic; modification of penicillin from fungi
ChyMax® (developed by Pfizer, Inc currently manufactured by Chr. Hansen Inc.)	enzyme that curdles milk for cheese production; cloned in bacteria
Cervarix, human papillomavirus vaccine (GlaxoSmithKline)	modified recombinant virus vaccine for the prevention of cervical cancer caused by oncogenic human papillomavirus (HPV) types 16 and 18
digitalis	organic chemical extracted from the foxglove plant to treat congestive heart failure and heart rhythm problems

Modeling the Research and Development of a Potential Product

The following sections of this chapter discuss the research and development (R&D) of a potential marketable product, amylase. **Amylase** is an enzyme produced by several organisms to break down the polysaccharide amylose (plant starch) to the disaccharide maltose (see Figures 6.3a and 6.3b). Maltose is, in turn, degraded to the monosaccharide, glucose, by maltase:

amylase (am•y•lase) an enzyme that functions to break down the polysaccharide amylose (plant starch) to the disaccharide maltose

$$\text{amylose} \xrightarrow{\text{Amylase}} \text{maltose}$$

$$\text{maltose} \xrightarrow{\text{Maltase}} \text{glucose} + \text{glucose}$$

Estimating Market Size   Amylase is an important commercial product with a large market because many industries need to easily and economically break down starch or produce sugar. The textile industry,

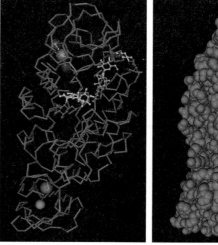

Figure 6.3a-b   These molecular models show the three-dimensional structure of a bacterial amylase. The pink amylase "tube" confirmation clearly shows where amylase's substrate, starch, binds in a groove at amylase's active site. The space-filling confirmation shows how tightly starch fits in the active site. Unwanted changes is amylase's structure could affect its ability to bind starch.

Photos courtesy of *Molecule World*, Digital World Biology. LLC.

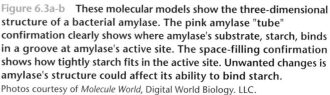

Identifying a Potential Biotechnology Product

183

for example, needs to quickly remove all the starch added to fabric during clothes manufacturing. Starch does not easily go into solution so just washing it does not remove it from fabric. Adding amylase to the wash process catalyzes the breakdown of starch to sugar, which then washes away easily. Papermakers also use amylase to remove excess starch from paper products. The difference in quality of paper products, such as toilet paper versus paper bags, is partly due to the amount of starch added and then removed during paper manufacturing. Other industries such as beer and juice manufacturers use amylase to remove starch and its cloudiness from their products.

Another industry that utilizes amylase in starch breakdown is cellulosic biofuel production. In cellulosic biofuel production, cellulose fibers from plant materials are broken down to glucose, which is, in turn, fermented to bioethanol fuel. The cellulose fibers are in the plant cell walls and they are impregnated with polysaccharides including starch. When amylase is added to biofuel plant substrates, it clears away the starch making it easier, and cheaper, for cellulases to breakdown cellulose to glucose.

Figure 6.4. **Sugar is used by many industries, including beverage makers and baking companies. It is much more expensive to use sugar grown on islands than to use amylase to convert cornstarch to sugar.**
© isitsharp/iStock.

Many industries require glucose production and starch is a good source of sugar. Beverage companies may require large amounts of sugar or high-fructose corn syrup, which are easily made by converting glucose to fructose. Traditionally, sugar was harvested from sugar cane or sugar beet plants (see Figure 6.4). However, the islands that grew sugar cane are losing land to development, so there is not enough sugar available. In addition, harvesting sugar cane contributes to air pollution. Amylase breakdown of starch (ie, cornstarch) to sugar is an economical alternative to island-grown sugar.

With all of these applications, amylase is a product that would likely draw interest from a biotechnology company. Since the 1980s, amylase has been one of the most important biotechnology products and the best-selling industrial biotechnology product. According to The World Journal of Pharmaceutical Research, in 2012, industrial amylases generated $2.7 billion in revenues. To produce enough amylase to market, though, a company needed a substantial source of this enzyme.

Identifying Product Sources   Humans produce amylase in the salivary glands and the pancreas. As amylase is secreted onto ingested food, the starch in the food is broken down into absorbable sugar, which body cells use as fuel.

Many microorganisms are also amylase producers. Several species of decomposing bacteria or fungi use amylase to break down plant molecules for food. The bacterium, *Bacillus subtilis*, a common soil bacterium, is a known amylase producer. This species of bacteria lives in soil and on decaying plants. If *B. subtilis* is to be the source of amylase, scientists must be able to grow it in large volumes and purify an active form of amylase. Unfortunately, *B. subtilis* does grow or produce amylase well in large volume cultures. To make amylase in large volumes, a better cell line was needed.

Genencor International, Inc. scientists first created recombinant amylase (rAmylase) and developed a production process. Researchers developed procedures to genetically engineer *E. coli* bacteria to produce recombinant alpha-amylase on a large scale because they saw the possibility of many marketable applications. They chose *E. coli* as a production host since it is easy to scale up *E. coli* broth cultures. Scaled-up broth cultures produce the recombinant amylase well, and procedures to purify the protein product from the cell culture have been developed.

Creating a Comprehensive Product Development Plan (CPDP)   In Chapter 1, you learned that many companies use a CPDP, or something similar to it, to decide whether to pursue development of a potential product. An example of how

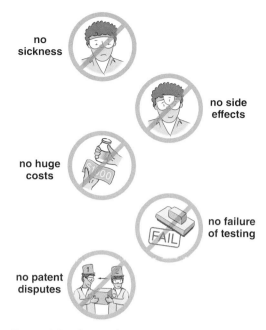

no sickness

no side effects

no huge costs

no failure of testing

no patent disputes

Figure 6.5. **Comprehensive Product Development Plan for a Pharmaceutical Product.** The CPDP for a pharmaceutical product would be slightly different from a CPDP for an industrial product that is not going to be used in humans. For a pharmaceutical, the potential product must be able to demonstrate safety and efficacy, as well as the potential for a large market.

Figure 6.6. Brock Siegel, PhD, was the vice president of Product Development & Manufacturing Operations at Applied Biosystems, Inc. (ABI) for 15 years. ABI (now part of Life Technologies Corp.) develops instruments and reagents that are used in DNA and protein research and manufacturing, including DNA sequencers and synthesizers. Brock's many responsibilities included ensuring that research ideas from ABI scientists met the needs of their customers, working with the marketing group to design each new product, and ensuring that new product ideas were rapidly and reliably converted into manufactured, shippable products. Each product must be profitable and must meet the customer's quality and performance expectations. Currently, Brock acts at Director and/or Chief Operations Officer at three California biotechnology companies. Photo by author.

an industrial product, such as amylase, would be evaluated using the CPDP is given below. Other products might be evaluated slightly differently, but still with the goal of demonstrating marketability (see Figures 6.5 and 6.6).

Amylase, as a potential product, can be evaluated in light of the CPDP. For an accurate evaluation, significant amounts of research would be conducted to identify the market and estimate the potential profit. A cursory evaluation results in the following assessment:

Recombinant Alpha-Amylase CPDP

1. **Does the product meet a critical need? Who will use the product?**
   Amylase is used in industry for several applications. Since amylase may be used to produce sugar, any industry needing large quantities of sugar would be interested in the product. The beverage industry, including manufacturers of soft drinks, such as sodas, is a substantial user of sugar. In addition, several industries including textiles and biofuels use amylase to remove starch from their products.

2. **Is the market large enough to produce sufficient sales? How many customers are there?** Sugar is a main ingredient in commercially produced beverages. Amylase can produce sugar at the site of demand. Amylase use has several applications at the present time and several potential industrial applications in the future including applications in biofuel production. The amylase industry has the potential to produce billions of dollars in sales annually for the foreseeable future.

3. **Does preliminary data support that the product will work? Will it do what the company claims?** Amylase decomposes starch molecules in the laboratory and in manufacturing. It can be assayed (tested) for activity and concentration with relative ease. The gene for amylase production can be engineered into host organisms that will grow well in fermentation tanks and produce significant qualities of amylase.

4. **Can patent protection be secured? Can the company prevent other companies from producing it?** Patent protection can be secured, and was, by Genencor International, Inc. for the genetically engineered version of bacterial amylase.

5. **Can the company make a profit on the product? How much will it cost to make it? How much can it be sold for?** Since sugar is mainly grown and harvested on isolated, tropical islands, the cost to manufacture and transport it is high. If sugar can be produced from cornstarch in the midwestern United States, manufacturing and transport costs decrease. This along with the the use of amylase for other applications translates into a demand for recombinant bacterial amylase. Recombinant amylase sales will generate billions of dollars in revenue.

Amylase can be synthesized, purified, and tested relatively easily in a small laboratory. Thus, it seems that amylase is an ideal product to enter the pipeline. Before amylase can be produced on either a small or large scale, methods must be developed to confirm the amylase in solution. The scientists and technicians must be able to know when they have amylase, how much they have, and how active it is. Tests for amylase and other molecules are called **assays**. Creating amylase assays is one of the first steps in the lengthy process of producing recombinant amylase for market. Assays and assay development are the focus of the next few sections and activities in this chapter and Chapter 7.

assay (ass•ay) a test

activity assay (ac•tiv•i•ty as•say) an experiment designed to show a molecule is conducting the reaction that is expected

---

Section 6.1   Review Questions

1. Why are antibiotics important biotechnology products?
2. What is the function of the enzyme, amylase?
3. Why might a company be interested in producing amylase as a product?
4. Summarize the criteria that a potential product must meet in a CPDP review.

6.2 The Use of Assays

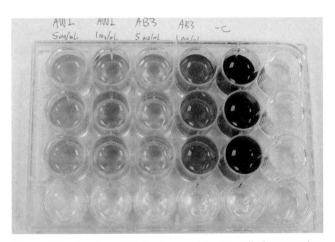

Figure 6.7. Amylase Activity Assay. The 24-well plate reveals samples with different amounts of amylase activity. Amylase breaks down starch to sugar. In this activity assay, as amylase breaks down starch, the dark color of a starch/iodine mixture becomes lighter. A dark blue color indicates no measurable amylase activity.
Photo by author.

Many biotech products headed for manufacturing are proteins. Yet, protein molecules are often colorless and submicroscopic. While conducting R&D, and throughout manufacturing, scientists must be able to quantify the amount and activity of a protein. If a substance is chosen as a potential product, researchers must be able to test for its presence, activity, and concentration. The protein must be "assayed."

The term "assay" is synonymous with the term "test." Many kinds of assays exist. Some assays are simple and straightforward, such as the assay for the presence of a compound or group of compounds. Indicator tests, such as the Bradford reagent, test for the presence of protein in solution. A polyacrylamide gel electrophoresis (PAGE) may also be considered an assay since it may be used to show the presence and concentration of a particular protein.

Another important type of assay is an **activity assay** (see Figure 6.7). Activity assays not only show that a compound is present, but that it is active or

Figure 6.8. **Decreasing amounts of protein are indicated by less blue color in the tube. The right-hand tube is a negative control that contains no protein. The tubes to the left have increasing concentrations of protein.**
Photo by author.

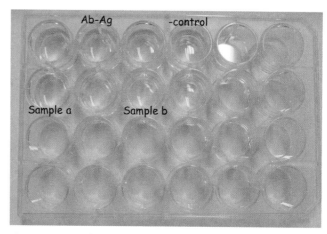

Figure 6.9. **In this ELISA, conducted in a 24-well plate, the antigen is bound to the bottom of a well. An antibody that recognizes it has an enzyme attached to it. The enzyme will change a colorless reagent from clear to blue and then to yellow with the addition of acid. The darker the yellow, the higher the concentration of the antigen protein in the sample.**
Photo by author.

functioning. These assays are necessary to demonstrate that an enzyme, for example, is active and conducting the reaction that is expected. In the case of amylase, an activity assay would indicate the degradation of amylose (starch). It would measure either how much starch is broken down by amylase or how much sugar is produced.

Scientists must be able to report the concentration of protein being used, produced, or sold. Several **concentration assays** exist for the enzyme, amylase. Using colored indicator solutions, such as a Bradford protein reagent or bicinchoninic acid (BCA) protein reagent, the concentration (amount per unit volume) of a protein in solution can be estimated (see Figure 6.8). Either by eye or, more accurately, using a spectrophotometer, a technician can determine the concentration of an unknown sample by comparing the unknown to known solutions. Bradford protein reagent is a nonspecific protein indicator and will show the presence of virtually any protein in solution. It can be used to set up a concentration assay to estimate the amount of the protein.

In a mixture of other molecules, it may be difficult to confirm the presence of just one kind of protein. Some concentration assays are very specific and will recognize a single type of protein in a mixture of others. An enzyme-linked immunosorbent assay (ELISA) has high specificity. An ELISA can be used to determine the presence and concentration of a specific protein utilizing antibody-antigen specificity. Since antibodies recognize and bind to only very specific molecules, they can be used to attach to and indicate these molecules. In an ELISA, an enzyme attached to an antibody binds to a specific protein (antigen) and causes a color change that can be measured when a specific substrate is added (see Figure 6.9). An ELISA is a very sensitive analytical tool that can be used to ensure the presence and concentration of a product throughout the entire research and manufacturing process. ELISAs are so common in biotechnology laboratories that many labs have their own ELISA plate washers and readers (see Figure 6.10 and 6.11). ELISAs are discussed in more detail in the next section.

The types of assays discussed above are used throughout process development as a product moves through the pipeline. If a product is a pharmaceutical,

concentration assay (con•cen•tra•tion as•say) a test designed to show the amount of molecule present in a solution

Figure 6.10. **An ELISA plate reader such as the one pictured here, automatically reads the absorbance and determines the concentration of up to 96 samples in a 96-well plate placed in the tray at the lower right. In Figure 6.11, a 96-well ELISA plate is shown with positive antibody antigen recognition (dark yellow color).**
Photo by author.

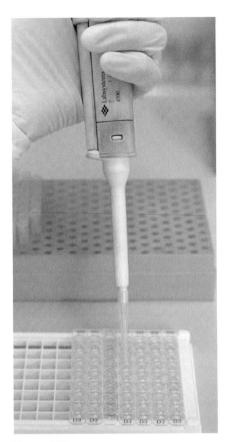

Figure 6.11.
© Science Photo Library/Photo Researchers.

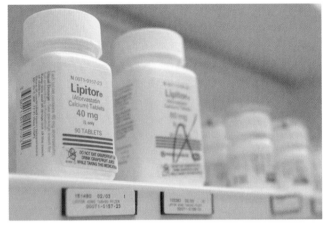

Figure 6.12. It is obvious why companies must perform stability and dosage assays on medications such as Lipitor® (Pfizer, Inc.), which is prescribed to lower cholesterol, and the various drugs used to control blood pressure.
© James Leynse/Corbis Documentary/Getty Images.

it must go through much more extensive testing. As a product gets closer to clinical trials and an **Investigational New Drug (IND) application** to the FDA, assays must be developed to prove a product's safety and efficacy. In addition to activity, concentration, and ELISA tests, multispecies **pharmacokinetic (PK) assays** and **pharmacodynamic (PD) assays** must be developed and conducted. These tests show the amount and length of activity of the protein in humans, as well as in other test organisms. These assays must demonstrate activity in monkeys, mice, rabbits, or other animals that will be used in testing.

Other assays include those for **potency, toxicology**, and **stability**. Potency assays are used to determine how the dosage of a drug affects its activity and how long it stays in the body. Toxicology assays show what quantities of the drug are toxic to cells, tissues, and model organisms. These studies help to determine the appropriate dosage for humans. Stability assays show the shelf life of a product (see Figure 6.12) and the proper storage conditions for the compound to maintain its activity. At what temperature, humidity, and light level should the product be stored? In what form should it be stored: liquid, powder, freeze-dried, capsules, etc?

Assays are performed at every step in the development of a product. Early in the R&D process, and especially if a company is small, assays are performed regularly at a researcher's lab bench. Small companies also send some samples out for testing by third-party companies. If a company grows large enough, as is the case at a company like Genentech, Inc., an entire department may be established for assay development and for conducting assays. Some companies have Assay Services and Quality Control Departments that specialize in testing company products.

The results of assays are important for the approval of drugs and products for the market. They are also important for patent acquisition and protection. Valid assays may be used to show the first development of a product or process.

Investigational New Drug (IND) application (In•ves•ti•ga•tion•al New Drug ap•li•ca•tion) a document to the FDA to allow testing of a new drug or product in humans

pharmacokinetic (PK) assay (phar•ma•co•ki•net•ic as•say) an experiment designed to show how a drug is metabolized (processed) in the body

pharmacodynamic (PD) assay (phar•ma•co•dy•nam•ic as•say) an experiment designed to show the biochemical effect of a drug on the body

potency assay (po•ten•cy as•say) an experiment designed to determine the relative strength of a drug for the purpose of determining proper dosage

toxicology assay (tox•i•col•o•gy as•say) an experiment designed to find what quantities of a drug are toxic to cells, tissues, and model organisms

stability assay (sta•bil•i•ty as•say) an experiment designed to determine the conditions that affect the shelf life of a drug

Section 6.2 Review Questions

1. What kind of assay would use Bradford reagent in the test?
2. For what purpose would a technician use an ELISA?
3. What does a stability assay measure?
4. In a large company, which department would have several employees developing and conducting assays?

One of the most important and frequently used molecular assays is the ELISA. An ELISA is very specific and will recognize a single type of protein or other antigenic molecule in a mixture of others. ELISA utilizes two important phenomenon: 1) antibody-antigen specificity to recognize only very specific molecules (see Figure 6.13), and 2) enzyme activity on colorimetric reagents for visualization purposes.

To understand how ELISAs work and how to conduct one, an understanding of antibody structure and function is needed. Antibodies, as discussed in Chapter 5, are complicated, four-chained proteins. The shape of an antibody is similar to a Y (see Figures 5.8 and 5.9). The tips of the Y recognize and bind only to certain molecules. The molecule that is bound by an antibody is called its antigen. Usually, an antibody recognizes one, and only one, kind of antigen, ignoring even closely related antigens. The bodies of mammals produce thousands of different antibodies (the tips of the Ys are different in each antibody) in response to all the foreign antigens entering the organism.

For example, all mammals have albumin protein in their blood sera. Each version of the albumin molecule is very similar in size, shape, and amino acid sequence. But they are not exactly the same. If a mouse is injected with cow blood, the mouse will produce antibodies that will recognize and bind to the cow blood albumin but not to the mouse's own blood albumin. These mouse antibodies are called mouse anti-cow albumin IgG. The abbreviation IgG stands for immunoglobulin G, a group of antibodies.

In an ELISA, the goal is to recognize the antigen and measure its concentration. The ELISA antibodies and their antigens are colorless and submicroscopic, so how is a technician to know if an antibody is present and bound to its antigen? To visualize the antibody and make the antibody-antigen binding recognizable, an enzyme that causes a color change in a specific substrate is attached to the antibody. If the antibody is specific to the particular antigen, the enzyme-carrying antibody will bind to the antigen. Then, when the enzyme-specific color substrate is added, a color change occurs at the site of the antibody-antigen binding. Thus, the presence and concentration of the antigen can be determined by the amount of color change (see Figure 6.14).

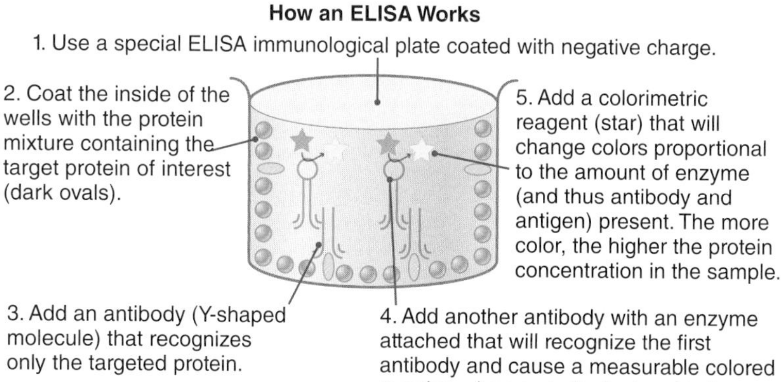

Two enzymes are commonly linked to the antibodies used in ELISAs and in Western blots (see Sections 6.4 for a discussion of Western blotting). One is horseradish peroxidase (HRP), which causes the colorimetric reagent tetramethylbenzidine (TMB) to change from clear to blue (see Figure 6.15). The blue color is not stable and the

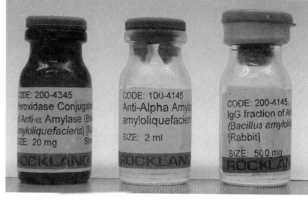

Figure 6.13. The anti-amylase antibody will recognize and bind to alpha-amylase produced by bacteria from the genus *Bacillus*. "Peroxidase-conjugated" means that the enzyme horseradish peroxidase is attached to the antibody. The HRP catalyzes a reaction that causes a color change, letting a technician know that the antibody is present. Look carefully at the three bottles. Each contains a different antibody. Be careful to order and use the correct antibody for the application.
Photo by author.

How an ELISA Works

1. Use a special ELISA immunological plate coated with negative charge.

2. Coat the inside of the wells with the protein mixture containing the target protein of interest (dark ovals).

5. Add a colorimetric reagent (star) that will change colors proportional to the amount of enzyme (and thus antibody and antigen) present. The more color, the higher the protein concentration in the sample.

3. Add an antibody (Y-shaped molecule) that recognizes only the targeted protein.

4. Add another antibody with an enzyme attached that will recognize the first antibody and cause a measurable colored reaction when an indicator is added.

Figure 6.14.

Figure 6.15. This 96-well plate shows an ELISA reaction near its endpoint, where the antigen-bound antibody HRP has turned TMB from clear to blue. The reaction must be stopped at some point with acid, turning the TMB from blue to yellow (see Figure 6.16).
Photo by Arun Asundi and Brian Woodall.

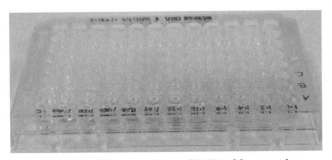

Figure 6.16.　The HRP conversion of TMB to blue must be stopped at some point with acid, turning the TMB from blue to yellow; otherwise all samples will eventually turn dark blue. The amount of yellow color, indicating the amount of antigen present, can be measured.
Photo by author.

HRP must be denatured to stop the reaction at some endpoint, so acid is added to the ELISA reaction. The acid turns the blue TMB to a semi-permanent yellow color (See Figure 6.16). The amount of yellow is dependent on the amount of antibody binding to the amount of antigen present.

A second enzyme that is commonly linked to ELISA antibodies is alkaline phosphatase (AP). Alkaline phosphatase catalyzes a combination of two substrates, nitro blue tetrazolium chloride (NBT) and 5-Bromo-4-Chloro-3-Indolyphosphate p-Toluidine salt (BCIP). The product is a dark purple-blue color. The amount of blue present indicates the amount of antigen bound by the antibody.

An ELISA is usually conducted in a plastic well plate coated to make the wells charged and attractive to proteins. These are called immunological plates. ELISA plates come with a variety of sample wells, but ELISAs are done most commonly in 96-well plates.

An ELISA may be used qualitatively (detecting the presence of a particular antigen) or quantitatively (measuring how much antigen is present). The presence of a protein can be discovered using an ELISA. A qualitative ELISA test is used in several forms of disease detection or screening. Let's say there are a number of blood samples to be tested for a specific virus antigen, for instance, "virus A." The procedures for the ELISA might be similar to the following:

- Blood samples are collected, and white blood cells are separated from the sample.
- The cells are lysed, and the cell lysates are added to a 96-well plate assay tray. The proteins from the sample, including any virus antigen proteins that are present, will stick to the plastic walls of a 96-well plate.
- The plates are washed with a buffer, such as PBS, to remove any excess protein.
- The plates are then washed with a blocking solution that contains some nonspecific protein, such as 5% bovine serum albumin (BSA) solution, to reduce nonspecific binding. The BSA protein attaches to any uncovered plastic. This decreases random antibody binding.
- The plates are washed again with a buffer to remove any excess BSA protein.

BIOTECH ONLINE

ELISA Technology in Diagnostic Kits
ELISA technology, in which enzyme-linked antibodies are used to detect other molecules, is used in several diagnostic kits. One of the most well-known uses of ELISA diagnostic kits is in home pregnancy testing.

Go to biotech.emcp.net/animpregtest and view the animated tutorial showing how a pregnancy test ELISA works. After watching the animated tutorial, answer the following questions:

1. What antigen does a home pregnancy test detect?
2. Monoclonal antibodies are used in home pregnancy tests. What are monoclonal antibodies and how are they made?
3. Polyclonal antibodies are also used in home pregnancy tests. What are polyclonal antibodies and where do they come from?
4. Describe how the two types of antibodies work together to give a positive pregnancy test result.

- The enzyme-tagged antibodies (antivirus A tagged with horseradish peroxidase) are added, and the plate is washed with buffer to remove excess antibody.
- The substrate to the enzyme, tetramethylbenzidine (TMB), is added to the wells. If antibody to the virus-A antigen has bound, the TMB will be oxidized and a blue color will become visible (see Figure 6.15). Any samples lacking the antigen of interest should have no antibody recognition and therefore should not turn blue.
- Acid is added to turn the blue TMB reactions to a yellow color, and it stops further reaction by denaturing the enzyme. The yellow color is stable and easier to measure (see Figure 6.16). Samples that are more yellow than negative control samples show the presence of the antigen.

In a quantitative ELISA, samples of know antigen concentrations are tested at the same time as unknown samples. The amount of yellow color in the known samples can be used to judge the yellow color (due to antigen concentration) in the unknown. Since the yellow color in an ELISA is hard to measure with the naked eye, spectrophotometers are used to measure the absorbance of the yellow ELISA product. Samples could be read individually, but to make the process more efficient, ELISA plate readers are available that use a spectrophotometer to read the absorbance of each of the 96 wells in a plate (see Figure 6.10 and 6.11).

Finding an antibody to recognize a specific antigen is one of the challenges of developing an ELISA. Generally, antibodies for ELISAs and Western blots are produced by injecting an animal with an antigen of interest and then waiting for the animal to produce antibodies to the antigen in sufficient quantity that it may be harvested from the animal's blood serum.

Antibodies that are produced in this way and recognize an antigen directly are called primary (1°) antibodies. If an antigen is of significant research or production interest, a recognition enzyme, such as HRP, may be conjugated to it and it may be marketed and readily available. An ELISA that uses a 1° antibody with a conjugated enzyme to directly recognize an antigen is called a **direct ELISA** (see Figure 6.17).

More commonly, 1° antibodies are not conjugated with an enzyme since the market for an antigen's antibody is not that great. Instead, the primary antibodies for a variety of antigens are recognized by a second group of "all-purpose" IgG antibodies. The all-purpose antibodies are called 2° antibodies because they are the second antibody to bind in the recognition complex (antigen - 1° Ab-2° Ab). The secondary antibody has the conjugated enzyme that produces the colorimetric reaction in this **indirect ELISA** (see Figure 6.17).

For commonly conducted ELISA screening, the plates can be ordered already coated with antigen or antibody. For example, human immunodeficiency virus (HIV) screening is conducted using a HIV ELISA that has antibodies already coated onto the plates. Blood samples that may contain a person's antibodies to HIV exposure are added to the wells. If the ELISA antibody recognizes an anti-HIV antibody in a person's blood, it means that the person has been exposed to HIV.

An ELISA is a very sensitive analytical tool. It can be used to ensure the presence and concentration of a product throughout an entire research and manufacturing process. When developing vaccines to protect against HIV infection, the recombinant version of glycoprotein 120 (gp120) is produced in large quantities for eventual vaccine testing. As the purification process of gp120 is worked out, samples must be

direct ELISA (di•rect E•LI•SA) an ELISA where a primary antibody linked with an enzyme recognizes an antigen and indicates its presence with a colorimetric reaction

indirect ELISA (in•di•rect E•LI•SA) an ELISA where a primary antibody binds to an antigen and then a secondary antibody linked with an enzyme recognizes the primary antibody—the antigen presence and concentration is indicated by the degree of a colorimetric reaction

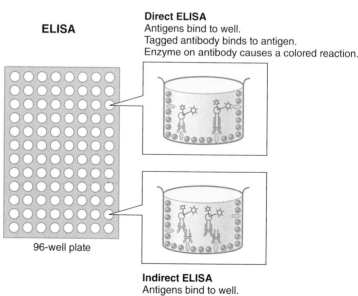

Direct ELISA
Antigens bind to well.
Tagged antibody binds to antigen.
Enzyme on antibody causes a colored reaction.

ELISA

96-well plate

Indirect ELISA
Antigens bind to well.
Primary antibody recognizes antigen.
Tagged secondary antibody recognizes primary antibody.
Enzyme on secondary antibody causes a colored reaction.

Figure 6.17. **ELISA.** ELISA antibodies can recognize antigens directly or by recognizing another antibody that recognizes the antigen.

Figure 6.18. **Molecular Biology Laboratory. Detecting pesticides in food (cereals) with ELISA (enzyme-linked immunosorbent assay) technique. AZTI-Tecnalia. Technological Centre specialised in Marine and Food Research. Sukarrieta, Bizkaia, Euskadi. Spain**
© age fotostock/SuperStock.

tested for the concentration of gp120 at every step in the process. Samples are sent to assay services, where ELISAs are conducted to determine the concentration of gp120 in solution. The degree of yellow color in the assay samples (absorbance of light) is measured on a spectrophotometer. The more yellow, the higher the gp120 concentration in the sample (the more enzyme-bound antibody to the gp120 antigen). The concentrations are determined by comparing the absorbance measurements with a set of standards.

ELISAs can also be used to look for contamination in samples. (see Figure 6.18) The ELISA reaction is so specific that the antibodies recognize the difference in variant forms of a protein. Let's say that a ground-beef sample is suspected to be contaminated with pork meat. Proteins from the ground-beef sample could be extracted and tested with anti-pig antibodies. Positive ELISA results would indicate that the sample contained pork. In fact, commercial kits are available for testing raw meat and poultry using this technology (**biotech.emcp.net/elisa-tek**).

ELISAs are routinely used in testing for allergens in foods. Having the ability to test for food allergens is a life-and-death matter for some people. Certain foods can cause mild reactions, such as a skin rash. Others can cause serious reactions, such as anaphylactic shock (swelling of body tissues, difficulty breathing, and possible loss of life). A food that causes a severe allergic reaction in a great number of people is peanuts. One of the major allergenic proteins in peanuts is Ara h1, a form of the protein arachine. Several peanut-detection kits are on the market, including one from R-Biopharm AG that tests for the presence of Ara h1 using antibodies. Their RIDASCREEN® FAST Peanut assay detects peanut contamination in food down to 1.5 ppm (parts per million). It is not an exaggeration to say that this kind of assay may save lives (**www.r-biopharm.com/products/food-feed-analysis/allergens/nuts/item/ridascreenfast-peanut**).

Section 6.3   Review Questions

1. Explain how antibodies and enzymes are used in ELISAs.
2. How can a technician know that an antigen is present during an ELISA?
3. How can a technician know the concentration of an antigen in an ELISA?
4. What is the difference between a direct and an indirect ELISA?

6.4 Western Blots

Running samples on PAGE gels can provide a great deal of information about the proteins in a sample. A technician may be fairly confident that a sample contains a protein of interest if it turns up as a band at the "right" molecular weight on a gel. If, for example, the enzyme amylase, with a molecular weight of about 60 kDa, is expected to be in a sample and a band is found on the gel at 60 kDa, one might think that the band is, indeed, amylase. Although it is likely that a band at the right spot on a gel is, indeed, the protein of interest, many different proteins have similar molecular weights. There are hundreds of proteins with molecular weights around 60 kDa. If the sample comes from a cell culture or a cell lysate, the chances of having several proteins of the same molecular weight is high. A technique is needed that can preferentially detect the protein of interest, and only the protein of interest, from other similar proteins on a gel.

To determine whether a protein band on a gel is actually the protein of interest, a technician can use a technique called Western blotting. In a Western blot, samples are run on a PAGE gel and then the protein bands are transferred to a blotting membrane (**PVDF** or **nitrocellulose**). The membrane is called a blot. The protein bands on a blot are colorless until they are visualized. Like in an ELISA, antibody recognition of an antigen (the protein of interest) is used to distinguish the protein of interest from other proteins that have transferred to the blot membrane. Using an enzyme linked onto an antibody, the blot protein bands can be colorized (see Figure 6.19). The visualization of a Western blot can confirm the presence of a particular protein at very low concentrations, often lower than concentrations visible on a Coomassie Blue stained gel.

Traditionally, a Western blot is conducted submerged in buffer in a "transfer cell." The transfer cell is placed in a gel box to utilize an electric field to force protein bands from the gel onto the blotting membrane (see Figure 6.20). A sandwich of gel, blotting paper or membrane, filter paper, and sponges is placed in the transfer cell (see Figure 6.21). The whole thing is submerged in buffer (it is called a "wet" blot) and run between 25 and 100 V for 1 to 3 hours, depending on the protocol. During that time, the peptide bands move from the gel to the blotting paper.

Once a blot has been completed and the sample is transferred to a blotting membrane, it is visualized. Like an ELISA, a direct visualization (1° antibody-enzyme) or indirect visualization (1° antibody-2°antibody-enzyme) may be used. A typical visualization includes using primary and secondary antibodies to recognize the protein of interest. The primary antibody binds to the protein of interest. The secondary antibody binds to the primary antibody. The secondary antibody has either a dye or an enzyme (HRP or

Figure 6.19. A blot is a transfer of molecules from a gel to a membrane or special paper. Often it is easier to test molecules on a blot than to test samples on a gel. On this blot, different concentrations of amylase have been recognized by an anti-amylase antibody conjugated with HRP. HRP converts TMB to a blue product that deposits on the protein sample in an amount proportional to the amylase concentration. Photo by author.

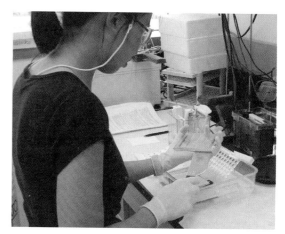

Figure 6.20. Several companies sell transfer cells that can be used in gel boxes for this purpose. This is called a wet transfer because the gel, blotting membrane, and sponges are all submerged in buffer. Recently, semidry and dry blotting techniques have been developed. Photo by author.

Western Blot Gel/Transfer Membrane Sandwich

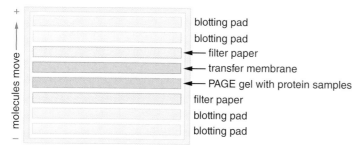

blotting pad
blotting pad
filter paper
transfer membrane
PAGE gel with protein samples
filter paper
blotting pad
blotting pad

molecules move

Figure 6.21. Western Blot Gel/Transfer Membrane Setup Diagram. During a Western blot, electrical current carries protein bands from the PAGE gel to the blot transfer membrane.

PVDF (P•V•D•F) polyvinylidene fluoride, a high molecular weight fluorocarbon with dielectric properties that make it suitable for Western blots because it is attractive to charged proteins

nitrocellulose (ni•tro•cell•u•lose) a modified cellulose molecule used to make paper membrane for blots of nucleic acids and proteins

Figure 6.22. **An image of a Western blot membrane is analyzed on a computer. Each black blot represents a protein sample that has been recognized by the primary antibody and colorized by an enzymatic reaction.**
© Vo Trung Dung/Getty Images.

AP) that can convert a colored substrate. In this way, the secondary antibody "colorizes" the whole complex, and a colored band can be seen on the membrane wherever the protein of interest is found (see Figure 6.22).

For example, once samples suspected of containing amylase are run on a gel, the gel proteins are blotted onto the membrane and they bind. In a dish, the membrane is covered with blocking solution (a mixture of buffer and the blocking protein BSA). The blocking solution coats the membrane with BSA protein. This ensures a minimum of nonspecific binding of antibody to the membrane itself. After blocking, the primary antibody to amylase (anti-amylase) is added and the membrane is incubated for about an hour. This gives the antibody time to find the amylase molecules on the membrane and bind with them. The membrane is washed with buffer several times to get rid of any unbound primary antibody. The "tagged" secondary antibody is then added to the membrane.

The secondary antibody (usually an IgG from an animal such as a goat, pig, etc) is tagged with an enzyme (such as horseradish peroxidase) that changes the color of a substrate. The secondary antibody recognizes and binds to the anti-amylase antibody. Again, the membrane is washed with buffer several times to rid it of any unbound antibody. The color substrate (TMB for HRP) is added. The TMB changes from clear to blue. The blue molecules drop onto the membrane at that point and "stain" the amylase molecules. If the conjugated enzyme is alkaline phosphatase, then when the antibody is bound, BCIP and NBT are added and the AP combines the BCIP and NBT, changing their color from yellow to purple-blue and the protein bands appear purple-blue.

An alternative method of completing a Western blot transfer is called "semi-dry transfer." A semi-dry transfer is conducted in a semi-dry electrophoretic transfer cell. Semi-dry electrophoretic transfer cells are set up like a gel box but without a buffer chamber. A Western blot sandwich of membranes, PAGE gel, and absorbent papers are set up in a similar orientation as in a traditional Western blot, minus the cassette holding it all together. The sandwich is laid down on the transfer cell platform and the excess buffer is wiped up. The top of the cell is attached and the Western blot transfer is run for 1–2 hours. Visualization of the proteins on the blot is essentially the same. A good video showing how a semi-dry blot is set up can be found on YouTube at **www.youtube.com/watch?v=krFKPwhiHJE**.

Blotting technology is used throughout the biotechnology industry for research and development (R&D) purposes. When manufacturing and purifying proteins, for example, blotting is a standard method of confirming the presence of the protein of interest anywhere in the process. For example, broth samples thought to contain a protein product may be collected from fermenters and tested using ELISA or Western blotting. Similarly, as protein product is purified after manufacturing, ELISAs and Westerns are performed to confirm protein concentration. These assays are run to verify that the manufacturing process meets its protein production objectives.

Section 6.4 Review Questions

1. What kind of molecule is being blot during a Western blot?
2. What causes molecules to move from a gel to a membrane in a Western blot?
3. How are blotted molecules visualized during a Western blot?
4. For what purpose are Western blots used in industry?

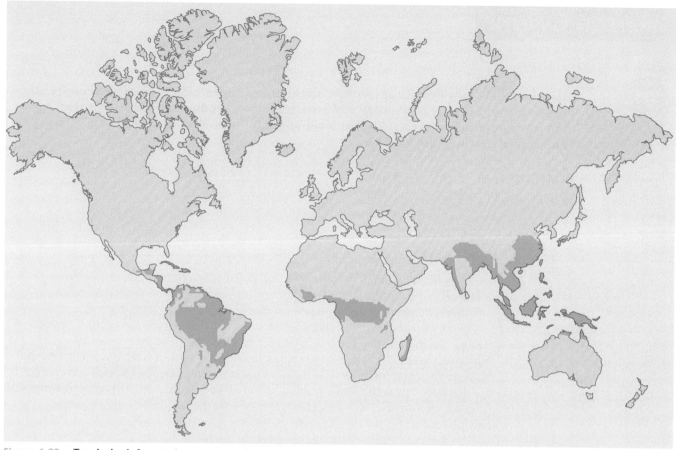

Figure 6.23. **Tropical rainforests (green areas of map) are dwindling so rapidly that some experts think they may be gone in the next 50 years. What cures for disease may be destroyed with the rainforests?**

6.5 Looking for New Products in Nature

You may have heard that the world's tropical rainforests are disappearing at an alarming rate of about 50 to 100 acres per minute (see Figure 6.23). Of major concern is the fact that we are not even sure what potential new drugs humans are destroying as vast areas of plant and animal diversity are wiped out.

Scientists estimate that there are a million different plant species, of which only 25% have been identified (see Figure 6.24). Most of these are found in the equatorial, tropical rainforests. Of those identified, only a very small percentage have been studied in any detail. How many of these unstudied plants may contain chemicals beneficial to humans as medicines, pesticides, herbicides, or other applications? Some plant biotechnology companies have recognized that rainforest plants must be studied in the hope of finding products helpful to mankind. In the 1990s, a small company in South San Francisco, California, Shaman Pharmaceuticals, Inc., focused on **herbal remedies** from the rainforest. Shaman sent scientists to the Amazon and found that, for centuries, the natives had been using an extract of the *Croton lechleri* tree as a treatment for diarrhea.

Diarrhea is a very serious ailment throughout the world. Water loss from diarrhea can be so severe

herbal remedies (herb•al rem•e•dies) the products developed from plants that exhibit or are thought to exhibit some medicinal property

Figure 6.24. **A tropical rainforest is composed of layers of plants blanketing equatorial regions. Each layer of plants houses hundreds of species of plants and animals that produce many unique compounds.**
© Wolfgang bogdanhoria/iStock.

Amazon Hide and Seek

Use the Pax Natura website at www.paxnatura.org/RainforestPreservationBenefits.htm to learn about efforts to identify and commercialize products while still protecting the Amazonian rain forests. Find at least one other websites that discuss the issues and record that website's URL.

1. How are scientists and government officials learning about potential rainforest products?

2. Identify five protein rainforest products and their uses or applications.

3. What is your opinion about future business opportunities in rainforest biotechnology product manufacturing?

that one can die from dehydration. In fact, diarrhea is a major killer of children in underdeveloped countries. Several agents, including bacterial or viral infection, radiation or chemotherapy, and several prescription drugs, cause diarrhea.

Scientists from Shaman Pharmaceuticals, Inc. extracted the sap of the *C. lechleri* tree for use in an herbal remedy called SB-Normal Stool Formula™. An active ingredient in the sap appeared to inhibit the flow of chloride ions and water into the bowels. Thus, the patient would lose less water into the bowels, thereby correcting the diarrhea. The result was a normalization of stool formation. Although the product is no longer available, it is a good example of how the rainforest might provide new pharmaceuticals.

Based on leads from native people or other scientists, botanists collect hundreds of plant, animal, and fungi samples from the rainforest looking for promising products or therapeutic agents. Each sample must be prepared, purified, and screened for its potential to meet some need. Scientists develop tests to determine the possible active ingredients in plants. A common method involves extracting soluble chemicals from plants using a solvent, such as distilled water, alcohol, or acetone, and then testing the extract to determine its activity. Hundreds of herbal remedies derived from a variety of natural sources are currently on the market (see Figure 6.25). Table 6.2 lists several conditions and some of the herbal therapies used to treat them.

nutraceutical (nu•tra•ceu•ti•cal)
a food or natural product that claims to have health or medicinal value

Recently there has been considerable interest in the possible antimicrobial activity of many plant extracts. Since bacteria cause so many illnesses and diseases, and so many species of bacteria are developing resistance to currently available antibiotics, companies are focusing their efforts on finding new, improved antibiotics. Antibiotics and antimicrobial agents are discussed in Chapter 12.

Many of the herbal remedies being developed or that have been marketed come from food plants and have been called nutraceuticals. The term **nutraceutical** is used by manufacturers to suggest that certain herbs or their extracts may have nutritional and/or pharmaceutical benefits. Nutraceuticals and their use and marketing have raised several controversies because nutraceuticals are not required to go through the same FDA testing and approval process as pharmaceuticals to demonstrate their safety or efficacy. Critics say there is no reliable evidence that these products do what they say they

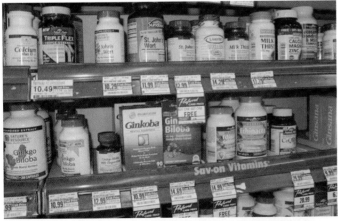

Figure 6.25. Most supermarkets, drug stores, and health food stores offer a large assortment of herbal products. Exercise caution when using these products, since the active ingredients are molecules that could have adverse effects if taken at the wrong concentration or in combination with some other compound.
Photo by author.

Table 6.2. **Herbal Therapies/Remedies for Some Common Ailments**   Note: Many of these remedies have not undergone clinical trials overseen by the Food and Drug Administration (FDA). Do not try to treat yourself with any of these products. They are listed for educational purposes only. Use therapeutics agents only as directed by a licensed professional or physician.

Condition	Herbal Therapy/Remedy
acne	lavender oil, tea tree oil, vitamins A and E
angina	hawthorne berry, Khella, choline, garlic
arthritis	chaparral, burdock root
asthma	*Ephedra sinica*
atherosclerosis	chromium, inositol
bronchitis	eucalyptus oil, licorice, *Echinacea*
burns	aloe, St. John's Wort, *Echinacea*
high cholesterol	lecithin, vitamin-B complex
cancer	Taxo®l by Bristol-Myers Squibb Co (paclitaxel from Yew trees)
common cold	vitamin C, *Echinacea* (recently shown to be ineffective)
colitis	cascara bark, psyllium, slippery elm
colic	fennel seed, chamomile, Valerian
constipation	cascara bark, papaya
cramps	blue cohosh, black cohosh, Belladonna, lobelia, Valerian
depression	pantothenic acid, magnesium, St. John's Wort, vitamin-B complex, ginkgo
diabetes	*Vinca rosea*, ginseng, geranium oil, plantain, knotgrass, cinnamon bark, patchouli, etc
diarrhea	mallein, yarrow, lotus stamins, cranes bill, oak bark, comfrey root
dry skin	aloe, St. John's Wort
endometriosis	vitamin E, goldenseal, dong quai, shatavari
low energy	bee pollen
fever	willow bark, feverfew, blessed thistle, oak bark, raspberry leaves, rehmannia
gastritis	aplotaxis amomum, aplotaxis carminative, yarrow
gas	carbo vegetables
hair loss	nettles, gotu kola, inositol
headache	feverfew
heart disease	hawthorne berry, foxglove, chromium, coq10, garlic, *Astragalus*, selenium
hemorrhoids	butcher's broom, oak bark, cascara bark, goldenseal, calc fluor
high blood pressure	cayenne pepper, barberry garlic, cascara bark, coq10, ginkgo, lecithin, Valerian
hypertension	hawthorne berry, lavender oil, skull cap, *Vinca rosea*, choline
indigestion	chamomile, nux vomica, ginger barberry, cardamon, parsley, yerba buena, slippery elm
injury	arnica hypericum, perforatum, xiao huo luo dan, *Echinacea*, arsenicum album, goldenseal
infection	ura ursi, thyme oil, tea tree oil, eucalyptus oil, barberry, garlic, lavender oil
leukemia	*Vinca rosea*, chaparral, colchicum, black cohosh, *Echinacea*, red clover
liver function	thisilyn, milk thistle, cascara bark, vervane, choline, vitamin B
muscle building	chromium picolinate
obesity	*calcarea carb*, coq10
periodontis	calcium, coq10, willow bark, St. John's Wort, colocynthis
premenstrual syndrome (PMS)	butiao, evening primrose oil
retinitis	coq10, vitamin A, black cohosh, sarsaparilla, chaparral, nettles
rheumatism	burdock root, calamus, cockle bur, willow bark, Valerian root
sex function	Damiana, hops, yohimbe, licorice, slippery elm, oak bark, *Apis mellifica* (royal jelly)
sore throat	juniper berries, hyssop, hawthorne berry
sunburn	aloe
sinus congestion	*Echinacea*, goldenseal, osha root
duodenal ulcers	juniper oil, birch oil, saimeian, frankincense oil
urinary infection	*Uva ursi*, shi lin tong, cranberry, *Echinacea*

do since they have not been tested properly. Supporters of nutraceutical products claim that the overwhelming anectodal evidence is enough to demonstrate safety and effectiveness. Currently, many nutraceutical manufacturers are addressing this issue by putting their products through the stringent testing regimes required for FDA approval.

Some herbal medical remedies and nutraceuticals have passed clinical trials and have been approved by the FDA. One such medication is vincristine, which is extracted from blue periwinkle plants. Vincristine is a chemotherapy used to treat many cancers, including some breast cancers, lymphomas, myelomas, leukemia, head and neck cancers, and soft tissue cancers. Vincristine sulfate was approved by the FDA in 2014 and is being manufactured by a a company called Hospira, which was just recently purchased by the large pharmaceutical company, Pfizer.

Section 6.5 Review Questions

1. From where do scientists expect that most of the remaining naturally occurring biotechnology products will come?
2. List a few herbal products that claim to have therapeutic value against depression.
3. How can molecules be extracted from plant samples for testing purposes?

6.6 Producing Recombinant DNA (rDNA) Protein Products

transfection (trans•fec•tion)
the genetic engineering, or transformation, of mammalian cell lines

formulation (for•mu•la•tion)
the form of a product, as in tablet, powder, injectable liquid, etc

Biotechnology became a common term when the first rDNA product was synthesized in the mid-1970s. Recall that recombinant refers to the process of combining DNA from two sources. Recombinant DNA containing a gene (or genes) for the production of a protein of interest is added to a laboratory cell line, such as *E. coli* cells or CHO cells (see Figure 6.26). When cells take up the foreign rDNA, the goal is that they start reading the DNA and making the new protein. The "transformed" cells, making the protein of interest, are scaled-up to large volumes, and the proteins are purified and sold as products (see Figure 6.27). Transformation that occurs in mammalian cells is called **transfection**. Hundreds of rDNA protein products, made in transfected laboratory cell lines, are currently on the market or in development.

Figure 6.26. **Chinese hamster *(Cricetulus griseus)* ovary (CHO) cells are grown in liquid media in bottles. Because they are easy to grow and transform with rDNA, they are one of the mammalian cells of choice for large-scale recombinant protein production. The media used in this lab has an indicator in it to show changes in pH as the cell culture grows. As more cells fill the culture, wastes builds up and the pH drops, turning the red indicator to yellow. At that point, the culture needs to be used or scaled-up to a larger volume.**
Photo by author.

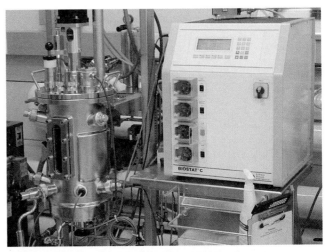

Figure 6.27. **This 10-L stainless steel bioreactor, or fermentation tank, contains CHO cell culture producing recombinant proteins for market. When the culture of genetically engineered cells reaches carrying capacity (and cells begin to get crowded), it is used to seed a larger-volume fermentation tank, such as a 100-L tank.**
Photo by author.

It takes approximately 15 years to develop, test, and market a typical rDNA protein product before it can be brought to market. Preliminary testing and designing a development process may take up to 10 years. During this time, a potential product must be identified. Then, assays to recognize its presence and activity must be developed. The next step is to determine genetic engineering and production methods for the particular protein. This can be complicated, and several protein products may be stalled as scientists try to determine what cells to transform and how to regulate the genes once they are in the new cells. Next, a purification process must be developed so that the recombinant protein product can be isolated from all other molecules that the production cells will synthesize. All these steps are part of the R&D of a protein product made through genetic engineering.

The R&D procedures are then scaled-up to manufacturing volumes, and it may take several years to produce volumes of protein large enough for the intended market. During manufacturing, pharmaceutical proteins must be produced under the supervision and rules of the Food and Drug Administration (FDA). Manufacturing must meet the FDA's Good Manufacturing Practices (GMP) guidelines to ensure safety and purity. The proteins produced during scale-up are purified and formulated into the final version to be used by customers (see Figure 6.28).

The final form a pharmaceutical protein product takes may be a tablet, an injectable liquid, an aerosol inhaler, a patch, or a cream, etc. The final **formulation**, or form of the product, must be tested during clinical trials. Clinical trials are also guided by FDA regulations and may take 3 to 5 years to complete. If a product demonstrates safety and efficacy (effectiveness) during the clinical trials, the FDA can approve it for marketing and sales.

Every pharmaceutical product goes through a similar product pipeline. Imagine the number of scientific and nonscientific staff involved in pipeline work and the number of years it takes to bring a product to market. It is no wonder that pharmaceuticals cost so much considering the expense of materials and the number of employees in the R&D, manufacturing, testing, legal, regulatory, marketing, and sales departments (see Figure 6.29). According to DiMasi and Grabowski of Tufts Center for the Study of Drug Development the R&D costs associated with the discovery and development of new therapeutic biopharmaceuticals (specifically, recombinant proteins and monoclonal antibodies) is over $600 million.

The cost of bringing a biotechnology crop to market is equally impressive. Crop Life International estimates the time it takes to develop and market a new biotechnology crop at about 13 years with a cost of approximately $136 million. In spite of these

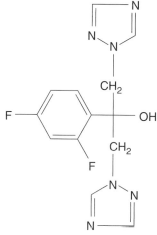

Figure 6.28. **Fluconazole Structure.** Fluconazole is the active antifungal agent in Diflucan® pills, which are used to treat yeast infections throughout the body. Prior to the availability of Diflucan®, yeast infections were treated with antifungal creams. Due to varying customer needs, a biotechnology company must make the best decisions for how a new product will be formulated and used.

Figure 6.29. **As a compliance specialist in the Quality Assurance Department at a pharmaceutical manufacturing company, Dina Wong spends most of her time writing audit reports and standard operating procedures (SOPs), training personnel, or in project team meetings to discuss and communicate any issues that need to be resolved.**
Photo courtesy of Dina Wong.

BIOTECH ONLINE

Quality ~~not~~ and Quantity

The Quality Assurance (QA) Department at a pharmaceutical company is responsible for meeting all the regulatory guidelines by the Food and Drug Administration. For most biotechnology products, ISO 9000, an international standard of policies for products and service, is used to guide QA.

Learn more about quality assurance in the manufacture of biotechnology products.

1. Find two websites that discuss what ISO 9000 is. Then create a summary description of ISO 9000.

2. Describe why ISO 9000 is important in biotechnology manufacturing. Give an example of how and where ISO 9000 standards would be important in the manufacture of a biotechnology instrument or therapeutic drug.

astronomical costs, between 2008–2012 over 6,000 new biotech crop products were in some stage of research and manufacturing.

Even with the cost of bring biotechnology products to market, the biotechnology industry is growing and substaining a global workforce. SelectUSA estimates that in 2012 more than 800,000 people worked directly in the U.S. biopharmaceutical industry and the products and services of biotechnology support nearly 3.4 million jobs in the U.S. economy. The numbers of global biotechnology employees is about double that amount and the future looks bright for those that hope to have a career in the science or business of biotechnology.

Section 6.6   Review Questions

1. What are CHO cells and what are they used for?
2. How long does it take to develop, test, and market a typical rDNA protein product?
3. What does GMP stands for and what does it cover?
4. Biotechnology products must be formulated before they can be marketed. Name two formulations for a pharmaceutical product other than tablet form.

Chapter Review

Summary

Concepts

- Biotechnology products come from many sources, including whole organisms, organs, cells, and molecules. Some products are found in nature, and some are synthesized in the laboratory in cells or test tubes.
- Antibiotic research and manufacturing are rapidly increasing as the challenge of combating antibiotic resistance increases.
- Approximately 20% of marketed biotechnology products are antibodies used in research and therapeutics.
- Amylase was one of the first recombinant protein products on the market. Amylase breaks down starch to sugar, a process important to many industries.
- Although amylase is made in nature, the increased volume needed for industry requires biotechnology manufacturing using transformable cells, such as *E. coli*.
- Many companies have a CPDP. A product is evaluated in light of the criteria in the CPDP. If the product does not appear to meet the criteria, production may be halted.
- The CPDP may include such questions as: Does the product meet a critical need? Who will use the product? Is the market large enough to produce enough sales? How many customers are there? Does preliminary data support that "it" will work? Will it do what the company claims? Can patent protection be secured? Can the company prevent other companies from producing it? Can the company make a profit on the product? How much will it cost to make it? How much can it be sold for?
- An assay is a test to detect some characteristics of a sample. Several assays are used to assess a biotechnology product, such as a protein. Assays might include tests for presence, concentration, activity, pharmacokinetics, potency, toxicology, and stability.
- ELISA stands for enzyme linked immunosorbent assay and is a technique to measure the presence and concentration of an antigen in solution.
- ELISAs visualize proteins using antibodies that recognize and bind to an antigen. Also attached to the antibody is an enzyme that can colorize a substrate. The amount of color is related to the amount of antibody that is bound to the antigen of interest.
- ELISAs are used in research and manufacturing of biotechnology products and are used in diagnostic kits.
- Specific proteins on a PAGE gel can be verified by doing a Western blot.
- In a Western blot, a PAGE gel is run, and the separated proteins are transferred to a blot membrane. The blotted membrane is probed with an antibody to the antigen conjugated with a visualization enzyme. The enzyme can colorize a substrate, depositing blue color on the protein of interest.

- The tropical rainforest is believed to be a great source of future biotechnology products. Since the rainforest is disappearing quickly, several ecological and scientific obstacles to discovering potential products need to be addressed.
- Natural products may have a historical record of some benefit. Scientists use local populations to lead them to prospective products to investigate. The active ingredients in the products must be identified, isolated, and tested for safety and efficacy to be approved by the FDA.
- Herbal products may have some beneficial value, but they have not been approved by the FDA.
- Antimicrobials are compounds that inhibit or kill microorganisms. Antimicrobials include antibiotics and antiseptics. The activity of different antimicrobial agents may be tested by growing plate cultures of certain bacteria in the presence of disks of each agent.
- If a mammalian cell is genetically engineered, it is called transfection. It is common for CHO cells to be used for transfection because they are easy to grow in cell culture.
- Recombinant proteins for pharmaceuticals must be produced following the FDA's GMP guidelines.
- Products are formulated into tablets, sprays, injectable fluids, inhalants, creams, or patches.

Lab Practices

- An ELISA is a type of concentration assay. Using an antibody that recognizes only one specific molecule, an ELISA causes a color change, the degree of which is dependent on how much of the ELISA antibody's antigen is present.
- Extracted proteins must be assayed through indicator testing, polyacrylamide gel electrophoresis (PAGE), ELISA, or Western blots.
- An amylase activity assay could test the decomposition of starch or the production of sugar.
- Iodine is the indicator of starch (amylose) and changes to blue-black in its presence.
- Benedict's solution is an indicator of aldose sugars and changes from blue, to green, to yellow, or orange, depending on the amount of aldose in the sample.
- Commercial test strips can be used to assay the concentration of glucose in a sample.
- A multichannel pipet can increase the accuracy of an experiment with many multiple replications.
- To test for antimicrobial activity of extracts, growing bacterial cultures are exposed to extract-treated filter-paper disks, so that zones of inhibition can be observed.
- The antibiotic, ampicillin, can be used as a positive control for the antimicrobial activity assay since E. coli is sensitive to it. Methanol is a known antiseptic agent and also is a good positive control.
- Bacteria and fungi that demonstrate amylase production can be found in nature. These organisms will produce clear halos around colonies when grown on starch agar.
- Naturally occurring amylase-producing bacteria or fungi could be used as commercial amylase producers, if they could be identified and characterized as safe, high-yield producers.
- It might be easier to transform E. coli into an amylase producer than to characterize an unknown bacterium.

Thinking Like a Biotechnician

1. Name a naturally occurring amylase producer that might be a source of an amylase gene.
2. A company is producing a diabetes medication with a new form of insulin. Part of the development includes designing multispecies PK/PD assays. What are multispecies PK/PD assays, and why do they need to be developed in a human insulin product pipeline?
3. Conducting an ELISA can be cumbersome because there is so much pipeting involved. Suggest some instruments that might make an ELISA easier to conduct.
4. Propose a method for testing the effectiveness of an antiseptic, such as rubbing alcohol, on inhibiting E. coli growth.
5. Amylase breaks down starch to sugar. What indicators could be used in an activity assay to see if a sample contains some active amylase?
6. If a cell extract is thought to contain a specific protein, how might a technician check to see if the protein is present?

7. An ELISA is done on a bacteria broth culture known to have a fairly high concentration of Protein A, based on PAGE analysis. When the ELISA shows color in all samples including the negative control, how should these unexpected results be interpreted?
8. Herbal therapies contain molecules that some people consider safe and effective. What argument can be made against herbal therapies being safe and effective?
9. A CPDP committee recommends that the development of a product should be halted because an injectable liquid formulation has a short shelf life. After many years of researching and developing the product, you are disappointed. What suggestion might you make?
10. Propose how antibody therapy could be used to treat someone who is allergic to peanuts.

Biotech Live

Exploring Potential Products

Activity 6.1

Before companies pick potential products to research and manufacture, scientists and technicians must become thoroughly knowledgeable about what is already known about the product or process. They spend many hours researching every resource related to their subject. They usually begin with a literature search, gathering all relevant books, journals, and reports. They must learn what is already known and what experiments have already been done before they can begin their own R&D.

As scientists gather information, they review it while considering the CPDP. A product that will be accepted into a company's pipeline must meet a market need, already have preliminary data that show it works, be able to be patent-protected, have enough customers over a long term to produce substantial sales, and be able to produce a substantial profit after it is on the market. A company will also review its research facilities and employees' skills before accepting a potential product in its pipeline.

Photo by author.

Gather enough information to recommend a potential product's placement in a company's pipeline.

Imagine that you work in a biotechnology lab. Select a product from the table below or from others presented by the instructor and gather data to complete a CPDP review.

HT (hyola triazine) tolerant canola	protects canola plants from insects
REVLIMID® (lenalidomide)	treatment for anemia and some blood cancers
Purafect® protease	protein-digesting enzyme
Endless Summer® tomato	slow-ripening tomatoes
bovine somatotropin	growth hormone for livestock
Shimadzu Protein Sequencer PPSQ-31B/33B	amino acid sequencer
ECOPULP®TX enzymes	paper processing treatment
besifloxacin	for "pink-eye" treatment

1. Conduct an exhaustive search using the usual reference resources (Internet, scientific journals, indexes, databases, etc) to access information. Use a search engine or a database, such as **biotech.emcp.net/bio**, to determine which company is currently producing the product or a similar product.

2. Construct a poster that addresses all of the criteria of the CPDP. Include the elements shown in the illustration below.
3. Prepare a 5-minute presentation offering evidence of how the potential product meets the criteria for acceptance into the pipeline. Make recommendations for accepting or rejecting the product into the pipeline.

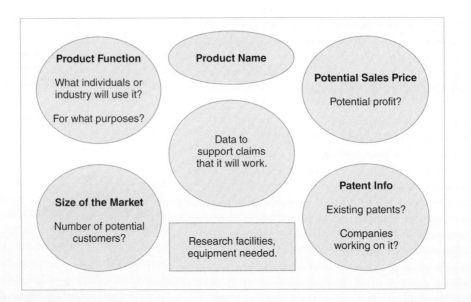

Activity 6.2

Amylase Three-Dimensionally

Development of this activity was done with the assistance of Sandra Porter, Ph.D., President, Digital World Biology LLC, www.DigitalWorldBiology.com

In earlier chapters you learned how the structure of DNA molecules is studied using the Molecular Modeling DataBase (MMDB). MMDB is actually just one of the databases of three-dimensional structures. The RCSB Protein Data Bank (PDB) is another. The PDB assigns a code of 4 letters and numbers to identify the data and images for a specific biomolecule. If a researcher finds a molecule that has been added to the MMDB and has a PDB code, its structure can be viewed by one of the molecular modeling applications such as Cn3D or Molecule World.

In this activity, you will learn how to view and study a protein (alpha-amylase from a species of bacteria) and compare it to an engineered version of the protein. You will use the Cn3D application, from NCBI on your computer. Alternatively, the Molecule World application, available through iTunes for the iPad or iPhone, may be used to view structural models.

 Use the Cn3D application to view a protein structural model and compare it to another similar protein's structure.

1. Make sure Cn3D application has been installed on computer. Click on the application icon to open Cn3D. Structures that have a PDB code will be imported and displayed by Cn3D.
2. Get the amylase structural data.

   - Go to www.ncbi.nlm.nih.gov. Change from "All Databases" search to "Structure."
   - Type in the name of the amylase molecule "**alpha amylase Bacillus licheniformis**" and hit "Search." Scroll down the page and find the entry for a 3D model of amylase with the PDB ID# "1BLI". When you click on the entry, the structural data for a wild-type "*Bacillus licheniformis* alpha-amylase" and its molecular interactions will appear on a MMDB Structure Summary page. The structural image is an alpha-amylase molecule with cofactor ions (3 Ca^{2+} and 1 Na^+). Scroll down the page to see what information is available. Record the URL for this website.

- Look on the right side of the page for the "View or Save 3D Structure" box. Click on "View structure." This will download the structural image to your computer and when it is opened, it will appear in Cn3D. Confirm it is the correct structure by checking that "1BLI" is in the title of the image file. When the file is open, it is in the most basic image style showing only the single polypeptide chain backbone of the amylase molecule. Below it, a "Sequence/Alignment Viewer" box showing the amino acid sequence should be visible. If the sequence is not visible, show it using the "Show" menu at the top. The amino acid sequence of this amylase can be decoded using the Table 2.2 in the text and the chemical nature of the amino acids can be determined using Table 5.1.

3. Cn3D gives options for studying the structure of this and other molecules. Explore what can be done using Cn3D, following the steps below. In the "View" menu, at any time, you can choose to restore the image to the original structure.

   A. <u>Changing the size of the image</u>. Using the "View" pull down menu, you can zoom in or zoom out.
   B. <u>Rotate the model</u>. To start the whole molecule rotating, pull down the View" menu and use "Animation" to select the "Spin" and "Stop" choices. Notice the shortcut keystrokes for these options.
   C. <u>View the structure</u>.

      - Go to the "Style" Menu. Choose "Rendering Shortcuts" and then choose "Ball and Stick."
      - Next go back to the "Style" Menu. Choose "Coloring Shortcuts" and then choose "Residue." You should now be able to see the color-coded amino acids. What are the colors of the glutamic acid and lysine? Look in both ball and stick style and space-fill styles (in Rendering Shortcuts). Can you confirm the sequence shown in the sequence viewer below the image? You can click on one or more amino acids in the sequence viewer to highlight them in the structural model image.
      - Go to "Worms" in the "Rendering Shortcuts" menu. Look for the secondary structure of the protein. Do you see alpha helices and beta-pleated sheets? Choose "Coloring Shortcuts" and then choose "Element." Do you see the Ca^{2+} and 1 Na^+ ions?
      - Change the style to "Space Fill" using the rendering shortcuts and to "Charge" using the coloring short cut. Rotate the protein stopping it as necessary. Look closely as it turns. Do you notice the two grooves in the long sides of the protein? The large one is the active site of the enzyme where its substrate, starch, binds to the amylase. This is important. If the active site is blocked or changes shape, it will not bind the substrate well and the enzyme's activity could be diminished.
      - Change the style to "Ball and Stick" using the rendering shortcuts. The blue and red amino acids are positively charged and negatively charged amino acids, respectively. Notice their location. Are they mostly on the inside or outside of the protein? Can you explain why their location makes sense for the environment in which this enzyme is usually found?
      - Reset the appearance to "Wire" (in rendering shortcuts menu) and "Rainbow" color and see how the sequence is shown from the beginning of the strand to the end following the colors of the rainbow. Rotate the structure so you can see the entire sequence.
      - Take the time to explore the molecular structure choosing other menu options. When you have an image of the structure that best represents explains to you how the molecule functions then, save or email yourself an image. Print that image and place it in your notebook. On the printed image, describe 3 features of the structural model that are important for its function.

4. Another analysis tool called, **Vector Alignment Search Tool Plus** (VAST+) allows you to align and compare two polypeptide (protein) sequences. With VAST+ you can look for differences in sequence and structure that may explain differences in function. Cn3D will take data from VAST+ and align the two sequences.

   - Go to **www.ncbi.nlm.nih.gov** and click on the box that says "Analyze." From the list of Analysis Tools, select "Vector Alignment Search Tool (VAST)."
   - Under "3D Macromolecular Structures" choose "VAST+."

- Input the original amylase PDB code "1BLI" and hit "Search." VAST+ will pull up the alpha-amylase from *Bacillus licheniformis* and other similar alpha-amylases.
- Look down the list for PDB ID # 1E3Z. This is a genetically engineered amylase that is more heat resistant than the original (native or wild-type). What percentage of its (1E3Z) sequence is identical to the sequence of amylase 1BLI?
- Click on the "+" beside 1E3Z and it will bring up a comparison of the 2 protein sequences.
- Next, click on "View aligned sequences" and see the blue-colored amino acids that denote different amino acids in the two molecules' sequence.
- Next, click on the "Visualize 3D structure superposition with Cn3D and it will download the Cn3D file of the molecules superimposed on each other.
- Rotate the 3D model of the superimposed proteins and see how the 2 chains of the 2 proteins are almost identical, but not quite. There are several changes but look for the deletions at amino acid 178, as well as the 4 amino acid changes at position # 264 and #290 in the chain. These each change the structure of the 1E3Z protein, making it more stable than the 1BLI protein.
- Change the model to "space-filling" and you can see that starch can still bind at the active site but there is extra rigidity in the mutated strand. Print that image and place it in your notebook. On the printed image, describe 3 features of the amylase structural models that are important for its function.

Activity 6.3

What's the Latest in ELISA and Western Blot Technology?

Learn more about advances in ELISA and Western blot techniques and how they are used to improve research and development.

Create a twelve-slide PowerPoint® presentation to teach others about these advances.

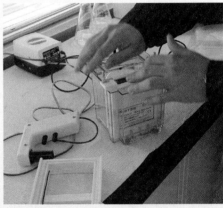

Photo by author.

Use the list below as a guide for what to include in your slide presentation.

Make sure each slide has text and images.

Be sure that each slide is easy to see and read from a distance.

Be prepared to present your PowerPoint® presentation to the class.

Slide 1	Title Page
Slide 2	Summary of ELISA procedure and applications
Slide 3	ELISA plate washer, what it does, and how it works
Slide 4	ELISA plate reader, what it does, and how it works
Slide 5	ELISA visualization methods, enzymes, and substrates (chromogenic, fluorescent, or chemiluminescent)
Slide 6	Examples of ELISA kits that are available to researchers or clinicians and what they detect
Slide 7	Summary of Western blot procedure (traditional "wet blot") and applications
Slide 8	Semi-dry Western blot and how it differs from a traditional wet Western blot
Slide 9	Western blot visualization methods, enzymes and substrates, fluorescence, luminography
Slide 10	Rapid Western blot transfer equipment (instruments/methods)
Slide 11	Sources of 1° and 2° antibodies for use in ELISAs and Western Blots, with examples
Slide 12	URLs, bibliographical references, and credits (number the images from the Internet for bibliographical reference.)

Create an information sheet on a common herbal remedy.

1. Pick an herbal remedy from the list below:
   lavender oil
   tea tree oil
   hawthorne berry
   burdock root
   St. John's Wort
   Echinacea
   lecithin
   cascara bark
   Psyllium
   slippery elm
   black cohosh
   belladonna
   Lobelia
   Ginkgo, yarrow
   goldenseal
   dong quai
   foxglove
   coq10
   Uva ursi
   Vinca rosea

***Eucalyptus camaldulensis*, also called the river red gum tree, is a type of eucalyptus. Many species of eucalyptus are sources of oil used in medications.**
© logorilla/iStock.

2. Gather as much information as you can about the herb and its uses. Include a photo if you can find one.
3. On an 8-1/2 in × 6-1/2 in sheet, report the following:
   a. Name(s) of the herb: common, scientific, trade, etc.
   b. Source(s): organism, country, and biome
   c. Function(s): uses, therapies, diagnostics, and applications, etc.
   d. Side effect(s)
   e. Three other interesting facts
   f. Two to three bibliographical references

   Be sure the fact sheet is neat and easy to read.

Product Pipeline Study

Most biotechnology companies fall into one of the following four categories of production/ development and sales:

- pharmaceuticals
- industrials
- agricultural products
- biotechnology instrumentation and reagents.

No matter what the product of a biotechnology company, the goal is to get the product to market as quickly as possible. Often there is a period of R&D where a great deal of money is invested in attempting to produce a product on a small scale. Much testing is conducted, as the protocols for small- and then large-scale production are determined. If the product is a pharmaceutical, it must undergo strict testing (clinical tests), under the guidance of the FDA, before it can be marketed. It takes anywhere from 10 to 15 years for a company to take a product through all these steps. The product pipeline is different at every company, but it follows the basic outline below:

- Identify a potential product.
- Complete R&D with assay development and quality control.

- Manufacture on a small scale.
- Continue testing for safety and efficacy (including Phase I, II, and III clinical trials for pharmaceuticals).
- Market the product.

 As a sales and marketing specialist would do, study and report on a product that has been recently, or is about to be, brought to market.

1. Review the products the instructor has identified as being appropriate for this study. Select one that you think would be interesting to extensively research. Create a folder, either a hard copy manila folder or an electronic computer file folder, into which you can collect research information. Title it with your name, the date, and the product name. This will be an exhaustive research folder into which all research documents will be placed.

2. Using search engines and searchable databases, find as many companies as possible involved in the research or production of the product. Save a copy of the Web pages that have information you feel will be helpful to completely understand the history, structure, and function of your product. Write down the URL information on the top or bottom of each page, and give each reference a number. Keep track of the information you collect about your product by making a bibliography, recording the reference number, title of the article or page, and the URL.

3. Determine the single company that is farthest along in the pipeline of your product. Look on the company's website for an "annual report." Annual reports usually include a basic product pipeline for the company's potential products. Most annual reports are available online (but require a lot of paper to print). Check with your instructor before printing a copy.

4. Once you have identified a product and a company that produces it, become an expert on the product. Gather information to answer the following questions using the Internet, databases, and scientific libraries, such as those at colleges or universities, or the ones at larger biotech companies. In addition, finding an expert (scientist, public relations person, etc) can provide you with much information.

   - What is the structure and function of the product?
   - What market will it serve?
   - How large a market will it serve?
   - How far along is it in the product pipeline?
   - What has been its history in the product pipeline?
   - When is the product first expected to reach the market?
   - What obstacles are there, if any, to reaching the market?
   - Are there any concerns or cautions about the production of the product?
   - How much money is it expected to produce for the company?
   - Are any other companies in serious competition to bring this product to market?
   - Is there any other interesting information?

5. Collect all the information you have gathered into your exhaustive research folder either in hard copy or electronic form. Use a bibliography as the table of contents. Make sure every bit of information has a bibliographical entry (standard or URL). Actively read all documents, highlighting the answers to all of the above questions. Include parts of the annual report in the folder.

6. Produce a Microsoft® PowerPoint® presentation (no more than 15 slides) that clearly describes the history, structure, function, and market for your product. Make sure each slide is interesting to look at, has an appropriate amount of information, and is easy to read from the audience's seating area. Include information on all the questions asked above. One slide should have a product pipeline diagram (timeline). Print copies of the pipeline at a size that fits into notebooks for each person in the class. Be prepared to make an oral presentation (10 minutes) to the class describing your product and its market. (See the deadlines below.)

Project Deadlines

End of Week 1   Identify product, set up an exhaustive research folder and bibliography, actively read, and record five articles/documents into the folder.

End of Week 4	Actively read and record a minimum of 15 articles/documents into the folder. Acquire the annual report.
End of Week 8	Submit a rough draft of your PowerPoint® presentation to the instructor.
End of Week 10	Be ready to present the final version of your PowerPoint® slide show and the exhaustive research folder.

7. After your presentation, turn in a copy of the entire PowerPoint® presentation to the instructor, along with the exhaustive research folder.

Ideas for Product Pipeline Products, similar to those below, can be found at company websites resourced through the Biotechnology Industry Organization's member site at **www.bio.org/articles/bio-members-web-site-links**.

Product	Description/Treatment of
Roundup Ready® soybeans (Monsanto Canada, Inc)	herbicide resistant soybeans
Recombinant human beta nerve growth factor (Neuromics)	stimulates nerve cell growth and reproduction
MerMade 192E Oligonucleotide Synthesizer (BioAutomation)	produces small strands of DNA
Pulmozyme® (Genentech, Inc)	cystic fibrosis medication
Purafect® protease (Genencor, International)	protein-digesting enzyme
Posilac® bovine somatotropin (Monsanto Co)	growth hormone for livestock
REVLIMID® (Celgene Corporation)	treatment of multiple myeloma
IndiAge® cellulose (Genencor, International Inc)	cellulose-digesting enzyme
Bollgard® cotton (Monsanto Company)	insect-resistant cotton
anti-IgE monoclonal antibody (Genentech, Inc)	allergic asthma
L-8900 amino acid analyzer (Hitachi High Technologies Corp)	determines amino acid sequence of polypeptides
Nutropin® (Genentech, Inc)	human growth hormone (hGH)
KALBITOR® (Dyax, Corp.)	plasma kallikrein inhibitor
Activase® (Genentech, Inc)	tissue plasminogen activator (t-PA)
Spezyme® starch enzyme (Genencor International, Inc)	starch-digesting enzyme
Herceptin® (Genentech, Inc)	breast cancer
Correlate-EIA™ (Assay Designs, Inc)	sex hormone immunoassay detection kits
BXN™ cotton seed (Monsanto, Inc)	herbicide-resistant cotton
FLAVR SAVR® tomatoes (Calgene, Inc)	slow-ripening tomatoes
Dual-Luciferase™ Reporter Assay System (Promega Corp)	gene-expression measurements
Rituxan® (Genentech, Inc and IDEC Pharmaceuticals, Inc)	treats some kinds of lymphomas
VISTIDE® (Gilead Sciences, Inc)	systemic treatment of cytomegalovirus (CMV) retinitis
Premise® 75 (Bayer Corp)	termiticide
EPOGEN® (Amgen, Inc)	RBC production for anemia
Kogenate® (Bayer Corp)	hemophilia
Glucobay® (Bayer Corp)	diabetes
INTEGRA® dermal regeneration template (Integra LifeSciences Corp)	artificial skin
Rigosertib (Onconova Therapeutics)	antibody therapeutic
RealTime HCV assay (Abbott Molecular)	PCR kit for hepatitis C
Avonex® (Biogen Idec)	relapsing multiple sclerosis
Humalog® (Eli Lilly and Company)	treatment for diabetes
Intron-A (Schering Corp)	hairy cell leukemia, genital warts, etc
Proleukin® IL-2 (Prometheus Laboratories Inc.)	kidney (renal) carcinoma, melanoma
NewLeaf™ Potato (Monsanto Co)	insect-protected potatoes
Enclomiphene (Repros Therapeutics, Inc.)	secondary hypogonadism
TZ101 (Targazyme, Inc)	cancer-fighting enzyme
Neupogen® (Amgen, Inc)	infection prevention for cancer/transplant patients
GS 4104 (Gilead Sciences, Inc and Hoffmann-LaRoche Inc)	influenza virus treatment and prevention

Bioethics

Limited Medication: Who gets it?

According to AVERT (**biotech.emcp.net/avert**), an international HIV and AIDS charitable organization, "The number of people living with HIV rose from around 8 million in 1990 to 34 million by the end of 2011. The overall growth of the epidemic has stabilized in recent years. The annual number of new HIV infections has steadily declined and due to the significant increase in people receiving antiretroviral therapy, the number of AIDS-related deaths has also declined. Since the beginning of the epidemic, nearly 30 million people have died from AIDS-related causes."

Many biotechnology companies are focusing product development on AIDS therapies and vaccines to prevent the symptoms of AIDS. Several challenges exist to the manufacture and distribution of these pharmaceuticals, including some of the following:

- The need is great. So many people have been infected and need treatment.
- Patients are not confined to a specific nation or region, but are spread throughout the world.
- The majority of HIV-infected individuals are extremely poor.
- The cost of producing these products is astronomical, and biotech companies need to make money.
- There are religious, societal, or political obstacles to administering medications.

Even if a company could produce substantial amounts of an HIV medication, there will likely never be enough at a cost low enough to treat everyone who needs it. This is a significant barrier to companies investing in R&D for these products.

Problem: How should a limited amount of an HIV therapeutic product be distributed?

Part I: What numbers of people are living with AIDS? How are they distributed throughout the world?

 Go to the AVERT website (biotech.emcp.net/avert) or any other site that publishes World Health Organization (WHO) HIV/AIDS statistics. Use the information on the website to fill in the following table, using data for the most recent year available.

Total Number of People Newly Infected with HIV in the Past Year	
Adults	
Women	
Children	
Total Number of People Living with HIV in the Past Year	
Adults	
Women	
Children	
In Sub-Saharan Africa	
In North Africa and Middle East	
In South and South East Asia	
In East Asia and Pacific	
In Latin America	
In the Caribbean	
In Eastern Europe and Central Asia	
In Western Europe	
In North America	
In Australia and New Zealand	
AIDS Deaths in the Past Year	
Adults	
Women	
Children	
AIDS Deaths Since the Beginning of the Epidemic	
Adults	
Women	
Children	

Part II: How should a limited amount of a medication that reduces the amount of HIV particles in the body be distributed?

Think of yourself as an administrator for the WHO. A biotechnology company has developed a therapeutic recombinant protein that halts the replication of the HIV in AIDS patients, giving the body a chance to get rid of the virus. Although the company can break even on its production investment, the science behind producing the therapeutic agent limits the amount that can be produced to 20 million doses per year. Since that amount is not enough for all the patients who need it, and since the AIDS epidemic has so many political and social implications, the company has asked the WHO to decide how the doses should be distributed.

1. With your colleagues' help, determine who (men, women, and/or children), living where (which geographic locations), should get these dosages, and when (first year, second year).
2. Make a flowchart or timeline that shows the distribution schedule.
   Write a position paper that specifically justifies the decisions you and your colleagues have made.

Biotech Careers

Photo by author.

Lab Technician

Jason Chang
CS Bio Company, Inc.
Menlo Park, CA

In a quality control lab, Jason tests the instruments (protein synthesizers) that CS Bio Company, Inc. produces and the products (peptides) made by the instruments. Here, Jason is using high-performance liquid chromatography (HPLC), which is hooked up to an ultraviolet light (UV) spectrophotometer. Peptides made at CS Bio Company, Inc. are separated on the HPLC and their purity is tested and reported using the spectrophotometer. This instrument is used in many applications to show the presence, purity, or concentration of molecules.

At CS Bio, for Jason's position, an Associate of Sciences (AS), or Associate of Arts degree (AA), a Biotechnician Certificate, or the equivalent is required. Many colleges have 1-year certification programs in combination with laboratory internships where students gain valuable experience. Many companies though require a 4-year degree in biotechnology, biology, biochemistry, or a related field for an entry-level laboratory technician. Jason was placed at CS Bio Company, Inc. through an internship program and continued to work in the facility after graduation.

7 Spectrophotometers and Other Analytical Tools

Learning Outcomes

- Describe how a spectrophotometer operates, compare and contrast ultraviolet and visible (white light) spectrophotometers, and give examples of their uses
- Determine which type of spectrophotometer is needed for a particular application and the wavelength to be used
- Explain the relationship between absorbance and transmittance in spectrophotometry and interpret the meaning of absorbance measurements
- Explain how protein indicator solutions are used with and without a spectrophotometer
- Describe how VIS and UV/VIS spectrophotometers are used to measure protein or DNA concentration.
- Use a best-fit standard curve to determine the concentration of an unknown protein sample and explain the usefulness of a protein absorbance spectrum when trying to isolate a specific protein
- In terms of how it works and what it measures, describe the difference between a spectrometer and a spectrophotometer.
- Distinguish between NMR, IR, and mass spectroscopy giving descriptions of what each measures.
- Give examples of how UV, IR and mass spectrometry each might be used in applications such as medical, agricultural, environmental or industrial biotechnology.

7.1 Using the Spectrophotometer to Detect Molecules

Molecules, mostly proteins and nucleic acids, are often the targets of biotechnology research and development (R&D). Molecules are too tiny to be seen, but confirming the presence of a molecule is important in virtually all aspects of R&D as well as during manufacturing. Molecular products must be assayed regularly to justify continued research and confirm that manufacturing is producing the product adequately. If the concentration or activity of a sample is too low, new protocols must be considered.

The quickest way to detect a molecule in solution is by using an indicator. Indicator solutions change colors when a molecule of interest is present, and they allow a scientist to quickly identify colorless molecules in solution. For example, one can quickly "see" if a solution contains protein by adding Bradford protein reagent or Biuret reagent. Other indicators, such as diphenylamine (DPA) or ethidium bromide (EtBr), make nucleic acids visible. The color of DPA changes to blue or green depending on whether deoxyribonucleic acid (DNA) or ribonucleic acid (RNA) is in a solution. When EtBr interacts with DNA, it absorbs **ultraviolet (UV) light** and emits light of a glowing orange color. Another indicator, iodine, quickly shows the presence of starch by changing from red-brown to blue-black. Several indicators can be used for sugars, including Benedict's solution and the copper-based indicators on glucose test strips.

Indicator solutions can show the presence of molecules of interest by showing a color change. The more color change, usually the more molecules in solution. However, to make more quantitative measurements, an instrument is needed to detect the amount of change in the indicator due to the number of molecules in the solution.

For most applications, quantifying the amount of a molecule is essential. One has to know how much of a given molecule is made or used in a reaction. During production or purification, a sample is measured at every step to ensure an adequate yield. To detect how much is present in a solution, the sample or molecule must be visualized and measured. A common way of detecting and measuring molecules is by using a **spectrophotometer**, or "spec" for short.

There are many different models of spectrophotometers, but all use some type of light to detect molecules in a solution. Light is a type of energy, and the energy is reported as wavelengths, in **nanometers** (nm). Different specs produce light of different wavelengths. Specs are classified as either **UV specs** (see Figure 7.1) or **visible light spectrum (VIS) specs** (see Figure 7.2). The UV specs use ultraviolet light (wavelengths

ultraviolet (ul•tra•vi•o•let) light (UV light) the high-energy light with wavelengths of about 100 to 350 nm; used to detect colorless molecules

spectrophotometer (spec•tro•pho•tom•e•ter) an instrument that measures the amount of light that passes through (is transmitted through) a sample

nanometer (nan•o•me•ter) 10−9 meters; the standard unit used for measuring light

visible light spectrum (vis•i•ble light spec•trum) the range of wavelengths of light that humans can see, from approximately 350 to 700 nm; also called white light

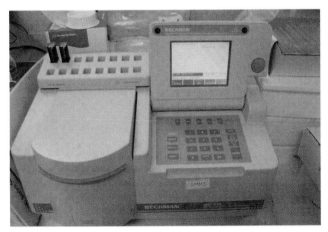

Figure 7.1. A UV spectrophotometer uses ultraviolet light to detect colorless molecules.
Photo by author.

Figure 7.2. A VIS spectrophotometer uses white light, composed of the visible spectrum, to detect colorful molecules.
Photo by author.

from 100 to 350 nm), and the VIS specs use visible light, also called white light (wavelengths between 350 and 700 nm). Whether a spec uses white or UV light, all specs work in a similar fashion.

All spectrophotometers shine a beam of light on a sample. The molecules in the sample interact with the light waves, and either absorb light energy, reflect the light, or the light transmits between and through the atoms and molecules in the sample (see Figure 7.3). The spectrophotometer measures the amount of light transmitted through the sample (**transmittance**). Although it measures the percent of light transmitted (%T), by using an equation it can convert the transmittance data to an **absorbance** value. By comparing the absorbance data to standards of a known concentration (the amount of molecules per unit volume), the concentration of an unknown sample can be determined. How a spectrophotometer takes readings and reports values is discussed below.

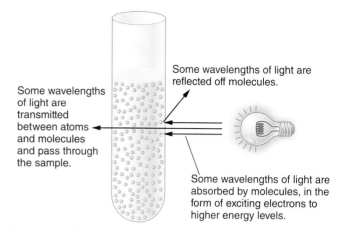

Figure 7.3. Absorbance, Transmittance, and Reflection. A spectrophotometer measures how light interacts with atoms or molecules in a sample.

Parts of a Spectrophotometer

Virtually every biotechnology lab has a spectrophotometer. Along with balances and pH meters, the spec is essential for nearly all research in molecular biology. Some spectrophotometers are very sophisticated and expensive. Some use only one type of light to detect molecules, while others have light source options. Some give digital displays or readouts, while others are hooked up to computers that gather, analyze, and print out the data collected. It is critical for a lab technician to have the ability to go into any lab and use any make or model of spectrophotometer. Understanding the basic hardware and operation will help you quickly learn to operate any spec you may encounter in the lab.

All specs have some common features, including a lamp, a prism or grating that directs light of a specific wavelength, a sample holder, and a display (see Figure 7.4). Knobs or buttons are used to calibrate the spectrophotometer to measure the designated molecule.

transmittance (trans•mit•tance) the passing of light through a sample

absorbance (ab•sor•bance) the amount of light absorbed by a sample (the amount of light that does not pass through or reflect off a sample)

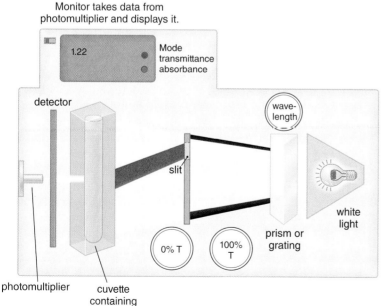

Figure 7.4. How a VIS Spectrophotometer Works. A sample in a cuvette is placed into the sample holder of a VIS spectrophotometer. White light is shone through a prism or hits a grating, which creates a spectrum of white light, essentially, a rainbow. The rainbow of light passes through a slit allowing only one wavelength (color) of light to be directed to the sample. The detector measures the amount of that wavelength of light that passes through the sample and calculates how much light the molecules in the sample absorb.

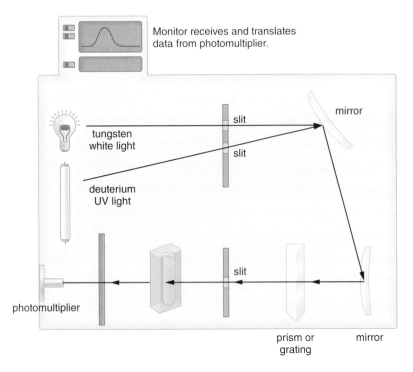

Monitor receives and translates data from photomultiplier.

Figure 7.5. **How a UV Spectrophotometer Works.** Similar to a VIS spectrophotometer, the UV spec shines ultraviolet light or visible light that reflects off mirrors until it passes through a prism or hits a grating, separating it into a spectrum of wavelengths. The light then passes through a slit which focuses just one wavelength on the sample. Light that is not absorbed by the sample passes through and a detector measures the amount of light transmitted.

tungsten lamp (tung•sten lamp) a lamp, used for VIS spectrophotometers, that produces white light (350–700 nm)

deuterium lamp (deu•ter•i•um lamp) a special lamp used for UV spectrophotometers that produces light in the ultraviolet (UV light) part of the spectrum (100–350 nm)

% transmittance (per•cent trans•mit•tance) the manner in which a spectrophotometer reports the amount of light that passes through a sample

absorbance units (ab•sor•bance un•its) (abbreviated "au") a unit of light absorbance determined by the decrease in the amount of light in a light beam

Spectrophotometers may contain only one lamp or several lamps. A VIS spec contains a **tungsten lamp** (see Figure 7.4). A tungsten lamp produces white light. A spec with a tungsten lamp is called a VIS spec because it produces light in the visible spectrum. A VIS spec is used to measure colored molecules such as chlorophyll, hemoglobin, or molecules that have reacted with indicators. A UV spec contains a **deuterium lamp** (see Figure 7.5), which produces light in the UV light part of the spectrum. Some molecules, such as colorless proteins and nucleic acids, are visualized on these types of spectrophotometers. Although not visible to the human eye, UV light is important in many biological processes and instruments. In most biotechnology facilities, the majority of specs are UV/VIS, containing at least one each of the tungsten and deuterium lamps.

How a Spectrophotometer Works

In a VIS spec, when the white light hits the prism or grating, it is split into the colors of the rainbow. The wavelength knob rotates the prism/grating, directing different colors of light through a slit that shines one wavelength of light toward the sample.

The wavelengths of light produced by the tungsten lamp range from about 350 nm (violet light) to 700 nm (red light) (see Figure 7.6). All of the other colors of the rainbow have wavelengths between these values. By turning the wavelength control knob, one can position any wavelength of visible light between 350 and 700 nm to interact with the sample. The molecules in the sample either absorb or transmit the light energy of one wavelength or another. The detector measures the amount of light being transmitted by the sample and reports that value directly (**% transmittance**) or converts it to the amount of light absorbed in **absorbance units (au)**.

To convert between transmittance and absorbance values, the spectrophotometer uses an equation based on Beer's Law. The equation is $A = 2 - log_{10}$ %T, and it shows that as the absorbance of a sample increases, the transmittance decreases, and vice versa. This makes sense, because if molecules are not absorbing all the energy from a light beam, then the remaining light beam penetrates (transmits) through the sample.

To be detected by a white light spectrophotometer, molecules must be colored or have a colored indicator added. Colorless molecules are not detected by a white light spec because they do not absorb one wavelength of visible light more than another. This is best illustrated by examining how a colored molecule is detected by the spectrophotometer.

Consider blue molecules. When white light shines on

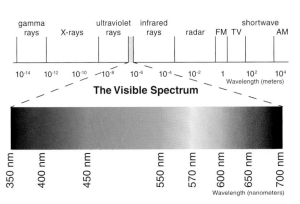

Figure 7.6. **Colors of Light in the Visible Spectrum.** Humans can see light with wavelengths of about 350 to 700 nm.

The wavelength of different colors of light energy

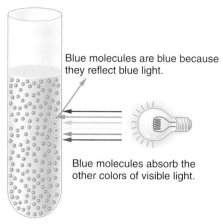

Figure 7.7. **Interaction of Light with Molecules.** Molecules are whatever color of light that they do not absorb. Blue molecules are blue because they absorb all the other wavelengths of the visible spectrum, except blue.

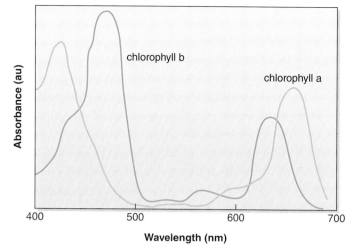

Figure 7.8. **Absorption Spectra of Chlorophyll a and b.** Chlorophyll a and chlorophyll b have slight differences in their molecular structure. This difference results in different interactions with light. Therefore, they appear as different colors.

a blue molecule, all the wavelengths of light are absorbed, except for the blue ones (see Figure 7.7). The blue wavelengths are transmitted or reflected off the molecules. If these blue wavelengths hit a detector (such as in the spec or the nerve cells in your eye), they appear blue. We then say that the molecules are blue. Molecules are whatever color of light that they do not absorb. Green molecules appear green because they absorb most wavelengths of visible light, except the green ones (around 540 nm). Red molecules appear red because they absorb most wavelengths of visible light, except the red ones (around 620 nm).

For light energy to be absorbed by a solution, there must be light-absorbing molecules present to do the absorbing. If colored molecules are present in a solution, they will absorb certain colors of light (certain wavelengths) and transmit other colors. In fact, a molecule reflects the light energy of the color that humans see. Remember, if a sample is green, it is because it is not absorbing green light; it is reflecting it. The spec can measure the amount of absorbance or lack of absorbance of different colored light for a given molecule. The graph of a sample's absorbance at different wavelengths is called an **absorbance spectrum** (see Figure 7.8).

The absorbance or transmittance of light at a given wavelength (the absorbance spectrum) is an indication of a molecule's presence in solution and, like a fingerprint, can be used to identify or recognize the molecule. If, for example, a green molecule, such as chlorophyll a, were in solution, you would expect the solution to transmit or reflect light of approximately 530 nm, that is, in the green wavelengths of light. You would also expect it to absorb lights of other wavelengths to some degree.

The concentration of molecules in a solution affects the solution's absorbance. If there are more molecules in one solution than in another, then there are more molecules to absorb the light. The amount of light that a sample absorbs indicates how many molecules (the concentration) are present. The more molecules in solution, the higher the light absorbance is for a sample. In fact, if there were twice as many molecules in solution, you would expect twice as much light absorbance. Likewise, half as many molecules absorb half as much light (see Figure 7.9). More information on the behavior of molecules at different wavelengths is presented in later sections.

■ ■ ■ ■ ■ ■ ■ ■ ■ ■ ■ ■ ■ ■ ■

absorbance spectrum (ab•sor•bance spec•trum)
a graph of a sample's absorbance at different wavelengths

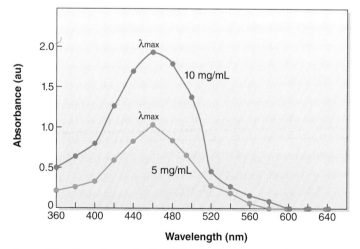

Figure 7.9. **How Concentration Affects Absorbance.** If a sample has twice as many molecules as another, it can absorb twice as much light. This is true at any wavelength. When measuring molecules in solutions, it is important to know a sample's wavelength of maximum light absorbance (λ_{max}), so that the difference in absorbance due to concentration is obvious.

lambda_max (lamb•da max) the wavelength that gives the highest absorbance value for a sample

Since spectrophotometers can detect molecules in solutions, they have several applications in the biotechnology lab setting. As described above, the presence of molecules and concentration of samples can be ascertained using a spec. The wavelength at which a molecule absorbs the most light is called **lambda_max** (λ_{max}). Commonly, scientists identify a molecule in a mixture using the lambda_max. Hemoglobin, for example, absorbs the most light at 395 nm, and if a sample gives an absorbance peak at 395 nm, it would suggest the presence of hemoglobin in the sample. Additionally, if one hemoglobin solution has a peak absorbance at 395 nm that is half as high as a second solution, then the second solution probably contains twice the hemoglobin.

In addition, scientists use specs to determine the purity of a sample. For example, DNA extracted from cells often contains protein molecules as contaminants. Since DNA and protein molecules have different absorbance spectra, measuring each at their λ_{max} will show the amount of each in a solution. By comparing the absorbance of both DNA and protein in a sample, a UV spec can determine the "purity" of DNA in the sample.

A spectrophotometer can also look at changes in samples over time (see Figure 7.10). Often, enzyme activity (kinetics) studies are conducted by monitoring the change in a colored product over time. The spectrophotometer can be used to measure the change in color. If, for example, a sample becomes a deeper yellow as more enzyme activity occurs, then the spec will show more absorbance at 475 nm, with more activity, until the reaction is complete.

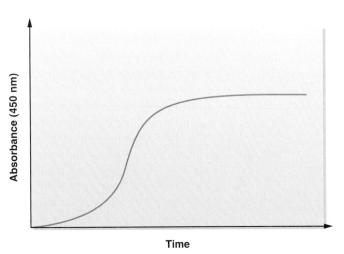

Figure 7.10. Absorbance of an Enzymatic Product. An enzymatic reaction can be monitored in a spectrophotometer. As a colored product is made, the absorbance of the reaction will change. When the maximum amount of product has been made, the absorbance of the sample will no longer increase.

BIOTECH ONLINE

Visual Spectrophotometry Virtually

Since molecules are submicroscopic, they cannot be seen directly by researchers. A spectrophotometer allows researchers to visualize molecules by how they interact with light waves. When shone on a sample, light of a selected wavelength will either pass between atoms and through a sample (transmittance) or they will hit electrons in atoms of a sample and absorb the light. The amount of light absorbed in a sample is dependent on the type and amount of atoms and molecules in the sample and the wavelength of light. Researchers and clinicians can determine the presence and concentration of a samples' molecules by measuring the light absorbance at specific wavelengths.

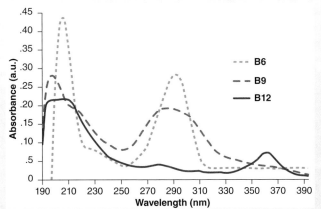

The Absorbance of B Vitamins at Different Wavelengths
From the Indian Journal of Pharmacology, 2013, 45:2.

Go to **web.mst.edu/~gbert/Color_Lg/color.html?963 to learn how a spectrophotometer "visualizes" molecules by measuring transmittance and absorbance.**

1. Use the website's Background and simulation to learn how a spectrophotometer works and how transmittance data is converted to absorbance data and concentration of a solution is measured.

2. Click on the link "Operation" to learn how to run the spectrophotometer simulation experiment.

3. Click on "Experiment" and follow the directions to produce an absorbance spectrum for each of two unknown samples. For each, sketch the absorbance spectrum in your notebook and identify the sample's lambda_max.

Section 7.1 Review Questions

1. What is measured in a spectrophotometer?
2. What is the difference between a UV spectrophotometer and a VIS spectrophotometer?
3. What happens to the absorbance of a sample as the concentration of a sample increases or decreases?
4. What color of light has a wavelength of 530 nm? If a molecule absorbs light at 530 nm, what color could it be? What color do we know that it is not?

7.2 Using the Spectrophotometer to Measure Protein Concentration

At a biotechnology company, protein concentration must be measured throughout the manufacturing of a protein product. Since proteins are too small to be seen, they must be visualized in a way other than direct observation. Several assays of protein concentration (using indicators, ELISA, and PAGE) and activity (activity assays) were introduced in Chapters 5 and 6. To measure protein concentration, a spectrophotometer may be used. A spectrophotometer can assess changes in light absorbance due to protein molecule concentration and give numerical values to proteins in solution.

Similar to other molecules, proteins interact with light waves. If a solution contains protein molecules, it will absorb or transmit light of certain wavelengths. A spectrophotometer measures the amount of light transmitted by the protein molecules in solution and can report the amount of light either transmitted or absorbed by the sample. If light of a certain wavelength is not absorbed, the light passes right through the molecules as if the molecules were not even there.

If a protein solution absorbs a substantial amount of light at one wavelength, but not at another, the data are useful in developing a method of detecting that molecule. A protein's absorption spectrum is determined by measuring the protein's light absorbance at different wavelengths. By measuring the absorption spectrum of a protein, we can learn how to detect it in solution. The wavelength that gives the highest absorbance (lambda$_{max}$), is the one that is "most sensitive" to the protein. It is the wavelength that interacts most with the protein, and it is used to measure the amount of that protein in solution (see Figure 7.11).

The absorption spectrum shown in Figure 7.11 is for a protein that absorbs visible light. Only colored molecules absorb in the visible light range. For example, hemoglobin is a protein that appears red in the presence of oxygen. Since hemoglobin is red, it would not be expected to absorb red wavelengths. Rather, hemoglobin would be expected to absorb non-red wavelengths such as blue light waves (on the opposite side of the visible spectrum). To detect hemoglobin using a spectrophotometer, set it to a wavelength of about 395 nm (its lambda$_{max}$) in the blue part of the spectrum to ensure that any hemoglobin molecules present will be detected.

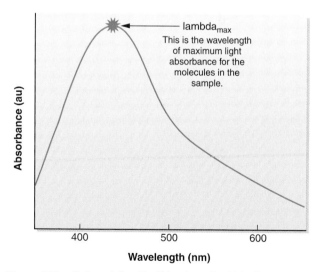

Figure 7.11. Determining the Wavelength of Maximum Light Absorbance. The lambda$_{max}$ for this sample is 440 nm. Once lambda$_{max}$ is determined for a molecule, the spec is set to that wavelength, and all readings for the molecule are made at that wavelength. The lambda$_{max}$ value is a characteristic of a molecule.

Hemoglobin is one of the few protein molecules that are colorful. Most proteins are colorless and cannot be detected by light in the visible spectrum. A solution containing colorless proteins is clear, and a white light (VIS) spectrophotometer cannot be used to detect them.

Colorless proteins do not absorb visible light; instead, they absorb in the UV range, at around 280 nm. This is because several of the amino acids in a protein absorb light at 280 nm. A 280-nm absorbance reading of a solution gives a value based on the total protein content and does not distinguish between types of proteins in solution. Even so, it is easy to quickly determine the total amount of protein in a sample by setting a UV spectrophotometer to 280 nm and measuring the absorbance of a sample as compared to some standard curve of protein absorbance. The absorbance reading reflects the presence and concentration of any or all proteins in a solution.

Since white light spectrophotometers do not detect in the UV range, the only way colorless proteins can be visualized with a white light spec is to colorize them. Several protein indicator solutions can be mixed with colorless proteins to change them to a visible color. Bradford reagent changes from brown to blue in the presence of protein. The degree of "blueness" of a Bradford-protein mixture can be determined by a spectrophotometer, and a concentration of the protein in solution can be calculated (see Figure 7.12). Bradford protein reagent is a universal protein indicator. The color reflects the presence and concentration of any or all proteins in the solution. Darker blue samples are at higher concentrations and, of course, absorb more light when set at an appropriate wavelength.

Bradford testing is often used to calculate the protein concentration of unknown samples. To do this, standard protein solutions of known concentrations are mixed with Bradford reagent. The absorbance of these known solutions is determined and plotted on a graph (see Figure 7.13). The line created by the known concentrations is called a **standard curve** (best-fit straight line). By comparing the absorbance of unknown mixtures with the absorbance of the known concentrations on the standard curve, one can estimate the number of protein molecules in a given volume of solution.

A standard curve is the most common and a fairly accurate method for determining the concentration of protein in a sample. However, since protein absorbance at 280 nm is dependent on the number of different amino acids (specifically tryptophan and tyrosine), comparable protein concentrations don't absorb light comparably, making protein determinations by this method slightly inaccurate.

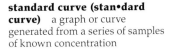

standard curve (stan•dard curve) a graph or curve generated from a series of samples of known concentration

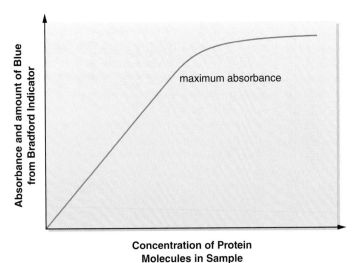

Figure 7.12. **Graph of Bradford Indicator versus Protein Molecules.** The absorbance of Bradford-protein mixtures increase with increased protein concentration until the maximum absorbance is reached.

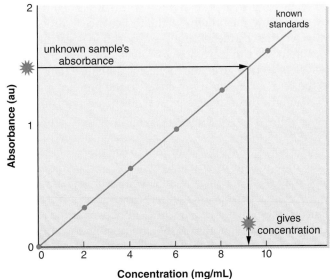

Figure 7.13. **Standard Curve Showing Absorbance versus Concentration.** The absorbance is proportional to the concentration.

Which Indicator is Indicated?

Since the majority of proteins are colorless in solutions, using indicators to colorize a protein sample is a valuable tool for protein researchers and lab clinicians. Although you have used Biuret and Bradford reagents to study proteins, several other protein indicators are available and for certain applications may be better suited for a procedure. The indicator to pick for a particular task is determined by the concentration of protein thought to be in a solution and the type of protein solution to be tested.

Gather information to compare some commonly used protein indicators with Bradford Reagent.

1. Create a chart similar to the one below to use to compare some common protein indictors. Use the Internet to fill in the chart with information to help distinguish when one indicator may be better for an application than another.

Protein Indicator Comparison Chart

Indicator/Reagent	Protein Detection	Wavelength Tested (nm)	How it works?	URL/Reference
Bradford	1 to 1500 µg/mL	595	Contains Coomassie Brilliant Blue G-250 which binds with protein and in acidic conditions, changes to blue	www.ruf.rice.edu/~bioslabs/methods/protein/bradford.html
BCA				
Lowry				
Biuret				

2. You are given a solution, thought to have 0.1 µg/µL of Protein A, to test for protein concentration. You have a VIS spectrophotometer available to use, which has a maximum wavelength setting of 660 nm. Which indicator would you select to use to complete the testing?

Sometimes, the protein of interest is the only one in a solution. Other times, the protein is in high concentration compared with other proteins in the solution. In either case, a universal protein indicator is enough to show the presence and amount of the protein. If the solution has many different proteins in it, then a Bradford protein reagent will not distinguish between them. An ELISA, which uses antibodies to recognize specific proteins, might be used to detect and measure the presence and concentration of a specific protein in a mixture. Or, an indicator specific to just one protein might be used. One example is Phadebus tablets (Pharmacia and Upjohn Co.), which change color in the presence of amylase. Because they are specific to amylase, they may be used to show the presence and amount of amylase in a mixture.

Spectrophotometers are often built into analytical instruments to measure the concentration of a molecule in solution while a reaction or process is occurring. For example, an ELISA may be used to measure the concentration of a specific protein in a cell culture. To quantify the amount of color change and therefore protein concentration in several samples, an ELISA plate reader with a built-in spectrophotometer can quickly read the concentration of colorimetric ELISA samples in each well of an immunological plate (see Figures 6.9, 6.12, and 6.13). Using a 96-well plate, a technician can run an ELISA with a set of protein samples of known concentration and with one or more samples of unknown protein concentration. The plate reader will automatically move each well in line with the spectrophotometer, measuring sample absorbance and comparing it to the absorbance of the known samples.

Section 7.2 Review Questions

1. What is lambda$_{max}$, and why is it important?
2. What is the lambda$_{max}$ for colorless proteins?
3. How is Bradford reagent used to detect a specific protein in solution?
4. Which graph is used to determine the concentration of unknown protein samples?

7.3 Using the Spectrophotometer to Measure DNA Concentration and Purity

Being able to quantify the amount of DNA in a solution is critically important in a biotechnology laboratory. Genomic (chromosomal) DNA and plasmid DNA are routinely isolated from cells for research or diagnostic purposes. DNA samples are also produced in laboratories through DNA synthesis or the polymerase chain reaction (PCR). The technician must know how much DNA (the mass or concentration) is in a sample before further analysis is done. In other words, DNA samples must be quantified.

Since DNA is clear in aqueous solution, DNA samples are commonly analyzed in a UV spectrophotometer. The lambda$_{max}$ for a pure DNA sample in a solution is 260 nm. The concentration of DNA in a sample can be determined by measuring the absorbance (or optical density, OD) of a sample at 260 nm. As a basis of comparison, a pure 50-µg/mL sample of double-stranded DNA in solution has an OD of approximately 1 au.

By measuring the absorbance of a sample at 260 nm and comparing it to the absorbance of a 50-µg/mL sample, the concentration of DNA in a solution may be determined using the simple ratio below:

$$\frac{50 \ \mu g/mL}{1 \ au} = \frac{X \ \mu g/mL}{Sample \ OD_{260} \ (au)}$$

or

DNA Concentration Equation

$$Concentration \ (\mu g/mL) \ = \ Abs_{260} \times 50 \ \mu g/\mu L$$

For example, a 1-mL sample of genomic DNA extracted from a yeast cell culture is placed in a calibrated UV spectrophotometer. The OD is read at 260 nm and is determined to be 0.255 au. What is the approximate concentration of DNA in the sample?

$$\frac{50 \ \mu g/mL}{1 \ au} = \frac{X \ \mu g/mL}{0.255 \ au} = 0.255 \times 50 \ \mu g/mL = 12.75 \ \mu g/mL \ of \ DNA \ in \ the \ sample$$

The absorbance indicates a yeast genomic DNA concentration of approximately 12.75 µg/mL. A concentration of 12.75 µg/mL is not enough to see on an agarose gel but it is enough for a PCR or sequencing reaction.

Determining DNA concentration is so common and important that most UV spectrophotometers come with A_{260} programs already programmed on the spec (see Figure 7.14). By selecting the A_{260} program, the spec will read samples at 260 nm and automatically calculate the concentration (based on the 50-µg/mL/1 au ratio).

$$\frac{50\ \mu g/mL}{1\ au}$$

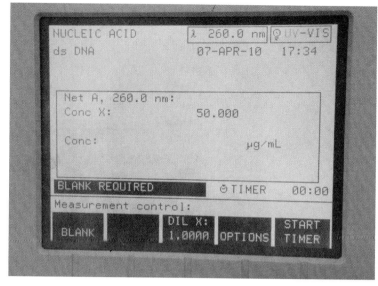

Figure. 7.14.   **Dialog box of a UV/VIS spectrophometer with the A_{260} program showing**
Photo by author.

It should be noted that the DNA concentration equation gives a good estimate of the actual concentration. However, the actual absorbance data, and thus the concentration determination, may vary from sample to sample depending on the length of the nucleic acid strands and the contamination in a sample. If a DNA sample is contaminated by RNA or DNA, the concentration of DNA in a sample may actually be higher or lower than the value calculated.

The amount of protein in cells and cell cultures is high. Therefore, protein contamination in DNA samples from cell extracts is common. Protein contaminants can interfere with many reactions including PCR and DNA sequencing. A technician needs to ascertain the relative amount of protein in a sample to decide if additional purification of the DNA sample is needed before proceeding with a reaction.

The UV spectrophotometer may be used to determine the relative purity of a DNA sample by assessing the relative amount of protein in the sample. Since protein is the most likely contaminant in a DNA sample, by measuring the absorbance of a sample at 280 nm ($lambda_{max}$ for most colorless proteins) and comparing it to the absorbance of DNA at 260 nm, a value for relative protein contaminant may be obtained. This is called an A_{260}/A_{280} reading or the OD_{260}/OD_{280} reading.

DNA Purity Equation

$$\frac{OD_{260}\ nm}{OD_{280}\ nm} = \frac{\text{relative amount of DNA}}{\text{relative amount of protein}}$$

For example, a 1-mL sample is placed in a calibrated UV spectrophotometer. The OD is read at 260 nm and is determined to be 0.365 au. The OD is also read at 280 nm and is determined to be 0.203 au. What is the A_{260}/A_{280} reading?

$$\frac{OD_{260\ nm}}{OD_{280\ nm}} = \frac{0.365\ au}{0.203\ au} = 1.798$$

What is the degree of contamination in the sample? The smaller the protein contamination is, the smaller the A_{280} reading and the larger the A_{260}/A_{280} value. If the A_{260}/A_{280} ratio is around 1.8, the DNA sample has a significant amount of DNA in it and is considered fairly clean or pure.

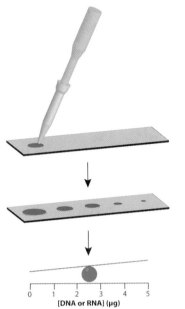

Figure 7.15. **Using a proprietary indicator, the Nucleic dotMetric™ assay can estimate the concentration of 1 μL DNA samples.**

Photo courtesy of G-Biosciences (Gene Technologies, Inc.).

As the amount of protein in the sample compared with the amount of DNA increases, the A_{260}/A_{280} ratio approaches 1.50 or even less, indicating a greater amount of protein contamination. Samples with A_{260}/A_{280} values less than 1.6 are commonly not used for DNA sequencing or PCR.

The closer the 260/280 ratio gets to 2.00, the more RNA contamination is suspected. RNA contamination may be a problem in a reaction, or it may not be. Proteases or RNase may be used to remove protein or RNA, respectively, from a sample. For both protein or RNA contamination, commercial kits are available to "clean up" a DNA sample.

In the example above, the sample is relatively pure, with little RNA or protein contamination. When determining the purity of a sample, the A_{260}/A_{280} reading will not be accurate if the concentration of DNA is too low in the sample. A technician must consider this when making inferences about the samples.

Sometimes the amount of sample is so small that measuring it in a spectrophotometer is not practical. New protocols have been developed to measure the concentration of DNA in volumes as small as a microliter. The NUCLEIC dotMETRIC™ quantitative nucleic acid indicator test, available from G-Biosciences (Geno Technology, Inc), will determine the concentration of 1 μL samples in the range of 1–10 ng/μL in 3 minutes (see Figure 7.15) For many applications, this relatively economical DNA concentration method saves time, product, and money.

Section 7.3   Review Questions

1. What are the lambda$_{max}$ values for DNA and colorless proteins in solution?
2. At what wavelength is DNA concentration determined?
3. What is the concentration of DNA in a sample if the OD at 260 nm is 0.43 au?
4. What is the relative purity of the sample in Question 3 if the OD at 280 nm is 0.29 au?

7.4   Other Specs and "Spec–Like" Instruments for Molecular Measurements

A critical instrument in the research and manufacturing of virtually all biotechnology products, a spectrophotometer is actually an instrument composed of two smaller instruments—a spectrometer and a photometer.

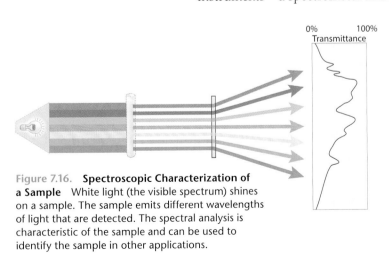

Figure 7.16.   Spectroscopic Characterization of a Sample   White light (the visible spectrum) shines on a sample. The sample emits different wavelengths of light that are detected. The spectral analysis is characteristic of the sample and can be used to identify the sample in other applications.

A spectrometer takes light and spits it into a spectrum of its component wavelengths. The spectrum of light shines on a sample and then the sample emits its own unique spectrum of light. That unique spectrum is like a fingerprint identifying the source and the spectrometer reports that data (see Figure 7.16). A photometer, one the other hand, measures the intensity of light that passes through or reflects off a sample. A spectrophotometer combines the two, separating light into its component wavelengths and then passing one of those wavelengths through a sample. The spec's light detector, or photometer, measures the amount of light transmitted through the sample. In this way, a spectrophotometer can measure the molecules in a solution.

Figure 7.17. Spectrometer can help solve crimes A Forensic scientist holding evidence box from a crime scene. Flakes of paint and other types of evidence (such as ash samples from a Picasso painting stolen in the Netherlands in 2012) may be analyzed in a spectrometer.
© Monty Rakusen/Getty Images.

Spectrometer Versus Spectrophotometer

A spectrometer is often confused with a spectrophotometer. Equally important in scientific research and diagnostics, a spectrometer measures the light emitted off or by a source, where as a spectrophotometer measures the light of a specific wavelength that passes through a sample (see Figure 7.17).

Use the Internet to learn more about how spectrometers are used in research and industry.

1. Find and record the website of articles, images, or videos on the Web that describe how a spectrometer is used to do the following. For each site, list one interesting thing you learned.

   a. Measure the amount of gold in coins, jewelry, etc.

   b. Look for chemical pollutants, such as greenhouse gases, in the environment

   c. Diagnosis skin cancer

2. Read the article at www.popularmechanics.com/science/news/a16270/tiny-spectrometer and describe how this new spectrometer can be so small but still measure the amount of light emitted from a source? List what it may be used for.

3. Go to www.popularmechanics.com/technology/gadgets/how-to/a8575/turn-your-laptop-into-a-spectrometer-14908361 to learn how to build your open inexpensive spectrometer. List 3 things you could study with a homemade spectrometer.

Mass Spectrometer

Spectrometers are used in other important biotechnology instrumentation and one of the most important is the mass spectrometer (mass spec). Mass spectrometry provides data on the molecular mass and elemental composition of a compound and that data can be used to recognize a compound in a mixture (see Figure 7.18).

A mass spec is very sensitive and is used in a variety of environmental, industrial, criminal and biomedical applications. For example, automated peptide synthesizers can produce short peptides (3–10 amino acids in length) for research or therapeutic purposes. After they are manufactured, the peptides are purified on an HPLC and then run through a mass spec to identify them by their molecular weight and to confirm the amount in the sample. In another example, forensic scientists use mass specs when analyzing biological samples from a crime scene. Mass spectroscopy can detect tiny amounts and differences in chemical and biological samples such as dyes, poisons, medications, paint flakes, glass splinters, and more.

Other spectrophotometers and spectrometers, some discussed below, use light of other wavelengths to measure molecules.

Infrared Spectrophotometer Infrared (IR) light is light with wavelengths of 700- 1500 nm. Similar to other specs, an infrared spectrophotometer produces a spectrum of infrared light and shines a selected wavelength on a sample. Depending on the bonds present in the compound being studied, the IR light is absorbed by the vibrating atoms' bonds. Different bonds, such as C-C, C-O, N-H and C-N bonds, vibrate

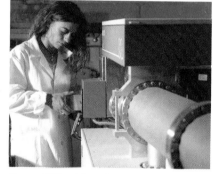

Figure 7.18. Mass Spectrometer Being Used by a Researcher Mass spectrometers are used to analyze the chemical composition of samples or materials. They work by converting the molecules in the sample into ions. The ions are propelled to form a beam, which is deflected by passing it through an electric or magnetic field. Each element has a different charge to mass ratio and so will be deflected by a different amount. This "spectrum" of ions can be analyzed, allowing very low concentrations of chemicals to be detected.
© Science Photo Library/Photo Researchers.

differently and absorb IR light differently. Light that is not absorbed passes through the sample and is measured. Analysis of IR absorption spectra reveals the types of bonds between atoms that are present in a sample, shedding light on the structure of the compound being studied

Fourier-transform Infrared Spectroscopy (FTIR)   FTIR spectroscopy also uses IR light to study chemical compounds. In FTIR though, an IR spectrum is shined on a sample and the whole IR wavelength range is measured at once time. A FTIR transmittance or absorbance spectrum reveals all the chemicals in a mixture. Analysis of the position, shape and intensity of peaks in the spectrum reveals details about the molecular structure of the sample. FTIR is able to determine the composition of complex mixtures of chemicals. In biotechnology applications, FTIR is used in research, manufacturing, and in quality control. It is used to confirm the purity and identity of reagent compounds and contaminants used in research and manufacturing as well as in products. FTIR is used to analyze food, feed, beverages, fuels, and industrial compounds (see Figure 7.19).

Figure 7.19. **IR Biomass analysis**   Using a Thermo Anteris II FT- NIR (Near Infrared Spectrometer) with an autosampler, a scientist analyzes corn stover, a crop residue used in biofuel production.
© Science Source/Photo Researchers.

Nuclear Magnetic Resonance (NMR) Spectroscopy
NMR uses the magnetic properties of atomic nuclei and how nuclei spin to determine the structure of a sample. NMR can study chemical reactions determining their mechanisms including kinetics, thermodynamics, and other physical organic properties. In biotechnology, NMR is used in chemical and biological molecule manufacturing and in protein quality control (QC) and quality assurance (QA) applications. While NMR is used to study molecules and reactions it also has several biomedical applications, where it is call MRI (magnetic resonance imaging). MRI is used to distinguish normal and abnormal tissues and organs in patients. Learn more about NMR and NMR spectrometers at: www.thermoscientific.com/en/products/nuclear-magnetic-resonance-nmr.html

Fluorescence Spectrometer (Spectrofluorometer, Fluorometer)   A spectrofluorometer measures the fluorescent light emitted from molecules when their electrons are excited by a light source. Some, but not all, molecules will fluoresce light when their excited electron release photons. Fluorescence is common in molecules with aromatic hydrocarbon rings, including several of the amino acids in proteins. A spectrofluorometer produces an excitation spectrum that resembles a UV absorbance spectrum and it is used to reveal structures and interactions in molecules.

One example of fluorescence spectrometry is the study of protein structural changes that may occur when a protein is in different environments. Since the amino acid tryptophan fluoresces differently dependent on the amino acids around it in a protein those differences are measured to understand if the amino acid is buried in the center of the protein or on the outside (see Figure 7.20). Fluorescence spectroscopy is used in many biochemical, medical, industrial, and chemical research applications for analyzing organic compounds.

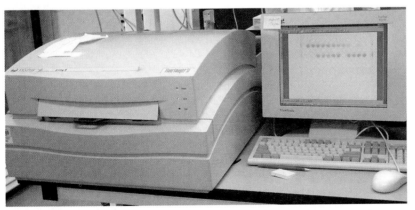

Figure 7.20. **Fluorometer Analysis of Protein Samples**   The fluorometer shown is being used to measure the difference in protein fluorescence in amylase samples that have undergone site-specific mutagenesis to change their structure to a more heat-tolerant one.
Photo by author.

1. What is the difference between a spectrometer and a spectrophotometer?
2. Which type of spectrometer is used to detect a compound by its molecular weight?
3. If a scientist is trying to determine the position of carbon to nitrogen bonds in a protein, which instrument might be used?

7.5 Applications of Spectrophotometry in Biotechnology

A detailed discussion of how VIS and UV/VIS spectrophotometers produce light of a specific wavelength and measure a sample's interaction with light was present in earlier sections. The importance of spectrophotometry in measuring cell cultures and protein and DNA concentrations during bioprocessing cannot be overstated. A UV/VIS spec is found in virtually every laboratory in a biotechnology facility.

Spectrophotometry is such an important diagnostic and research technique, that a spectrophotometer is built into other instruments to measure samples as they are studied, manufactured or in clinical testing. In a research facility, for example, UV specs are often built into or attached onto chromatographic columns used for protein purification (see Figure 7.21). High-performance liquid chromatography (HPLC) separates proteins in a mixture into small fractions to purify and characterize them from each other. A UV spectrophotometer is either built into or attached to the HPLC column and gives a quantitative analysis of the compound of interest after it comes off of the separation or purification column.

Clearly, spectrometry is an important and multipurpose technique, critical to the study and monitoring of molecules in biotechnology research and manufacturing. In this section, examples of how spectrometry is used in different biotechnology fields are given. This is just a survey of the important uses of spectrometry grouped using the *Domains of Biotechnology* graphic in Figure 1.13.

Figure 7.21. **HPLC with Attached UV Spectrophotometer** The HPLC/UV spectrophotometer analyzes peptide samples for purity. The HPLC column is in the small metal cylinder on the side of the unit. On the bottom of the unit the technician programs the UV spectrophotometer to measure the absorbance of the sample at different wavelengths as it comes through the column.
Photo by author.

Industrial and Environmental Biotechnology

Food and Beverage Production and Safety It is not an exaggeration to say that the use of mass spectrometry (MS) as a detection technique has completely revolutionized the analysis of food and beverage products. For example, in an attempt to monitor and characterize different varieties of wine, the industry uses MS to test wines for dozens of compounds that impact flavor, color, aroma, stability, and more. A MS can detect a wide range of compounds and identify them at very low concentrations. Testing foods for nutritional content, allergenic compounds, veterinary drugs, toxic residues, and food-processing contaminants are now routinely done using liquid chromatography (separation of compounds) and mass spectrophotometry (analysis of compounds).

Environmental Monitoring   Results obtained from UV spectrophotometry and MS can confirm the presence of different chemical and biologic residues in soil samples as well as ground, standing or flowing waters at very low concentrations. Government agencies and private industry do routine soil and water sampling in residential, agricultural and industrial areas to monitor the presence of contaminant molecules that may have been used or discharged into water supplies or onto plots of land. Examples of environmental contaminants include pesticides, herbicides, fertilizers, human or veterinary medications, trash, industrial processing products such as heavy metals or solvents. By monitoring the level of contaminant molecules, agencies can quickly respond to and remediate new sources of soil and water pollution that can impact the health of individual human and nonhuman organisms and communities.

Biofuels

Biofuels are fuels that originated from non-petroleum organic sources. These include ethanol from cellulose fermentation or oils from algae, seeds or other plant products (see Figure 7.22). Infrared spectrophotometry is extremely effective in monitoring oil production for biofuels as well as the oil content of food crops such as soybeans, corn, and sunflower seeds. See Chapter 14 for more on biofuels.

Medical Biotechnology

Pharmaceuticals   Traditional pharmaceutical companies, as well as companies that produce recombinant protein for medical purposes, must monitor and control the production of all molecules they manufacture. They rely on spectrophotometers and spectrometers for many of the steps in the process. Initially, all ingredients used in making pharmaceuticals are highly regulated. Prior to use, ingredients that go into pharmaceuticals are tested for purity and concentration using MS, IR, UV, or VIS spectroscopy. As a product is made and before packaging, each pharmaceutical is tested again in a series of quality control steps to ensure that patients are receiving safe and correct dosages of compounds that could be dangerous if not dosed correctly.

Clinical/Disease Diagnostics   Blood and urine research and diagnostics require several different methods and instruments, and spectrophotometric analysis is used to a large extent (see Figure 7.23). In routine, blood tests or urine testing, glucose concentrations are determined using spectrophotometry and indicators. This can provide important information about glucose blood levels that are a measurement of diabetic risk.

The presence, concentration, and activity of several blood proteins and other organic compounds are also measured using spectrophotometry. One example is the measurement of urea in blood or urine. Urea is a waste product from protein metabolism. Too high or too low of a urea concentration in blood or urine can indicate issues with kidney function.

Spectroscopy is now being used to diagnose disease. Recently infrared spectrometry has been used to differentiate between different kinds of arthritis by measuring differences in the synovial fluid (fluid in and around joints) of arthritis patients. Being able to distinguish between different types of arthritis allow doctors to prescribe the most appropriate medication.

Figure 7.22.   Algae Fermentation Bioreactor for BioFuel Production   Algae convert carbon dioxide into sugars during photosynthesis. Some strains of algae convert the sugars to oils that may be used as biodiesel. Solazyme, Inc. grows algae in fermentation bioreactor tanks and feed them sugar directly. This increases the algae's biodiesel production. Biodiesel oils are quantified using infrared spectrophotometry.
© Science Photo Library/Photo Researchers.

Figure 7.23.   HPLC analysis of a blood sample
Clinical pharmacology researcher studies the HPLC/UV spectrographic data of a blood sample analysis. The researcher is working to better monitor or quantify a pharmaceutical compound in the blood. A protocol such as this may be used to check a drug's stability or efficacy in the blood stream.
© Science Photo Library/Photo Researchers.

Another example of diagnostic spectrometry is the use of fluorescence spectroscopy to detect viruses in pathological samples. Tryptophan, in viral surface proteins, fluoresces and reveals information about viral infection. Fluorescence spectroscopy is in early development but is used for HIV studies and is expected to, in the future, be available as a routine diagnostic tool for microorganisms in daily medical practice. www.translational-medicine.com/content/7/1/99

Substance Abuse Testing   According to the National Institute on Drug Abuse at www.drugabuse.gov/publications/drugfacts/nationwide-trends, illegal drug (and alcohol abuse) has been increasing. In 2013, an estimated 24.6 million Americans aged 12 or older—9.4 percent of the population—had used an illicit drug in the past month. This number is up from 8.3 percent in 2002. Since drug abuse is closely correlated with crime, forensic scientists need sophisticated methods, such as IR spectrometry, to analyze samples collected from crime scenes for the dozens of illicit or illegal drugs (see Figure 7.24).

Although driving under the influence of alcohol has declined in recent years, drunk driving is still a significant problem in society with over 28 million arrests in 2013. Obviously, impaired driving is dangerous to the driver and the community. Testing for alcohol intoxication is one of the most common drug tests. Law enforcement can quickly and accurately measure blood alcohol content with breathalyzers that use indicator testing and intoxilizers that use infrared spectroscopy.

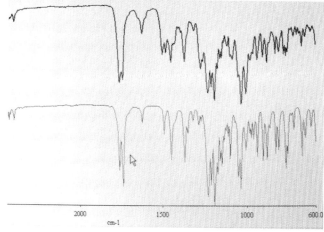

Figure 7.24.   Heroin Test   Heroin test.   Infrared (IR) spectra of the illegal drug heroin (blue) compared to that of an unknown sample (black). The samples match nearly perfectly, indicating that the sample contains a high percentage of heroin. Spectral analysis such as this can identify unknown compounds in mixtures or from samples taken from clothing or equipment. As such, the technique is widely used in forensic science.
© Science Photo Library/Photo Researchers.

Agricultural Biotechnology

Crop Analysis   Monitoring the growth, harvest, and formulation of agricultural crops require fast and accurate measurement of cells and molecules. Spectrophotometry provides a method to detect infections (including a coffee rust fungus), measure biomass, analyze nutritional factors, measure moisture and chemical content and provides quality control to ensure an excellent product.

Infrared spectroscopy has been used for many years to analyze both living plant tissue and dried plant materials. Scientists can quantify protein, amino acids, lipids, and carbohydrate contents in many crops including corn, canola, and soybeans. According to ASD, Inc., several types of spectrophotometry are used to analyze agricultural, timber and biofuel substrates for many of their constituent molecules including lutein, cellulose, hemicellulose, lignin, protein, moisture, ash, glucan, xylan, carbohydrates, and chlorophyll. www.asdi.com/applications/additional/agriculture-and-soils/commercial-agriculture

Figure 7.25.   Soil and Herbicide Analysis   A laboratory technician loads soil sample extracts (lower center) into the automatic sampling tray (center right) of a gas chromatograph-mass spectrometer. This will be used to test the samples for herbicide residues. Her research is being carried out at the Agricultural Research Service of the US Department of Agriculture.
© Science Source/Photo Researchers.

Soil Analysis   Analysis of soil nutrients and contaminants is a big business. Both field and laboratory UV/VIS and IR specs are used to quickly and economically measure organic and inorganic compounds in the soil. Based on spectrographic nutrient analysis, recommendations can then be made for fertilizer application. The mass spec can also analyze soil samples for toxic residues from organic or inorganic pesticides, herbicides, or fertilizers (see Figure 7.25). The build up of these residues can harm the crop product or the consumer.

Soil, Just a Bunch of Dirt?

Soils vary great from site to site and from time to time. In addition to rock and water, soil is a complex mixture of small animals, decomposing plants, microorganisms, and minerals in various concentrations. Since high quality soil increases the chances of successful crop production, being able to measure and compare soil samples is of critical importance. Spectrophotometry plays a big role in soil management.

Go to www.hunterlab.com/blog/color-measurement-2/methods-soil-analysis-using-spectrophotometric-technology and read about methods of soil analysis using spectrophotometric technology. Then answer the questions below.

1. What role does infrared spectrophotometry play in soil analysis?
2. What is "organic farming" and how does spectral analysis benefit organic farmers?

Then go to www.hunterlab.com/blog/color-chemical-industry/applications-of-spectrophotometry-in-agriculture-quantitative-analysis-of-fertilizer-properties to learn about fertilizers and how spectrometry is used to makes sure that the nutrient level in crop soils is optimal. Discuss how phosphates and nitrates are tested.

Section 7.5   Review Questions

1. Which type of spectrometry is used to measure alcohols in drug testing or in testing of biofuels?
2. Which type of spectrometry is primarily used to measure chemical contamination of soil or ground water?
3. Urea concentration is an indicator of kidney failure. What type of spectrophotometry is used to measure urea in urine samples?

Chapter Review

Summary

Concepts

- Since molecules can absorb the light of certain wavelengths, spectrophotometers may be used to detect molecules. Light energy that is not absorbed by molecules is reflected off the molecules or transmitted through molecules.
- When a sample is placed into a spectrophotometer, the light that is not absorbed by a sample is measured at the detector. The value is reported on the display as either % transmittance or the amount of absorbance in absorbance units (au).
- The amount of transmittance is related to the amount of absorbance by Beer's law. A spectrophotometer will perform the Beer's law conversion automatically to determine absorbance from transmittance.
- UV spectrophotometers have a deuterium light that produces light from 100 to 350 nm. These are wavelengths that are absorbed by many colorless molecules.
- VIS spectrophotometers have a tungsten light that produces light from 350 to 700 nm. These are wavelengths that are absorbed by most colored molecules.
- UV/VIS spectrophotometers have both a deuterium lamp and a tungsten lamp, and they can be set to wavelengths between 100 and 700 nm.
- When a sample's absorbance is measured at different wavelengths, an absorbance spectrum is produced. The wavelength of peak light absorbance is called lambda$_{max}$. The lambda$_{max}$ for a molecule is the wavelength that should always be used to study the molecule, since at lambda$_{max}$, the molecules are most strongly detected.
- If a sample has an absorbance peak that is the same as a molecule's characteristic lambda$_{max}$, then the presence of the molecule will be suspected.
- A spectrophotometer can monitor changes in colored reactions and is used commonly to assay for enzyme activity.
- The concentration of protein in samples can be estimated by comparing them to solutions of known concentration. Estimations can be done using indicator solutions with or without spectrophotometry. Using a spectrophotometer, a technician can determine the absorbance of standards of known concentration and plot them on a standard curve. The absorbance of the unknown samples are determined and plotted on the standard curve to estimate the concentration.
- Using standard curves is common practice in protein manufacturing.
- To determine the concentration of DNA in a sample, measure the absorbance of the sample at 260 nm and use the ratio:

$$\frac{50 \ \mu g/mL}{1 \ au} = \frac{X \ \mu g/mL}{Sample \ OD_{260} \ (au)}$$

- The relative purity of a DNA sample may be determined by obtaining an A_{260}/A_{280} value.

- A spectrometer and a spectrophotometer are similar in that each splits light into a spectrum. A spectrophotometer measures the way a sample interacts with light of one color or wavelength. A spectrometer measures the way a sample interacts with light of the entire spectrum that has been shone on the sample.
- Mass spectrometers are used to analyze the chemical composition of samples of materials and can recognize compounds by molecular weight. Characterizing molecules in pharmaceuticals and environmental monitoring of toxins is two examples of how MS technology is used.
- IR spectrometry identifies compounds based on the number and type of bonds present. IR spectrometry is used in many applications including soil analysis, alcohol testing, oil measurements, and more.
- NMR (also called MRI) uses magnets to study the properties of atomic nuclei. The data reveals information on the structure on molecules, cells, tissues, and organs.
- A spectrofluorometer may be used to detect small concentrations of DNA and protein. A sample produces a fluorescent excitation spectrum after a sample is exposed to light that may be used to understand the structure and concentration of molecules in a sample.
- Spectrometry is important in virtually every aspect of biotechnology research and manufacturing as well as diagnosis and remediation. UV/VIS spectrophotometry, IR spectrometry, mass spectrometry, and fluorometry are used to study and solve problems in agriculture, medicine, industrial products manufacturing, and environmental studies.

Lab Practices

- Spectrophotometers create light of different wavelengths. The colors of each wavelength of visible light can be observed by reflecting the light off a piece of filter paper in a cuvette.
- Wavelengths are measured in nanometers (nm).
- The visible spectrum (VIS) includes wavelengths between approximately 350 and 700 nm, and contains the colors of a rainbow.
- To use a spectrophotometer, fully warm it up and calibrate it.
- When a sample is placed in the spectrophotometer and light of a certain wavelength is directed on it, the molecules in the sample can absorb, transmit, or reflect the light energy.
- The spectrophotometer detects the light that is not absorbed by the sample and converts transmittance to absorbance.
- Absorbance is measured in absorbance units (au). On a Spec 20 D, absorbance values are most reliable between 0.1 and 1.1 au. Solutions should be prepared or diluted to given absorbance values in this range.
- Transmittance is measured in % and can be between 0% and 100%.
- Absorbance is also called optical density (OD).
- Calibrating a spectrophotometer requires use of a blank. A blank contains everything in a sample except the molecule of interest. By using the blank, and setting the transmittance to 100%, the spec is virtually ignoring all the other molecules in the sample except for the molecule of interest.
- The lambda$_{max}$ is determined by plotting a sample's absorbance data for different wavelengths on a line graph called an absorbance spectrum.
- The lambda$_{max}$ is needed for estimating the concentrations of protein (or DNA) samples.
- Amylase is made colorful by using a general protein indicator such as Bradford reagent. Bradford reagent turns from brown to blue in the presence of protein. The more blue, the more protein. A lambda$_{max}$ can be determined for an amylase-Bradford mixture.
- Once the lambda$_{max}$ is determined, it can be used to measure the absorbance of unknown amylase samples. This is done by comparing the absorbance of solutions of *known* amylase concentrations with the absorbance of solutions of *unknown* concentrations. This is possible by the creation of a best-fit standard curve.
- A linear regression equation, $y = mx + b$, representing a best-fit standard curve of the absorbance of known protein solutions, can be used to mathematically determine the concentration of unknowns.
- Lambda$_{max}$, standard curves, and protein-concentration determinations can be made for colorless proteins using a UV spectrophotometer. The methods employed are similar to those used on a VIS spectrophotometer, except that a smaller amount of sample is needed, and the protein sample can be retrieved for further use.

Thinking Like a Biotechnician

1. What color is light of the following wavelengths: 600 nm, 525 nm, and 475 nm?
2. A colorless protein is purified from a cell extract. What kind of spectrophotometer should be used to detect its presence and concentration?
3. A molecule has a lambda$_{max}$ of 475 nm. What wavelengths would probably not be good to use for testing samples for the presence of the molecule?
4. A set of standards is prepared by diluting a stock sample in a 1:2 ratio. If the stock solution has an absorbance of 1.2 au, and the 1:2 dilution has an absorbance of 0.6 au, what would be the expected amount absorbance of the 1:4, 1:8, and 1:16 dilutions? If the absorbance of the dilutions is not as expected, what might be the reason?
5. You are growing algae as a film on a glass slide in a biofuel research lab. To understand more about how the algal film is interacting with light, shall you chose a spectrometer or a spectrometer to test the sample?
6. A dog has been poisoned and a sample of saliva has been collected and brought into the lab to be tested. What instrument can be used to test the saliva for compounds by their molecular weight?
7. As a forest and timber researcher, you are studying the effects of a prolonged drought on the growth of pine trees. You have collected samples that have been properly stored from each of the past 10 years. What type of spectroscopy would be appropriate to use to analyze these samples for differences in their chemical composition, including celluloses, lignins, and other carbohydrates?
8. Since food allergies can be life threatening it is critical that food manufacturers measure and report each ingredient in a processed food. Facilities that process foods must test their products for traces of food allergens including peanuts, milk, wheat, soy and more. What spectroscopic instrument is used to test foods for allergenic proteins?
9. Soil samples are taken from a farm that claims to grow their food adhering to strict "organic" standards of no added herbicide, pesticide, or inorganic chemical fertilizers. What spectroscopic instruments might be used to analyze the samples to look for pesticide residues?
10. An HPLC is commonly used in conjunction with a MS. The HPLC separates compounds based on size, charge, or chemical nature prior to the sample being analyzed in the mass spec. Why is it important to use the HPLC when trying to characterize the amino acids in a mixture?

Biotech Live

Activity 7.1 ## DIY Spectrophotometry

Build your own spectrophotometer. Find instructions at The Royal Society of Chemistry website.

www.rsc.org/education/eic/issues/2007Sept/BuildYourOwnSpectrophotometer.asp

Once it is built, use your homemade spec to create absorbance spectra for 3 different fruit juices. Make sure the juices are clear but colored. Dilute them down until the absorbance data can be measured. See Laboratory 7b in the lab manual for hints how to set up the analysis.

Activity 7.2 ## Why do Proteins Absorb UV Light?

Colorless proteins absorb light in the UV spectrum. This characteristic may be used to recognize and measure them in solution.

Use the Internet to learn more about UV spectrophotometry of proteins. Record the URL of the sites that helped you.

1. Find a website that explains why colorless proteins usually absorb light at 280 nm. Summarize your findings.
2. Not all amino acids in a protein absorb light at 280 nm. Which amino acids do and how does that affect the overall absorbance of a colorless protein?

Activity 7.3 ## Diagnostic Spectrometry is Spot on!

A new type of spectroscopy called diffuse reflectance spectroscopy is being used with great success to identify malignant breast tumors.

Go to www.medicalnewstoday.com/articles/254502.php to learn more about diffuse reflectance spectroscopy and how it is used in tumor diagnosis. Read through the article and answer the questions below.

- What is diffuse reflectance spectroscopy and how is it used to distinguish between malignant and benign breast tumors?
- How many samples were tested in this study? And how were the different tumor results grouped?
- What is an advantage to using diffuse reflectance spectroscopy during breast tumor biopsies versus other methods?

Bioethics

Test Results: Who Should Get Access to Them?

The Hippocratic Oath

I swear by Apollo the physician, by Æsculapius, Hygeia, and Panacea, and I take to witness all the gods, all the goddesses, to keep according to my ability and my judgment, the following Oath.

"To consider dear to me as my parents him who taught me this art; to live in common with him and if necessary to share my goods with him; to look upon his children as my own brothers, to teach them this art if they so desire without fee or written promise; to impart to my sons and the sons of the master who taught me and the disciples who have enrolled themselves and have agreed to the rules of the profession, but to these alone the precepts and the instruction. I will prescribe regimens for the good of my patients according to my ability and my judgment and never do harm to anyone. To please no one will I prescribe a deadly drug nor give advice that may cause his death. Nor will I give a woman a pessary to procure abortion. But I will preserve the purity of my life and my art. I will not cut for stone, even for patients in whom the disease is manifest; I will leave this operation to be performed by practitioners, specialists in this art. In every house where I come I will enter only for the good of my patients, keeping myself far from all intentional ill-doing and all seduction and especially from the pleasures of love with women or with men, be they free or slaves. All that may come to my knowledge in the exercise of my profession or in daily commerce with men, which ought not to be spread abroad, I will keep secret and will never reveal. If I keep this oath faithfully, may I enjoy my life and practice my art, respected by all men and in all times; but if I swerve from it or violate it, may the reverse be my lot."

Although outdated and not really "politically correct," the Hippocratic oath is presented to physicians upon graduation from medical school. For doctors, genetic counselors, clinical scientists, and other healthcare professionals, the underlying overall message of the Hippocratic oath is valuable. In essence, it says that healthcare professionals will not knowingly do harm to their patients and that they will maintain a patient's privacy. But is privacy the best policy? Are there any circumstances in which a person's medical condition or test results should be revealed to persons other than the patient? When does a patient's right to privacy outweigh the public's right to knowledge and protection, and vice versa?

To Think About: Who should have access to an individual's medical records and under what circumstances?

Read each of the scenarios that follow. Using the chart displayed, decide which of the following persons or organizations should be given access to "private" medical information.

Scenario 1: A male fetus, still in the mother's womb, is diagnosed with HIV infection through antibody testing. The mother has been living with AIDS for over a year. The father (HIV clear) does not live in the same state as the pregnant mother. Who should be informed of the fetus' condition?

Persons/Organizations	Should be informed? Yes, No, or NA (not applicable)	Why/Why not? Any special conditions/qualifications?
Parent(s) of Patient		
Day Care Workers/Parents		
Healthcare Workers of Patient		
Educators of Patient		
Military		
Insurance Companies		
(Health, Car, Life)		
Department of Motor Vehicles		

Scenario 2: A male patient is diagnosed with generalized epilepsy characterized by grand mal seizures during which he may lose consciousness. He is a first-year kindergarten teacher responsible for 30 5-year-old children. His school is 35 miles away from his home. Who should be informed of his condition?

Persons/Organizations	Should be informed? Yes, No, or NA (not applicable)	Why/Why not? Any special conditions/qualifications?
Patient's Parents		
Spouse		
Adult Children of Patient		
Employer of Patient		
Parents of Students of Patient		
Military		
Insurance Companies (Health, Car, Life)		
Department of Motor Vehicles		

Scenario 3: A female pilot, who has just completed flight training, has been recently diagnosed with sickle cell disease through a genetic test and blood cell analysis. She has trained for 10 years to be an airline pilot and this is her "big break." She was hoping to serve in the Air National Guard as a pilot before entering her civilian job. She has never had any of the symptoms, including the most serious one, weakness due to lack of oxygen. Who should be informed of her genetic condition?

Persons/Organizations	Should be informed? Yes, No, or NA (not applicable)	Why/Why not? Any special conditions/qualifications?
Patient's Parents		
Fiancé		
Spouse		
Children of Patient		
Employer of Patient		
Educators of Patient		
Military		
Insurance Companies (Health, Car, Life)		
Department of Motor Vehicles		
Federal Aviation Administration		

Biotech Careers

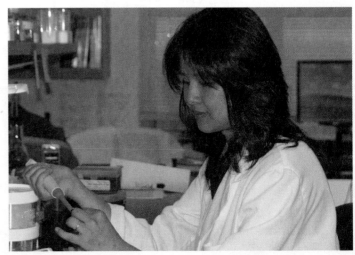

Photo courtesy of Ping Chen.

Molecular Biologist

Ping Chen, PhD
Emory University School of Medicine
Atlanta, GA

Dr. Chen's research team studies the sensory organ for hearing deep in the human ear. The team applies procedures used in molecular biology to study the molecules and genes that are important to the formation of the ear during embryonic development. These genes can then serve as potential targets for research with the aim of rejuvenating an aging or faulty hearing organ in patients with hearing impairment.

Dr. Chen directs a research team of students, research associates, and postdoctoral fellows. Her team works on DNA and protein analysis both in cells (in vivo) and in the lab (in vitro). DNA isolation, genetic engineering, PCR, and protein recognition using biomarkers are common techniques used in Dr Chen's lab. She also teaches graduate and medical school courses. Dr. Chen performs laboratory tasks and supervises others in their research. As do many scientists, Dr. Chen participates in scientific conferences, stays current on scientific literature, and contributes to science journals.

8 The Production of a Recombinant Biotechnology Product

Learning Outcomes

- Outline the fundamental steps in a genetic engineering procedure and give examples of genetically engineered products
- Describe the mechanism of action and the use of restriction enzymes in biotechnology research and recombinant protein production
- Discuss techniques used to probe DNA for specific genes of interest
- Explain the steps of a bacterial transformation and various selection processes for identifying transformants
- Differentiate transformation, transfection, and transduction
- Discuss the considerations for scaling up the production of transformed or transfected cells, the general cell culture protocol for scale-up, and the importance of complying with standard manufacturing procedures
- Explain the usefulness of plasmid preparations, how they are performed, and how the concentration and purity of plasmid samples can be determined

8.1 An Overview of Genetic Engineering

One of the most important methods in biotechnology is genetic engineering (GE), which includes all the techniques and technologies of modifying and manipulating the genetic information in cells. Genetic engineering encompasses any deliberate change made in the genetic code of an organism. The engineering can be as simple as modifying a single nucleotide in a DNA sequence or transferring a single gene to a cell, to being as expansive as changing large sections of chromosomes that code for multiple characteristics. Even a single nucleotide change can cause drastic changes in protein structure and function for better or worse (see Figure 8.1). Many changes in genetic information can lead to new or improved products.

The specifics of each genetic modification procedure are unique to the particular organism being engineered and to the characteristics of the desired product. In general though, the steps of genetic engineering to produce a protein product use the processes of recombinant DNA technology, **transformation**, cloning, and protein purification. The general procedure for a genetic engineering procedure is listed below. Figure 8.2 provides a visual representation of these steps.

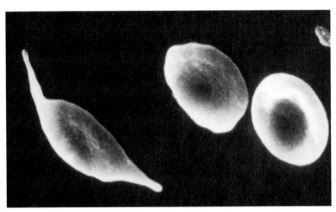

Figure 8.1. The dramatic results of a single nucleotide change in the DNA code for hemoglobin are seen in the abnormal, sickled red blood cells (RBCs) of a sickle cell disease patient. A single nucleotide replacement in the gene for hemoglobin can change a glutamic acid to valine leading to the sickling shape in the cells in low oxygen conditions. This is an example of how changes in DNA can have adverse effects. Genetic engineers, however, try to improve phenotypes (characteristics) by modifying DNA. ~2,000X
© Custom Medical Stock Photo.

**transformation
(trans•for•ma•tion)** the uptake and expression of foreign DNA by a cell

General Protocol—Genetic Engineering to Produce a Protein Product

1. Recombinant DNA Technology - The genetic code (DNA) for the desired characteristic or protein is identified and isolated from a donor cell, confirmed by restriction digestion and/or sequenced, and pasted into a vector, producing a recombinant DNA [rDNA] plasmid, that can carry the desired DNA code into a recipient cell.

2. Transformation - Genetically engineered cells are produced when the recombinant vector carrying the gene of interest is transferred into new host cells. If the cells express the new DNA, transcribing and translating it into a new (recombinant) protein, the cells are then said to be "transformed." Assays confirm that the transformation has happened by testing for the gene product.

3. Cloning - The transformed cells, producing their recombinant protein, are grown in culture (cloning). First, they are grown on a small scale in Petri plates and small broth cultures up to 10 L (fermentation). If the transformed cells grow well and are making sufficient amounts of the recombinant protein they are scaled up to larger volumes, up to 30,000 L (manufacturing).

4. Purification - Finally, the recombinant protein product, being produced in manufacturing, must be isolated and purified from the cells and other proteins in the cell culture. Protein purification is a highly technical procedure that is specific to each protein product. Before purified recombinant protein product is sent to market, it is tested for purity and may have to go through governmental approvals. The final formulation of the product as a liquid, powder, tablet, etc., depends on what is best for delivering a good product to the user.

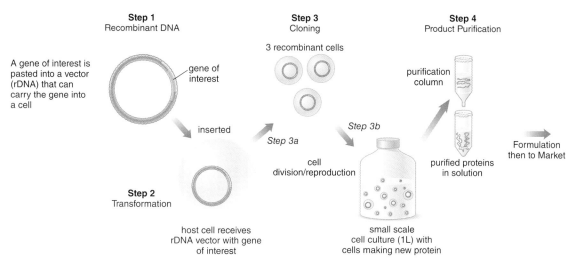

Step 1
Recombinant DNA

A gene of interest is pasted into a vector (rDNA) that can carry the gene into a cell

gene of interest

inserted

Step 2
Transformation

host cell receives rDNA vector with gene of interest

Step 3
Cloning

3 recombinant cells

Step 3a

Step 3b

cell division/reproduction

small scale cell culture (1L) with cells making new protein

Step 4
Product Purification

purification column

purified proteins in solution

Formulation then to Market

Figure 8.2. **The Genetic Engineering Process.** There are four main steps in the genetic engineering process used to make a recombinant protein.

By altering the genetic code, genetic engineers can produce cells or organisms with new or modified proteins that result in new, improved characteristics. The genetically modified organisms (GMOs) are capable of synthesizing new products potentially beneficial to humankind. There are hundreds of genetically engineered products currently on the market representing all the domains of biotechnology.

The majority of genetically engineered products are pharmaceuticals made using recombinant DNA technology, combining the DNA from two or more organisms. For these products, the DNA code for existing proteins with important medical value is modified and moved to cells that can produce the proteins in large amounts. An example is the pharmaceutical, Neupogen® (Amgen, Inc.). The protein in Neupogen® is recombinant filgrastim (rFilgrastim). Naturally made in small amounts in the human body, filgrastim is a protein that stimulates the production of certain white blood cells (granulocytes). rFilgrastim is produced by engineering the gene for human filgrastim into a plasmid vector. That recombinant plasmid is then transferred into *E. coli* bacteria, transforming them into filgrastim producers. The rFilgrastim is harvested from the genetically engineered bacteria and available to market. As the key ingredient in Neupogen®, rFilgrastim encourages white blood growth and reproduction, helping patients fight infection (see Figure 8.3).

Another example of a recombinant pharmaceutical is a vaccine called CERVARIX® (GlaxoSmithKline). CERVARIX® is a recombinant DNA variation of a human papillomavirus (HPV). HPV infection is known to significantly increase the risk of cervical cancer in women and genital warts in men and women. To produce rHPV, the genetic code for the virus is altered so that it cannot cause cancer. When rHPV is injected into a person as a vaccine, antibodies against HPV are produced in large amounts with the expectation that it will stimulate the body to recognize and destroy virulent HPV in the future. Removing HPV before it can infect cells reduces the likelihood that it can cause cancer in a person.

Several genetically engineered products that are used as industrial and agricultural products are on the market and many more are in the pipeline. IndiAge® cellulase and Purafect® protease, both by Genencor International, Inc. are recombinant enzymes that are produced by genetically engineering selected bacteria and fungi. IndiAge® cellulase breaks down cellulose fibers in plant materials such as cotton and is used to make denim jeans softer. Purafect® protease is recombinant enzyme product that is used in detergents to better remove protein stains.

Some examples of genetically engineered agricultural products are Roundup Ready® soybeans and Bollgard II ™ cotton both by Monsanto Co. For these products, selected genes are modified in target plant cells and engineered plants are grown from the

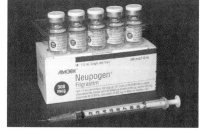

Figure 8.3. **Recombinant Filgrastim** Neupogen (rFilgrastim) is a granulocyte colony-stimulating factor (G-CSF) analog protein used to stimulate the production of white blood cells when chemotherapy has lowered their production. Neupogen is one of dozens of recombinant pharmaceuticals manufactured by Amgen, Inc., which is one of the largest biotechnology companies. Learn more about Amgen and their products at www.amgen.com/products.
© Leonard Lessin/Photo Researchers.

genetically engineered cells. In Roundup Ready® soybeans, the modified gene is one that makes the soybean plant resistant to the herbicide Roundup®. When Roundup® is sprayed on Roundup Ready® soybean fields, weeds die but the Roundup Ready® soybeans do not.

In another example, cotton and some other agricultural crops, are damaged by bollworm infestations. To grow cotton economically, bollworms must be controlled. Before Bollgard II™ cotton, farmers sprayed their fields with large amounts of pesticide to kill bollworms. Pesticide runoff though can contaminate water sources so scientists looked for other ways to control bollworm. Genetic engineering cotton, with two genes from the bacterium *Bacillus thuringiensis* inserted into it, confers resistance to these cotton-devouring insects. These genes produce the insect-killing compounds Cry1Ac and Cry2Ab that are toxic to bollworms when eaten. Bollgard II™ cotton, a genetic engineering product, has increased U.S. cotton yield significantly.

Genetically engineered products require a significant investment in research and development (R&D). The rationale for investing money and human resources in developing these products rests on whether a product can be more successfully produced through genetic engineering than through conventional manufacturing methods. One example of a process improved through the genetic engineering of a protein is cheese production (see Figure 8.4).

Chymosin is the genetically engineered version of the naturally occurring protein, rennin. Rennin, an enzyme purified from calf stomachs (veal), is a protease that cleaves casein, the protein in milk, into fragments. The casein fragments fall out of solution (precipitation) and appear as curds in milk. Curds are pressed and aged into cheese. Rennin is used in cheese making to speed cheese production. The price and availability of rennin have always been problems for cheese makers because of the availability of veal. As veal production goes up, rennin is more readily available, which makes the cost more reasonable. However, when there is a decrease in the veal supply, the price of rennin skyrocket

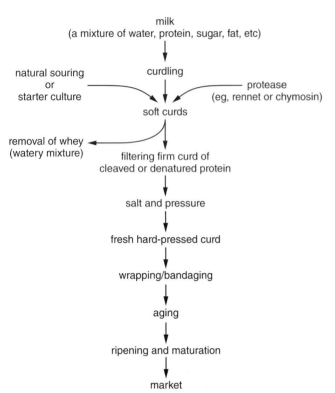

Figure 8.4. Cheese Production. Cheese is composed of the curds of degraded milk protein. The milk might curdle naturally from contamination by random bacteria. Sometimes "good" bacteria fall into the milk and produce "good cheese." However, often, "bad" bacteria fall into the milk and result in bad-looking, -tasting, or -smelling cheese. Also, some "bad" bacteria can produce toxic compounds. An enzyme such as rennin or chymosin (genetically engineered rennin) can be added to speed the curdling process and avoid "bad spoiling."

Biotechnologists at Genencor International, Inc. recognized the need for a constant source of rennin. They isolated the rennin gene from calf cells and transferred it into fungal cells. The fungal cells, not distinguishing between their own DNA and foreign DNA, began transcribing the calf DNA and producing the rennin. The transformed fungal cells were grown in very large amounts, and the genetically engineered rennin was purified from them.

Producing genetically engineered rennin is cheaper and more reliable than obtaining it from cows. Think about the ease of growing fungi in a lab compared with raising cows. To distinguish calf rennin from genetically engineered rennin, Genencor researchers renamed the enzyme chymosin. Chymosin (marketed as ChyMax®) has been on the market since the mid-1980s.

Genes for Recombinant DNA

If a potential product is to be made through genetic engineering, it must have originated from an organic compound already produced in some type of cell. With a small amount of information about its structure, a scientist can locate cells that make the compound of interest and then locate and isolate the genes (DNA instructions) for it's synthesis. If the genetic code can be isolated and placed into a plasmid vector, the recombinant vector can transfer the genetic code into cells and hopefully the synthesis of the target compound will begin.

Isolation and purification of target DNA depend on the type of cell. However, no matter what the source is, the first step in retrieving DNA is to explode or lyse the cells. Some cells have walls that must be eliminated. All cells have membranes and cytoplasmic constituents that must be removed. When burst open, the cells dump all of their molecules into the collection vessel. Centrifugation steps and special buffers allow the DNA molecules to be separated from all of the cell's other molecules (see Figure 8.5).

Target gene sequences may come from any kind of cell. Due to the differences in cell structure, the actual protocol (what reagents to use and in what order) for isolating animal, plant, fungal, or bacterial genomic DNA varies. There are many steps and reagents required for the isolations. In the early days of biotechnology, researchers prepared all their own reagents and solutions for DNA isolation and did everything from scratch. Now, it saves time and money to purchase kits for genomic or plasmid isolation (See Figure 8.6). The kits speed the retrieval and purification of all types of nucleic acids from all kinds of cells.

Once the target DNA has been removed from cells and purified from contaminant molecules, the search for specific genes of can begin. Locating and isolating the DNA code for a specific protein is no trivial task. A cell contains so much DNA, and so many genes, that searching for a single gene is often like "looking for a needle in a haystack."

One method of locating a gene for a particular protein is by using a **probe**. A probe is a DNA or RNA molecule that is complementary to the DNA sequence being sought (see Figure 8.7). One method of probing DNA involves the use of restriction enzymes (DNA chopping enzymes) and gel electrophoresis. Restriction enzymes digest the DNA isolated

Figure 8.5. This high-speed microcentrifuge is used to help separate extracted DNA from other cell wastes. At about 12,000 × g ("g" is a unit of pull due to centrifugal force), cell wastes precipitate to the bottom of microtubes, and the DNA stays suspended in the supernatant.
© Andrew Brookes/Getty Images.

probe (probe) a single-stranded DNA or RNA molecule that is complementary to the DNA sequence being investigated, often bound to some kind of "reporter" molecule, used when looking for a gene or nucleic acid sequence

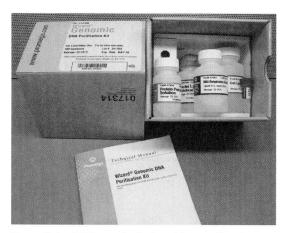

Figure 8.6. DNA may be isolated from cells using a genomic DNA purification kit such as this one by Promega Corp.
Photo by author.

Step 1

genomic DNA isolated from cell samples, loaded and run on gel

Step 2
DNA probe with radioactive label
The probe's DNA sequence is complementary to the DNA sequence of the gene of interest.

Probes are washed over the gel.

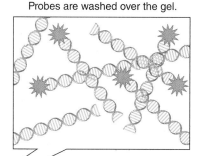

Probes bounce around until they find their complement. They bind and emit radiation that can be seen on x-ray film.

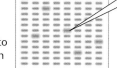

Figure 8.7. **Probing DNA.** A probe is used to identify certain DNA sequences that are hidden among the billions of nucleotides in a genome.

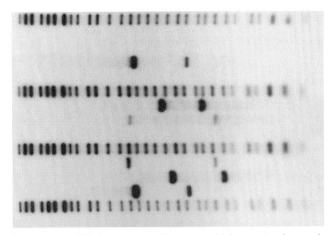

Figure 8.8. This is an autoradiogram, which is a visual record of a gel analysis created on film by the reaction of small amounts of radioactivity present in each sample of DNA. The samples have been processed using the polymerase chain reaction (PCR), a technique in which small sections of the DNA are amplified from very small starting amounts. Each band on the autoradiogram represents thousands of identical fragments of amplified DNA. Genomic (chromosomal) DNA must be extracted from cells for this type of lab work.

© 2005 Custom Medical Stock Photo.

hybridization (hy•brid•i•za•tion) the binding of complementary nucleic acids

autoradiogram (au•to•ra•di•o•gram) the image on an x-ray film that results from exposure to radioactive material

Southern blotting (south•ern blot•ting) a process in which DNA fragments on a gel are transferred to a positively charged membrane (a blot) to be probed by labeled RNA or cDNA fragments

blot (blot) a membrane that has proteins, DNA, or RNA bound to it

cDNA (cDNA) abbreviation for "copy DNA," cDNA is DNA that has been synthesized from mRNA

primers (pri•mers) the small strands of DNA used as starting points for DNA synthesis or replication

thermal cycler (therm•al cy•cler) an instrument used to complete PCR reactions; automatically cycles through different temperatures

from the source cells. The restriction fragments are then split into single-stranded pieces of DNA and run on an agarose gel. Once the DNA pieces are separated, a solution containing probes is flushed over the gel. The probe molecules are "tagged" with either a radioactive marker or a fluorescent label. The probes bounce around the gel. If they bump into a complementary sequence, they bind to it. This is called **hybridization**. The hybridized "tag" shows the location of the DNA sequence of interest. If the tag is a radioactive label, then an x-ray film can be exposed, resulting in a permanent record, called an **autoradiogram**, of the hybridized complex (see Figure 8.8). Although radioactive methods are still used, other methods of visualizing probes without radioactive labels are routinely being used. Some of these methods are inherently easier and safer than using radioactivity.

DNA probes are also used to identify DNA during **Southern blotting**. In a Southern blot, a gel with the DNA fragments is transferred to a positively charged membrane (the **blot**). The blot is then probed. Once the DNA sequence of interest is identified on the blot, its can be located on the gel and removed for further processing. Chapter 13 has a detailed discussion about Southern blots.

Probes can also be made from RNA molecules. Some cells, like the pancreas, make an abundance of certain proteins (insulin). To produce large amounts of insulin, pancreas cells, in turn, must contain large amounts of the mRNA that codes for insulin. The insulin mRNA code is complementary to the insulin gene (DNA) code. When RNA is extracted from pancreas cells, insulin mRNA is in the highest concentration. The RNA molecules can be separated, and the insulin mRNA can be isolated for use as a probe. When RNA is used as a probe, the process is called mRNA/DNA hybridization.

One drawback of mRNA probes is that they commonly degrade at high temperatures. To avoid this, scientists often produce copy DNA (**cDNA**) probes. Using reverse transcriptase and DNA polymerase, copies of RNA, complementary to the original DNA sequence of interest, can be synthesized. The process is "reverse transcription" because the production is from RNA to DNA, not the "normal" process of DNA to RNA.

If probe technology is successful and a gene of interest is located and isolated, the gene must be pasted into a vector plasmid and inserted into host cells. If the host cells, express the new genes by making new proteins, then the genetic engineering is a success.

Another method of finding a gene of interest utilizes the polymerase chain reaction (PCR). During PCR, two **primers** (similar to probes) recognize two target DNA sequences that are some distance apart, flanking a gene if interest. Using a polymerase enzyme, PCR can replicate the primer-delineated, targeted gene sequence to millions of copies within a few hours. A machine called a **thermal cycler** (see

Figure 8.9. A thermal cycler is used to make millions of copies of DNA fragments in just a few hours. Underneath the cover is a heatblock that holds 96 sample tubes that are cycled through different temperatures. Using primers with a specific DNA sequence, PCR can recognize and amplify a target gene sequence that later may be inserted into a vector for genetic engineering purposes.

© Alcuin/iStock.

Figure 8.9) is required for a PCR. Assuming you know the approximate size of the piece you are looking for, the PCR product (DNA of interest) can be confirmed by gel electrophoresis and used to create a recombinant plasmid for genetic engineering. Chapter 13 presents much more on PCR technology and its widespread applications.

By whatever method, once the gene of interest is isolated, it can be pasted into a vector plasmid and the recombinant DNA can be transferred into new host cells. If the recipient cells express the genes and produce a nonnative protein, the cells will exhibit new traits. These cells then, have been transformed through genetic engineering. The ability to transform cells into protein factories was the driving factor in creating the biotechnology industry. Genentech, Inc. was the first biotechnology company to market a product (human insulin) manufactured in transformed cells (see Figure 8.10).

Figure 8.10. DNA Way at Genentech, Inc., South San Francisco, CA. In 1982, Genentech was the first biotechnology company to market a rDNA product, recombinant human insulin (HumalinR®). Its headquarters in South San Francisco is primarily responsible for the R&D of pharmaceutical products. Wedged between three buildings, DNA Way is visible from planes as they fly into and out of the San Francisco International Airport. Photo by author.

BIOTECH ONLINE

Some Say Genetic Engineering Is a Fishy Business
For more than 10 years, scientists have been genetically modifying species of fish. These genetically modified organisms are called GMOs. Genetic engineering of fish used for food is a big business. Bigger fish that can better resist disease or environmental changes are very desirable for obvious reasons.

Using information posted at the Food and Agriculture Organization of the United Nations website at biotech.emcp.net/fao_genfish, learn how genetic engineering has modified several aquatic species.

1. Study the table showing fish species and the target genes that have been added to them. List four different species of fish that have each received the GH gene. What traits does the GH gene confer to the GMO fish? List a potential pro and a potential con of engineering this gene into these fish.

2. List two species of fish that have each received the AFP gene. What traits does the AFP gene confer to the GMO fish? List a potential pro and a potential con of engineering this gene into these fish.

3. Scroll down the website. Find the discussion of the techniques used to induce the transgenic (transfer of genes to produce GMOs) fish. Describe two methods of how genes are transferred into fish species.

Section 8.1   Review Questions

1. What is the name of the genetically engineered rennin molecule? Why is it desirable to produce genetically engineered rennin instead of harvesting rennin from nature?
2. What is the name of the process that occurs when complementary pieces of DNA or RNA locate and bind to each other? How is that technology used in the laboratory?
3. What is a Southern blot, and how is it used?
4. What does a thermal cycler do?

8.2 Using Recombinant DNA for Transformation

Transformation is a general term describing the uptake and expression of foreign DNA by a cell. If a transformation is successful, the recipient cell acquires the characteristics coded for on the incoming DNA, and a new phenotype arises. When this happens, we say that the cell has been "transformed."

One of the first transformations was the transfer of the human insulin gene into *E. coli* cells. In nature, insulin is only made in mammalian pancreas cells. Genetic engineers working at Genentech, Inc. revolutionized medicine when they isolated the human insulin gene and inserted it into bacteria cells. The bacteria cells read the human DNA, transcribed it into mRNA, and produced the human insulin protein (see Figure 8.11).

Under the "right" conditions, *E. coli* bacteria grow well in a lab. As the transformed cells grow and multiply, they create large vats of insulin-producing bacteria. The insulin is purified from the bacteria cells and sold as a therapy for diabetes. In this way, bacteria cells are manipulated to produce large amounts of a protein they normally do not make.

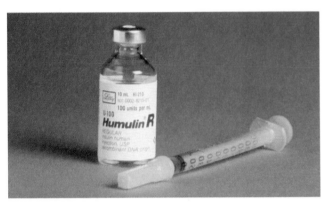

Figure 8.11. Humulin R® contains recombinant human insulin. It was the first insulin analog approved for use in insulin-pump therapy.
© Custom Medical Stock Photo.

Scientists first conceived of doing bacterial transformations when they observed that some living, naturally occurring bacteria have the ability to take up DNA from dead cells. The cells taking up DNA showed new characteristics; they were transformed. However, when scientists tried to add individual DNA gene fragments into cells, only a very few cells were transformed. The transformation efficiency was very low. It was soon observed that plasmid DNA, small rings of extrachromosomal bacterial DNA, was taken up rather easily and better expressed by certain bacteria cells.

Scientists proposed using plasmid DNA as vectors to carry genes of interest into cells. They learned how to introduce genes of interest into plasmids, producing what is now known as rDNA. The rDNA plasmids, carrying the genetic code from at least one different species, were actually doing the transforming. Since the first transformations were conducted in the 1970s, scientists have also learned how to use viral DNA as vectors of genetic information (see Figure 8.12). When viruses are used to transform cells, the process is called **transduction**.

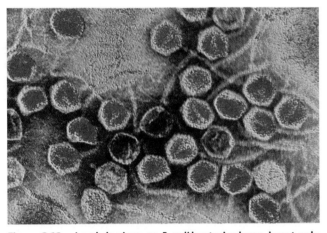

Figure 8.12. Lambda virus, an *E. coli* bacteriophage, is not only a common cloning vector, it is also the source of DNA for the lambda/*Hind*III sizing standards used on agarose gels. 18,000X
© Science Photo Library/Photo Researchers.

Soon after scientists discovered how to produce rDNA from bacteria, they learned how to transform mammalian cells. The term "transfection" is used to describe mammalian transformation. Transfection is a little trickier than transformation of bacteria cells because mammalian cells are so much more complex than are bacteria. Often, additional regulatory sections of DNA must be used to ensure transfection success. Also, it is common to use transcription inducers to induce expression of transformed genes.

Several kinds of mammalian cells are used as host cells, including Chinese hamster ovary cells (CHO) and mouse kidney cells. These cells have good transfection efficiencies and grow well in broth culture. Viruses are often used as vectors to carry genes of interest into the mammalian host cells. Viral DNA is cut with restriction enzymes and a gene of interest is inserted. Cosmids are also used as vectors. Cosmids are like very large plasmids. It is also common to inject DNA directly into the nucleus of a mammalian cell using a microinjection syringe.

transduction (trans•duc•tion)
the use of viruses to transform or genetically engineer cells

BACs versus YACs

To transform cells with large pieces of DNA, larger vectors are needed. Scientists have developed BACs and YACs to host pieces of foreign DNA up to 500 kb (kilobases) in size. These have been critical for transformation, sequencing, and genome projects.

TO DO

Using at least two different websites, learn more about BACs and YACs, and their uses in biotechnology. Create a chart like the one below to record the information you gather.

Characteristics	BAC	YAC
meaning of acronym		
primary uses		
circular or linear		
typical size of insert		
another interesting fact		
reference 1		
reference 2		

Transformation Begins with Recombinant DNA

As you know, a plasmid is a small, circular, double-stranded DNA molecule that is distinct from a cell's chromosomal DNA. Plasmids naturally exist in many different kinds of bacterial cells. Usually, the genes carried in a plasmid DNA sequence provide the bacteria with an environmental advantage, such as antibiotic resistance or a reproductive factor.

Some plasmids are well suited for recombinant DNA vectors. Many have been isolated, modified, and replicated and are available for purchase from biological supply companies. To carry a gene(s) into a cell, a plasmid or other vector must first be "spliced" or cut open. The gene(s) of interest is then pasted into the open plasmid. This produces another circular piece of DNA containing DNA from two different species. It is called a rDNA plasmid since the DNA pieces have been recombined from two different sources. The recombinant plasmid may act as a vector, carrying the gene of interest into a new cell (see Figure 8.13).

Two sets of enzymes cut and paste the DNA to produce a piece of rDNA. The cutting ones, restriction enzymes (also called **endonucleases**), recognize specific A, C, G, and T sequences within DNA molecules, and cut the DNA strands at those recognition sites. Some restriction enzymes cut like scissors, straight across the DNA strand, to produce "blunt ends." However, the most valuable restriction enzymes cut to produce "**sticky ends**."

Sticky ends have one side of the DNA strand that is longer than the other. These overhangs allow for complementary matches between two DNA pieces cut by the same enzyme. A plasmid can be cut with a restriction enzyme (or two), and a gene of interest can be cut by the same restriction enzyme(s). Their sticky ends will match, and pasting may occur to produce an rDNA molecule. The pasting is done by a second enzyme called DNA ligase.

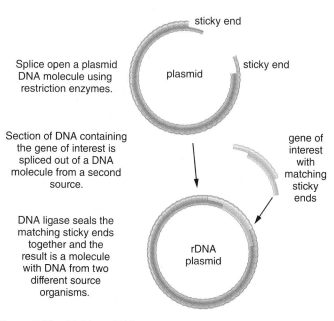

Splice open a plasmid DNA molecule using restriction enzymes.

Section of DNA containing the gene of interest is spliced out of a DNA molecule from a second source.

DNA ligase seals the matching sticky ends together and the result is a molecule with DNA from two different source organisms.

Figure 8.13. **Making rDNA.**

endonucleases (en•do•nu•cle•a•ses) the enzymes that cut RNA or DNA at specific sites; restriction enzymes are endonucleases that cut DNA

sticky ends (stick•y ends) the restriction fragments in which one end of the double stranded DNA is longer than the other; necessary for the formation of recombinant DNA

restriction fragments (re•stric•tion frag•ments) the pieces of DNA that result from a restriction enzyme digestion

restriction fragment length polymorphisms (re•stric•tion frag•ment length pol•y•mor•phi•sms) "RFLP" for short, restriction fragments of differing lengths due to differences in the genetic code for the same gene between two individuals

More than 4000 restriction enzymes recognizing over 300 different sequences have been discovered and isolated from the entire group of existing bacterial species. It is thought that restriction enzymes evolved in bacteria as a defense against invading viral DNA chopping it up as it tried to infect the cells. Restriction enzymes are named based on their origin. Consider the restriction endonuclease *Eco*RI, which is very commonly used in cloning and DNA studies. The "R1" shows that *Eco*RI was the first restriction enzyme isolated from a strain of *E. coli,* called *E. coli* RY13.

When *Eco*RI finds the DNA sequence GAATTC in the 5' → 3' strand, it cuts between a guanine and an adenine. Since the complementary side has the GAATTC sequence going in the opposite direction, the EcoRI also cuts on that side and a staggered cut across made in the DNA. This produces the sticky ends needed for making rDNA, as illustrated below.

5'...G|AATTC...3'	*Eco*RI restriction digestion	5'...G	AATTC...3'
3'...CTTAA|G...5'	⟶	3'...CTTAA	G...5'

Notice that the restriction enzyme sequence is palindromic, meaning it reads the same forward and backward. For example, the word "radar" is a palindrome. *Eco*RI's restriction site nucleotide sequence is a palindrome since each 5' end reads GAATTC.

In making rDNA, it is important that the two pieces of DNA that will be joined, come together in the correct orientation. Remember that the two antiparallel sides on a DNA strand have different codes on them. In one direction the code reads for the gene of interest and on the other side it does not. To ensure that the pieces come together correctly, two different restrictions enzymes are used on both the plasmid vector and the donor DNA to open or cut each producing distiguishable sticky ends (see Figure 8.14). Once the fragments are brought together in the correct orientation, the enzyme DNA ligase is used to seal them and the rDNA molecule is ready to use.

Using restriction enzymes and DNA ligase means that potentially any combination of genetic information is possible. Theoretically, DNA from any number of organisms can be pasted together and incorporated into the cells of another organism.

Restriction enzymes are critically important for producing rDNA molecules used in genetic engineering. In addition, restriction enzymes are important tools for studying DNA. For example, restriction enzymes are used to cut DNA into pieces of manageable size for analysis. The resulting pieces are called **restriction fragments**. Studies of the differences in fragment lengths that result from restriction digestion can reveal information about the difference in DNA sequences. The technique is called **restriction fragment length polymorphism** (RFLP) analysis.

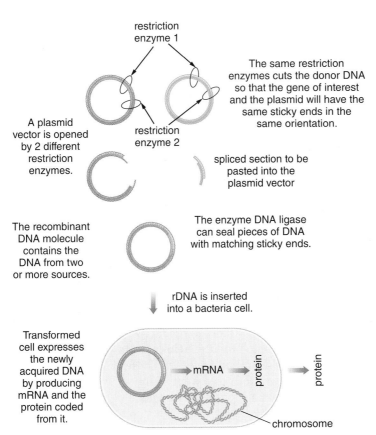

Figure 8.14. **A Big Picture View of Genetic Engineering.**

Endonucleases: Real "Cut-ups"

TO DO

Go to biotech.emcp.net/nebecomm and click on the link for the restriction endonucleases (enzymes) *Enzyme Finder* webpage available from New England BioLabs. Duplicate the chart below and fill in the DNA recognition site for each restriction enzyme. Determine if the recognition site is a pallindrome (stereotypical). Also, determine if the restriction enzyme digestion produces sticky or blunt ends.

Restriction Enzyme	DNA Recognition Site (Show both strands of the DNA sequence.)	Does the endonuclease produce sticky ends?
Alu I		
Bam HI		
Bgl I		
EcoRI	5'...G|AATTC...3' 3'...CTTAA|G...5'	Yes
Hind III		
Hae III		
Pst I		
Sma I		

Applications of RFLP Analysis

The first DNA fingerprinting used RFLP analysis (see Figure 8.15). A DNA fingerprint is the unique pattern that results from the DNA analysis of an individual. A section of DNA that is known to differ in individuals is subjected to restriction digestion analysis. When the restriction digestion fragments are run on a gel, a unique banding pattern, or fingerprint, is seen. The only two individuals who might share an exact DNA fingerprint are identical twins or asexually produced clones.

If DNA fingerprints are made using a particular restriction enzyme, then any variations in restriction fragments will be due to differences in the DNA sequence. The different RFLPs are easily seen on a gel electrophoresis. In recent years, though, DNA fingerprinting technology has changed; it now uses PCR almost exclusively. The PCR process targets specific sections of genetic information and reproduces it millions of times. Variations in the DNA information can be deduced by how pieces are targeted and duplicated. Chapter 13 presents more information on PCR fingerprinting.

Another application of RFLP is in evolutionary studies, where banding patterns help scientists understand DNA mutations that lead to differences in species. Recognizing these differences helps to identify related organisms and allows scientists to better understand how to address the needs of threatened or endangered species. In fact, RFLP studies, along with other methods, are used to identify specific individuals of endangered species for breeding purposes.

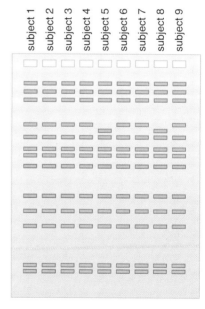

All samples except subjects 5 and 8 show the same restriction digestion pattern after being cut by a single restriction enzyme. This indicates a difference in DNA sequence in these individuals due to a mutation.

Figure 8.15. DNA Fingerprint. In these samples, a mutation in a gene is recognized by a difference in banding patterns. In DNA fingerprinting, RFLP analysis gives unique banding patterns because each person's DNA code is unique.

Before DNA sequencing techniques drastically improved in the 1980s, small DNA sequences could be determined by identifying which restriction enzymes cut a molecule, and at what location, relative to another. This is possible because scientists know the sequences at which each restriction enzyme recognizes and cuts. For example, if after digestion, a *Hin*dIII cut occurs right next to a *Bam*HI cut, then one can deduce that the sequence at that spot on a DNA strand is AAGCTTGGATCC. Determining the order of restriction sites of enzymes in relation to each other is called **restriction enzyme mapping**. At one time, restriction enzyme mapping was the primary method by which plasmids and other small DNA sections were sequenced.

restriction enzyme mapping (re•stric•tion en•zyme map•ping) determining the order of restriction sites of enzymes in relation to each other

BIOTECH ONLINE

RFLPs Can Reveal Disease Mutations

Huntington's disease (Huntington's chorea) is a horrible and fatal disease in which symptoms may not present until an afflicted person reaches his or her 40s or 50s. Recently, scientists have used *Eco*RI, and a few other restriction enzymes, to identify several unique RFLPs in people who possess this genetic disorder.

TO DO

Create a single Microsoft PowerPoint slide that describes the cause, symptoms, and treatments for Huntington's disease. Include graphics and references.

Section 8.2 Review Questions

1. What is the name of the process in which bacteria receive and express a) recombinant plasmid DNA, and b) recombinant viral DNA? What is the name of the process in which mammalian cells receive and express rDNA?
2. Which two types of enzymes are needed to produce a rDNA molecule?
3. What is the name for the differences in gel banding patterns in DNA samples that result from a restriction enzyme's activity?

8.3 Transforming Cells Using rDNA

You might be surprised to learn that transformations do occur in nature. In fact, this is how some species of bacteria quickly acquire new traits. One particular kind of "natural" transformation has caused recent concern in the medical industry. Due to the widespread use of antibiotics and antiseptics and the selective pressure to survive, several species of bacteria have undergone transformations from antibiotic sensitivity to antibiotic resistance. Antibiotic-resistant strains are occurring so quickly that scientists are having difficulty developing new treatments against them.

In the lab, scientists have exploited the ability of bacteria to readily take up foreign DNA, be transformed, and express new traits. The basic transformation protocol is basically the same for different cell strains, with only a few modifications. The general steps of a transformation are outlined below and illustrated in Figure 8.16.

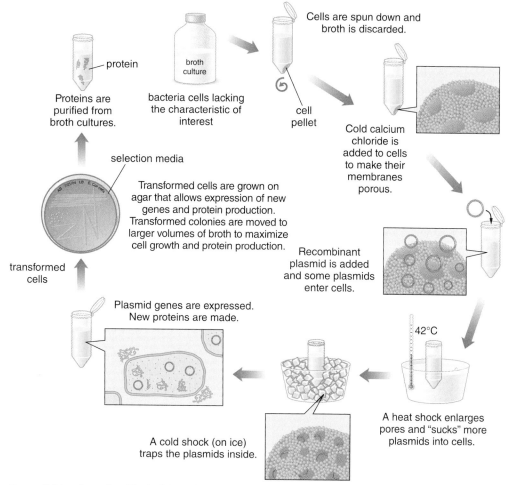

Figure 8.16. **Steps in a Typical Transformation.**

General Steps of a Transformation

- Grow the host cells in broth culture.
- Keeping the cells on ice, make them **competent** (ready to take up DNA) with a treatment of calcium chloride (CaCl$_2$) or magnesium chloride (MgCl$_2$).
- Add the rDNA plasmids to the competent cells.
- Heat shock the cells by rapidly moving them from ice to a hot water bath (37°C or 42°C for 20 to 90 seconds, depending on the strain of cells used), and then quickly place them back on ice. This is called a heat shock/cold shock.
- Add a nutrient broth for cell recovery and gene expression at some optimum temperature, such as 37°C for *E. coli*.
- Plate out the cells on some kind of **selection** media/agar that shows that the cells are producing the new protein.

competent/competency (com•pe•tent/com•pe•ten•cy) the ability of cells to take up DNA

selection (se•lec•tion) the process of screening potential clones for the expression of a particular gene; for example, the expression of a resistance gene (such as resistance to ampicillin) in transformed cells growing on ampicillin agar

Although transformations occur in nature, in the biotechnology lab, they occur too slowly or inefficiently to make marketable products. Researchers have learned "tricks" to induce bacteria to take in foreign DNA more efficiently. One trick is to create large pores or channels in the outer boundaries of the cell. This may make it easier for plasmids to enter the cell. This is called competency. Although the mechanism of competency is not completely understood, one theory suggests it may occur as follows: Cells have a membrane around them that blocks the movement of large molecules (ie, DNA) into or out of the cell. Embedded in the membrane are proteins that are arranged to make intramembrane channels or pores. Adding cations, such as Ca^{2+} or Mg^{2+}, covers the membrane channel proteins with a positive charge. It is thought that the proteins in the channels may repel each other, thereby producing expanded channels. As the channels in the cell membrane become larger, it is easier for DNA molecules to move into the cell. When cells are treated with the ions in this way, they are said to be "competent." Competency increases **transformation efficiency** by thousands of times. Using a molecule like $CaCl_2$ to induce competency is called chemical competency. Cells may be made competent by exposing cells to an electric field. This is called electroporation and requires an expensive instrument called an electroporator.

A second trick is to give the cells heat and cold shocks after they have been treated with the rDNA. When cells are transferred from cold to hot, they swell rapidly, pulling in DNA at or near the membrane. When quickly transferred back to the cold, the cells shrink rapidly and trap the DNA inside. A distinct heat shock followed by a distinct cold shock markedly improves transformations.

Once cells have been transformed, they must undergo a **recovery period**. During this time, the cells are given nutrients in a sterile broth and allowed to repair their membranes. The cells that survive the whole process grow and divide. They are spread on Petri plates containing selection media and grown in incubation ovens. Selection media has a specific ingredient that makes it easy to tell if transformed cells are growing on the plate. A selection ingredient may be an antibiotic, a nutrient, or a type of chemical.

Figure 8.17. Stacks of Petri plates containing cells spread on selection media. Only those cells that have been transformed will grow into colonies. The plate on the top shelf is white due to starch in the agar. Starch digestion is an indicator that a transformed colony contains the amylase gene.
Photo by author.

Often, the selection media kills off or slows the growth of nontransformed cells. For example, if ampicillin-sensitive cells are transformed with a plasmid carrying an ampicillin-resistance gene, the transformed cells can be selected for on an ampicillin-containing agar. Only those cells that are transformed will be able to grow in the presence of ampicillin. The process, called selection, screens the cells to see if they are making the new proteins from the newly acquired gene (see Figure 8.17). As transformed cells grow and reproduce, colonies of identical transformed cells arise. If conditions are suitable, the transformed cells begin expressing their new genes. The colonies are **clones** of transformed cells. The clones are all expressing the same genes and producing the same proteins.

Transformation technology has greatly improved transformation efficiencies in recent years. Several commercial laboratory kits are available to simplify or speed the process. New strains of bacteria have been developed to improve the transformation results.

Sometimes it is difficult to select for the desired gene directly. In that case, additional genes can be added to rDNA plasmids for the sole purpose of selecting for transformants. Often, antibiotic-resistance genes are added onto plasmids purely for screening purposes. The kanamycin-resistance gene (Kan^R) is frequently used so that transformants can be seen on a plate containing kanamycin agar.

The **beta-galactosidase (β-gal) gene** is commonly inserted into recombinant plasmids as a selection gene. In cells, the β-gal gene produces beta-galactosidase, an enzyme which converts the carbohydrate X-gal to a blue product. If cells that are transformed with a recombinant β-gal plasmid are grown on X-gal agar, the colonies will turn blue (see Figure 8.18). Blue bacteria colonies are indeed an unusual sight.

Biotechnicians can add other genes of interest to the recombinant β-gal plasmid and use the blue colonies (or, actually, the lack of blue colonies) to determine whether cells are transformed. They actually insert the gene for the desired gene product into the

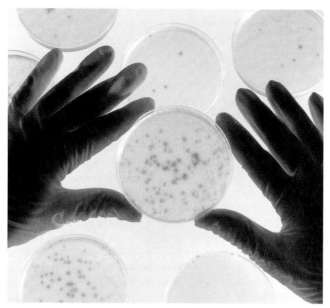

Figure 8.18. **Transformed cells expressing the β-gal gene.**
© Science Faction/Superstock.

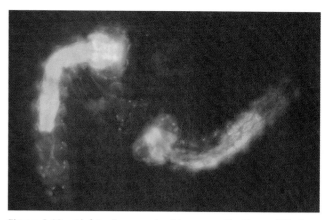

Figure 8.19. **Light micrograph of two genetically modified** *Anopheles* sp mosquito larvae glowing under ultraviolet light. Adults of this mosquito carry the disease, malaria. A gene for green fluorescent protein (GFP) from the jellyfish *Aequorea victoria* has been introduced into these mosquitoes. Genetic engineering could be used to introduce a gene into the mosquitoes to make them unable to carry the *Plasmodium* sp protozoa that cause malaria, thus saving millions of lives. ~40X
© Science Photo Library/Photo Researchers.

middle of the β-gal gene. This disrupts the code for the β-gal enzyme that makes the blue product. In this way, two types of colonies can grow on selection plates with X-gal. Blue colonies are those that have been transformed with plasmids, but they do not have a gene of interest in the plasmid. The genetic engineers do not want these. The scientists are looking for white, transformed colonies on the selection plates. These are the colonies that have cells containing the gene of interest, right in the middle of the β-gal gene.

A very popular selection gene is the **green fluorescent protein** (GFP) gene. The GFP gene, and the protein it codes for, is found in nature in a species of deep ocean jellyfish, *Aequorea victoria* (see Figure 8.19). The fluorescence of the protein causes a green glow in certain wavelengths of light. When a recombinant vector plasmid containing the GFP gene is used, transformed cells in UV light will glow a fluorescent green color. When a gene of interest is also present on the vector plasmid, a technician can be confident that the gene of interest got into the target cells if the cells glow after transformation.

A transformation is considered a success if even one transformed colony is seen on a selection plate. However, a few transformed colonies on a Petri dish will not make enough of the protein of interest to be purified and marketed. Huge volumes of transformed cells, in the tens of thousands of liters, are needed to manufacture the amounts of protein required to make a marketable product. Large volumes of transformed cells are produced during **scale-up** as part of the manufacturing process (see Figure 8.20).

green fluorescent protein (green fluor•es•cent pro•tein) a protein found in certain species of jellyfish that glows green when excited by certain wavelengths of light (fluorescence)

scale-up (scale-up) the process of increasing the size or volume of the production of a particular product

Figure 8.20. **During scale-up, transformed cells in broth culture are grown in progressively larger fermentation tanks. The goal is to scale-up to the tens of thousands of liters. Early in the scale-up process, the glass bioreactor/fermenter, shown here, has several liters of mammalian cell broth culture producing GVAX cancer vaccine proteins.**
Photo courtesy of Cell Genesys, Inc.

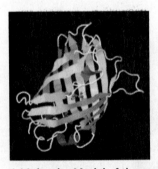

A Molecular Model of the GFP Protein

Courtesy of *Molecule World*, Digital World Biology, LLC.

A Glow in the Dark Cat? What's Next?

The green fluorescent protein (GFP) is a beautiful protein. It is not only pretty to look at but it is "beautiful" in what it is allowing scientists to visualize. Use the Internet to learn more about GFP's role in genetic engineering, cloning, and other scientific research.

1. Go to www.biotech-now.org/health/2011/09/doggonn-it-this-kitty-may-be-researchers'-new-best-friend and read about how a glow-in-the-dark cat is being used in cancer research. Describe in 2–3 sentences how researchers at the Mayo Clinic are using GFP as a reporter molecule.

2. Read through the GFP article at the RCSB Protein Data Bank website (www.rcsb.org/pdb/101/motm.do?momID=42). Describe what GFP is, what it is used for, and what scientists have done to "improve" GFP. Describe how the GFP chromophore colorizes GFP.

3. Go to www.rcsb.org/pdb/explore/explore.do?structureId=1ema and view the structure of the green fluorescent protein, PBL ID# 1EMA. Click on "View in 3D: JSmol" (another molecular modeling program) and view the GFP structure with its chromophore in 3D. Use the app to rotate the structure and view its primary and secondary structure. Describe its unique shape based on what you see.

4. Find another article about how GFP is used. What other organism(s) has it been added to and for what purpose? Cite the reference.

Section 8.3 Review Questions

1. List two techniques used to increase transformation efficiency.
2. Why do transformed cells need a recovery period?
3. What is GFP and why is important in genetic engineering work?

8.4 After Transformation—Cloning for Manufacturing Purposes

Transforming cells is a technology that attracts a lot of attention. Scientists create new and unique cells and organisms that do not exist in nature. Glowing plants, blue bacteria, cloned sheep, and new types of cells growing in the lab generate excitement about genetic engineering. But a few transformed cells on Petri plates produce only a tiny amount of product, not enough to harvest and sell. At a biotechnology company, the goal is to produce enough of a product to sell and make a profit. The profit is reinvested into the company for more R&D.

Figure 8.21. Cell cultures in four 2-L spinner flasks. Each flask has a spinner apparatus (propeller blade) inside to keep the cells suspended and aerated. The amount of oxygen the cells are receiving is the most critical factor in growing the culture at a maximum rate.
Photo by author.

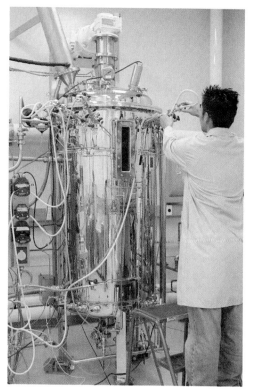

Figure 8.22. A technician prepares a bioreactor/fermentation tank for inoculation with cell culture. It must be cleaned thoroughly and sterilized before any culture is added. If even one contaminating bacterial or fungal cell enters the culture, thousands or millions of dollars of work could be destroyed.
Photo by author.

The Scale-Up Process

To produce enough volume of a product, the selected transformed cells are grown into ever-increasing amounts, in larger and larger containers. This scale-up process begins with the transfer of a colony of transformed cells to liquid media. The first scale-up is to a small volume, perhaps 50 mL of broth. The nutrient broth allows cells more room and more nutrients. As the volume of cells increases, so does the amount of product. If the culture does well, producing enough cells and product, the culture will be scaled-up to 1- or 2-L **spinner flasks**, then increased to 10 L, 100 L, 1000 L, and even up to 10,000-L or more (see Figures 8.21 and 8.22).

Increasing the volume of broth provides more nutrients and space for more cells. During each scale-up, the cell growth rate, product concentration, and product activity are measured. There must be assurance that the cells are growing as rapidly as possible and producing as much protein as possible. If the protein is an enzyme, then it must continue to show maximum activity.

Throughout the entire transformation, selection, and scale-up periods, the transformed cells must be monitored. The goal is to encourage maximum cell reproduction and protein synthesis. The requirements for each cell system are unique. Each cell culture has an optimum temperature, pH, nutrient concentration, and oxygen content. For certain strains of *E. coli* cultures, maximum growth is achieved in liquid media, in the dark, at 37°C, at a pH of 7.5, with glucose in the broth, and while the broth is being stirred and aerated.

Early in the scale-up process, when the batches of transformed cells are relatively small, technicians monitor "by hand." Samples are taken at regular intervals and tested for each factor. As necessary, sugar or other nutrients are added to maintain optimal growing conditions. When batches are large enough, 10 L or more, computers automatically monitor and adjust the batches in the **fermenters** (bioreactors). To maintain sterile conditions, bioreactors that hold large volumes have pipes going into

spinner flasks (spin•ner flask)
a type of flask commonly used for scale-up in which there is a spinner apparatus (propeller blade) inside to keep cells suspended and aerated

fermenters (fer•men•ters)
the automated containers used for fermentation, or growth of micro-organism cultures designed to be easily monitored and controlled

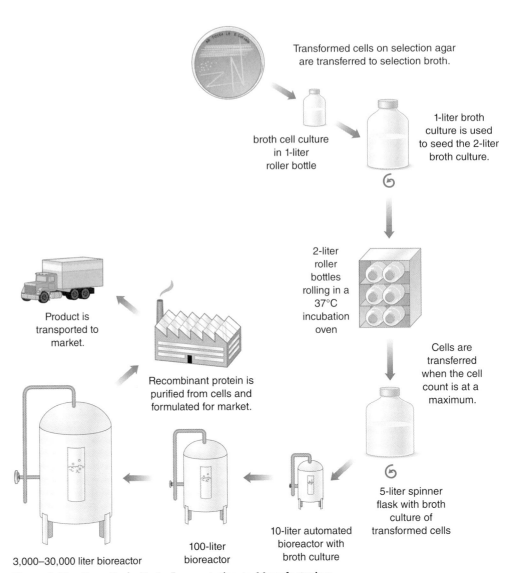

Transformed cells on selection agar are transferred to selection broth.

broth cell culture in 1-liter roller bottle

1-liter broth culture is used to seed the 2-liter broth culture.

2-liter roller bottles rolling in a 37°C incubation oven

Cells are transferred when the cell count is at a maximum.

5-liter spinner flask with broth culture of transformed cells

Product is transported to market.

Recombinant protein is purified from cells and formulated for market.

3,000–30,000 liter bioreactor

100-liter bioreactor

10-liter automated bioreactor with broth culture

Figure 8.23. From Scale-Up to Fermentation to Manufacturing.

Figure 8.24. Booties, lab coats or smocks, caps, gloves, and goggles are standard attire when working in fermentation and manufacturing facilities. "Gowning up" is critical to protect the product from contamination.
© Royalty-Free/Corbis.

and out of them for monitoring and adjustment. As the culture moves from scale-up to fermentation to manufacturing (see Figure 8.23), the process becomes more automated. There are a few different types of bioreactors. Learn about them at **bioprocessing.weebly.com/types-of-fermenters.html**. Fermentation is discussed in greater detail in the next section.

Not only are cultures checked for their growth rate and protein production, they are monitored to ensure that they have not been contaminated with unwanted bacteria, fungi, or viruses. The expense of scaling-up is enormous. If even one unwanted bacterium or fungal cell entered the culture, the entire batch could be ruined. Depending on the batch size and the point in the scale-up process at which the contamination occurs, the lost product could cost a company millions of dollars, as well as significant amounts of time and labor.

Ensuring sterile conditions throughout scale-up requires following careful procedures. There are strict protocols for cleaning and sterilizing equipment. Workers must maintain sterile technique and "gown up" when working with fermentation tanks (see Figure 8.24). In many situations, scale-up is done in "clean rooms," in which the environment is kept sterile through use of HEPA (high-efficiency particulate air) filters and monitored to ensure that microbes do not enter the facility in reagents, on equipment, or on the bodies of employees.

Using Assays during Scale-Up

It does not matter how quickly transformed cells are growing if they are not producing adequate amounts of the product. Throughout scale-up, assays measure protein concentration and activity. Early in R&D, assays must be developed to recognize and monitor the protein of interest. In general, this is the job of scientists in the Assay Services department. Initially, some assays may be done on an individual researcher's lab bench, but as more and more time, money, and materials are invested in a product, technicians who specialize in testing perform the assays. Assay Services develops and conducts testing and verification of products as they move through the pipeline. Technicians work full-time testing samples from throughout the company's R&D labs and manufacturing facilities.

There are many different types of assays, some of which were discussed in earlier chapters. Certain assays are enzymatic; they test for the presence and activity of an enzyme. In the case of the recombinant α-amylase produced by transformed cells, an enzymatic activity assay may be used to measure the degree to which amylase's substrate, starch (or a synthetic substrate), is converted to a detectable product (maltose). Although maltose is an aldose and is indicated by Benedict's solution and other copper-based indicators, the colored reactions are not quantitative.

A quantitative α-amylase test (Enzymatic Assay of α-AMYLASE (EC 3.2.1.1) has been developed by Sigma, Corp.). The assay comes with all the reagents (starch substrate and colored indicator and control samples) to test samples during manufacturing. The color indicates the amount of maltose produced by amylase breakdown of starch. This quantitative assay produces a stable colored maltose product that is measured at 540 nm in a VIS spectrophotometer. The higher the absorbance at 549 nm, the more amylase activity. The product and the protocol for how to run the assay are available at **www.sigmaaldrich.com/technical-documents/protocols/biology/enzymatic-assay-of-a-amylase.html**.Using the α-amylase assay, scientists can detect the presence of their product and measure its activity throughout the manufacturing process. In fermentation and manufacturing, technicians take samples regularly and perform several assays similar to this one. If the assay results show a low concentration of an enzyme or low activity, the production protocols may be reviewed or even scrapped.

Another commonly used protein assay (discussed in Chapter 6) is an enzyme-linked immunosorbent assay (ELISA). To detect the presence of a specific protein in a mixture, an ELISA uses antibodies to recognize the protein molecule's unique three-dimensional structure. The ELISA can determine whether a specific protein is, indeed, in a mixture and, if so, at what concentration.

For example, an ELISA is used during the production of HIV glycoprotein 120 (gp120) for use in HIV vaccine trials. Throughout the product purification process, the presence and concentration of the gp120 protein must be ensured. Samples are collected and sent to Assay Services to have ELISAs run. To "run" an ELISA, a sample (in this case, thought to contain gp120) is added to a test plate. Proteins in the sample stick to and coat the bottom of each sample well. Then, the samples are washed with antibodies (immunoglobulin G [IgG]) that recognize the gp120 antigen. Next, a second antibody, one that recognizes IgG, is added. Linked to the other end of the anti-IgG antibody is an enzyme, such as alkaline phosphatase, which can cause a color change if given the right substrate.

The goal is for the antibodies and antigens to find each other and bind. After some time, excess antibody-enzyme

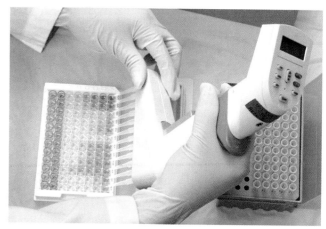

Figure 8.25. Here an ELISA detects HIV antibodies that the body makes when someone is infected with HIV. A second, more specific test, called the Western blot, is used to check the results. Together, these two tests detect 99.8% of HIV-positive blood. In manufacturing, the same antibodies can be used in ELISAs to ensure that the protein product (gp120) is being produced at a sufficient concentration.
© Lester Lefkowitz/Corbis.

complex is washed off the sample. A substrate is added that will cause a color change by the linked (conjugated) enzyme proportional to the amount of antigen-antibody-antibody-enzyme complex. An ELISA plate reader can measure the color change. In this way, the concentration of gp120 can be determined. ELISAs are one of the most basic and commonly used assays in a protein lab. Much time is spent at companies developing and using ELISAs for research, manufacturing, and diagnostic testing (see Figure 8.25).

To support manufacturing, another group, the Quality Control (QC) department, conducts testing and provides verification. The QC department performs additional assays and more extensive testing. Often, the test results are required by the Food and Drug Administration (FDA) or other regulatory agencies, and this information must be reported to them.

BIOTECH ONLINE

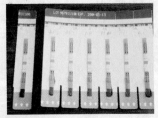

An HIV Test Kit.
© Gideon Mendel for The International HIV/AIDS Alliance/Getty Images.

You Can Try This at Home
Several diagnostic antibody-based testing kits are now available for home use.

Go to biotech.emcp.net/homeaccess and click on the link "Products" to learn about the diagnostic tests available from Home Access Health Corporation. Find three test kits available from Home Access. List the name of each diagnostic test kit, the disease/disorder it tests for, and whether or not the test kit is FDA approved.

Section 8.4   Review Questions

1. What is a spinner flask and where is it used?
2. What type of environmental conditions must be monitored in cultures as they are scaled-up?
3. How are ELISAs utilized during manufacturing?
4. How is a QC department involved in the manufacturing process?

8.5 Fermentation, Manufacturing, and GMP

When one thinks of fermentation, images of beer brewing and winemaking come to mind (see Figure 8.26). In the broad sense of the word, fermentation is the process by which cells utilize glucose under anaerobic (lack of oxygen) conditions. Several different species carry out one of two kinds of fermentation, **alcoholic fermentation,** and **lactic acid fermentation**.

Yeast cells and many bacteria carry out alcoholic fermentation. Alcoholic beverages are produced when certain yeast or bacteria cells convert the sugar in grapes or grains into alcohol and carbon dioxide:

$$glucose \rightarrow carbon\ dioxide + ethanol$$

alcoholic fermentation (al•co•hol•ic fer•men•ta•tion) a process by which certain yeast and bacteria cells convert glucose to carbon dioxide and ethanol under anaerobic (low or no oxygen) conditions

lactic acid fermentation (lac•tic ac•id fer•men•ta•tion) a process by which certain bacteria cells convert glucose to lactic acid under anaerobic (low or no oxygen) conditions

Figure 8.26. Men tread on grapes during the blessing at the Sherry Vintage Festival in Jerez, Spain. Treading (crushing) grapes releases the juice (cytoplasm) from the grape cells. This allows the yeast *Saccharomyces* sp, which occurs naturally on the grape skins, to ferment the sugar in the juice to alcohol. Red wine is produced from fermenting the grape pulp and skins, while white wine is produced from fermenting only the grape pulp.
© Ted Streshinsky/Getty Images.

Figure 8.27. Yogurt is very easy to make at home. All that is needed is a scoop of starter culture (commercial yogurt with *Lactobacillus* sp bacteria) and a way to incubate the culture at about 37°C (hot water bath in a styrofoam cooler). The *Lactobacillus* bacteria perform lactic acid fermentation. The lactic acid lowers the pH of the milk and causes the milk proteins to denature. The protein precipitates out of the solution as yogurt curds. This brand of yogurt contains *Lactobacillus bulgaricus (L. bulgaricus)*, *Streptococcus thermophilus (S. thermophilus)*, and *Lactobacillus acidophilus (L. acidophilus)*.
Photo by author

Some bacteria and, under certain conditions, some animal cells undergo lactic acid fermentation. Yogurt, cheeses, buttermilk, sauerkraut, and soy sauce are produced when certain bacteria cells convert the sugar in milk, cabbage, and soybeans into lactic acid (see Figure 8.27). The lactic acid lowers the pH of the milk, for example, which degrades the milk protein molecules into curds of cheese:

$$glucose \rightarrow lactic\ acid$$

Many biotechnology products are manufactured by alcoholic and lactic acid fermentation as described above. In addition, though, the term "fermentation" has come to have a different meaning in the biotechnology industry. In biotechnology, fermentation refers to growing cells (bacteria and fungi) under optimum conditions for maximum cell division and product production.

Many biotechnology companies have fermentation departments that grow their cell lines in highly controlled fermentation tanks or bioreactors to ensure the highest productivity. Usually, transformed cells move from R&D labs to fermentation for initial scale-up, and then to manufacturing.

An earlier section of this chapter provided an overview of the scale-up and manufacturing phases. This section offers a more in-depth look at the processes. Recall that scale-up begins with the transfer of transformed cells to broth. Volumes of broth cultures begin as small quantities (eg, 50 mL of broth), in flasks. A large colony, growing well on Petri plates and producing product, is used as the **seed**. The seed colony may have several million identical transformed cells in it. With the ideal temperature, pH, aeration, and nutrients, the cells will grow and multiply as quickly as every 20 minutes. Hence, 1 million cells become 2 million in 20 minutes. In 40 minutes, there are close to 4 million transformed cells. After an hour, there are 8 million cloned cells. It is called **exponential growth** when a cell culture doubles in cell count with every cell cycle. As long as the conditions are optimal, the culture will maintain exponential growth (see Figure 8.28).

When a broth culture is first inoculated, it appears clear, since there are only a few million cells in a relatively large volume of broth. However, as the culture becomes

seed (seed) the initial colony or a culture that is used as starter for a larger volume of culture

exponential growth (ex•po•nen•tial growth) the growth rate that bacteria maintain when they double in population size every cell cycle

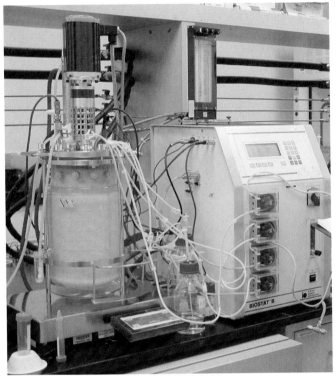

Figure 8.28. **A computer monitors the temperature, pH, glucose concentration, and other important growth and production factors of the cell culture in this 5-L fermentation tank. As long as the cells are in exponential growth, they are kept in the 5-L tank. As cell division slows because of overcrowding, the entire 5-L culture is used to "seed" a 10-L or larger tank.**
Photo by author.

■ ■ ■ ■ ■ ■ ■ ■ ■ ■ ■ ■ ■

lag phase (lag phase) the initial period of growth for cells in culture after inoculation

stationary phase (sta•tion•ar•y phase) the latter period of a culture in which growth is limited due to the depletion of nutrients

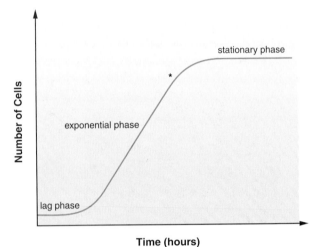

Figure 8.29. **To ensure healthy cells that produce protein at a maximum rate, cultures are grown to exponential growth, measured as an OD_{600} of about 0.6 au. Before the growth rates decrease* below the level of exponential growth, the sample is harvested or used to seed a larger vessel.**

more concentrated with cells, it turns cloudy. After the initial inoculation, it takes some time (the **lag phase**) before the presence of cells in the culture can be seen (see Figure 8.29). Eventually, the culture becomes cloudy, indicating the presence of cells. The concentration of cells in the culture can be monitored using a spectrophotometer. Exponential culture growth is easily seen when the culture absorbance rises in a linear fashion on a graph.

Over time, the cells in culture begin to use up the broth nutrients, which slows their growth and cell division. When this happens, the culture is said to be at **stationary phase**. The culture must be transferred into a new, larger broth culture before it reaches stationary phase. If the transfer to a larger volume occurs before the culture reaches the stationary phase, it ensures that the dividing cells always have optimum conditions. Under optimum conditions, the cultures grow quickly and are used to seed, or start, larger volumes every few days. The larger volumes, 2 L and more, may be grown in spinner flasks and, eventually, in fermentation tanks. In medium- to large-size companies, the Manufacturing department takes over the job of scaling-up transformed cell lines.

The fermentation protocols are different for different cell lines and for different products. Each cell line has specific temperature, pH, and nutrient requirements. A comprehensive discussion of a typical cell culture protocol for scaling-up transfected CHO cells is presented below. It is very similar to the fermentation protocols for bacterial or fungal cultures. The objective of mammalian cell culture is to grow transfected cells as quickly as possible with maximum protein production. Remember, ultimately, a company must produce enough protein for purification and marketing.

Eventually, the purified protein product must be formulated, or prepared, for delivery and storage. If the product is a pharmaceutical, the route of drug administration must be considered: orally, transdermally (through the skin), subcutaneously (under the skin), intramuscularly (into the muscle), or intravenously (into the blood). Some products are sold as liquids, some are lyophilized into a freeze-dried powder, some are granulated, and some are put into pumps or patches. There is a considerable amount of testing and market analysis performed before a formulation is chosen. Some companies have a Formulations department focused on the best way to prepare the final product.

General Cell Culture Protocol, Transfected CHO Cell Scale-up (Mammalian Cell Culture)

1. A transfected cell line (CHO cells containing newly acquired DNA) must be obtained. Transfections are usually done in R&D. The transfected cells are grown in Petri plates, small tissue culture flasks, or small spinner flasks containing selection media.

2. Selection and assay protocols for testing transfected cell growth and protein production must be developed. These assays may be developed and used right at the lab bench of the researcher who needs them. However, if a company is large enough, they may have a separate department, Assay Services, in which the assays are created and tested.

3. When the transfected cell lines show enough growth and production, they are divided into three to five samples and grown in progressively larger flasks or plates. This may be done in the R&D or Fermentation departments.

4. The broth cultures are grown until there is a large enough volume to "seed" a spinner (100 to 300 mL) suspension. A spinner has a blade in it, which keeps the culture mixed and aerated. In the spinner, a concentration of about 2 to 5 million cells/mL is desired so that scale-up to larger fermentation tanks can occur (see Figure 8.30).

5. Cultures undergo passage (seed) from one vessel to the next every 3 to 4 days. Cultures are moved from 500-mL spinners to 5-L spinners. Next, spinner cultures are transferred to 10-L bioreactors. The goal is to maintain a minimum of 300,000 transfected cells/mL in suspension. Of course, the cells must be expressing their new genes and making the target protein at a sufficient concentration. Also, the protein of interest must be functional, and if it is an enzyme, it must show activity.

6. From the 10-L bioreactors, 80-L bioreactors are seeded. Then, 400-L bioreactors, 2000-L bioreactors, and ultimately 10,000-L or more bioreactors are seeded and grown (see Figure 8.31). This all takes place in the Manufacturing department (in fermentation for bacteria, and in cell culture for mammalian cells). Throughout manufacturing, an unclumpy, single-cell suspension that continues to produce the protein of interest is desired.

7. Cultures are harvested from bioreactors after they have reached production goals. From bioreactors, cultures are piped through huge filters or into centrifuges where the cells are separated from the broth. Sometimes, the cells retain the target protein, and they must be burst open to begin the protein purification process. Often, the proteins are not stored in cells, and they are released to the extracellular fluid. In this case, the cells are discarded, and the broth is processed to separate the protein of interest from all other proteins in the broth.

Figure 8.30. At Genencor International, Inc., a research associate collects a broth culture sample from a 10-L bioreactors/fermenter for testing. Cell density, as well as product concentration and activity, is determined. Photo by author.

Figure 8.31. A 400-L bioreactor is cleaned and sterilized for a 100-L "seed" culture. Photo by author.

During the manufacture of a pharmaceutical product, the biotechnology company follows current good manufacturing practices (**cGMP**), which are set and monitored by the FDA (see Figure 8.32). cGMP are outlined in Title 21, Parts 210 and 211, of the Code of Federal Regulations, which can be found at: **biotech.emcp.net/fdaregs**. This monstrous document details the quality management and organization, device design, buildings, equipment, purchase, and handling of components, production and process controls, packaging and labeling

cGMP (cGMP) abbreviation for current good manufacturing practices

controls, device evaluation, distribution, installation, complaint handling, servicing, and record keeping. Site inspectors regularly visit production facilities to audit for cGMP compliance. Maintaining cGMP is a main focus of a pharmaceutical production facility.

Figure 8.32. **A technician cleans a 5-L fermenter following current cGMP and validates it for clean-in-place effectiveness. Meticulous records, available for review by the FDA, must be kept.** Photo by author.

Section 8.5 Review Questions

1. Distinguish between the processes of alcoholic and lactic-acid fermentation. Give an example of a product made by each process.
2. Manufacturing teams want to keep cell cultures in exponential growth. What is exponential growth?
3. Place these events in order, from early in the product pipeline to late in the pipeline.

Transformation	Manufacturing
Fermentation	R&D
Assay Development	QC
Scale-up	

4. Which federal agency is responsible for setting cGMP guidelines?

8.6 Retrieving Plasmids after Transformation

preparation (prep•ar•a•tion)
the process of extracting plasmids from cells

miniprep (mi•ni•prep) a small DNA preparation yielding approximately 20 µg/50 µL of plasmid DNA

midiprep (mid•i•prep) a DNA preparation yielding approximately 100–300 µg/mL of plasmid DNA

maxiprep (max•i•prep) a DNA preparation yielding approximately 500 µg or more of plasmid DNA

After a successful transformation, it is often necessary to extract the transforming plasmid back out of the transformed cells. The process of extracting plasmids from cells is called a **preparation** (called "prep," for short). When the amount of cells and cell culture is small, the amount of plasmid recovered from the cells is relatively small. A **miniprep** is a plasmid isolation that yields about 20 to 30 µg of DNA (usually 20 µg in a 50 µL sample).

When larger amounts of DNA are needed, larger volumes and concentrations of cell cultures may be used. A **midiprep** isolates plasmids from 15 to 25 mL of culture with the goal of isolating several hundred micrograms of DNA. To complete a midiprep, a large, refrigerated centrifuge is needed to handle the larger volumes of sample. For larger volumes, a **maxiprep** starts with 100 mL or so of culture, giving yields of over 500 µg of plasmid DNA. For an even larger yield of plasmid (in the milligram amounts), megapreps and gigapreps can be done if the necessary equipment is available.

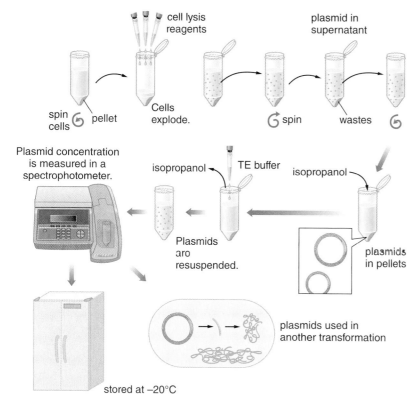

Figure 8.33. **Miniprep Procedure.** Transformed cells are spun down, and enzymes and other reagents are added to explode the cells. Salts precipitate proteins, which will be discarded. Alcohol washes and centrifugation clean to precipitate a pellet of plasmids on the bottom of the tube. The plasmid pellet is resuspended in TE buffer and stored at –20°C until ready for use.

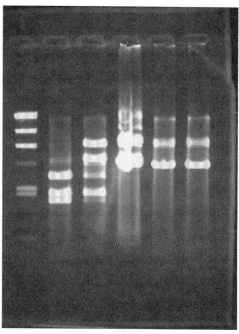

Figure 8.34. **An extracted plasmid sample (Lanes 4 through 6) is treated with restriction enzymes (Lanes 2 through 3) to confirm its size and sequence. The size of the DNA fragments is determined by comparison with the DNA sizing standards in Lane 1.** Photo by author.

The miniprep process is similar to extracting chromosomal DNA from cells except that plasmids are much smaller, and they are not attached to the cell membrane. There are a few different versions of the basic miniprep procedure. In any prep procedure, though, the cells must be exploded and the DNA separated from all the other molecules of the cell. An alcohol precipitation is one of the final steps (see Figure 8.33). A technician can prepare all the reagents for a miniprep from scratch or purchase a ready-to-use kit from such companies as G-Biosciences, QIAGEN Inc., or Zymo Research Corp.

Most laboratories now use commercially available kits for plasmid preparations. When kits are not available, a technician can prepare the solutions needed. In a miniprep "from scratch," cells are grown in an overnight culture and then spun out of the broth using a centrifuge. Pellets of cells are resuspended in a cell lysis buffer that usually contains SDS to dissolve cell membranes and NaOH (high pH) to degrade and precipitate cell walls, protein contaminants, and genomic DNA. Potassium acetate, or another salt, is added in solution to bring the pH down to a more neutral level. Adding potassium acetate also increases the precipitation of proteins.

The mixture is spun in a centrifuge again. The pellet of cell debris is discarded. The supernatant contains small plasmid rings and RNA. Isopropanol is added to precipitate these nucleic acids, and the mixture is spun a third time. Washes with alcohol (usually ethanol) remove everything except the pelleted plasmid DNA. RNase may be added at one or more steps to decrease RNA contamination. To resuspend the plasmids, TE buffer is added. This is a time-consuming method, but it often produces 20 µg of plasmid, which is plenty for more transformations. After a miniprep, the plasmid sample must be checked by restriction enzyme digestion to ensure that the size and fragments are correct for the target plasmid (see Figure 8.34).

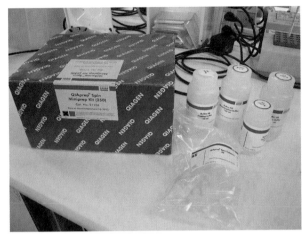

Figure 8.35. Many companies sell kits for plasmid preparation. Most kits require some kind of vacuum pump or centrifuge.
Photo by author.

Commercial kits conduct preparations in much the same way, although they may contain one or more columns that bind DNA or contaminants at different steps in the procedure (see Figure 8.35). Some new prep procedures contain plates or manifolds to handle dozens of samples in purification columns at one time. An advantage of commercial kits is that the buffers and solutions needed are provided, so the technician does not have to prepare these from scratch. This saves time and may in the long run save money.

Reasons for Performing a Prep

There are several reasons one might want to do a prep. A miniprep ensures that a transformation has actually occurred as planned. Extracting and analyzing the transforming plasmid confirms that a new phenotype (characteristic) seen in the transformed cells is due to the acquisition of the planned new genotype (DNA). In other words, this verifies, on a molecular level, that a cell has received and is processing the new genes carried in on the transforming plasmid.

Another reason minipreps are done is to collect more plasmids for future transformations. Transformed cells may not store well for long periods of time. Plasmid DNA is easily stored, almost indefinitely, at very low temperatures. At some future time, plasmid samples can be retrieved and used to transform other competent cells.

Once a plasmid preparation is completed, the resulting sample in TE buffer appears clear, as if it were in water. How do you know if you have plasmid DNA in the tube, and if there *is* DNA in the tube, how can you be sure it is actually the plasmid you hoped to extract?

Testing for the Presence of DNA

The presence of DNA in a sample must be demonstrated. The concentration and purity of the sample must be determined. It is also important to confirm the structure of the plasmid through restriction digestion.

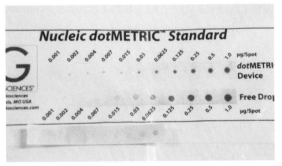

Figure 8.36. G-Biosciences NUCLEIC dotMetric™ Assay kit.
Photo by author.

A quick and easy test to confirm the presence of DNA in a sample is to use a DNA indicator test. In Chapter 4, ethidium bromide (EtBr) dot tests were described. EtBr is an indicator that glows a pinkish-orange color when DNA is present.

Recently, the company G-Biosciences has developed a DNA indicator test (NUCLEIC dotMetric™ Assay, Geno Technology, Inc.) to quantify small amounts of DNA (from about 0.1–1.0 µg) in solution (see Figure 8.36). This is particularly useful in determining DNA yields in minipreps since the volume of miniprep samples is often only 50 µL, making it challenging to use a spectrophotometer to measure the sample.

A common way to determine the concentration of DNA in a sample is to use a UV spectrophotometer to visualize the DNA molecules (see Figure 8.37). For use in transformation, a concentration of approximately 0.005 µg/µL is preferred. For restriction digestion, a minimum concentration of 0.1 µg/µL is required. Scientists have determined that double-stranded DNA molecules absorb a maximum amount of light at 260 nm and can detect DNA in a sample by setting a UV spec to this wavelength.

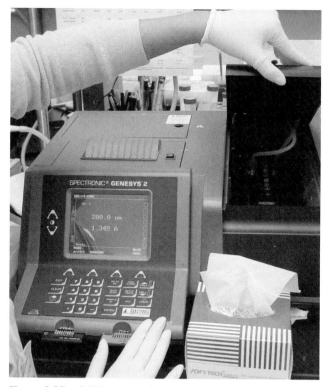

Figure 8.37. A UV spectrophotometer is used to quantify the amount and determine the purity of DNA samples. A sample is placed in a cuvette, and a deuterium lamp shines 260-nm light on the sample. The amount of 260-nm-light absorbance is related to the concentration of DNA in the sample.
Photo by author.

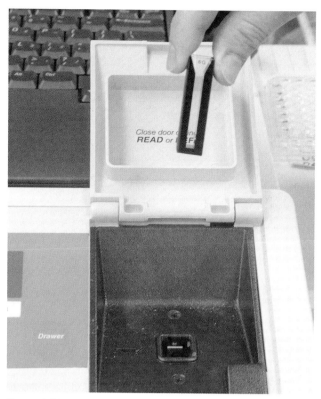

Figure 8.38. A quartz cuvette has a narrow slit that allows UV light to pass through in only one direction. The cuvette must be placed in the spectrophotometer in the correct orientation.
Photo by author.

To use a UV spectrophotometer, locate either quartz cuvettes or specially treated plastic cuvettes (see Figure 8.38). Quartz cuvettes are small and very expensive. They must be very clean and handled correctly so they are not damaged. Like visible specs, a UV spec must be zeroed (calibrated) using a blank containing the buffer in the sample.

To calculate the concentration of DNA in a sample, use a simple ratio. It is known that 50 µg/mL of pure, double-stranded DNA absorbs approximately 1 au of light at 260 nm. One can determine the concentration of an unknown DNA sample using the following ratio:

$$\frac{50\ \mu g/mL}{1\ au\ at\ 260\ nm} = \frac{X\ \mu g/mL}{the\ absorbance\ of\ sample\ 260\ nm}$$

By measuring a sample's absorbance at 260 nm, you can cross-multiply and solve for X (the concentration of the sample). To use the equation, a sample should have an absorbance in the range of approximately 0.1 to 1.0 au. Samples that are either too concentrated or too dilute must be diluted or concentrated to a readable absorbance level.

Ultraviolet spec data are useful in many ways. For example, if a plasmid sample absorbs 1.05 au at 260 nm, then its concentration would be approximately 52.5 µg/mL. This is a very good yield for a miniprep, but this concentration is so high that the sample would need to be diluted considerably for use in a transformation. This is assuming that the plasmid sample is pure and not contaminated by RNA or protein. Often, though, RNA and proteins contaminate plasmid preps.

By measuring a sample's absorbance at 280 nm (the wavelength of maximum light absorbance for a colorless protein) and comparing that value to the absorbance at 260 nm, one can establish a ratio of nucleic acid (DNA or RNA) to protein. Use the ratio below to estimate the purity of a DNA sample:

$$\frac{\text{absorbance (au) at 260 nm}}{\text{absorbance (au) at 280 nm}} = \text{the purity value of the sample}$$

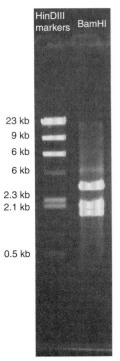

23 kb
9 kb
6 kb
6 kb
2.3 kb
2.1 kb

0.5 kb

Figure 8.39. **Plasmid Digestion by *Bam*HI.**
Photo by author.

A purity value of at least 1.8 is expected for pure double-stranded DNA. A purity value between 1.8 and 2.0 is preferred when conducting plasmid preparations. If the purity value is over 2.0, the sample is probably contaminated with RNA. Depending on what the plasmid sample is to be used for, RNA contamination could be a problem. Using RNases during or after the preparation can reduce RNA contamination.

If the purity value is below 1.5, the sample has substantial contamination. Depending on the intended use of the plasmid sample, protein contamination could be a serious problem. As the purity value reaches 1.0, the sample becomes virtually worthless for transformation or restriction digestion purposes. Using proteases or column chromatography during or after the preparation may reduce protein contamination.

When a sample's DNA purity and concentration are determined, a plasmid sample can be used for restriction digestion to confirm that the plasmid that has been purified is, indeed, the one desired. For example, if one conducts an extraction of the pAmylase plasmid from transformed cells, the plasmid can be confirmed by digestion using *Hind*III and *Bam*HI. *Hind*III has one restriction site on the pAmylase plasmid and, thus, opens the 6600 bp plasmid into a single, linear piece. *Bam*HI has three restriction sites on the plasmid and, thus, cuts the plasmid into three pieces totaling 6600 bp. The results of the digestions can be visualized on an agarose gel (see Figure 8.39). Once the plasmid has been retrieved from transformed cells and its identity is confirmed through restriction digestion, it may be used in subsequent transformations.

Section 8.6   Review Questions

1. What is the name of the procedure in which plasmids are extracted from cells?
2. How is plasmid DNA precipitated in the final steps of a plasmid prep?
3. Once plasmid is extracted from a cell, how can a technician know that it is the "correct" plasmid?
4. If a DNA sample gives a 260-nm reading of 0.8 au and a 280-nm reading of 0.5 au, what are its concentration and purity? Is this purity acceptable?

Chapter Review

Summary

Concepts

- Genetic engineering is the addition, subtraction, or modification of genetic information in an organism. Genetic engineering has led to many new products since the 1980s.
- Isolation of DNA from cells is required for a source of genetic information. Cells have to be burst open using enzymes, solvents, or mechanical means. Contaminant molecules must be removed using solvents, enzymes, and centrifugation. Alcohol precipitation is used to pellet and isolate DNA from solution.
- Finding a gene of interest on a piece of DNA is challenging. Labeled probes are used to find complementary sequences in a DNA sample. The probe will hybridize to a complement and the probe and the complement become visible. Radioactive and fluorescent probes are common.
- Sections of DNA can be recognized and reproduced through PCR technology. A thermal cycler and primers (similar to probes) are needed for a PCR. A PCR results in the production of billions of copies of a DNA sample in just a few hours. A PCR product can be used in both genetic engineering and diagnostic testing.
- Transformation is the insertion and expression of foreign DNA into a cell. If the transformation is of a mammalian cell, it is called a transfection. Transformations are usually done using recombinant plasmids or recombinant viral DNA.
- To make a recombinant plasmid, one or more restriction enzymes are needed to splice the donor and vector (plasmid) DNA molecules. Recombinant enzymes that produce matching sticky ends allow for the matching of pieces from different DNA sources. To seal the sticky ends, DNA ligase is needed.
- There are hundreds of different known restriction enzymes. Each recognizes a specific sequence and will make a cut in the double-stranded DNA at or near the recognition sequence.
- Knowing the recognition sequence of a restriction enzyme helps a technician plan how to cut DNA molecules. Recognition-site information also helps to recognize plasmids and determine sites of interest.
- The pattern of restriction fragments on a gel is called restriction fragment length polymorphisms (RFLPs). The RFLPs are useful in diagnostics, DNA fingerprinting, and restriction mapping.
- For greater transformation efficiencies, cells are made competent with calcium chloride or

magnesium chloride solutions. Electroporation may be used to induce competency. Cells are also given a hot and cold shock when mixed with plasmids to increase the number of plasmids that become trapped inside the recipient cells.

- Transformed cells are given a good environment in which to grow and reproduce exact copies, or clones. Clones are grown on Petri plates with selection media. This results in cells that demonstrate new, desired phenotypes. Often, additional selection genes, such as GFP, β-gal, ampicillin, or tet are added to a recombinant plasmid so that transformation can be recognized.

- Cloned cells are scaled-up to broth cultures and put through a series of larger volumes and bigger tanks. The goal is usually tens of thousands of liters of broth culture, producing high concentrations of active recombinant protein.

- Fermenters and bioreactors are automated culture vessels designed to monitor the conditions of the cell culture they hold. Technicians set up the fermenters and monitor their operation. Sterile conditions must be maintained for all work performed with reagents and equipment used with the fermenters and cell cultures.

- During R&D, fermentation, and manufacturing, assays ensure the presence, concentration, purity, activity, stability, and potency of the cells and proteins from a culture.

- During fermentation and manufacturing, cells are encouraged to grow and reproduce exponentially. During exponential growth, the number of cells in a culture doubles with each generation.

- Cultures that start to use up their resources, and are overcrowded, are said to be in stationary phase. The high-density cultures, in exponential growth, are used to "seed" larger volumes of broth that contain more nutrients and room for growth.

- Once cultures have filled the largest fermentation vessel, the cells are separated from the broth. The proteins are purified from either the broth or the cells using several separation techniques. The goal is a high concentration of purified, active protein.

- Once the protein is pure, it must be formulated before it is marketed. Some formulations that might be considered include injectables, inhalers, patches, creams, tablets, etc.

- Throughout the manufacturing process of pharmaceuticals, guidelines written and enforced by the FDA are followed. These are called current Good Manufacturing Practices (cGMP).

- Plasmids responsible for transformation can be retrieved from transformed cells during a plasmid preparation. A "prep' is similar to other DNA isolations in that the cells are burst and, then, through a series of isolations and extractions, plasmids are separated from other molecules in the transformed cells. Mini-, midi-, and maxipreps are all similar in their procedures, but they result in respectively greater yields of plasmids.

- A UV spectrophotometer is used to determine the concentration and purity of DNA samples extracted from cells. The concentration of DNA can be determined using the following equation: 50 µg/mL/1 au at 260 nm = X µg/mL/sample's absorbance at 260 nm.

- The relative purity of a DNA sample can be determined using the following equation: absorbance at 260 nm/absorbance at 280 nm and then comparing that number to accepted values.

Lab Practices

- Lambda bacteriophage is the source of DNA used in many sizing standards.
- One of the most common sizing standard is the lambda + HindIII ladder, produced when the restriction enzyme HindIII cuts lambda DNA into eight pieces, from 125 bp to 23,130 bp in size.
- Restriction digestion can be used to verify characteristics of a given DNA sample. The plasmid pAmylase2014 has a total size of close to 4300 bp. When the circular plasmid is cut with HindIII, one cut linearizes the plasmid, and the 4300-bp piece can be seen on a gel.
- When EcoRI cuts pAmylase2014 it also cuts in only one place, although it is a different spot than the HindIII cut, and linearizes the plasmid. The 4300-bp piece can be seen on a gel. When mixed together the 2 enzymes cut the plasmid in 2 places, resulting in a characteristic 2-band fingerprint.
- E. coli cells are not normally amylase-producers. E. coli can be transformed into amylase-producers by inserting a recombinant plasmid (pAmylase) carrying the amylase gene and an

amp-resistance gene.

- By growing *E. coli* cells on starch/amp LB agar plates, selection of pAmylase-transformed *E. coli* cells is accomplished. Cells that can thrive with ampicillin in the agar and also digest starch are the desired transformed cells.
- Transformation efficiency is increased through competency plus hot and cold shocks.
- Transformed colonies are composed of transformed cells. The identical cells are called clones.
- Petri plates of transformed colonies do not contain sufficient amounts of marketable product. During manufacturing, samples are scaled-up in progressively larger volumes of broth.
- Cells growing well in broth culture exhibit exponential growth and produce large amounts of protein. Cell culture growth is checked using the UV spec as well as other instruments and techniques, which include monitoring the cell culture environment.
- The growth of cells at stationary phase slows because nutrients become scarce and the environment becomes polluted.
- Cultures in scale-up, fermentation, and manufacture must be assayed for protein production, concentration, and activity.
- A common type of plasmid miniprep is an alkaline lysis miniprep. Minipreps should produce yields that are high enough to see the retrieved plasmids from a transformation on a gel and provide enough plasmid for future transformations. Mini-, midi-, and maxiprep kits are available for purchase from biological suppliers.

Thinking Like a Biotechnician

1. In DNA isolation, what is the function of each of the following enzymes?
   salt   RNase   protease   cellulase
2. A technician is attempting to transform cells with a rather large plasmid and the transformation efficiency is very low. What might he or she do to increase the transformation efficiency?
3. How can a technician obtain the current GMP regulations?
4. A plasmid has a gene of interest that you would like to transfer to another plasmid that is a better vector for transformation. How would you know which restriction enzymes to use to make the rDNA?
5. A technician has found a section of DNA responsible for an antifreeze phenotype in an ocean fish. She wants to make several copies of the gene for genetic-engineering purposes. What technique(s) and instruments could be used to make multiple copies of the gene?
6. You are interested in finding the insulin gene in some human chromosomal DNA. Propose a method by which you could accomplish this.
7. A cell culture is in a 2-L spinner flask. Overnight, the growth rate of cells slowed, and the culture is no longer in exponential growth. The density of cells in the culture is not high enough to "seed" another flask. What might the technician check?
8. A DNA sample gives a 260-nm reading of 0.85 au and a 280-nm reading of 0.65 au. What is its concentration and purity?
9. In a given transformation, only a relatively low number of cells are transformed. How can a technician tell if a transformation, such as transforming *E. coli* into amylase producers, actually occurs?
10. What are the advantages and disadvantages of purchasing plasmid isolation kits?

Biotech Live

Activity 8.1

Lambda (λ) Phage: A Good Virus to Have

Bacteriophages are viruses that infect bacteria cells. Like other viruses, phages can take over a cell and trick it to reproduce and release more viruses. Some viruses, including phage lambda (λ) can remain dormant in a cell for a long time before it eventually inserts itself into a chromosome and takes over the cell. The phage can actually cut a chromosome and insert its DNA into the DNA of a chromosome. In this way, a phage's genes can be read and expressed.

Lambda phage (λ) is useful to genetic engineers for several reasons. One reason is that scientists know the entire DNA code for the virus. They know how to cut it and put it back together once new genes have been inserted. They can use lambda phage to transform cells to produce new proteins.

Lambda phage's DNA is often cut with restriction enzymes, and the resulting pieces are used as DNA sizing standards on agarose gels.

Using Internet resources, learn more about the characteristics of bacteriophage lambda (λ) and its use in biotechnology. Create a paper model of the virus that teaches others about the phage's characteristics.

1. Find a website that has a diagram of a lambda (λ) bacteriophage. Using the diagram, design a three-dimensional model of the virus that will demonstrate its important characteristics (those that help it infect cells and reproduce).
2. Gather materials to build your phage model. Include something to represent the "head" of the phage, the tail of the phage, and the DNA molecule inside the virus.
3. Label the outside of the "head" with important facts about lambda phage (λ), including the organism it infects, how it reproduces, the length of its DNA strand, and how biotechnologists use it. Include at least two websites that you used as references.

Activity 8.2

Restriction Enzymes: Protein Scissors

Restriction enzymes recognize certain sequences of base pairs in a double strand of DNA and cut the DNA at or near that point. Without restriction enzymes, creating pieces of rDNA (DNA from two organisms) would not be possible.

Use GenBank (at NCBI) and www.restrictionmapper.org to determine the number of restriction sites in the lambda bacteriophage sequence for the following restriction enzymes, *MseI*, *Bam*HI, *Eco*RI, and *Sma*I.

1. Go to the National Center for Biotechnology Information at www.ncbi.nlm.nih.gov.
2. Under Search, scroll down and choose "Nucleotide."
3. Then type "Enterobacteria phage lambda, complete genome" in the search box.
4. Hit "Search." About a dozen sequences will be listed that contain all or part of lambda DNA sequence.
5. Select the sequence entry for the entire lambda sequence of 48,502 base pairs (JO2459.1). It will open an entry for a sequence (GenBank: J02459.1) submitted by reference authors Jeong, H., Kim, J.F. and Studier, F.W.
6. Scroll to the bottom of the entry to the DNA sequence. Highlight and copy the entire sequence.
7. Go to www.restrictionmapper.org and paste in the lambda nucleotide sequence.
8. For the restriction enzymes *MseI*, *Bam*HI, *Eco*RI, and *Sma*I, find the number (frequency) of each enzyme's restriction sites in the sequence. Choose the enzyme, and hit "Map Sites."
9. Record the number of "Cut Positions" that result from the digestion of the lambda DNA molecule by each enzyme.

10. Go back to the GenBank: J02459.1 entry. What other information can be found on this page besides the lambda genome sequence?
11. Now go to www.neb.com, choose "NEB cutter" and use this program to find the restriction sites for lambda DNA. The lambda sequence is pre-loaded under "standard sequences."

Animals, Animals, Animals, Animals...

Activity 8.3

1. Use the Internet to find 2–3 definitions of the term "clone". From them, create a definition for "clone" that describes it best.
2. Visit the cloning information website hosted by the NIH's National Human Genome Research Institute at www.genome.gov/25020028. Answer the questions below:

   • What is a natural clone? Give an example.
   • How is an animal cloned?
   • Has a human been cloned?
   • Give an example of a benefit derived from a cloning procedure.
   • Give an example of a possible drawback to cloning animals.
   • Why might someone question the ethics of cloning animals?

3. Several animals have been cloned. Using the information at www.businesspundit.com/20-animals-that-have-been-cloned, give the year that each was cloned and then list one interesting thing that you learned about each cloned animal.

   • "Dolly" the sheep
   • "Mira" the goat
   • "Millie" and her family of pigs
   • "Ombretta" the mouflon
   • "Copy cat"
   • "Snuppy" the Dog

GMP: It Makes Good ~~Sense~~ Cents

Activity 8.4

To ensure that pharmaceuticals are properly identified and free of contamination, the federal government has developed very strict guidelines for their production. These guidelines are described in detail on a website produced and maintained by the FDA. Inspectors from the FDA regularly check companies to verify that they are maintaining cGMP.

In groups of two to four students, produce a poster that outlines and illustrates one of the sections of the Code of Federal Regulations for cGMP.

1. Go to biotech.emcp.net/cGMPregs for the cGMP regulations. Look it over and be sure to review the definitions of terms used on the website.
2. For one of the following sections (assigned by the instructor), create a poster that lists/summarizes/outlines the guidelines of the assigned cGMP section.

Subpart B	Organization and Personnel
Subpart C	Building and Facilities
Subpart D	Equipment
Subpart E	Control of Components and Drug Product Containers and Closures
Subpart F	Production and Process Controls
Subpart G	Packaging and Labeling Controls
Subpart H	Holding and Distribution
Subpart I	Laboratory Controls
Subpart J	Records and Reports

3. Find or draw illustrations that would help a user understand the guidelines.
4. Hang the poster in the laboratory to allow frequent use by lab technicians.

Activity 8.5 Standing on the Shoulders of Others

In 1962, James Watson, Francis Crick, and Maurice Wilkins shared the Nobel Prize in Physiology and Medicine for their determination of the structure of DNA. DNA's twisted ladder/double helix structure explained how DNA could replicate by unzipping and making copies of each side. One copy of each chromosome then moves into daughter cells during cell division. Their understanding of how the nucleotides came together to make the double helix has opened up research avenues that did not exist prior to their determination of DNA structure.

Just as Watson, Crick, and Wilkin's discovery was dependent on the work of others (Erwin Chargaff's A-T, G-C pairing rule and Rosalind Franklin's x-ray crystallography of DNA, for example), their work has been the basis of new research and discoveries on how genetic information is stored and expressed in organisms. These discoveries have lead to the understanding and application of genetic modification to produce new products and organisms, as well as a better understanding of how organisms function, disease mechanisms, and therapies.

 Learn more about the significant discoveries and the scientists that made them in DNA, molecular biology and biotechnology.

1. Go to the Dolan DNA Learning Center's Nobel Laureate website at biotech.emcp.net/dnalcNobel for a list of scientists whose work was so important that they were recognized by a Nobel Prize.
2. Click on the image for each of the scientists or team of scientists listed below to learn more about their work.

> Francis Jacob, Jacques Monod (1965)
> Marshall Nirenberg (1968)
> David Baltimore, Howard Temin (1975)
> Frederick Sanger (1958 & 1980)
> Barbara McClintock (1983)
> Michael Bishop, Harold Varmus (1989)
> Richard Roberts, Phil Sharp (1993)
> **Edward Lewis, Christiane Nusslein-Volhard, Eric Wieschaus (1995)**

3. Read through the animated explanations of their work and record the following in your notebook:

> Their major discovery (discoveries)
> The organisms and/or molecules they studied
> The importance (or applications) of their discoveries

Bioethics

Designer Babies: When Should New Technologies Be Used to Change the Gene Pool?

Advances in cell and molecular biology, embryology, and biomarker technology allow scientists to identify phenotypes (traits) in embryos and even in sperm and egg cells. The power of the technologies can be applied in a variety of ways.

It is obvious why couples would want to check for "healthy" sex cells and embryos. Being able to test for enzyme deficiencies (such as PKU) or chromosome abnormalities (such as Down syndrome) allows parents to decide how they might handle a pregnancy with a child that has one of these conditions. But now, less life-threatening phenotypes can also be identified early in the reproductive process. For example, the sex of an embryo is easily determined. With new techniques, an embryo of a specific gender can be selected or aborted. This is called gender selection. Gender selection is highly controversial.

Many parents say that as long as their baby is healthy, they do not care whether it is a boy or a girl. But, some parents do care. In certain cultures, having a baby of one particular gender is celebrated. In fact, in some countries, babies that are "the wrong gender" may be given away or killed.

How do you feel about gender selection? Is it ever justifiable? If so, when? Would you want to be able to select the gender of your babies? How much would you be willing to pay for it?

Using the Internet, find information about additional gender-selection techniques. Examine your own beliefs regarding gender-selection technologies.

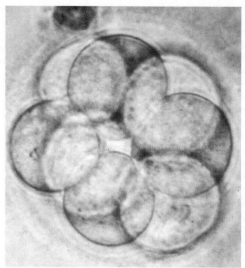

An eight-celled human embryo (blastocyst).
Photo courtesy of Joe Conaghan, PhD.

Using the Internet, find information about additional gender-selection techniques. Examine your own beliefs regarding gender-selection technologies.

1. Go to biotech.emcp.net/newsweek_gender and read the article on gender selection.
2. Next, go to biotech.emcp.net/PGD. The Fertility Institutes claim "PGD" (preimplantation genetic diagnosis) has taken sex selection to the next and most successful level ever (greater than 99.9%). Read through the website to learn more about how technology is not only being used in gender selection but in the selection of other factors important to prospective parents. Finally, watch the *60 Minutes* video clip available on the website.
3. Write a position statement as to whether or not you would want the right to predetermine the gender of your child. Address the following points:

   If you believe couples should have the right to choose, explain your reasons. List a reason some other couple might not want the right to be able to choose. Conversely, if you do not want the right to be able to select the gender of your offspring and you think others should not have that right, explain your reasons. Give one reason why other couples might want to select the gender of their child.

4. In addition to choosing the gender of your child, are there specific traits that you would want your child to possess? List three other characteristics a couple might want to predetermine in their offspring and why.
5. Gender selection is expensive. Find a website for a gender-selection service. How much do they charge? Do you think your medical or health insurance should pay for these kinds of services? Why or why not? How much would you personally be willing to spend? How did you come up with that amount?

Biotech Careers

Photo by author.

Research Associate, Fermentation

Meng Heng
Genencor International, Inc.
Palo Alto, CA

On a production line in biomanufacturing, Meng Heng, a research associate at Genencor International, Inc. monitors fermentation cultures. These 10-L bioreactors grow broth cultures of bacteria-producing enzymes used in industrial applications such as paper softening or detergent additives.

The bioreactors are automated and computer controlled. Factors that influence cell growth are monitored including pH, sugar concentration, and especially oxygen content. Maintaining these and other factors keep the cells growing, dividing, and producing protein at a maximum rate. Meng tests samples for protein concentration and activity and also sends samples to **Quality Control (QC),** where very sensitive assays are run. The company must always know exactly how much product they have and the condition of each culture. When the cells become crowded in the cultures and protein production is at a maximum, the cultures are either harvested or transferred to larger bioreactors. When the cultures in these 10-L bioreactors reach a maximum growth rate and product concentration, they will be used to seed 30-L bioreactors.

Quality Control (QC)
(qual•i•ty con•trol) a department in a company that monitors the quality of a product and all the instruments and reagents associated with it

9 Bringing a Biotechnology Product to Market

Learning Outcomes

- Outline the major steps in bring a genetically engineered protein product from a selection plate through biomanufacturing to marketing
- Compare and contrast the methods for harvesting intracellular and extracellular proteins
- Define chromatography and distinguish between paper, thin-layer, and column chromatography, giving examples of each procedure
- Discuss the variables used to optimize column chromatography
- Explain how product quality is maintained for key types of biotechnology and pharmaceutical products
- Describe the clinical testing process for pharmaceuticals
- Discuss the final marketing and sales considerations in bringing a product to market

9.1 Introduction to Biomanufacturing

Cells produce thousands of different proteins in hundreds or thousands of copies each. Isolating a specific protein product in large quantities from a sample of cells requires techniques that apply an understanding of the biology, chemistry, and physics of the molecules involved. Through a series of molecular separations, scientists purify protein product in marketable volumes from other molecules in a sample.

A company that has cloned a protein in a transformed cell line wants to manufacture that particular protein in large quantities without other cellular proteins contaminating the sample. This is called biomanufacturing. Biomanufacturing is the process by which commercial volumes of a biotechnology product are produced, purified, and formulated. The specific biomanufacturing procedures are different for each protein product but are similar in the general process steps.

Biomanufacturing starts with broth cultures of transformed, transfected, or native cells. The cells in the culture have been selected for the

harvest (har•vest) the method of extracting protein from a cell culture, also called recovery

recovery (re•cov•er•y) the retrieval of a protein from broth, cells, or cell fragments

purification (pu•ri•fi•ca•tion) the process of eliminating impurities from a sample; in protein purification, it is the separation of other proteins from the desired protein

intracellular (in•tra•cell•u•lar) within the cell

extracellular (ex•tra•cell•u•lar) outside the cell

expression of the specific protein of interest and then scaled-up from agar selection plates to broth selection media. The liquid culture of cells producing the target protein are moved through a series of processes including:

- fermentation or cell culture (upstream manufacturing)
- **harvest/recovery** (downstream manufacturing)
- protein **purification** (downstream manufacturing)
- formulation (downstream manufacturing)
- filling/packaging (downstream manufacturing)

For pharmaceutical and industrial proteins, biomanufacturing starts with cells in liquid culture (see Figure 8.23). Using scale-up and fermentation or cell culture methods, discussed in Chapter 8, the protein of interest is produced in progressively larger cultures. Specifically, when a cell line in research and development is shown to produce a protein product (usually on selection plates or through protein assays), it is used to inoculate broth cultures in flasks or spinner bottles. Monitoring of the culture's cell density (OD_{600}), pH, nutrient levels, and protein content reveal when the culture should be used to "seed" the next larger volume, usually in 1-L and then 5-L spinner flasks to 10-L fermentor/bioreactors. Bioreactors holding up to a few dozen liters of cell culture are used to inoculate or seed larger bioreactors, for example, a 100-L bioreactor. Then at the "right time" the cultures are moved from the 100-L bioreactor to a 1000-L bioreactor to a 5000-L bioreactor, and then finally to a final large-scale (30,000-L) bioreactor (see Figure 9.1). All along the way, the cultures are monitored and assayed to ensure uncontaminated, highly concentrated sample. When protein concentration is high enough, the cultures are moved to recovery.

Downstream Manufacturing: Recovery and Purification

In recovery, the protein of interest is separated or harvested from the production cell cultures and purified from other proteins using a series of centrifugations (see Figure 9.2), filtrations, and chromatography. The method of harvesting a protein of interest from cloned cells depends on the specific protein and whether that protein is found within the cell (**intracellular**) or outside of the cell (**extracellular**). Intracellular proteins are made and function inside the cell. To harvest intracellular proteins from cells, the cells must be burst open. Several methods can be used. One way is to use detergents, such as sodium dodecyl sulfate (SDS), to solubilize the cell membranes, which results in cell explosion. A problem with using SDS is that it denatures many proteins and is hard to remove after use. **Sonication** (use of high-

Figure 9.1. Technicians monitor the operation of bioreactors used in biomanufacturing protein products.
© Michael Rosenfeld/Science Faction/Getty Images.

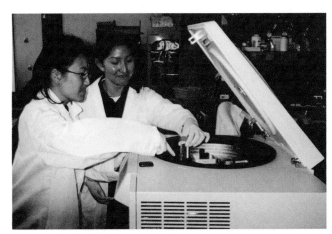

Figure 9.2. Using a large-volume, refrigerated centrifuge, technicians spin down transformed cells, separating them from broth. The broth or the cells may contain the protein of interest.
Photo by author.

frequency sound waves) is another way to release proteins from cells. Alternatively, cells may be mechanically burst open using a tissue homogenizer such as the Tissuemizer by IKA Works, Inc., a mortar and pestle, or blender. A disadvantage of bursting cells open is that all of a cell's thousands of intracellular proteins are released. This makes the job of isolating and purifying the protein of interest from other proteins substantially more difficult.

Although they are fewer in number, extracellular proteins are easier to isolate. An example of an extracellular protein is amylase. In humans, salivary gland cells produce salivary amylase and secrete it into the salivary gland ducts. Many bacteria, including *Bacillus subtilis*, also produce amylase and secrete it outside the cell to digest starch into absorbable sugar units. As you know, commercial recombinant amylase has been made by genetically engineering bacteria to make it in large volumes. To harvest extracellular proteins such as rAmylase, the transformed cells must be separated from the rAmylase in the growth media and other extracellular proteins must be separated from the rAmylase.

Recovery from the "broth" begins with separating the protein from cell debris, either by filtering or centrifugation. The "broth" is piped into a large-volume centrifuge, which is spun at a sufficient speed (revolutions per minute [rpm]) to pull cells and cell debris to the bottom of the centrifuge. The proteins, other molecules, and low-density fragments continue to float in the supernatant. After centrifugation, the supernatant is passed through one or more filters to eliminate other large contaminants that may not have been removed by centrifugation and to adjust or change buffers, if necessary.

At this stage, the protein has been harvested from the cell culture but not from the broth suspension. Next, the protein has to be separated, or purified, from other molecules in the broth suspension in sufficient quantity and purity. Purification is usually accomplished through a process called **column chromatography**. In column chromatography, a sample is passed through a column packed with microscopic resin beads (see Figures 9.3 and 9.4). The beads act as a molecular sieve, separating molecules based on their characteristics, such as size, shape, charge or solubility. Several columns in series may be needed to remove each group of contaminant proteins.

sonication (son•i•ca•tion) the use of high frequency sound waves to break open cells

column chromatography (col•umn chrom•a•tog•ra•phy) a separation technique in which a sample is passed through a column packed with a resin (beads); the resin beads are selected based on their ability to separate molecules based on size, shape, charge, or chemical nature

BIOTECH ONLINE

I only want THAT protein.

Isolating a sample of a particular cellular protein is a tricky task since there are so many different kinds of protein in a cell. The use of buffers in protein separation technology makes it possible to separate molecules based on their size, shape, or charge.

Review how and why proteins are purified from cells and how and where buffers are used in protein purification from cell extract samples.

1. Go to **biotech.emcp.net/techtvprotein**.
   A. List the procedures that are used to separate and study protein samples from cells.
   B. Consider how and where buffers might be of importance in the separation technologies.
2. Go to **biotech.emcp.net/icyoucolumnchrom**.
3. Describe how buffers are used to purify proteins from cell extractions using column chromatography.

Figure 9.3. **Chromatographic columns, filled with microscopic beads of resin, are used to separate proteins based on their size, charge, shape, or solubility. Small volume columns, such as the one pictured here, are used early in R&D to determine separation protocols. A small amount of a particular resin is added to the gravity-flow column (the white resin in the bottom one-third of the column). The technician tests different resins that might separate molecules based on the size, shape, charge, solubility, or some other special property of the protein of interest.** Photo by author.

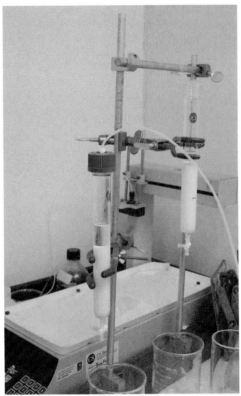

Figure 9.4. Larger columns, such as these, run too slow if only relying on gravity. Instead, the column is attached to a pump (through the tube) that creates pressure to force the sample through the resin bed. Once a protocol for a chromatography is determined, it is scaled-up to manufacturing volumes.
Photo by author.

Figure 9.5. Students from the University of Massachusetts–Lowell BioManufacturing Center learn how to set up and use computers to control large pressure-pumped chromatographic columns used in large volume protein purification.
© Science Faction/SuperStock.

gravity-flow columns (grav•i•ty-flow col•umns)
column chromatography that uses gravity to force a sample through the resin bed

pressure-pumped columns (pres•sure-pumped col•umns)
a column chromatography apparatus that uses pressure to force a sample through the resin bed

ultrafiltration (ul•tra•fil•tra•tion) a type of filtration that separates proteins or other large organic molecules from low molecular weight solutes

tangential flow filtration (tan•gen•tial flow fil•tra•tion)
a type of ultrafiltration where small molecules diffuse out of a solution as they pass by a filtering membrane, resulting in a solution with a higher concentration of the molecule of interest—also called TFF

Once R&D has designed chromatography protocols in **gravity-flow columns** on a small scale, the chromatography is scaled-up to **pressure-pumped columns** (see Figure 9.5) large enough to treat the large volumes of protein solutions coming from fermentation or manufacturing. Large-scale columns, each filled with a specific resin to separate molecules based on their differing characteristics, are run in a series until the protein product is of sufficient concentration and purity. Column chromatography is discussed in greater detail in the next section.

The final chromatographic product is then run through a series of filters to concentrate the product and to remove small contaminate molecules. This is called **ultrafiltration** (see Figure 9.6). The method of filtration is important since standard "through a membrane" normal flow filtration may result in decreased yield of protein because the filters plug up with accumulated protein. Commonly, a method of ultrafiltration called **tangential flow filtration (TFF)** is used, in which protein solution is fed into a tube that has membranes lining it. As the sample flows by the membranes, small molecules diffuse out of the solution, leaving proteins in the solvent. In some cases the solution recycles through the TTF tube so that additional filtering and concentrating occurs. After ultrafiltration, what remains is a highly concentrated, relatively pure protein of interest, ready for formulation (into liquid, powder, tablets, etc) and filling into the desired packaging.

Whether intracellular or extracellular, the protein of interest must be recovered from the broth, cells, or cell fragments and then purified, validated, and put into its final formulation. Each protein presents new challenges for recovery. The size, shape, charge, concentration, and location of the protein affect its ability to be purified and

Figure 9.6. Ultrafiltration is used in the pharmaceutical industry to isolate a protein of interest from contaminant molecules and to produce protein concentrate. Each column holds a series of filter stacks.
© Science Faction/SuperStock.

recovered. Many hours of R&D go into devising new purification and recovery techniques; the goal is to recover the highest concentration of the purest protein possible. Figure 9.7 summarizes the main steps in biomanufacturing and the commercial production of proteins.

Section 9.1 Review Questions

1. When harvesting broth cultures, how are cells separated from the broth?
2. Describe why it may be more challenging to purify an intracellular protein versus an extracellular protein.
3. In column chromatography, what accomplishes the separation of molecules in a mixture?
4. Summarize the differences in the objectives of the upstream and downstream processes during biomanufacturing.

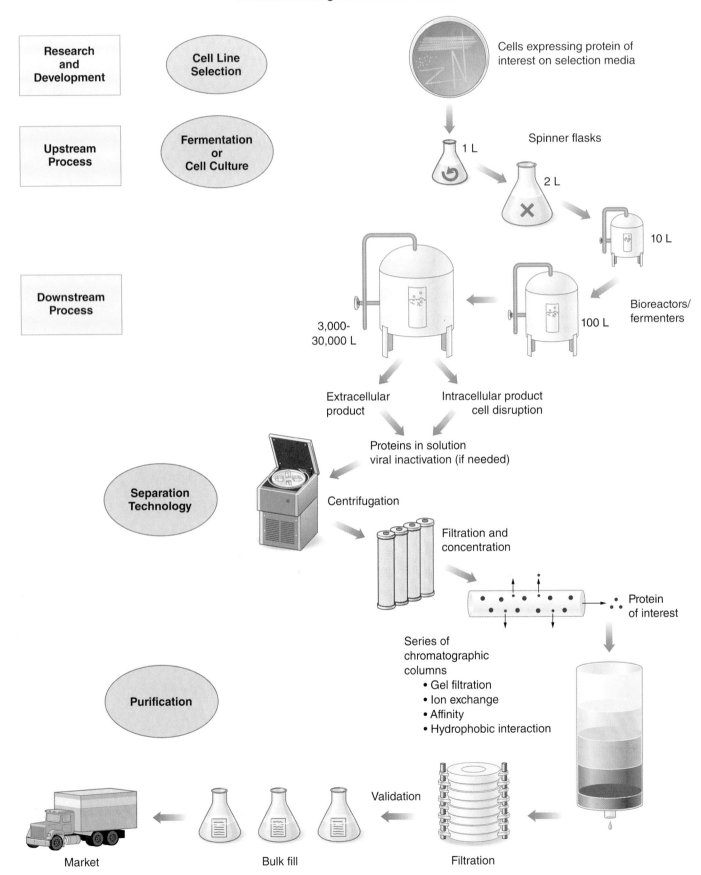

Figure 9.7. **Biomanufacturing and Protein Purification**

9.2 Using Chromatography to Study and Separate Molecules

Any method of separating molecules in some kind of buffer or solvent through a solid phase that acts as sieving material is called a chromatography. Separating molecules on columns is only one kind of chromatography. In biotechnology applications, there are three common types of chromatography: **paper chromatography**, **thin-layer chromatography**, and column chromatography. This section compares the three types.

Paper Chromatography

You probably have separated ink samples or plant pigments using paper chromatography. In paper chromatography, filter paper is used as the solid phase. A sample is placed on the filter paper, near the bottom. The filter paper is then placed in a chromatographic chamber with solvent. The solvent creeps up the paper by capillary action, carrying soluble molecules along with it. Depending on their size and solubility, molecules are deposited on the paper, at different distances from the starting point, as the solvent moves (see Figure 9.8). If the molecules are colorless, they might have to be visualized with developer.

The solvent for paper chromatography determines how a molecule behaves on the **chromatograph**. Some molecules are only soluble in water-based solvents. Some are only soluble in nonpolar, organic solvents, such as acetone or petroleum ether. Many lipids, proteins, and amino acids require this kind of solvent. To allow molecules to move and separate on the solid phase, one must choose an appropriate solvent.

Paper chromatography may be used to ascertain the amino acid composition of a protein. A sample is digested with proteases. The resulting amino acid mixture is run on a chromatograph. The amino acids travel up the paper to different heights. Ninhydrin is sprayed on the paper, which causes the amino acid spots to colorize. The distance the amino acid spots travel, compared with the solvent front, is measured and reported as an R_f value. Each of the 20 amino acids has a characteristic R_f for a given solvent. By determining the R_f value for the sample, scientists can learn which amino acids are found in a given protein. Although some labs still use paper chromatography, most have replaced it with HPLC (to be discussed later in this chapter).

paper chromatography (pa•per chro•ma•tog•ra•phy) a form of chromatography that uses filter paper as the solid phase, and allows molecules to separate based on size or solubility in a solvent

thin-layer chromatography (thin-lay•er chro•ma•tog•ra•phy) a separation technique that involves the separation of small molecules as they move through a silica gel

chromatograph (chro•mat•o•graph) the medium used in chromatography (ie, paper, resin, etc) through which the molecules of interest move and separate

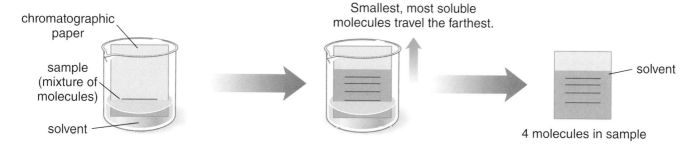

Molecules on the dried chromatograph are recognized by the distance they move in the solvent.

Figure 9.8. Paper Chromatography. Molecules separate as they move up the paper. The distance that the molecules travel depends on their size and solubility in the solvent.

Thin-Layer Chromatography

Thin-layer chromatography is similar to paper chromatography in several aspects. A glass plate is spread with a thin layer of silica gel, which serves as the stationary phase. A sample is added at one end. A buffer creeps from the sample side to the opposite

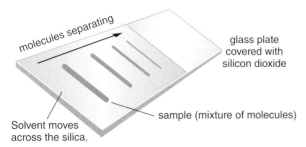

Figure 9.9. Thin-Layer Chromatography. Molecules separate as they move through the silica gel. Thin-layer chromatography is used to separate small molecules, such as amino acids.

frit (frit) the membrane at the base of a chromatographic column that holds the resin in place

fraction (frac•tion) a sample collected as buffer flows over the resin beads of a column

dialysis (di•al•y•sis) a process in which a sample is placed in a membrane with pores of a specified diameter, and molecules, smaller in size than the pore size, move into and out of the membrane until they are at the same concentration on each side of the membrane; used for buffer exchange and as a purification technique

end, carrying soluble molecules along with it. Molecules are deposited based on their size and solubility in the buffer (see Figure 9.9). If the molecules are colorless, they might have to be visualized with developer.

Column Chromatography

How a Column Works A chromatographic column is basically a tube with a membrane (**frit**) near the base (see Figure 9.10). Resin beads of a particular type (in buffer) are poured into the column. The resin beads settle onto the frit and form a matrix through which molecules can pass. When the sample is added to the top of the resin bed, gravity will pull it into the matrix. Depending on the resin selected, the molecules in the sample will either bind to the beads or pass through them. As the sample drips out, it is collected at regular intervals into vessels of manageable volumes. These are called **fractions**.

Before a sample is passed through a column, it must be in the "right" buffer. This can be accomplished either by buffer exchanges during centrifugation or through a technique called **dialysis**. In dialysis, a sample with a mixture of desirable and undesirable proteins is placed into a dialysis tube composed of dialysis membrane; the tube is then tied off at each end (see Figure 9.11). The membrane is made up of a cellophane-type of material with microscopic holes (pores), which are tiny enough to keep the protein of interest inside the bag. However, smaller molecules, including water, salts, sugars, and other constituents of the broth, can diffuse in or out. The broth-filled dialysis tube/bag is placed into a large volume of the desired buffer. By diffusion, buffer goes into the dialysis bag and unwanted molecules diffuse out of the bag. After several hours and several buffer changes, the broth is exchanged for chromatographic buffer, by which time many unwanted molecules have been removed.

For large volumes, such as those in manufacturing recovery, the dialysis process described above is impractical. Instead, a process called **diafiltration** is used. In diafiltration (similar to TFF), the sample flows by a dialysis membrane. At right angles to the flow, pressure is applied to speed the diffusion of buffers, solvents, and/or contaminants. Large molecules, such as proteins, stay

The sample is added to the resin at the top of the column. Gravity pulls the sample through the resin bead matrix. Depending on the resin, molecules can be separated based on size, shape, or charge.

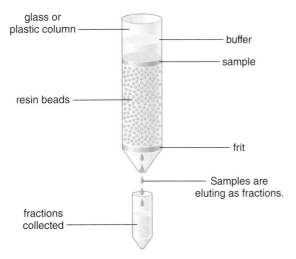

Figure 9.10. How a Column Works. Samples that pass through a column are collected in small volumes called fractions. Under ideal circumstances, molecules that separate in the column are collected in different fractions.

Figure 9.11. A technician transfers a small amount of sample in a given buffer into a soaked, tied-off, dialysis bag. The dialysis bag is then placed into the desired buffer and left to exchange for several hours. The buffers exchange because of differences in concentration (diffusion). Several rounds of dialysis are required for a complete exchange.
Photo by author.

in the flow while small molecules, such as buffer salts, are exchanged across the membrane.

Once dialyzed in the appropriate buffer, the protein mixture is loaded onto the top of the resin bed in the chromatographic column. Buffer pushes the sample through the resin. As buffer and sample drip out of the column, small subsamples (fractions) are collected.

Since most proteins are colorless, their separation on a column is not visible. Likewise, as samples are collected into fractions, proteins are not visible. To visualize which proteins separate into which fractions, samples are run via polyacrylamide gel electrophoresis (PAGE) and identified by molecular weight. By comparing the bands in the original column **load** to the bands in the fractions, technicians can determine the amount of separation (see Figure 9.12).

Often, it takes more than one kind of column chromatography to separate a protein of interest from all the other proteins in a mixture. At the end of each chromatography, the scientist checks the separation of the proteins from each other using an ultraviolet (UV) spec, indicators, ELISA, and/or gels. The goal is to have the desired protein completely separated from other proteins in the mixture but still in a high concentration. An isolated protein at a high concentration would be visible as the only dark band of a sample in a lane on a gel. When isolation is confirmed in this way, the batch is considered pure, and its concentration is determined. If enough of the pure protein product can be harvested and purified, it can be sold in the marketplace.

In industry, large columns are attached to pumps that force large volumes through the columns. Several kinds of column chromatography are routinely used in manufacturing. The most common types are as follows:

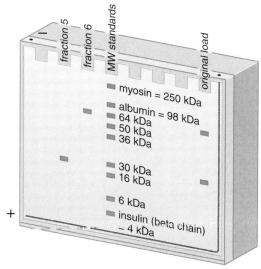

Figure 9.12. Running Fractions on a PAGE. Lane 8 shows the original sample load. Fractions 5 and 6 show the separation of two of the proteins from the load.

- **Gel-filtration** (also called size-exclusion) **chromatography** separates molecules based on size.
- **Ion-exchange chromatography** separates molecules based on charge.
- **Affinity chromatography** separates molecules based on shape or unique functional groups.
- **Hydrophobic-interaction chromatography** separates molecules based on hydrophobicity (not soluble in water).

Each method of column chromatography may be used to identify the characteristics of molecules. Each is also used to purify or separate molecules in a mixture.

Gel-Filtration (Size-Exclusion) Chromatography In gel-filtration chromatography, a sample is processed through a size-exclusion resin (see Figure 9.13). Size-exclusion beads possess tiny channels. In a sample, some molecules will be tiny enough to go into the channels, and some will be too big to enter the channels. It takes a longer time for molecules to go through the channels than to go around them (think of a maze). As a result, larger molecules move through the column faster than smaller molecules. Fractions are collected as the sample travels down the column. The early fractions contain larger molecules than do the late fractions. A sample from the fractions can be loaded onto a polyacrylamide gel and their molecular weight determined. By comparing fractions to the original sample loaded on a column, one can determine the degree of separation.

Several factors affect the separation of molecules within the sizing column. The height and width of the resin bed can affect the resolving power of the column. The longer the column, the greater the separation between molecules of different sizes (see Figure 9.14). The width of the column affects how much sample can be loaded and how quickly different sizes of molecules will separate. Resin beads of different pore size can be selected to include or exclude molecules of a certain size.

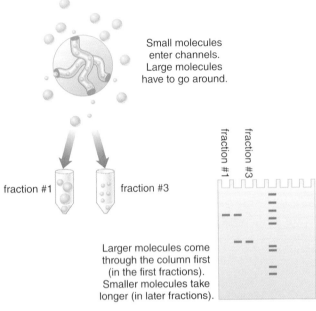

Small molecules enter channels. Large molecules have to go around.

fraction #1   fraction #3

Larger molecules come through the column first (in the first fractions). Smaller molecules take longer (in later fractions).

Figure 9.13. Gel Filtration Resin. When starting protein purification, technicians often use a gel-filtration (size-exclusion) column first. They know the molecular weight of their protein, so they can often eliminate several contaminant proteins by a quick run through a sizing column.

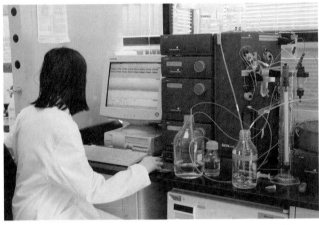

Figure 9.14. A technician runs a fast-performance liquid chromatography (FPLC) sizing column. Pumps that push buffer through the resin beads characterize FPLC. The column (on the right-hand side) is long and narrow. Most sizing columns are rather long to give molecules time to separate completely. Bottles supply buffers to the column. Pumps run the buffers through tubing from the bottles to the column. Samples coming out of the column are run through a spectrophotometer (one of the black boxes) for a concentration reading. Data are processed on the computer. Photo by author.

Gel-filtration chromatography can be the first step in sorting the proteins in a large mixture. Such is the case when cells are genetically engineered to produce a particular protein. The cloned cells are exploded, and hundreds of proteins are released, including the desired protein. The mixture of proteins is initially separated into fractions using a sizing column. The fractions may be further treated and identified by other chromatographic techniques, as well as by PAGE.

Ion-Exchange Chromatography A second type of column chromatography is ion-exchange chromatography, which separates molecules based on their overall charge at a given pH (see Figure 9.15). Many proteins have a net charge, either positive or negative, due to the kinds of amino acids present on the polypeptide (recall that amino acids can have no charge, a positive charge, or a negative charge, depending on their molecular structure and the pH). If there are more positively charged than negatively charged amino acids, the protein will carry a net positive charge.

Ion-exchange columns are packed with resin beads carrying a charge opposite the charge of the protein of interest (see Figure 9.16). Positive and negative resin beads are available for purchase. An ion-exchange column resembles a gel-filtration column, but it runs differently. The primary difference is in the types of resin and buffers used.

In ion-exchange chromatography, the resin bed is equilibrated with a buffer of a certain pH. This charges the resin beads, giving them a positive or negative charge. The sample is loaded and pushed through the column with more of the equilibration buffer. Molecules of a charge opposite to that of the beads will attach to the beads. All other molecules wash through the column and are collected in "wash" fractions. The bound molecules will stay attached to the beads until they are knocked off. This is called "eluting the sample." The **elution** buffer either contains a high salt concentration, or it has a high or low pH. As the elution comes off, the sample is collected. All fractions are then run on a PAGE to visualize what was separated and collected.

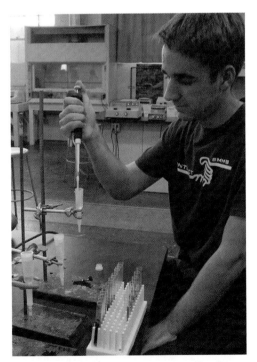

Figure 9.15. A technician runs a small-volume, gravity-flow, ion-exchange column. To get an idea of the molecule's charge and how it will behave in a larger column, such as those used in manufacturing, he or she can load a pure sample of protein and see which fraction it ends up in. Photo by author.

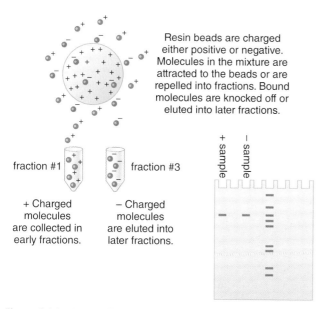

Resin beads are charged either positive or negative. Molecules in the mixture are attracted to the beads or are repelled into fractions. Bound molecules are knocked off or eluted into later fractions.

fraction #1

+ Charged molecules are collected in early fractions.

fraction #3

− Charged molecules are eluted into later fractions.

+ sample

− sample

Figure 9.16. **Ion Exchange Resin.** Resins are manufactured with ions attached. The ions present a certain degree of positive or negative charge, depending on the buffer pH.

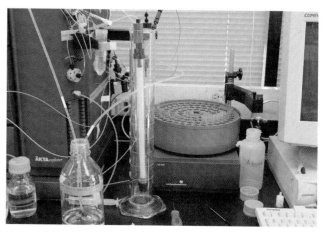

Figure 9.17. Since so many fractions are collected off a running column, a fraction collector with collection tubes (right side of the photo) can be set up to collect each fraction automatically.
Photo by author.

Sometimes, there are 50 to 100 fractions from a column run. Fraction collectors are used so that technicians do not have to stand for hours changing collection tubes every few minutes (see Figure 9.17).

A protein's charge is the separating factor in an ion-exchange chromatography. The column is equilibrated with buffers of a specific pH to ensure that the resin beads and the protein in the sample have the "correct" charges. Depending on the pH of the solution in the column, a protein's charge, and its behavior in the column, may change. If a protein's charge is unknown, then its behavior on an ion-exchange column will reveal its overall charge. If a protein's charge is known, it may be separated from other proteins of a lesser or opposite charge.

The technician selects a particular ion-exchange resin, depending on the goal of the experiment. If one knows that the molecule of interest is a positively charged protein, then negatively charged resin is used. This is called **cation exchange** since positively charged ions (cations) are bound to and released from the resin. If one knows that the protein of interest is negatively charged, then positively charged resin is used. This is called **anion exchange** since negatively charged ions (anions) are bound to and released from the resin. Once the conditions for protein separation in a column are determined, they can be applied in manufacturing to a large-volume column for large-scale purification of protein (see Figure 9.18).

Affinity Chromatography A protein's shape (due to amino acid composition, side groups, and folding) is critical to its structure, function, and purification. Several purification methods rely on the shape of proteins. Affinity-column chromatography is an isolation method based on shape or molecular configuration. The most common methods use antibodies to recognize and bind a protein based on, in this case, Ab-Ag interactions. Antibodies can be coupled to resin beads. As sample flows through a column, the antibodies will bind with their complementary antigenic epitope on the protein of interest and pull it out of the solution. It is in this manner that a molecule of a given shape or molecular configuration can be separated from other molecules (see Figure 9.19).

Since antibody-antigen interactions are very specific, affinity chromatography can be used to isolate and remove a single type of protein from a mixture of hundreds. The challenge in affinity chromatography is finding a complementary molecule/antibody to attach to the resin beads. Sometimes these already exist in nature. Sometimes these can be made in the lab.

elution (e•lu•tion) when a protein or nucleic acid is released from column chromatography resin

cation exchange (cat•i•on ex•change) a form of ion-exchange chromatography in which positively charged ions (cations) are bound by a negatively charged resin

anion exchange (an•i•on ex•change) a form of ion-exchange chromatography in which negatively charged ions (anions) are bound by a positively charged resin

Figure 9.18. These large-volume, ion-exchange columns are used in manufacturing to separate proteins produced during biomanufacturing. The resin is the white material at the bottom of the column.
Photo by author.

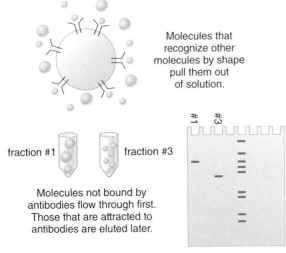

Figure 9.19. Affinity Chromatography Resin.
Separating molecules based on shape or molecular configuration is often done using antibody resin. Antibodies recognize only certain antigens and will bind those and pull them out of solution (fraction #3). High salt buffer may be used to elute the bound protein.

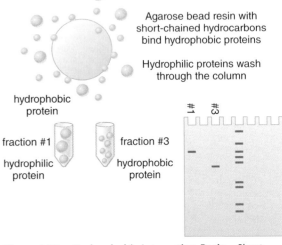

Figure 9.20. Hydrophobic Interaction Resin Short-chained hydrocarbons are bound to agarose bead resin. Hydrophobic proteins in a high salt buffer are added to the column and bind to the hydrophobic hydrocarbons on the HIC resin. Hydrophilic proteins wash through the column. Low salt buffer releases the protein of interest, here in Fraction #3.

Hydrophobic Interaction Chromatography Due to the distribution of different amino acids along a protein's primary structure, each protein molecule has regions that are more or less hydrophilic (soluble in water) or hydrophobic (not soluble in water). As a result, the more hydrophilic groups on the surface of a protein, the more soluble the protein is in water. Conversely, the more hydrophobic groups on the surface of a protein, the less likely it is that it will go into solution. In a solution, the concentration of salts will affect how the protein folds exposing more or less of the hydrophobic groups on the surface of the protein. Protein manufacturers may utilize this phenomenon to separate proteins in a mixture.

Hydrophobic interaction chromatography (HIC) separates protein molecules using the differences in their hydrophobicity in buffers of high salt concentration. In a typical HIC chromatography, resin beads made of agarose with hydrophobic groups (short carbon chains) covalently bound to them are added to a column (see Figure 9.20). A mixture of proteins, containing the protein of interest, in a high salt buffer, such as 2 M or 4 M ammonium sulfate or sodium sulfate, is added to the column. The ammonium sulfate molecules have a strong attraction to the water molecules in the solution so there are less water molecules to interact with the hydrophilic regions of the target protein. The proteins will change shape exposing more hydrophobic regions on the surface and these are available to bind to HIC resin beads. Hydrophilic proteins and other contaminant molecules are repelled by the hydrophobic groups on the HIC resin and pass through the column. Once the contaminant molecules are washed out, a neutralization buffer containing a relatively low salt concentration is added to the column and the target protein, changes back to its original shape, is released from the beads on the column and collected in a vessel.

open-column chromatography (o•pen-col•umn chro•ma•tog•ra•phy) a form of column chromatography that operates by gravity flow

Section 9.2   Review Questions

1. What is the solid phase for each of the following types of chromatography?

    Paper chromatography   Ion-exchange chromatography
    Thin-layer chromatography   Affinity chromatography
    Gel-filtration (size-exclusion) chromatography   Hydrophobic interaction chromatography

2. If a molecule is the smallest in a mixture, will it be the first or last molecule to come off a size-exclusion column?
3. Diethylaminoethyl (DEAE) sepharose is a type of ion-exchange resin. At a pH of 7.5, it has a positive charge. What would be expected if a sample containing one positively charged protein and one negatively charged protein were put on a DEAE column? Where should the proteins end up?
4. What is the value of a fraction collector?

9.3 Column Chromatography: An Expanded Discussion

As a separation technology, column chromatography is so important, and so complex, that this entire section is devoted to expanding on the basic concepts of chromatography described in the previous section.

There are two ways to run a column. One way is to allow gravity to draw the sample and buffers through the column resin. A second, more efficient, way is to use pumps to push a sample and buffers through a column.

Open–Column Chromatography

Also called gravity-flow chromatography, **open-column chromatography** is simple and can be conducted with a minimum of equipment. With open columns, a plastic or glass column is packed with resin and the technician adds samples and buffers, by hand, to the top of the resin bed. Open columns work well for small samples and small column volumes. Open columns are great for working out a column chromatography process on a small scale, as in R&D (see Figure 9.21). Since they are tiny and the volumes used are tiny, these columns can be set up and a run can be completed fairly quickly (ie, in an afternoon). In these columns, though, the buffer runs rather slowly, and, if large volumes are required, the quantitation (collection of numerical data) is relatively poor. Molecules do not separate as well as they do in more sophisticated methods of column chromatography. An advantage of open-column chromatography is that it gives the technician an idea of how to scale-up the purification process to larger, more effective columns.

Figure 9.21. A technician uses a gravity column to determine the right parameters for a column chromatography that will eventually be scaled-up to large purification volumes. With small columns, several factors such as resin volume, concentration of samples, and pH of the samples and reagents maybe quickly tested and optimized.

Photo by author.

Fast–Performance Liquid Chromatography (FPLC)

In the 1970s, as more biotechnology companies began to devise procedures to synthesize and purify protein products, the need for faster, more effective separation technologies arose. Technicians began attaching pumps to columns to push samples through the tightly packed resin. Pressure pumping gave faster and better separation of similar compounds. The column's

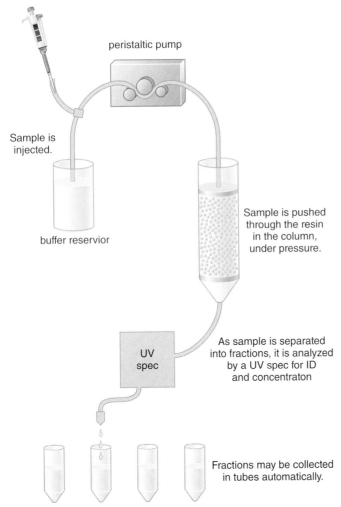

Figure 9.22. **Fast-Performance Liquid Chromatography.** Pumps push the buffer or sample through tubing, into and through the column. As fractions come off the column, they are run through a spectrophotometer that determines the protein concentration of each fraction.

peristaltic pump

Sample is injected.

buffer reservior

Sample is pushed through the resin in the column, under pressure.

UV spec

As sample is separated into fractions, it is analyzed by a UV spec for ID and concentraton

Fractions may be collected in tubes automatically.

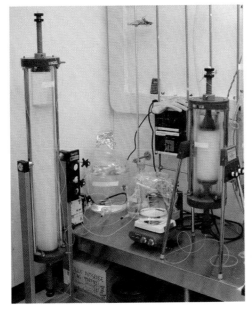

Figure 9.23. **The technician equilibrates an FPLC column with buffers at a certain concentration and pH. Several columns remain from past separations. When needed, a column can be connected to buffer reserves and a pumping unit.**
Photo by author.

output was attached to UV spectrophotometers and a chart recorder for immediate and accurate detection of the molecules in the system. This type of column chromatography, using pressure pumps, is called **fast-performance liquid chromatography**, or, **FPLC** (see Figure 9.22).

A typical FPLC apparatus includes the column with a top gasket that seals the column and allows for increasing pressure. Tubing carries the buffer from the reservoirs to the column and, then, carries the fractions away from the column. The tubing is fed through a pump mechanism that pushes the sample into the column and through the resin at different rates. Computers and pumps are used to set the flow rate of both the sample and the buffers to achieve maximum interaction with the resin beads (see Figure 9.23). As the buffer flows through the resin, the beads separate the molecules based on their characteristics. The fractions leave the column through the frit membrane; they pass into exit tubing, through the sample detector (UV spec) and chart recorder, or computer, and finally land in a fraction collector.

FPLC can be scaled-up to very large volumes. FPLC separation technology is used when thousands of liters of protein product in broth are harvested from bioreactors (see Figure 9.24).

Figure 9.24. **Several large-capacity FPLC columns, packed with resin, are stored in cold rooms until needed. These are used for larger volume protein purifications such as from 10-L bioreactors. Each is labeled with the type of resin and buffer.**
Photo by author.

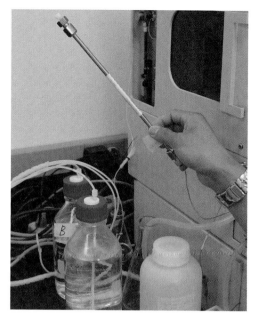

Figure 9.25. The HPLC columns are encased in metal and constructed to withstand the high pressure created as samples are pushed through the resin inside.
Photo by author.

Figure 9.26. The HPLC units are sophisticated, and it takes time to learn how to use them correctly. This instrument actually has four components, including pumps on the bottom and a spec on the top. In the center are two components holding a sampler and the column cartridge, a smaller version of the one shown in Figure 9.25. The technician is programming the HPLC unit through a computer interface. The buffer flow rate and fraction size are just two of the variables that can be controlled. As fractions come off the column, they are run through the spectrophotometer, which gives information about the presence and concentration of protein in the sample.
Photo by author.

High-Performance Liquid Chromatography (HPLC)

With the development of **high-performance liquid chromatography (HPLC)**, researchers have greatly improved their ability to separate, purify, identify, and quantify samples. The HPLC instrument uses tiny microcolumns (see Figure 9.25). These resin-containing columns are made of metal that can withstand very high pressures. The columns are packed with minute resin beads, which provide increased surface area for better separation. Using HPLC, technicians can study tiny amounts of proteins, DNA, and RNA. The HPLC instrument is highly sophisticated, and it requires specialized training to run the computer programs and optimize the running conditions (see Figure 9.26).

No matter what type of column chromatography is used, there are a number of different ways to set up the running conditions for a column. Several factors affect the resolving power of a column apparatus. The type of resin and buffers used will determine how molecules will be separated in the resin. The flow rate and pressure also affect the binding of the sample to the resin beads. In maximizing the separation of molecules, the concentration of both the sample and the buffer is critical. Each chromatographic process is unique to the protein being studied, and the procedures must be optimized for each product.

Resins Used in Column Chromatography

Several types of resins are available for use in column chromatography. For size-exclusion (gel-filtration) columns, the diameter of the channels in the resin beads determines which molecules can be separated from others. For example, the channels in Sephacryl-100 resin beads are smaller than those in Sephacryl-200 beads, and much smaller than those in Sephacryl-300 beads (all three types are made by GE Healthcare). Molecules of 60 kDa can be separated from 250 kDa proteins by using Sephacryl-100 or Sephacryl-200. The 60-kDa molecules travel into the beads and through the channels. The 250-kDa proteins cannot enter the beads so they flow around them. However, if Sephacryl-300 resin were used, both proteins would be small enough to enter the beads, and there would be no separation of these proteins.

fast-performance liquid chromatography (FPLC) (fast-per•for•mance liq•uid chro•ma•tog•ra•phy) a type of column chromatography in which pumps push buffer and sample through the resin beads at a high rate; used mainly for isolating proteins (purification)

high-performance liquid chromatography (HPLC) (high-per•for•mance liq•uid chro•ma•tog•ra•phy) a type of column chromatography that uses metal columns that can withstand high pressures; used mainly for identification or quantification of a molecule

equilibration buffer (e•quil•i•bra•tion buf•fer) a buffer used in column chromatography to set the charges on the beads or to wash the column

elution buffer (e•lu•tion buf•fer) a buffer used to detach a protein or nucleic acid from chromatography resin; generally contains either a high salt concentration or has a high or low pH

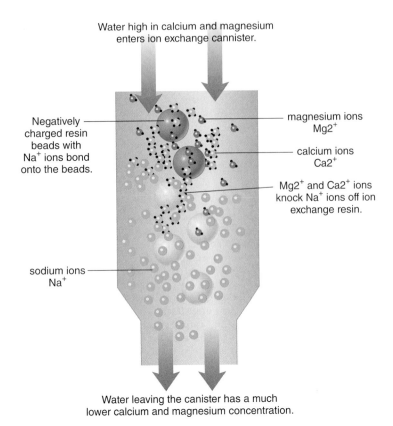

Water high in calcium and magnesium
enters ion exchange cannister.

Negatively charged resin beads with Na⁺ ions bond onto the beads.

magnesium ions Mg2⁺

calcium ions Ca2⁺

Mg2⁺ and Ca2⁺ ions knock Na⁺ ions off ion exchange resin.

sodium ions Na⁺

Water leaving the canister has a much lower calcium and magnesium concentration.

Figure 9.27. How Ion-Exchange Water Softeners Work. Canisters of ion-exchange resins can be connected to domestic water sources, softening the water as it enters a home. Mg^{2+} and Ca^{2+} ions knock Na^+ ions off ion-exchange resin and bind tightly to the resin beads, effectively removing the ions from the water.

For ion-exchange chromatography, resins have either positive or negative charges at a given pH. Anion-exchange resin has positive charges on the beads and attracts negatively charged molecules. Cation-exchange resin has negative charges on the beads and attracts positively charged molecules. For example, the protein, lysozyme, can be separated from the protein, albumin, at a pH of 7.2 using an anion resin called DEAE Sepharose (GE Healthcare). The lysozyme is positively charged at this pH, so the beads repel it. The lysozyme flows through in the wash. The albumin attaches to the positively charged beads until it is eluted with a high-salt buffer.

Ion-exchange chromatography has many practical uses outside of a biotechnology lab. For example, homeowners and water-treatment companies have used ion-exchange chromatography for many years to soften "hard" water. Hard water has a high concentration of dissolved calcium or magnesium ions, which are a problem because they stick to pipes, appliances, and tiles, and form a scaly buildup. Hard water tastes bad and does not allow soaps and detergents to work well. Ion-exchange water softeners work by exchanging sodium (Na^+) ions on resins with the magnesium (Mg^{2+}) ions and calcium (Ca^{2+}) ions in the water (see Figure 9.27). This takes the offending ions out of solution, making the water "softer." The Na^+ ions coming off the resin do not build up on surfaces, and they rinse away with the water.

Buffers Used in Column Chromatography

Depending on the kind of column chromatography, the type of buffer to be used is an important consideration. For gel-filtration columns, the purpose of the buffer is to carry the sample down the column. The buffers used to dissolve the sample are often used as the gel-filtration buffer.

In ion-exchange columns, **equilibration buffer** is used to set the charges on the beads and proteins in the sample. A common equilibration buffer is a sodium monophosphate prepared at a specific pH. At an appropriate pH, some of the molecules in a sample with opposite charge to the beads will bind. The elution buffer in ion-exchange chromatography has to be able to knock a bound protein off the charged beads. An **elution buffer** with a high salt concentration can be used to elute molecules off a column. For the above example, the elution buffer might have 0.5 *M* NaCl added to the equilibration buffer. The chlorine ions can knock the negatively charged proteins off the beads, eluting them from the column.

When a sample is to be loaded onto a column, it must be in an appropriate buffer. Often, the sample must undergo buffer exchange, where the buffering compounds in the sample solution are removed, and new buffering compounds take their place. Dialysis may be used for buffer exchange. For example, if a sample has been prepared in a buffer containing SDS, it usually needs to be removed before chromatography. Although SDS is good for denaturing proteins for electrophoresis, it interferes with most column chromatography. Before the sample is loaded onto a column, the buffer may be exchanged for another buffer, such as sodium monophosphate buffer.

A sample suspended in an "inappropriate" buffer is placed in a dialysis tube. The dialysis tube is an artificial membrane with submicroscopic pores small enough for buffer molecules to pass through, but too small for the protein sample to escape. The ends of the tube are tied. The entire dialysis bag is placed in buffer for several hours. During that time, the "old" buffer moves out and the "new" buffer moves in by the process of diffusion (see Figure 9.28).

Resin Bed Volume Versus Sample Concentration

Another important factor in column chromatography is the relationship between the resin bed volume and the sample concentration. The amount of resin beads must be sufficient to interact with the sample. In a gel-filtration column, the bed must be long enough for a given concentration of molecules to be trapped in the bead channels and separated from the larger molecules. In an ion-exchange column, there must be enough charged beads to bond all of the desired protein. In an affinity column, there must be enough beads with bonding groups to bind all of the desired protein.

The concentration of the sample may be unknown. If the concentration of the sample is known, it is usually adjusted to between 0.1 mg/mL and 1 mg/mL. Typically, a sample volume of 3% of the bed volume is loaded on a column.

Often the best conditions for conducting chromatography are discovered through trial and error. Samples are run under various conditions, and the fractions are checked using PAGE.

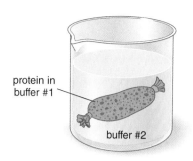

A dialysis bag with the old buffer is submerged in the new buffer. Buffer #1 diffuses out of the bag while buffer #2 diffuses into the bag and tiny undesirable molecules diffuse out.

Figure 9.28. Dialysis Buffer Exchange. Typically, dialysis is conducted using 10X the volume of buffer outside the bag as that inside the bag. Also, the buffer is changed several times after several hours. This ensures the complete exchange of buffers. Sometimes the volume of the sample increases substantially from the influx of buffer. If this happens, the sample can be concentrated using concentrators or centrifuge filters.

BIOTECH ONLINE

Got Gas?

Gas chromatography was one of the first types of chromatography used in biology and chemistry labs. Although it is not quite as common a practice in biotechnology now as it was in the past, many companies still use gas chromatography for R&D.

Go to the University of Colorado Boulder website on organic chemistry at biotech.emcp.net/orgchemCO to learn about gas chromatography.

In general terms, describe what a gas chromatograph separates, how it works, and what it is used for.

Forensic scientists use a gas chromatography/ mass spectrometer to separate (GC) and analyze (MS) molecules in a sample from a criminal investigation. Applications of GC-MS include drug detection, peptide analysis, fire sample investigation, environmental analysis, explosives investigation, and identification of other unknown samples.
© BSIP/Getty Images.

1. A technician wants to quickly determine if an antibody affinity resin will bind a particular protein for purification. Which type of chromatography should he or she use to test the resin?
2. Which instrument, FPLC or HPLC, is used for large-scale protein separations/purifications?
3. Why are spectrophotometers hooked up to most FPLC or HPLC units?
4. You are to dialyze 10 mL of protein extract in PAGE running buffer into sodium monophosphate buffer before running an FPLC ion-exchange column. Into what volume of sodium monophosphate buffer should you place the dialysis bag?

9.4 Product Quality Control

Quality Assurance (QA)
(qual•i•ty as•sur•ance) a department that deals with quality objectives and how they are met and reported internally and externally

A company must be certain that its manufacturing process is capable of synthesizing a product of high quality (see Figure 9.29). As a protein product is scaled-up, manufactured, and purified, technicians must monitor its concentration, purity, and activity to ensure that it meets certain standards. The QC and **Quality Assurance (QA)** departments monitor the characteristics and performance of the company's products.

The roles of the QC and QA departments may vary from one company to another. Usually, the QC department handles product quality and testing during product development, well before a sample is close to marketing. The QA department usually deals with quality objectives, such as how certain objectives are met and reported, both internally and externally, especially as a product is closer to marketing. At a pharmaceutical company, such as Genentech, Inc., QC receives samples from fermentation, cell culture, or other manufacturing areas. These samples undergo the same types of assays that were performed during R&D. In addition, testing may be conducted on animals, including mice, rats, monkeys, and chimpanzees. Animal testing is necessary to safely bring therapeutic drugs to the human market. Some of the common assays for a pharmaceutical product are listed in Table 9.1. These assays confirm the presence and performance of the protein product at each step of manufacturing.

Figure 9.29. Several biotechnology companies develop pharmaceuticals that target cancer, including cancer vaccine proteins. As proteins are produced in these fermenters, samples are taken and tested to ensure their high quality.
Photo courtesy of Cell Genesys, Inc.

Table 9.1. **Pharmaceutical QC/QA Testing**

Type of Assay	Function (determines:)
ELISA	presence and concentration of the protein
enzyme activity	degree of protein activity
multispecies pharmacokinetic	behavior of protein in nonhuman test animals
toxicology	quantities of the drug toxic to cells and organisms
potency	effect of dosage on drug activity
stability	length of time the product remains active

At an instrumentation company, such as Life Technologies Corp., the products may be instruments rather than engineered proteins. Instruments and the reagents used on the instruments are the company's products, so the QC department measures different variables for these products. There are several QC departments; some test reagents and some test instruments. For example, technicians might use the HPLC and the mass spectrometer, DNA synthesizers, and/or DNA sequencers to validate their chemical reagents' purity and performance (see Figure 9.30).

If a protein product is pure and properly formulated, the company can launch steps to bring the product to market. If the product is not a pharmaceutical, but rather an industrial enzyme or research instrument, the sales staff can begin advertising and selling the product as long as all of the requirements of the Environmental Protection Agency (EPA) and the US Department of Agriculture (USDA) have been met. It takes many years to move from the initial period of R&D to marketing.

If the protein product is a pharmaceutical, an **Investigational New Drug (IND) Application** must be filed with the FDA. An IND Application describes the structure, specific function, manufacturing process, purification process, preclinical (animal) testing, formulation, and specific application of the proposed pharmaceutical. The company submits an IND Application in anticipation of clinical testing, a requirement before the FDA will approve a therapeutic drug for market.

Clinical testing, also called clinical trials, may take from 2 to 5 years to complete. Clinical trials conducted on human subjects are used to determine the safety and efficacy of the therapeutic product. The trials are designed based on a set of protocols that include the following:

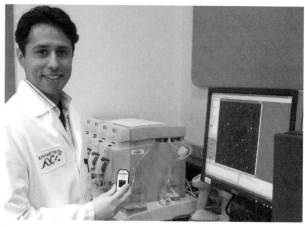

Figure 9.30. Joaquin Trujillo, a Research Associate in the Quality Control Department at Affymetrix, Inc., performs functional and analytical testing of reagents using the Affymetrix GeneChip probe microarray assay. Joaquin follows implemented QC procedures to optimize manufacturing and regulatory requirements. Here, Joaquin works with a fluidic system that dispenses the reagents used with the array assay. He is holding an array. Microarrays are used to study gene expression and genetic diversity.
Photo by author.

- types of people accepted into trial
- procedures
- dosages
- schedule of trials
- medications
- study duration

Throughout the clinical trial, companies continue to develop and perform assays on human serum and plasma to determine the dosage that will provide the greatest efficacy with the fewest adverse effects.

Clinical trials are performed in four phases. In Phase I trials, a small sample of high-risk patients test a new drug therapy for safety, dosage range, and side effects. In Phase II trials, the study group is expanded to several hundred subjects, and additional safety, dosage, and efficacy testing is completed. In Phase III, the study group is expanded to several thousand people. Safety and efficacy continue to be monitored. In addition, the treatment's effectiveness and its safety are compared with those of existing drug therapies.

Clinical trials are most often **double blind**. In a double-blind protocol, neither the researchers nor the study subjects know which treatment the subjects are receiving: the new test drug or a placebo. A **placebo** looks exactly like the test drug, but it contains only the inactive ingredients, not the actual drug. Double-blind trials decrease the risk for prejudices during the study, which could result in biased data.

After a product has been on the market for a period of time, the FDA may require Phase IV trials. These trials provide better safety and efficacy data based on larger, more diverse populations.

Once a company proves the safety and efficacy of a product through clinical trials, the appropriate department composes a marketing application, which includes all of the product descriptions and clinical trial results. The marketing application is sent to the FDA for review and approval. The FDA approval requires site visits to the manufacturing company to check for Good Manufacturing Practices (GMP) and to meet with key company administrators and researchers. The FDA approval may take 1 to 2 years.

Investigational New Drug (IND) Application (in•ves•ti•ga•tion•al new drug ap•pli•ca•tion) an application, filed with the FDA for the purpose of testing and marketing a product, that describes the structure, specific function, manufacturing process, purification process, preclinical (animal) testing, formulation, and specific application of a proposed pharmaceutical

clinical testing (clin•i•cal test•ing) another name for clinical trials

double-blind test (dou•ble-blind test) a type of experiment, often used in clinical trials, in which both the experimenters and test subjects do not know which treatment the subjects receive

placebo (pla•ce•bo) an inactive substance that is often used as a negative control in clinical trials

Products "in the Pipeline"

TO DO Go to Genentech, Inc.'s product development pipeline Web page at: www.gene.com/medical-professionals/pipeline. Find an example of a drug product in each of the three phases (Phases I, II, and III) of clinical trials. Identify each drug, its stage in the product pipeline, and its potential application.

For every product, regardless of the target market, a large team of people is responsible for ensuring that the product's quality is high before it reaches the marketplace (see Figure 9.31).

Figure 9.31. **Atticus Rotoli works in the Business Excellence Department as part of the Global Operations and Services organization for Life Technologies Corp. in Foster City, CA. Atticus is responsible for teaching, mentoring, and certifying internal staff on the proper application of Lean Principles to reduce waste variation in any process. Atticus is also a project manager and teaches basic project management principles and more advanced statistical applications to Life Technologies staff. Atticus's time is divided among five places: at his desk (30%), in meetings (25%), training (20%), travel (15%), and in the lab or field (10%). He has a Bachelor's degree in microbiology and bacterial genetics, and he is a member of the International Society of Six Sigma Professionals (ISSSP).**
Photo courtesy of Atticus Rotoli.

Section 9.4 Review Questions

1. What type of biotechnology product undergoes clinical testing/clinical trials?
2. How many people (subjects) are usually involved in Phases I, II, and III of a clinical trial?
3. In which phase of a clinical trial, Phase I, II, or III, is product safety tested?

9.5 Marketing and Sales

In a biotechnology company, the business departments develop operating plans and strategic objectives that promote the growth of the company and the research and development of new products. An assortment of administrative, financial, legal, and scientific staff work together to assess and develop the company's long- and short-term goals. A long-term plan outlines a company's R&D scheme, as well as business strategies, with the intent of maximizing the company's value. Obviously, a company needs to continue to have investment and sales income to continue to research, manufacture, and market its products.

The goals of R&D include decisions about which products, and how many, should be in the product pipeline at one time. Some products will require accelerated or expanded development. Decisions to change R&D plans are made, if appropriate. The plan outlines goals for product sales and services, and describes the appropriate allocation of company resources. The plan also delineates ways to improve financial returns.

Bringing a Product to Market

Bringing a product to market is a challenging task with uncertain outcomes. Many promising products may not reach the market for reasons that may, or may not, be under a company's control (see Figure 9.32). Some factors that may impede a product reaching the marketplace include the following:

- A product may be found to be ineffective during preclinical or clinical trials.
- During testing, a product may be shown to have harmful side effects.
- Production may turn out to be uneconomical.
- A product may fail to receive necessary regulatory approvals, such as from the FDA.
- Competing products may already control a large portion of the market.
- Patent protection for the product may be unobtainable, or another company may hold proprietary rights.

Marketing

As soon as a product receives all of the necessary approvals, it can be sold. Depending on the product in question, size of the company, and its operating budget, a staff with diverse talents is responsible for marketing the product. The sales force's job is to advertise and publicize the product to an appropriate audience. For pharmaceuticals, the audience is typically physicians. For industrial enzymes, such as subtilisin, the market is the laundry detergent industry. The marketing/sales staff must convince potential buyers that the cost of adding subtilisin to their detergent will be offset by the increase in profits from a greater number of sales of the "new, improved detergent."

In the case of a pharmaceutical, such as the blood-clot buster, t-PA, manufacturer Genentech, Inc. must convince doctors, hospitals, and health insurance providers of the safety and efficacy of the product. Competitive products have regularly challenged its cost-effectiveness relative to its degree of clot clearing. There has been a lot of "press" about these heart attack treatments. Marketers of t-PA must educate consumers about the possible life-saving capability of their product.

Product Sales

Factors that affect a company's product sales include the following:

- effectiveness of the marketing team
- pricing decisions made by the company
- degree of patent protection afforded a product
- use of alternative therapies or products for the product's target population
- timing of FDA approval of competitive products
- rate of market penetration for competitive products. For example, Retavase, by Centocor, Inc., is a competitor of Genentech, Inc.'s Activase. Both therapeutic agents are used to treat acute myocardial infarction (heart attack), and Retavase sales have adversely affected Genentech's market share.

 A product cannot have short-term side effects.

 A product cannot have long-term side effects.

 A product cannot cost the producer or the consumer too much.

 A product cannot fail to pass all the FDA testing and requirements.

 A product cannot have patent disputes.

Figure 9.32. Potential Product Problems. Unexpected results in R&D, as well as in testing, may delay or stop a product's marketability or sales.

Proprietary/Patent Rights, and Community and Government Relations

**proprietary rights
(pro•pri•e•tar•y rights)**
confidential knowledge or
technology

**patent protection (pat•ent
pro•tec•tion)** the process of
securing a patent or the legal rights
to an idea or technology

A company may invest hundreds of millions of dollars in the development of a product. At any time, another individual or company could steal protocols or product information, and begin producing or marketing the product. Doing so is unfair and illegal, and is regarded as "intellectual theft." To protect against intellectual theft, companies proceed in two ways: First, most companies require their employees to sign **proprietary-rights** contracts. In doing so, an employee agrees to keep secret the R&D of the company's products. Second, as soon as it is possible, a company will move to secure **patent protection** for a product under development.

Having strong patent protection is of utmost importance to a company's sales and royalty revenue. A company must gain proprietary or patent rights to protect against other groups using or selling a product or technology. Gaining and retaining patent protection is of constant concern and may involve enormous costs. Patent disputes could entail complex legal issues, and necessitate the use of patent attorneys in and out of a courtroom (See Figure 9.33).

Often, patent disputes interfere with bringing a product to market. Sometimes patent investigation or litigation reveals that another company holds a patent for a process or a product. If this is true, then a third-party license must be obtained to continue production or R&D, and development of the potential product may be terminated.

Almost as disastrous to product development is a bad corporate image. The government, clients, and citizens must trust that a company is acting in good faith. Several departments and many employees in a company focus their work in legal, community, or corporate relations.

Figure 9.33. J. Peter Parades is a patent attorney for Rosenbaum & Silvert, PC, in Northbrook, IL. His primary responsibility is to draft and prosecute domestic and international patent applications in the fields of biotechnology, nanotechnology, and medical devices; to develop patent portfolios for early state companies; and to conduct due diligence investigations, clearance and freedom-to-operate opinions, patentability searches and opinions, and intellectual property litigation. Peter worked as a research associate in a biotechnology laboratory facility for eight years before deciding to pursue a career in biotechnology law.
Photo courtesy of author.

Product Applications

Once a product is being synthesized and has been approved for use for one application, it makes sense for a company to look for other applications for that particular product. If other uses are found for a product that is already being manufactured, a company can save thousands of dollars in R&D costs.

Consider the many applications of the enzyme, cellulase, originally manufactured by Genencor International, Inc. Cellulase was first produced for use in paper making. Cellulase breaks down the cellulose fibers in paper, resulting in softer paper products. At a later date, scientists realized that cellulase could be used in other applications as well. Now, it is used to break down denim fibers in the production of "stone-washed" jeans. It is also used to break down fruit cells to increase juice production. The apple juice industry uses significant amounts of cellulase. The sale of cellulase increases considerably each time a new application for the enzyme is found. Finding new applications for products is so important that often a biotech company will have a separate applications department.

Sometimes a company changes a product on a molecular level, resulting in different versions and different applications. An example of this is megakarocyte growth and development factor (MGDF), which Amgen, Inc. originally cloned in 1994. Thirty different versions of MGDF were produced. Many of these are being studied for the treatment of different types of cancer, and some are used in bone marrow transplantation. Each product version still must complete rigorous testing, but the R&D costs are substantially reduced once the safety has been proved for the first application. This translates into important profits for the company.

BIOTECH ONLINE

Approved Biotechnology Drugs

Go to biotech.emcp.net/fda-approvals to learn about drugs recently approved by the FDA.

1. Find two pharmaceuticals, that have gained FDA approval in the past 5 years, produced by different companies for treatment of the same disease or disorder, such as arthritis. Record the product name, company name, and the application or use of each drug.

2. Use the Internet to learn more about each drug. Record an advantage or a disadvantage of the use of each.

Section 9.5 Review Questions

1. What are some of the reasons that a product in development may not make it to the marketplace?
2. What is covered in an "employee's proprietary-rights contract"?
3. Why must a company gain patent protection on a product?

Chapter Review

Summary

Concepts

- Biomanufacturing of a specific protein product includes growing cells that express the protein in liquid culture and moving the cell culture through fermentation or cell culture where increasing volumes of cells and protein are produced. Then, cultures are harvested and the protein is recovered through a series of centrifugations, filtrations, and column chromatographies. Purified protein is formulated and then sent to filling and packaging.
- Recombinant protein products must be harvested from cell cultures. Cells can be separated from broth by centrifugation or filtration. If the proteins of interest are inside the cells, they must be burst open and purified from all the other proteins of the cell. If the proteins are extracellular proteins, they must be purified from the broth proteins.
- Recovery of the protein of interest begins with dialysis of the mixture into a buffer for column chromatography. Dialysis is performed in dialysis tubing, in two to three rounds of 10 times the buffer, compared with the sample being dialyzed.
- Chromatography is the separation of molecules along a stationary phase due to solubility, size, charge, shape, hydrophobicity, or other special properties of molecules of interest. Common types of chromatography include paper, thin-layer, and column chromatography.
- Size-exclusion columns are usually the first to be used in protein purification. Gel filtration resin is composed of resin beads with tiny channels. Resin size is selected based on the proteins to be separated. Gel-filtration columns are usually long to allow space for molecules to separate. Fractions from columns are collected and analyzed.
- Ion-exchange resin has either positively or negatively charged groups on the beads. Ion-exchange beads bind proteins of the opposite charge to the beads. The column resin has to be equilibrated to a certain pH by equilibration buffer. This ensures that the resin has the "right" charge. A sample is loaded and run. Molecules that bind to the resin are knocked off (eluted) with either high-salt buffer or high- or low-pH buffer.
- Affinity chromatography uses resin beads with antibodies to recognize very specific molecules and to pull them out of solution.

- Column chromatography is conducted on open columns, or on FPLC or HPLC. Open-column chromatography is easy to set up and run, but if large volumes of samples are run, FPLC must be used. HPLC is designed more for analytical purposes, to separate molecules to check their purity and concentration.
- FPLC and HPLC units include a pumping system to push samples through resin beads. This improves yield and purity. Fractions are checked on a UV spectrophotometer.
- QC and QA departments ensure high-quality product development through testing and reporting. QC departments utilize assays and testing developed in R&D.
- Some of the reasons that a product in development may not make it to the marketplace include the following: ineffectiveness, harmful side effects, uneconomical production, failure to gain FDA approval, or failure to win patent protection.
- A company must gain patent rights for a product to protect against other groups using or selling their product or technology.

Lab Practices

- Protein purification from transformed cells begins with harvesting cells or broth from the fermentation or manufacturing broth culture. This is accomplished by spinning the cells down in a centrifuge or by separating the cells from broth through filtering.
- If cells contain the protein of interest, they must be burst open to harvest the protein prior to purification. If the protein is extracellular and already in the broth, the cells are discarded to reduce the starting number of proteins in the harvest.
- Column chromatography is the principal method of purifying proteins from cell extracts or harvested broth cultures. After a column protocol for chromatography has been determined on a small scale in R&D, FPLC is used to purify large amounts of protein mixtures; HPLC is used to characterize samples.
- The columns used in all chromatography work in a similar fashion. Each is filled with some kind of resin bead that interacts with proteins in a mixture and causes them to separate. Fractions come off the column and are collected for analysis using UV spectrophotometry and PAGE.
- Proteins are separated on columns based on size (gel filtration), charge (ion exchange), or shape (affinity).
- Dialysis tubing, with a certain molecular-weight pore size, is used for buffer exchange. The dialysis membrane traps protein molecules in the tube, but allows other buffer components and water to exchange. A sample must be in the appropriate buffer for column chromatography. It takes several rounds of incubating a dialysis tube with 10 times the amount of buffer for a complete buffer exchange.
- After a dialysis, the volume of a sample is usually higher, and the concentration is lower. It is often necessary to concentrate the sample.
- Recombinant amylase from transformed *E. coli* broth cultures is released to outside of the cell. Centrifugation can be used to separate transformed cells from broth containing amylase. For ion-exchange chromatography, amylase may be dialyzed into a sodium monophosphate buffer, pH 7.35.
- To run a column, the column must be equilibrated with running buffer. For an ion-exchange column, the equilibration buffer is at a specific pH or salt concentration. Once proteins have bound to the resin in an ion-exchange column, elution buffer is used to knock the proteins off the resin and into a fraction. Elution buffer usually has a high salt concentration, or a high or low pH.
- Amylase molecules are negatively charged at pH 7.35 and bind to DEAE Sepharose resin (GE Healthcare) in an ion-exchange column. Lysozyme molecules are positively charged at pH 7.35 and do not bind to DEAE Sepharose resin in an ion-exchange column, but flow through in column washes.
- To determine the charge on amylase molecules at pH 7.35, amylase standards are run on a positively charged DEAE Sepharose column. If the amylase sticks to the resin, it is assumed to be negatively charged at pH 7.35.
- If amylase or other molecules are separated during a column chromatography, PAGE can be used to visualize the proteins in the fractions, and UV spectrophotometry can be used to determine their concentrations.

Thinking Like a Biotechnician

1. What are the advantages of pressure-pump chromatography (FPLC and HPLC) compared with open-column chromatography?
2. What is the difference between clinical trials Phases I, II, and III?
3. A gel-filtration column, containing resin with 60 kDa pore size, is run with two proteins: one is 48 kDa and one is 100 kDa. At the end of the column run, one of them is found in fraction 10, and one is found in fraction 15. Which protein would be in which fraction? Explain why.
4. A small sample of valuable protein, MW = 58 kDa and positively charged at pH 7.5, needs to be confirmed on a column. Which kind of column chromatography should be used?
5. A mixture of proteins was separated on a sizing column. One 2-mL fraction has two proteins left in it, and the technician wants to run an ion-exchange column. First, he or she has to dialyze it into the appropriate buffer. How much ion-exchange buffer is needed for the dialysis?
6. A technician sets up a negatively charged ion-exchange column and puts a small volume of a mixture of proteins on it. What should happen to positively charged proteins in the mixture? What should happen to negatively charged molecules in the mixture?
7. Propose a method of purifying recombinant amylase from transformed *E. coli* cells and their proteins.
8. How can you tell if the protein purified in item 7 actually is amylase?
9. As a sample elutes from a column, the technician wants to determine the concentration of protein in the sample. What instrument can be used to determine the concentration of the sample, and at what setting(s) should it be operated?
10. How do the QC and QA departments in a company differ from each other?

Biotech Live

Protein Manufacturing

Create a poster that diagrams the major steps in the R&D, biomanufacturing, and marketing of a rDNA protein product that is destined to become a pharmaceutical.

The poster should look like a big flowchart with annotated photographs and diagrams demonstrating all of the major steps in the protein production process. Each diagram or photo should have reference/website information. Include all of the following items on the poster:

Research and Development

Product Identification	Examples of products and how they are found.
Assay Development	Examples of how assays (tests) are developed to identify and quantify the protein product.
Genetic Engineering	Examples of how cells are genetically engineered to produce the protein product.
Protein Synthesis	Description of how, and where, protein synthesis (transcription and translation) in the genetically engineered cells occurs.

Manufacturing

Fermentation/Cell Culture	Descriptions of how small amounts of genetically engineered cells are grown in increasingly larger volumes under strict regulations (scale-up).
Recovery/Harvest	Examples of how a protein product is purified on a large scale from cells in culture and other proteins, using centrifugation, filtration, and chromatography.
GMP	Discussion of the use of "Good Manufacturing Practices."
Product Formulation	Examples of the variety of formulations possible for a protein product.

Marketing

Product Testing/Clinical Trials	Examples of the type of testing required before patients or customers use a product.
FDA Approval	The process by which a pharmaceutical agent is judged safe for use and distribution.

Setting the Standard in Biomanufacturing

In a biotechnology company, the Quality Control and Quality Assurance departments work to ensure that biomanufacturing meets the necessary standards to produce the high-quality biotechnology product. Several agencies, such as the FDA, USDA, and EPA have rules that must be followed, and the application of the rules must be documented. In addition, many facilities apply for ISO certification.

To Do: Use the Internet to find answers to the following questions and then create a one-page flyer that could be used to convince a small biotechnology company to apply for ISO certification. List the URLs you used as references at the bottom of the flyer.

What is ISO and why is it important to the biotechnology industry?
What is ISO 9000? Who should use ISO 9000 guidelines?
What is ISO 22000? Who should use ISO 22000 guidelines?
What is ISO 13485? Who should use ISO 13485 guidelines?
What are some other ISO guidelines that might be of interest to biotechnology companies?

Reading Annual Reports

All companies publicly traded on the stock market, as well as some privately owned companies, produce annual reports to inform their investors. An annual report describes the present state of the company's scientific and business ventures, as well as its plans and expectations for the future.

An annual report can be obtained by writing, telephoning, or e-mailing a company. The reports include a wealth of information about the company's products and/or services. In annual reports from biotechnology companies, most of the pipeline products are described. Often, information about the application of a product, clinical trials, and marketing is included.

Obtain information about a company's business and scientific interests through an annual report.

1. Obtain an annual report from one of the biotechnology companies listed below (or its parent company) or a different one approved by your supervisor. Check out the Yahoo Finance website at **biz.yahoo.com/ic/515_cl_all.html** for a list of other companies. Write, phone, or e-mail the company to request a copy of the annual report. Often, annual reports can be downloaded from the Internet. The website for the Biotechnology Industry Organization (**biotech.emcp.net/bio**) may provide the address, phone number, and/or e-mail address for a company.

Abbott Laboratories	Bayer Corp.	Biogen Idec, Inc.
Pfizer, Inc.	Nektar Therapeutics	Onyx Pharmaceuticals, Inc.
Geron Corp.	Gilead Sciences, Inc.	Elan Corp. plc
Charles River Laboratories	ACADIA Pharmaceuticals, Inc.	Telik, Inc.
Amgen, Inc.	GlaxoSmithKline, plc	Affymax, Inc.
Sangamo Biosciences, Inc.	Baxter International, Inc.	

2. On a small (11 in × 17 in) poster board, report the answers to the following questions in an interesting and informative way. Include drawings, diagrams, photos, etc. Assume that potential investors will be reviewing the information on the poster. It is your goal to inform them so that they may make a decision as to whether or not they should invest in the company.

   - What are the stated long-term goals for the company's R&D and manufacturing?
   - What are the company's total revenues (if they have revenues)?
   - What is the total amount of product sales in dollars (if they have product sales)?
   - What are the company's R&D expenses?
   - On a chart, list the company's marketed products (generic and trade names) and their applications.
   - Pick one of the company's marketed products. Determine the total amount of net sales for the product over each of the past 3 years (if it has been on the market that long).
   - Briefly, discuss any pending legal matters that may affect the company's resources.

Bioethics

Should Genes be Patentable?

In June of 2013, The U.S. Supreme Court issued a ruling that naturally occurring genes may not be patented. This ruling threw out existing patents that were applied for and granted to Myriad Genetics for the human genes BRCA1 and BRCA2. Attorneys for Myriad Genetics had applied for patents for the DNA sequence for the BRCA1 and BRCA2 genes. Myriad had spent a significant amount of time and money sequencing these genes and developing the genetic test that recognizes them. Using their sequencing technology, Myriad also provides patients with important information about their risk of developing aggressive breast or ovarian cancer. Learn more about Myriad Genetics at **www.myriad.com** and by watching the video at **vimeo.com/126497053**.

Many industry representatives and consumers applauded the Supreme Court decision since, if a company owns the patent to a naturally occurring gene, then other scientists could not work on it without paying a royalty. This would slow research that might lead to important discoveries. On the other hand, if other original research can gain patent protection, a company like Myriad Genetics will be able to reap the financial reward of the time, money and risk of their research investment.

How do you feel about it? Should genes be patentable?

1. What exactly is a patent? Find a definition that makes sense for use when describing a patent for a biotechnology product. Record the definition and the source of the definition.

2. The Supreme Court ruled that new genes produced through biotechnology should be patentable. This is called "biopatenting." Do you agree? Go to **debatewise.org/debates/3693-the-human-genome-should-be-patented** and read through the arguments for and against gene patenting. Find and record 3 points on each side of the argument that you believe could make a good case either way.

3. Read through the discussion of the important "takeaways" from the Supreme Court ruling at **news.nationalgeographic.com/news/2013/06/130614-supreme-court-gene-patent-ruling-human-genome-science**. Which sway you to support or not support gene patenting?

4. Before making a decision about the ethics of biopatenting, read the blog discussion at **ipkitten.blogspot.com/2015/01/the-ethics-of-biotech-patenting.html**.

5. Write an 8–10 sentence "post" that you could submit to a blog or a discussion that outlines your position on biopatenting. Give specific reasons and examples for why you support or do not support the patenting of engineered genes.

Unit 3
Current and Future Applications of Biotechnologies

Scientist uses a thermal cycler to recognize and amplify DNA fragments in the process called Polymerase Chain Reaction (PCR). PCR has revolutionized molecular diagnostics and genetic engineering.
© Monty Rakusen/cultura/Corbis.

It is an exciting time to be in the science and business of biotechnology. New products are being developed everyday that are improving the quality of life for the billions of people who live on our crowded planet. In 2004, the Biotechnology Industry Organization said that biotechnology's mission was to heal, fuel and feed the world but that mission has expanded to include protecting our planet and the public.

As humans all over the globe are living longer, the need for addressing the impact the growing population has on the earth increases. How do we grow more and better food without harming the environment? How do we address the need for better medications to provide, not just longer lives but a better quality of life?

In Unit 3, the biotechnology advances to meet these goals from the past few decades, that are being made now, and those expected in the future are discussed.

In Chapters 10 and 11 the practices and advances in agricultural biotechnology are presented. Research and development of new biotechnology crops for food, traditional medicines, biopharming (the production of human pharmaceuticals in plants), food security and fuel alternatives continues to expand opportunities to feed, fuel, heal, and protect the world in a healthy way.

Chapter 12 focuses on the use of current biotechnology practices in medicine. From diagnostic testing and tools to therapeutic medications including personalized medicines, biotechnology research and development brings new and better products from nature and the laboratory to market almost daily.

Chapter 13 presents how DNA synthesis, including PCR, and DNA sequencing technologies are used in research and development of products and services in all domains of biotechnology. The ability to synthesize DNA fragments of a few to several hundred nucleotides allows researchers and clinical technicians to target genetic codes for identification, engineering, and therapeutic purposes.

In Chapter 14, the promise of biotechnology advances being worked on now and expected in the near future is explored. DNA computer chips, genome editing, biosensors, bioplastics, biodefense products, and nanobiotechnology tools are starting to create better and smarter biotechnology products to address some of our greatest challenges.

From advances in stem cell research and therapies to creating new cells, tissues, or organs using synthetic biology, to new biomedical products using tools such as 3D bioprinters to deactivating genes using CRISPR technology, some studies suggest that the social and technical impact of biotechnology in this century may surpass that of the industrial revolution of the 1800s and the information revolution of the 1900s.

Biotech Careers

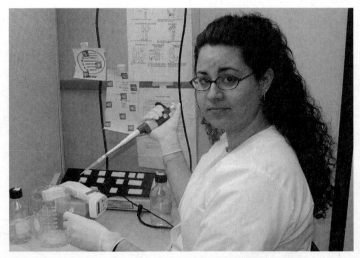

Photo by author.

Plant Biologist

Jennifer Costa, Research Associate
Mendel Biological Solutions
Hayward, CA

Mendel Biological Solutions, Koch Agronomic Services, LLC, focuses on controlling gene expression to create new opportunities to improve plant growth for applications in plant biotechnology, plant breeding, horticulture, and forestry. Their scientists direct research toward a large class of genes called transcription factors. Transcription factors control the degree to which each gene in a cell is activated.

Jennifer, like many Mendel employees, works with *Arabidopsis thaliana*, a plant that is a model organism for genetic studies and biotechnology research. In a laminar flow hood, Jennifer sterilizes *Arabidopsis* seeds prior to planting. Seeds are placed in 1.7-mL tubes and are incubated with a 30% bleach solution for 30 minutes. To completely remove all of the bleach, seeds are rinsed a total of six times with sterile, deionized water. Sterile seeds are suspended in 0.1% agarose before planting on media. *Arabidopsis* is a good lab plant because it can be grown indoors under grow lights, and because it has a short generation time of about 2 months from seed to flower.

See some of Jennifer's projects at **patents.justia.com/inventor/jennifer-m-costa**.

10 Introduction to Plant Biotechnology

Learning Outcomes

- Describe mechanisms of plant pollination and differentiate between haploid and diploid cells and their role in sexual reproduction
- Identify various natural and artificial ways to propagate plants to increase genetic variety or maintain the genetic composition
- Discuss the function and composition of different plant structures, tissues, and organelles and give examples of foods that are derived from various plant organs
- Describe the processes of seed germination and plant growth
- Perform the calculations to predict expected plant phenotypes for specific genotypes, using Punnett Square analysis in a plant breeding experiment
- Describe the role of meristematic tissue in asexual plant propagation
- Explain the role of plant growth regulators, as well as the advantages and disadvantages of plant tissue culture

10.1 Introduction to Plant Propagation

Plant biotechnology has a long history dating back to the origins of agriculture and includes several methods of modifying or improving plants or plant parts. Early humans discovered that seeds could be collected and planted the following season to produce a new crop. They also quickly learned that new seeds were created when insects, animals, or the wind pollinated flowers. Attempts were then made to control which plants were pollinated with others (see Figure 10.1). These early ancestors were the first plant propagators attempting to control plant reproduction.

One of the first advancements in plant propagation occurred when farmers discovered that they could not only control **pollination**, but the resulting seed crop as well. Pollinating and fertilizing parent plants, showing one or more desired characteristics, produced new varieties of crops. Wheat, corn, rice, and oats have been cultivated for thousands of years in just this fashion. More recently, the cultivation and breeding of plants has resulted in a vast assortment of fruits, vegetables, grains, flowers, timber trees, ornamentals, and agricultural plants (see Figure 10.2).

pollination (pol•lin•a•tion)
the transfer of pollen (male gametes) to the pistil (the female part of the flower)

307

Figure 10.1. **An African honeybee,** *Apis,* **pollinates a Cosmos,** *Cosmos bipinnatus,* **flower.** Early propagators observed insects spreading pollen and tried to mimic them in an attempt to control breeding. To increase honey production, African bees, also called "killer bees," were introduced into the northern hemisphere in the 1950s to crossbreed with native honeybees. Now, the imported bees have escaped into the wild, and countries are concerned about their particularly aggressive behavior.
© akova/iStock.

Figure 10.2. **The large number of lettuce varieties is a result of selective breeding, a process in which scientists or farmers "control" sexual reproduction.**
Photo by author.

breeding (breed•ing) the process of propagating plants or animals through sexual reproduction of specific parents

sexual reproduction (sex•u•al re•pro•duc•tion) a process by which two parent cells give rise to offspring of the next generation by each contributing a set of chromosomes carried in gametes

zygote (zy•gote) a cell that results from the fusion of a sperm nucleus and an egg nucleus

embryo (em•bry•o) a plant or animal in its initial stage of development

Plant **breeding** involves **sexual reproduction**, in which two parent cells give rise to offspring of the next generation. Pollen is produced in the anther of the stamen (the male part of the flower) and is carried to the pistil, the female part of a flower (see Figure 10.3). Within each pollen grain is a sperm nucleus. The chromosomes of the sperm nucleus contain half of a set of the genetic information describing the characteristics of a plant.

If a pollen grain from an appropriate flower reaches the stigma of the pistil of a flower, it grows a tube toward one of the ovules in the ovary of the pistil. Within the ovule is an egg nucleus. The chromosomes of the egg nucleus also contain one-half of a set of genetic information describing characteristics of the plant.

When the sperm nucleus finds the egg nucleus, they combine to form a **zygote**, the first cell of the next generation. The zygote grows and divides within the ovule to produce a multicellular **embryo** (the offspring plant). Many ovules may be fertilized and develop in the ovary. The rest of the ovule matures into a protective and nutritive seed. As the ovules mature into seeds, the ovary swells into a fruit. Under the right

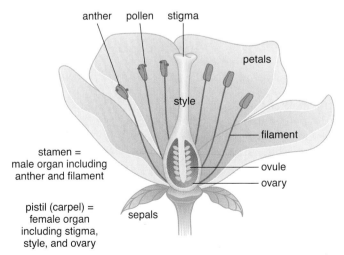

Figure 10.3. **Flower Structure.** Most flowers are "complete" with both male (stamen) and female (pistil) sex organs. A seed develops when a pollen sperm nucleus reaches and fertilizes an ovule.

Figure 10.4. This beautiful *Hibiscus* flower demonstrates the results of breeding flowers to be large and showy. Notice the large pistil with attached stamen protruding from the center of the flower. Once the ovules in the flower's ovary are fertilized, the ovary develops into a "fruit." The fruit of the *Hibiscus* is small and not edible, but it contains several seeds.
Photo by author.

conditions, the ovary bursts, releasing seeds, and the seeds grow into the next generation of plants. The plants show characteristics determined by the genes they have received through the sex cells (sperm and eggs). The goal of a plant breeder is to ensure that the seeds contain certain "desired" characteristics by carefully selecting the parent plants (see Figure 10.4).

BIOTECH ONLINE

These genetically engineered seeds contain a gene that protects plants from Round-up® (Monsanto Corp.) , a common herbicide (weed killer).

Seeds: The Next Generation of Biotech Products

Seeds carry the next generation of a flowering plant. Biotechnology companies that engineer plants with new characteristics most often grow those plants to flowering, pollination, and seed production. The seeds of transformed plants must carry the new genes and express the new characteristics to be of commercial value. Dozens of plant biotechnology companies produce seeds for the marketplace that have desired characteristics gained through either traditional breeding or genetic engineering.

Learn more about biotechnology companies that produce seeds and their products.

Go to **biotech.emcp.net/Purduehort** and find three companies that produce seed for agricultural or landscaping purposes. For each company, report the following:

- the company name and location
- the kinds of seeds they sell—specifically three seeds that are available and the plants into which they grow
- other kinds of plant products that are available from the company.

Breeding involves two parents each contributing genetic information. The **gametes** (sperm and egg cells) are produced by a special kind of cell division called **meiosis**. During meiosis, chromosomes in a developing sex cell break apart and recombine with partner chromosomes. Because of the recombination of chromosomes during meiosis, it is random whether a sex cell receives one version or another of a particular gene (see Figure 10.5). Imagine the infinite combination of all the genes a plant may pass in gametes. The result is that each sex cell contains a unique assortment of genes. When a unique sperm nucleus fuses with a unique egg nucleus, a unique zygote is formed.

Due to gene shuffling, a plant can produce an infinite variety of seeds with an infinite variety of gene combinations during sexual reproduction. New varieties

gametes (gam•etes)   the sex cells (ie, sperm or eggs)

meiosis (mei•o•sis)   a special kind of cell division that results in four gametes (N) from a single diploid (2N) cell

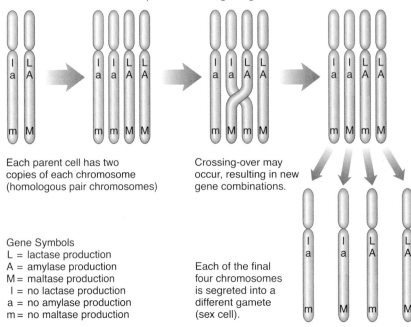

The chromosomes replicate at the beginning of meiosis.

Each parent cell has two copies of each chromosome (homologous pair chromosomes)

Crossing-over may occur, resulting in new gene combinations.

Each of the final four chromosomes is segreted into a different gamete (sex cell).

Gene Symbols
L = lactase production
A = amylase production
M = maltase production
I = no lactase production
a = no amylase production
m = no maltase production

Figure 10.5. New Gene Combinations. Crossing-over and gene shuffling during meiosis (sex cell division) create new combinations of genetic information on chromosomes. The new combination of genes is carried in sex cells to the zygote of the next generation. What genetic information would a sex cell be carrying if it got one or another of the final chromosomes?

selective breeding (se•lec•tive breed•ing) the parent selection and controlled breeding for a particular characteristic

of plants are commonly produced during **selective breeding**; pollen from plants with desired traits is purposefully crossed with other plants exhibiting desirable traits. The majority of agricultural crops on the market today are the result of selective breeding. Several agricultural biotechnology companies, including Monsanto Co., Archer Daniels Midland Co., and Novartis AG employ hundreds of workers who create new breeds of plants and animals through selective breeding.

Although breeding increases variety in plant populations, plant biotechnologists sometimes do not want variety, but, rather, many identical plants. Suppose you have a rare orchid worth more than $10,000. If you breed orchids through sexual reproduction, you risk producing plants that have lost desirable traits and have gained unwanted ones. To ensure production of plants identical to the original, a biotechnologist will make more plants through asexual plant propagation.

Asexual reproduction, also called cloning, is the production of offspring from a single parent. The offspring are identical to the parent and are considered clones (see Figure 10.6). In fact, clones are common in many species. Bacteria colonies are actually thousands or millions of identical bacteria cells (clones) originating from a single parent cell.

Depending on the species of plant, there are several kinds of asexual plant propagation techniques that plant scientists use to create clones (see Figure 10.7). One of the oldest methods of asexual propagation is the use of

Figure 10.6. Cloning. The cloning process creates exact copies of the parent.

cuttings. Cuttings are pieces of stems, leaves, or even roots, placed in an appropriate media, such as vermiculite, sand, perlite, potting soil, or water. Under the right conditions, existing cells in growing tissue will develop into missing roots, stems, or leaves. Many houseplants, annuals, and perennials, including geraniums and *Coleus* plants, are easily "started" from cuttings. Even some woody shrubs and trees can be started from cuttings by treating them with a plant hormone rooting compound (see Figure 10.8).

Another kind of asexual reproduction commonly used to propagate plants is **tissue culture**. In tissue culture, one or a few cells, under sterile conditions, are excised from the parent plant and placed in a medium containing all the nutrients necessary for growth. Under the right conditions, the cells will grow and divide, resulting in a mass of cells called **callus**. If callus is placed in a suitable medium, it produces roots, stems, and leaves. The resulting plant is identical to the plant that was the source of the original cells. Tissue culture has several advantages, including the ability to grow hundreds or thousands of plants from just a small sample of the parent. Another advantage is that growing tissue culture specimens under sterile conditions, in a controlled environment, reduces pest problems. African violets, orchids, and other exotic plants are commonly cloned through tissue culturing (see Figure 10.9).

Often, the goal of a biotechnician is to produce whole plants, as in breeding or asexual reproduction. However, sometimes biotechnicians want to produce or retrieve plant products. Rubber is a plant product that has been harvested from rubber trees for centuries. Other plant products include cotton fibers, wood fiber, and many different medicinal compounds, such as the precursor to aspirin.

Agricultural biotechnologists have improved the productivity of plants. The result is increasing amounts of food and industrial crops. Sometimes

Figure 10.7. Valuable and unusual plants, such as these orchids, can be cloned into thousands of identical offspring through asexual propagation. The most common method of orchid reproduction is the use of protocorms and plant tissue culture. Protocorms are tiny, bulb-like branches excised from a very young orchid. Protocorms are cultured in sterile broth and subdivided every few months. Some mature protocorms are transferred to sterile agar to grow true leaves and stems, and develop into adult orchids. Photo by author.

cuttings (cut•tings) the pieces of stems, leaves, or roots for use in asexual plant propagation

tissue culture (tis•sue cul•ture) the process of growing plant or animal cells in or on a sterile medium containing all of the nutrients necessary for growth

callus (cal•lus) a mass of undifferentiated plant cells developed during plant tissue culture

Synthetic auxin in a commercial product stimulates cell division and root growth.

Figure 10.8. Some undifferentiated cells in certain parts of plants retain the ability to specialize into new plant tissues. Rooting compounds contain hormones (plant growth regulators, such as auxin) that encourage cell division and cell specialization.

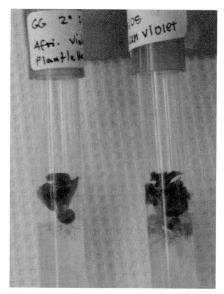

Figure 10.9. Several African violet plantlets (clones) grow in tissue culture from a piece of parent plant leaf. Each clone, developed from a few cells of the leaf disk, has the same genetic code and the same characteristics as the original parent. Photo by author.

Figure 10.10. The pigeon pea legume is a source of protein in the tropics and has for the first time been genetically modified to incorporate the insect resistant gene (Bt). Developed at the International Crops Research Institute for the Semi-Arid Tropics (ICRISAT), the transformed plant is resistant to attack by the dreaded bollworm, which, according to ICRISAT, causes crop loss of about $475 million to India, despite the use of insecticides worth $211 million.
© Pallava Bagla/Getty Images.

these improvements are due to selective breeding. Recently, scientists have created techniques for transferring deoxyribonucleic acid (DNA) into plant cells (genetic engineering). Plant cells can be coaxed to synthesize a variety of new plant products or to exhibit new phenotypes coded for on an imported DNA segment (see Figure 10.10). The possibilities are limitless. Theoretically, a plant could be genetically engineered to synthesize virtually any chemical or to have almost any trait.

Section 10.1 Review Questions

1. How many parents are necessary for an offspring to be produced by sexual reproduction? How many parents are necessary for an offspring to be produced by asexual reproduction?
2. Which two cells fuse to make a zygote? From where do the chromosomes of a zygote come?
3. How does a cutting become a functioning, independent organism?
4. What is the smallest number of cells required to clone a plant through tissue culture?

10.2 Basic Plant Anatomy

Plants are multicellular organisms composed of organs and tissues. The organs of plants include stems, leaves, and roots, and, depending on the type of plant, flowers or cones. Each plant organ has a specific function. Leaves are specialized for food production (through photosynthesis) and food storage. Stems are specialized for water and food transport. Roots conduct water absorption for plants. Roots may also serve as food storage areas and anchorage for the plant. Flowers and cones are reproductive organs specializing in sperm and egg production, as well as seed production and dispersal.

There is wide diversity in plant organ structure, including several examples of elaborate plant organs specialized for specific environments. Cactus stems, for example, are hollow to create more water storage space than is found in most stems. The leaves of a Venus flytrap act as a tiny jail, with barred doors that close to trap unsuspecting insect prey (see Figure 10.11).

Figure 10.11. The leaves of the Venus Flytrap plant are specialized to trap insects and dissolve them for mineral nutrients.
© lovleah/iStock.

Figure 10.12. The base (ovary) of the cucumber flower develops into the cucumber fruit that we eat. Inside the fruit are tiny seeds.
© syaber/iStock

Figure 10.13. Seedless watermelons, developed in the late 1980s, are the result of breeding watermelons of different chromosome numbers. Since they are seedless, these watermelons are sterile and must be rebred with each generation.
© Royalty-Free/Corbis.

Plant organs are often grown as food or commercial crops. Cut flowers, grown for bouquets, are a huge business. Flowers are grown in fields or greenhouses, and are shipped to markets around the world. Breeding bigger, more diverse flowers is the goal of many horticultural biotech companies.

Many food crops are flowers or flower parts. Grains are the flower or seed clusters of plants in the grass family. Examples are oats, wheat, rice, and rye. Some other flowers grown as food are broccoli flower buds and cauliflower heads.

Interestingly, the ovaries of flowers are the most common fruits or vegetables. Tomatoes, zucchini, watermelon, bananas, cucumbers, apples, and cherries are the ovaries of their respective plants (see Figure 10.12). In a plant ovary, seeds are produced through sexual reproduction (the fusion of sex cells or gametes). Seeds contain an embryo, developed from the zygote, which is the next generation of a plant. Food seeds include peas, beans, peanuts, corn, soybeans, and all varieties of nuts.

Many stems and leaves are grown for food as well. Asparagus stems, celery stalks, bok choy, and chives are a few examples. Potatoes, yams, and onions are stems that grow underground. Spinach, lettuce, cabbage, and parsley leaves are also popular food crops.

Agriculture is the largest industry in the United States. More people work in food and crop production than in any other industry. Developing specific breeds of plants, appropriate fertilizers, and safe, effective pest control for crops is an ever-expanding area of agricultural biotechnology (see Figure 10.13).

BIOTECH ONLINE

Pros and Cons of Fertilizer Use

TO DO

Search the Web to find an article that discusses the benefits and concerns of using fertilizers on commercial agricultural land. Print the article. Actively read it. Place an asterisk next to three interesting or important facts. Cite references.

All-purpose fertilizers contain, at a minimum, the compounds nitrogen, phosphorus, and potassium. The percentage by weight of each compound is shown on the package as three numbers, in this case, 12, 5, and 7, for N, P, and K, respectively.
Photo by author.

Table 10.1. **Plant Tissues and Their Functions**

Tissue	Function	Location
epidermis	covering, protection, and gas exchange	on the surface of plant organs
meristem	cell division	in shoot and leaf buds, and root tips
cortex	food and water storage	filling stems and roots
xylem	water and mineral transport	roots, stems, and veins of leaves/flowers
phloem	food and sap transport	roots, stems, and veins of leaves/flowers
parenchyma	food and water storage	filling stems and roots
collenchyma	support	thick-walled cells of plant organs

Plant Tissues

Plant organs contain specialized tissues and cells. Tissues are groups of similar cells with a specific function. Different plant tissues working together allow the plant to complete all of its basic life functions, including food production, food storage, food and water transport, growth, reproduction, and gas exchange. Table 10.1 shows some different kinds of plant tissues and their functions.

Some plant tissues are more interesting to biotechnologists than others. Of particular interest is **meristematic tissue**, the regions of cell division. When plants are grown in tissue culture, a piece of plant is cut to include some meristematic tissue (see Figure 10.14). With the use of special media, the tissue is stimulated to produce new shoots and roots. Without these dividing and differentiating cells, no other new plant tissues or organs can be made.

Plant Cells

Plant tissues are made up of plant cells. Plant cells have all the organelles found in other cells (nucleus, mitochondria, vacuole, Golgi bodies, etc), plus special organelles, including chloroplasts. Chloroplasts make food molecules by converting carbon dioxide and water to glucose and oxygen. This process, photosynthesis, is of interest to biotechnologists. Chloroplasts have been found to contain their own DNA. Biotechnologists find this especially interesting for several reasons, including the potential to manipulate the food-making process.

Plant cells are surrounded by a cell wall composed of cellulose molecules (see Figure 10.15). Cellulose molecules are long and fibrous. They are the major component in many commercial products, such as paper, cotton, and hemp. Biotechnologists have produced enzymes that break the bonds holding cellulose molecules together. These enzymes, called cellulases, shorten cellulose molecules. When used commercially, the enzyme produces softer paper and cotton products. Cellulases are also used to break down cellulose plant fibers to produce sugar that is converted to ethanol for use as biofuel.

Plant cells, like all eukaryotic cells, contain DNA in their nuclei. The genetic instructions for all of a plant's products are held within the DNA code. Plant biotechnologists manipulate the DNA code to alter a plant's growth and chemical production. Adding more DNA, removing some DNA, or changing the code may modify the existing DNA sequence and thus the genetic code. Calgene, Inc. produced the FLAVR SAVR® tomato by modifying the fruit-ripening genes in the tomato chromosomes. The goal was to create tomatoes that could stay on the vine longer, taste better, and arrive at the market retaining more flavor.

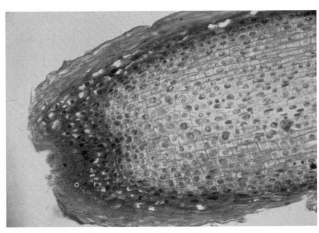

Figure 10.14. This is a longitudinal micrograph through an onion (*Allium* sp.) root tip. The growing tip, or root meristem (cubic, golden cells with dark blue nuclei), is visible just under the root cap (pink). Meristems are found at all growing tips (stems, branches, and roots) and in a thin circle, longitudinally, around the stem and root. In plants, meristematic cells are the only cells that undergo cell division. Meristematic cells eventually elongate and specialize into cells with specific functions, seen higher up on the root.
© Lester V. Bergman/Getty Images.

■ ■ **Chapter 10**

The ability to modify the DNA of plants is an exciting and rapidly growing area of biotechnological research. Many companies are employing plant-gene engineers to explore the genetic manipulation of their plant crops. Pest resistance (such as pest-resistant Roundup Ready® soybeans), rapid growth, increased chemical production, and the potential to grow plants in less than desirable conditions are major areas of plant biotechnology research.

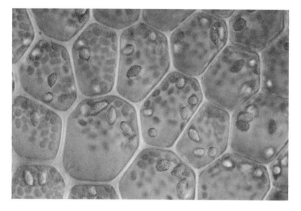

Figure 10.15. **Lettuce leaf cells with cell walls containing different amounts of cellulose. 400X**
© Clouds Hill Imaging Ltd./Corbis Documentary/Getty Images.

BIOTECH ONLINE

Whatever Happened to the FLAVR SAVR® Tomato?

In 1994, Calgene, Inc. of Davis, CA, brought the first genetically engineered food to market. But, if you go to the supermarket today, you will not find any FLAVR SAVR® tomatoes. What ever happened to the FLAVR SAVR® tomato?

Overripe tomatoes (left) compared to genetically engineered or modified tomatoes (right). By altering the DNA in the genes of the tomato, different properties can be introduced to the plant. These are FLAVR SAVR® tomatoes, the first genetically modified food to be sold (in 1994).
© Science Photo Library/Photo Researchers.

Go to the website at biotech.emcp.net/allbusiness and read about the history and future of genetically engineered foods. Summarize the article, including the traits of the FLAVR SAVR® tomato, why the tomato is not currently on the market, other new crops that are showing promise, and some of the controversies surrounding genetically modified food crops. Next go to biotech.emcp.net/tomatocasual to learn about two new tomatoes engineered to fight important disorders. Describe them.

Section 10.2   Review Questions

1. List some foods that are examples of the following plant organs: stems, roots, leaves, flowers, fruits (ovaries), and seeds (fertilized ovules).
2. Which plant tissue type is the source of cells for tissue culture?
3. Give an example of a plant that has been modified by genetic engineering.

10.3 Plant Growth, Structure, and Function

In many respects, plant growth is similar to animal growth. Like animals, plants become bigger primarily by adding cells. However, plant and animal growth differs in many other respects. Unlike animals, in which most organs and tissues can produce new cells, plants have specific regions where cell division can occur. These are called **meristems**, and they are composed of meristematic tissue (see Figure 10.16).

meristems (mer•i•stems)
regions of a plant where cell division occurs, generally found in the growing tips of plants

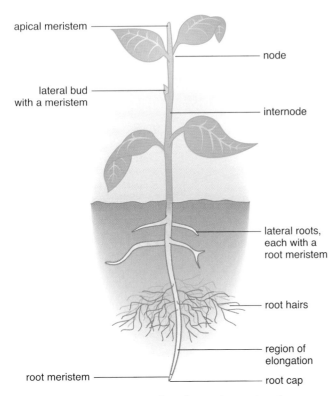

apical meristem

node

lateral bud
with a meristem

internode

lateral roots,
each with a
root meristem

root hairs

region of
elongation

root meristem

root cap

Figure 10.16. Meristems are found at each growing tip.

Meristems are found in the growing tips of the plant. Root tips (root meristems), shoot tips (apical meristems), and branch tips (lateral meristems) have meristems. Flower and leaf buds have meristems. There is even a thin layer of meristematic cells in the center of a plant's trunk that is responsible for plants growing in width. This is called the vascular cambium.

In the meristematic regions, cells undergo **mitosis** (cell division) to make more body cells (see Figure 10.17). Mitosis is a continuous process, but it is often described as four steps: prophase, metaphase, anaphase, and telophase. Just prior to mitosis, all of the DNA in the cell replicates, resulting in a set of doubled chromosomes. Mitosis begins when the doubled chromosomes tighten and shorten. The chromosomes line up in the middle of the cell and pull apart into new daughter cells. Thus, one cell makes two identical ones, each containing a copy of the chromosomes found in the parent cell. Since hundreds of cells are found in each meristem, each cycle of cell division results in adding more and more tissue to the growing tip. The tip continues to grow in length, producing longer stems, roots, and leaves (see Figure 10.18).

Cell division in meristematic regions is the main process involved in seed germination. Seed **germination**, also called sprouting, occurs when a seed absorbs water, triggering an embryo to start mitosis. As the embryo grows, its root pushes out and ruptures the seed coat.

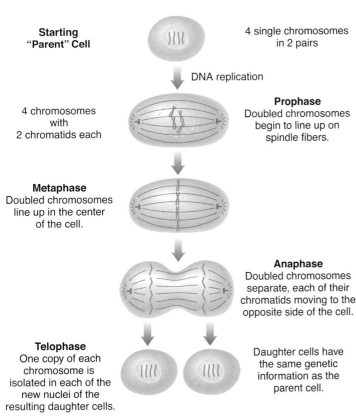

Starting "Parent" Cell

4 single chromosomes in 2 pairs

DNA replication

4 chromosomes with 2 chromatids each

Prophase
Doubled chromosomes begin to line up on spindle fibers.

Metaphase
Doubled chromosomes line up in the center of the cell.

Anaphase
Doubled chromosomes separate, each of their chromatids moving to the opposite side of the cell.

Telophase
One copy of each chromosome is isolated in each of the new nuclei of the resulting daughter cells.

Daughter cells have the same genetic information as the parent cell.

Figure 10.17. During mitosis, cells make exact copies of themselves. The tightly wound mitotic chromosomes are visible using a microscope.

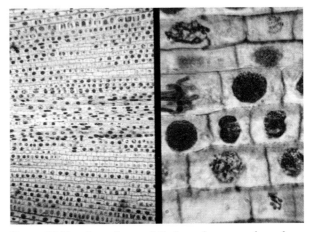

Figure 10.18. An onion root tip has a large number of cells undergoing mitosis. Mitosis adds length or width to plant parts. At a higher magnification (right), the chromosome strands in mitosis are visible.
© Custom Medical Stock Photo.

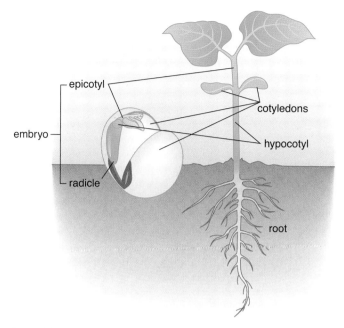

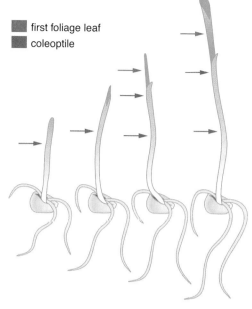

☐ first foliage leaf
☐ coleoptile

Figure 10.19. **Dicot Seed Germination.** The diagram shows a germinating bean seed with an enlarged radicle growing down. The epicotyl will grow up into the leaves and stem. The two cotyledons emerge from the seed as a food source, but they wither after a short time.

Figure 10.20. **Monocot Seed Germination.** Four germinated corn seedlings. The coleoptile protects the emerging leaves. Corn plants are monocots since they have only one section to the seed.
© Custom Medical Stock Photo.

The seed's root, called a **radicle**, grows downward due to gravity, and the seed's shoot grows upward, bringing the seed's leaves, called cotyledons, above the soil (see Figure 10.19 and Figure 10.20). The cotyledons store starch and lipids, which the plant uses for energy until it is aboveground and can photosynthesize.

Different cells are found in different plant tissues and organs. Roots, stems, and leaves have specialized functions because of their unique cells and tissues. It is the process of **differentiation** that changes newly produced meristematic cells into specific tissues. During differentiation, certain genes are turned off or on, resulting in the production of specific **plant hormones**. The hormones diffuse to other locations and, in certain concentrations, turn on and off other genes that trigger certain cell specialization and organ formation in that region of the plant. Hormone regulation will be discussed in more detail in later sections.

Some cells become storage cells, some become vascular cells (water transport and food transport), and others become structural cells giving support to the plant. Some cells of leaves and stems specialize into photosynthetic cells (see Figure 10.21).

A new seedling is green and tender. Some plants stay in this tender state for their entire life. They are called herbaceous. **Herbaceous plants** usually live for only one (annual) or two (biennial) seasons. Some examples are cornflowers, violets, baby blue eyes, squash, and tomatoes (see Figure 10.22).

Some plants grow to be thick, add wood, and become strong and hard. These are called **woody plants**. Woody plants add many cells horizontally by increasing cell division at the vascular cambium in the center of the stem and roots. Woody plants can better survive changes in weather and the effects of gravity. They last many seasons (perennial), and some live hundreds of years. Woody plants include shrubs and trees (see Figure 10.23).

radicle (rad•i•cle) an embryonic root-tip

differentiation (dif•fer•en•ti•a•tion) the development of a cell toward a more defined or specialized function

plant hormones (plant hor•mones) the signaling molecules that, in certain concentrations, regulate growth and development, often by altering the expression of genes that trigger certain cell specialization and organ formation

herbaceous plants (her•ba•ce•ous plants) the plants that do not add woody tissues; most herbaceous plants have a short generation time of less than one year from seed to flower

woody plants (wood•y plants) the plants that add woody tissue; most woody plants have a long generation time of more than one year from seed to flower; most woody plants grow to be tall, thick, and hard

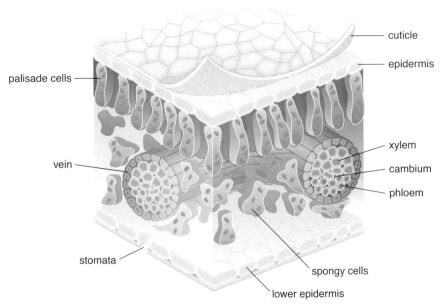

Figure 10.21. In this leaf diagram, the diversity of plant cells is shown. Xylem and phloem carry water and food, respectively. Cambium, in the center of the vein, is actively dividing cells. Palisade and spongy cells conduct photosynthesis. The epidermis protects inner cells from dehydration.

Figure 10.22. *Nemopila insignis*, commonly called Baby Blue Eyes, is an herb that grows in many meadows.
© motorolka/Shutterstock.

Figure 10.23. Teddy Roosevelt and a group pose in front of a giant redwood. The redwood tree is one of the largest organisms. The tree's woody vascular tissue gives it substantial vertical support.
© Royalty-Free Corbis.

Section 10.3   Review Questions

1. Name the parts of the plants that contain actively dividing cells.
2. After a mitotic division, how many chromosomes do daughter cells have compared with the parent cell?
3. When a seed germinates, what is the first plant part to emerge from the sprouting seed?
4. There is a vast diversity among plant cells and plant tissues. What are the chemicals called that trigger much of the cell and tissue specialization in plants?

10.4 Introduction to Plant Breeding

In many respects, sexual reproduction in plants is similar to that in animals. For instance, plants produce sperm and eggs. Plant sperm fertilizes a plant egg and a zygote is formed. The zygote divides by mitosis and becomes an embryo with differentiated tissues and organs. The organs (roots, stems, leaves, and flowers) have specialized functions. The flowers have structures that produce more eggs and sperm (through meiosis) for the next generation.

Plants are **diploid** organisms represented by the symbol "2N" (see Figure 10.24). Diploid (2N) refers to the fact that each cell of the organism (except the sex cells) has two (2N) sets of homologous (matching) chromosomes. One set (1N) comes from the female parent and is carried in the egg cell. The other set (1N) comes from the male parent and is carried in the sperm nucleus. These sex cells are **haploid**, with only one set of chromosomes, or half the diploid amount. When the sperm fertilizes the egg, the two sets of chromosomes come together in the zygote's nucleus, reestablishing the diploid number in the parent cells. The rest of an organism's cells arise from mitotic cell division of this original zygote (2N).

In humans, the diploid number is 46. All human cells, except the gametes (sperm or eggs), have 46 chromosomes. This is because one set of 23 chromosomes comes from the mother, in her eggs, and one set of 23 chromosomes comes from the father, in his sperm. The chromosomes in the egg and sperm of each species are homologous, meaning that they contain all the same genes in the same order.

In flower plants, the sperm and eggs are produced in the anther and pistil of the flower, respectively. Each sex cell is the result of meiosis, sex-cell division. In the anther and pistil, regular diploid cells, called "mother cells," undergo two sets of cell divisions resulting in thousands of haploid (1N) sperm or egg cells, respectively (see Figure 10.25).

Figure 10.24. A Wisconsin Fast Plant, the *Brassica rapa* plant, has 2N = 30. Therefore, each cell has 15 pairs of chromosomes, except for sperm and egg cells, which contain 15 chromosomes each. In this photo, there are three height mutants of wild-type (normal) plant. From left to right, the plants are rosette mutant, petite mutant, normal wild-type, and tall mutant.
© Wisconsin Fast Plants Program, University of Wisconsin-Madison.

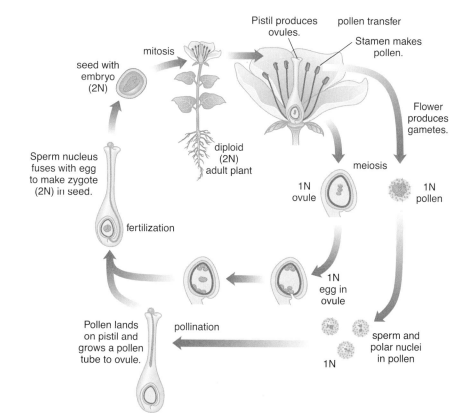

Figure 10.25. Alternation of Generations. Each sex cell gets one copy (1N) of each chromosome and, therefore, one copy of each gene. Most genes exist in one of two or more forms (alleles). When the zygote (2N) forms, it receives both sets of chromosomes (and genes) from the two sex cells. Depending on what was carried in the sex cells, the zygote could receive two matching alleles or two different alleles for a particular trait. The alleles of an organism are its genotype and, ultimately, determine the traits expressed (phenotype).

In the life cycle of a plant, the genetic information is transferred from one generation to the next when sex cells fuse and contribute their chromosomes in the zygote nucleus. Each chromosome has hundreds of genes, most of them coding for some protein's structure. Since there are two copies (2N) of each chromosome, one from the egg and one from the sperm, there are two copies of each gene.

Alternate forms of a gene are called **alleles**. There are different alleles for different genes. For example, *Brassica rapa* contains two alleles coding for one of the chlorophyll production genes. A plant can receive either two alleles for normal chlorophyll production (dark green leaves), two alleles that result in lessened chlorophyll production (yellow-green leaves), or one each of an allele for normal chlorophyll production and lessened chlorophyll production (see Figure 10.26).

Genotypes and Phenotypes

The alleles an organism possesses are called its genotype, and they determine the plant's characteristics, or phenotype. The genotype for a particular trait is represented by allelic symbols. In this example, the capital letter "G" could represent the normal chlorophyll production allele, and the lower case "g" could represent the lessened chlorophyll production (yellow-green) allele. These symbols are selected because, when the G allele is present in an organism with the g allele, a plant will have enough normal chlorophyll production to be dark green. We say that the G allele is **dominant** over the g allele, and that the g allele is **recessive**.

If a plant inherits two G alleles, it has the genotype, GG. The DNA represented by these alleles can be transcribed into messenger ribonucleic acid (mRNA) and when translated into the proteins cause the plant to have the dark-green leaf phenotype. These plants are called **homozygous** because they have two of the same alleles for the gene. In this case, they are called **homozygous dominant** because the two alleles are for the dominant version of the gene.

If a plant inherits two g alleles, it has the genotype, gg. These alleles are either poorly transcribed, which results in less chlorophyll production, or the transcribed mRNA is not translated into the proteins "correctly." Either way, the result is that the plant has less chlorophyll in its leaves than a plant that is Gg or GG. The gg plants are called **homozygous recessive** because they have two of the same alleles for the recessive version of the gene.

If a plant inherits both a G and a g allele, it has the genotype, Gg. The G allele can be transcribed into mRNA and translated into the proteins that result in a plant with dark-green leaves. The g allele does not interfere with the G allele. These plants are called **heterozygous** because they have two different alleles for the gene (see Figure 10.26).

Many phenotypes result from the expression of multiple genes. The hairy (leaf hairs) phenotype of *Brassica rapa* is due to the expression of several different genes, each with alleles for hairiness (see Figure 10.27). A trait coded in this way is called **polygenic**. Since there are several genes expressed in a polygenic trait, there are several possible phenotypic outcomes of the genes.

In this example, the allelic symbols used to represent the hairless, or hairy condition, might be the letter "H." The H represents alleles that code for no hairs. The h allele codes for the production of stem and leaf hairs. The H allele is strongly expressed, and when present with an h, it suppresses the hairy allele expression. The H is dominant over the h. So, a plant can be HH, which results in no hair production; hh, which codes for hairs; or Hh, which codes for a small amount of hair production. Table 10.2 shows the graduation of phenotypic expression in a trait expressed, for example, as the sum of three genes.

Figure 10.26. The wild-type plants on the left side of the quad have dark green leaves due to at least one dominant chlorophyll production allele. Their genotype could be represented as either GG or Gg. The yellow-green mutant plants on the right side of the quad have two recessive alleles and could be represented by a gg genotype.
© Wisconsin Fast Plants Program, University of Wisconsin-Madison.

Figure 10.27. The degree of hairiness depends on the number of recessive "hairy" alleles a plant receives.
© Wisconsin Fast Plants Program, University of Wisconsin-Madison.

Figure 10.28. Plant breeders try to control the result of crosses by collecting the pollen of one flower and transferring it to the pistil of another. The pistil of this flower is visible in the center of five stamen. Using a beestick or pollination wand, a breeder can collect pollen (containing the male gametes) from the anther (tip) of the stamen.
Photo by Timothy Wong.

Table 10.2. An Example of How a Polygenic Trait Might Be Expressed

No. of HH or Hh Genes Present	No. of hh Genes Present	Phenotype
0	4	large amount of stem and leaf hairs
1	2	large amount of stem hairs and some leaf hairs
2	1	small amount of stem hairs
3	0	no leaf or stem hairs

Breeding Plants for Desired Phenotypes

Plant breeders try to study, predict, and manipulate crosses between flowers in an attempt to produce plants of desired phenotypes (see Figure 10.28). Since meiosis results in a recombination of genetic information in sex cells, breeding introduces variety into the phenotypes of plants. For genes of interest, plant breeders have to consider all the combinations of alleles that could be in the gametes being crossed.

For example, consider a *Brassica rapa* cross used to study the inheritance of mutant short plants. Breeders have seen that when a short plant (genotype = tt) is crossed with a tall plant (genotype = TT), all of the offspring are tall (phenotype). This indicates that the allele responsible for shortness is recessive to the allele for tallness. We can assign the symbols "T" for the tall allele and "t" for the short allele. To produce all tall plants from tall and short parents, the parents' genotypes had to be TT and tt, respectively. How this is determined is shown below:

Parents' genotypes   TT   tt

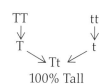

Alleles possible in gametes   T   t
Genotype of offspring   Tt
Phenotype(s) of offspring   100% Tall

Punnett Square (pun•nett square)   a chart that shows the possible gene combinations that could result when crossing specific genotypes

monohybrid cross (mon•o•hy•brid cross)   a breeding experiment in which the inheritance of only one trait is studied

Using Punnett Square Analysis

To make predictions and to study crosses of specific traits, plant breeders use a chart called a **Punnett Square**. A Punnett Square shows the possible gene combinations that could result when crossing specific genotypes. Preparing a Punnett Square analysis allows breeders to determine the probability of having offspring with certain genotypes and phenotypes. The steps are similar to those in the example above.

Consider a cross between *Brassica rapa* parent plants known to be heterozygous tall and homozygous short. Using the same allelic symbols as above, Table 10.3 shows the resulting gene combinations:

Table 10.3.   **Punnett Square of the Cross**

Parents' genotypes	Tt	tt
Alleles possible in gametes	T or t	t

Punnett Square of possible reproductive combinations

	t	
T	Tt	offspring possibility 1
t	tt	offspring possibility 2

Expected genotypic results of crossing these gametes:

1/2 of offspring are expected to be = Tt
1/2 of offspring are expected to be = tt

Expected phenotype(s) of offspring:

1/2 of offspring are expected to be = tall
1/2 of offspring are expected to be = short

This means that each seed has a 50% chance of growing into either a short plant or a tall plant. In this example, the inheritance of only one trait is being studied. This kind of cross is called a **monohybrid cross**.

Notice:

- The parents have two symbols each to represent each genotype.
- The gametes have one symbol each to represent each genotype.
- The offspring have two symbols each to represent their genotypes.

Thus, if these plants were crossed, and the result was 30 offspring, a technician would expect 15 tall and 15 short plants. It is unusual, though, for the actual results to be the same as those expected. Of course, to check the results of the cross, one must harvest the seeds from the parent plants and grow them for a sufficient period of time to check the phenotypes.

Have you ever thought about why, in some families, there are more girls than boys, or vice versa (see Figure 10.29)? For both the plant and human examples, the difference between the observed and expected

Figure 10.29.   **In this family, the parents had five girls and a boy. This is not what is expected if each sperm cell carries either an X or a Y chromosome, and the number of X sperm is equal to the number of Y sperm available. Can you explain why?**
Photo courtesy of author.

results is due to the fact that many sperm and egg cells of each type (X and Y chromosomes, or T and t alleles) are present, and which sperm find which eggs is random. So, even though there are thousands of T sperm, there are also thousands of t sperm. Over thousands of crosses, though, there is an equal chance that either type of sperm could find any egg. Table 10.4 displays the results of an XY and XX chromosome cross.

Table 10.4. **Punnett Square of the Cross**

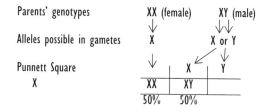

Expected genotypic results of this cross: 1/2 each XY and XX
Expected phenotype(s): 1/2 each boys and girls

Frequently, more than one trait is studied at the same time. It is called a **dihybrid cross** when two traits are studied at the same time. Using the allelic symbols given above, consider a cross of a heterozygous, tall, green plant with a short, yellow-green plant. Table 10.5 displays the results.

Table 10.5. **Punnett Square of the Cross**

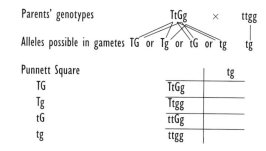

Expected genotypic results of this cross: 1/4 each TTGg, Ttgg, TtGg, and ttgg
Expected phenotype(s): 1/4 each tall/green, tall/yellow-green, short/green, short/yellow-green

How well are you following the symbols used in the cross shown above? Check out the genotypes:

- Does it make sense that the parents would have four symbols to represent their genotypes for this cross?
- Does it make sense that the gametes would have two symbols each to represent their genotypes for this cross?
- Does it make sense that the offspring would have four symbols to represent their genotypes for this cross?

Thus, if these plants were crossed and the result was 100 offspring plants, a technician would expect 25 of each phenotype.

Statistical Analysis of Data

average (av•er•age) a statistical measure of the central tendency that is calculated by dividing the sum of the values collected by the number of values being considered

mean (mean) the average value for a set of numbers

When crossing *Brassica rapa* plants (also called Wisconsin Fast Plants, or WFPs), thousands of seeds are produced, so there is a large amount of data to collect and analyze. This is true for many types of scientific experiments, especially genetic studies. How does a scientist know if the data collected support or refute the hypothesis of an experiment? How does one know whether the numerical data are meaningful and significant? How close do numbers have to be to what is expected?

Whenever possible, scientists collect numerical (or quantitative) data. Quantitative data allow for statistical analysis because numbers are easy to compare and evaluate. On the other hand, descriptive, or qualitative, data are difficult to analyze, compare, and appraise because they rely on subjective interpretation. For example, suppose scientists were studying the concentration of hemoglobin in different serum samples. Assessing the samples by recording that one was a little more or a little less red than another could provide inaccurate or misleading data. One observer's judgment of "a little more or a little less red" likely would differ from another observer's perspective. It would be difficult to study accurately the range of redness and the average redness using these types of observations.

Instead of using words to describe data, scientists try to make all observations of experiments in numerical form. In the example of the hemoglobin samples, it is simple to measure the absorbance (optical density) of each sample with a spectrophotometer (see Figure 10.30). From such data, we can determine the concentration of one sample compared with another sample. The numerical data allow us to accurately evaluate the samples.

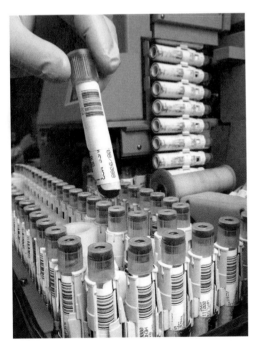

Figure 10.30. Blood serum samples are stored in freezers until ready for use. If not stored properly, the cells may degrade, and the concentration would be affected. Absorbance and concentration are measured using a spectrophotometer.
© Lester Lefkowitz/Corbis.

Using Multiple Replications to Determine Averages

When conducting experiments, it is important to repeat trials enough times to ensure that the results reflect what really happens. These are called "multiple replications." The value of multiple replications becomes clear when you consider the example of a baseball player's batting average. At the beginning of a baseball season, every player has a .000 batting average. If a batter gets a "hit" his first time at bat, then he is one for one, batting 100%, or as baseball reports it, 1.000. Another batter strikes out at his first at bat. He has batted zero for one and still has a .000 batting average. Does this mean that the first batter is necessarily a better baseball player than the second player? Of course not. What is important is a batter's average over the entire season (162 games). By the end of the season, a typical player will bat about 500 times. If a batter's average over the entire season is .333, which is outstanding, he gets a hit 33.3% of the time. A second player's average over the entire season is .250. He is only getting a hit about 25% of the time, or on average, one out of four times at the plate. Which player would you want on your team?

The most basic form of statistical analysis is that of determining the **average** for a group of samples. The average value for a set of numbers is also called the "**mean**." To determine an average value, add all the values and divide by the number of values being considered (the sample size). The larger the sample size, the better the average will represent the true value.

Experiments are conducted with multiple replications so that an average can be determined. However, there can be a great deal of variation in the individual trials data that are used to determine an average value, due to measuring errors, timing, or other human errors. The average value helps negate large variations and rare or erroneous data (see Figure 10.31).

Figure 10.31. A lab technician at Mendel Biotechnology, Inc., in Hayward, CA, prepares hundreds of seeds for germination screening assays. To get valid data, the numbers of seeds must be very high.
Photo by author.

Evaluating the Validity of Data

Scientists often want to evaluate the validity of individual pieces of data. Consider 10 juice extraction measurements with an average juice volume of 10 mL. If most of the individual trials give a volume of between 9.1 and 11.3 mL, is a value of 8.8 mL a "good" piece of data? Should the 8.8 mL value be accepted as collected under proper procedures?

The 10% Error Rule   Often a technician looks at data with the goal of making a rapid determination of whether or not to continue a procedure. For many applications, a quick calculation of 10% of the expected value gives you an idea if the data make sense or not. We call it the "10% Error Rule." Considering 10 juice measurements with an average volume of 10 mL, 10% of 10 mL is 1.0 mL. Using the 10% error rule, values of + or – 1.0 mL would be considered valid and representative. This means that samples with juice production between 9 and 11 mL would be considered valid and acceptable. Based on the 10% rule, a value of 8.8 mL would be considered not acceptable, and possibly erroneous.

Standard Deviation   Another way of looking at the validity of data is by determining the **standard deviation (SD)**. The SD is a value that describes the range on either side of the mean (average) where data are considered valid. It describes how tightly the data are clustered around the mean. The SD takes into account both the sample size (number of data entries) and the range of samples (how many are high or low, and how high or how low). The SD is time consuming to calculate, so using a calculator or a spreadsheet program, such as Microsoft® Excel®, is helpful. For many experiments, a +/- 2D rule (2 SD) is used. If a value falls within 2 SD of the average, then it is considered a valid piece of data.

Consider the data in Table 10.6, which resulted from a size determination of a DNA fragment produced by a DNA synthesizer.

The SD reveals that the measurements of the gel fragments within 15.28 bp above or below the average of 759 bp are considered valid. Values outside of 2 SD must be questioned as erroneous. The errors that cause deviations in measurements could be trivial (ie, mismeasurement) or large (eg, poor experimental design). The SD also reveals the validity for the whole set of data (see Figure 10.32). A large SD means that data are very widely spaced and not very similar.

standard deviation (stan•dard de•vi•a•tion)   a statistical measure of how much a dataset varies

Table 10.6.   Average Length of DNA Fragments in Base Pairs (bp)

Sample No.		Fragment Length (bp)
1	750	
2	765	
3	760	
4	765	
5	750	
6	775	
7	760	
8	765	
9	760	
10	750	
11	755	
12	755	
Average	759.17	
SD	+/-7.64	
2 SD	+/-15.28	

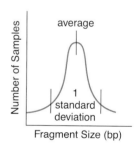

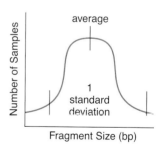

Figure 10.32.   **Standard Deviation.**   If the SD is twice as large (bottom graph) as another (top graph), then we would know that the data for the bottom graph are very dissimilar, and we would have less confidence in the data collection.

Using Goodness of Fit (Chi Square Analysis) to Test the Hypothesis

In every experiment, there are expected results (the hypothesis). The experimental procedures are designed to test the scientific question and the hypothesis for the question. When data are collected and analyzed, researchers must determine how well the data support a hypothesis.

Among the several ways to analyze how well data support a hypothesis, **Chi Square** analysis ("Goodness of Fit") is commonly used in genetics and breeding experiments. A Chi Square analysis gives a numerical value that determines whether the actual data are close enough to the expected data to support the experimental hypothesis.

Consider a fruit fly breeding experiment. When crossing *Drosophila* fruit flies, long wing length is dominant over short wing length. A cross of two heterozygous, long-winged parents should result in 3:1 long-winged to short-winged offspring, as shown in Table 10.7. Since several random events can occur, the number expected does not always match the number observed in a cross.

From the Punnett Square, one can determine that each fertilized egg has a 75% chance of producing flies with long wings. In addition, if the sample size is large enough, for every four offspring, three should have long wings and one should have short wings, a 3:1 ratio. If a cross of two flies results in 80 eggs, three out of four (three-fourths) of the offspring (60) are expected to have long wings and one out of four (one-fourth) are expected to be short-winged. In fact, though, when all the offspring flies are observed and counted, 54 have long wings and 26 have short wings. The scientist will determine, using Chi Square analysis, if 54:26 is close enough to 60:20 to confirm a valid cross of correctly identified parents. This analysis is as follows:

Chi Square (Chi square) a statistical measure of how well a dataset supports the hypothesis or the expected results of an experiment

Table 10.7. **Results of Crossing Two Heterozygous Fruit Flies**

Gametes	L	l
LLL	Ll	
lLl	ll	

How To Calculate the Chi Square Value (χ^2)

Use the equation below to calculate the χ^2 value for a set of data:

$$\chi^2 = \Sigma \left[\frac{(O\text{-}E)^2}{E} \right]$$

The terms represent the following:

O = observed number for a phenotypic group
E = expected number for a phenotypic group
Σ = the Greek letter sigma, which represents "sum of"

The χ^2 equation relates each phenotypic group's observed data to the expected data. The difference is the deviation. Then, Σ sums up all the phenotypic group's deviations. So, when you get the final χ^2 value, the number represents the sum of all the deviations in the experiment. Then you check a χ^2 probability table to see whether the value is small enough to accept your experimental hypothesis (see Table 10.8).

From the *Drosophila* example, there are two phenotypic groups, long wings and short wings. Each phenotypic deviation is calculated as follows, based on 54 long wings being observed when 60 were expected, and 26 short wings being observed when 20 were expected:

$$\frac{(54-60)^2}{60} + \frac{(26-20)^2}{20} = 0.6 + 1.8 = 2.4 = \chi^2$$

The χ^2 for the *Drosophila* cross is 2.4. However, is this amount of deviation from the expected results small enough to accept the hypothesis that there is no significant difference between the expected and observed values and that a valid heterozygous cross was done? To answer the question, look at a χ^2 probability table.

Table 10.8. Chi Square Probability Table

df*	0.95	0.90	0.70	0.50	0.30	0.20	0.10	0.05	0.01	0.001	←Probability (P)
1	0.004	0.016	0.15	0.46	1.07	1.64	2.71	3.84	6.64	10.83	
2	0.10	0.21	0.71	1.39	2.41	3.22	4.61	5.99	9.21	13.82	
3	0.35	0.58	1.42	2.37	3.67	4.64	6.25	7.82	11.35	16.27	
4	0.71	1.06	2.20	3.36	4.88	5.99	7.78	9.49	13.28	18.47	
5	1.15	1.61	3.00	4.35	6.06	7.29	9.24	11.07	15.09	20.52	
6	1.64	2.20	3.83	5.35	7.23	8.56	10.65	12.59	16.81	22.46	
7	2.17	2.83	4.67	6.35	8.38	9.80	12.02	14.07	18.48	24.32	
8	2.73	3.49	5.53	7.34	9.52	11.03	13.36	15.51	20.09	26.13	
9	3.33	4.17	6.39	8.34	10.66	12.24	14.68	16.92	21.67	27.88	
10	3.94	4.87	7.27	9.34	11.78	13.44	15.99	18.31	23.21	29.59	

← Accept the experimental hypothesis | Reject →

The data are close enough to the expected | Data are not close enough to the expected

*df: degrees of freedom

To use the χ^2 probability table, you must determine a value called the "**degrees of freedom**" (df). The df are the number of phenotypic groups −1. For this cross, since there are only two phenotypic groups (long wings and short wings), there is 1 degree of freedom (because 2 − 1 = 1). Looking at the 1 df row, you find that the calculated χ^2 (2.4) at 1 df falls between the probabilities (P) 0.20 and 0.10.

The location where the χ^2 value falls on the probability table determines whether or not the deviation in data is significant. Look at the probability (P) value under which the χ^2 value lies. In general, if the P value is greater than 0.05 (P > 0.05), the **null hypothesis** that there is no significant difference between the observed and expected values is accepted. If the P value is less than 0.05 (P < 0.05), the null hypothesis is rejected, meaning that there are enough discrepancies in the data to reject the hypothesis. If P values are 0.05 and higher (P ≥ 0.05), it means that your data are "good enough" to accept.

Since the χ^2 value of 2.4 is to the left (between 0.10 and 0.20) of the 0.05 P value (3.84), we can accept the data as being within an acceptable amount of deviation. The χ^2 value supports the hypothesis that there is no difference between these data and the expected data. The results indicate that a valid heterozygous cross was conducted.

The bold print values across the top of the chart are the probabilities that the deviations are due to random chance and not to error. Think of it this way: If the χ^2 is at 0.20, then 20% of the time that this experiment is duplicated, results will be equal to or worse than those in this experiment. Usually, the smaller the χ^2 value, the "better" the data are at supporting the experiment's hypothesis.

In the case of a dihybrid, heterozygous cross, the data is expected to give a 9:3:3:1 ratio of individuals in 4 phenotypic group. After conducting a dihybrid, heterozygous cross, and after counting the number of offspring in the four phenotypic groups, an χ^2 is calculated. If the calculated value of χ^2 exceeds the critical value of χ^2 (7.82 at 3 df), then we have a significant difference between the observed data and the expected data (9:3:3:1). A probability of 0.05 or less (P≤0.05) (χ^2 = 7.82 or greater) shows an extreme departure from expectation. We would know that the data show a significant difference between expected and observed data. The data would not support a valid dihybrid, heterozygous cross. To summarize, in the Chi Square analysis, we test whether the deviation from expected results support a null hypothesis that there is no difference between the observed and expected results.

degrees of freedom (de•grees of free•dom) a value used in Chi Square analysis that represents the number of independent observations (eg, phenotypic groups) minus one

null hypothesis (null hy•poth•e•sis) a hypothesis that assumes there is no difference between the observed and expected results

1. A plant has purple flowers and hairy stems. Is this description its genotype or its phenotype? Propose some allelic symbols for purple flowers and hairy stems.

2. Consider a cross between two *Brassica rapa* parent plants known to be heterozygous tall. Using the same allelic symbols as in the text, show the entire cross (problem), including the chances of this breeding resulting in short plants.

3. A set of plant DNA extractions is measured on the UV Spec. There are 13 samples, and the average concentration is 18.7 μg/mL. One sample has a value of 17.2 μg/mL. Using the 10% rule, is the sample's value valid and acceptable?

4. An experiment is conducted to determine the number of kindergarten children with attention deficit disorder (ADD) in the United States. The result shows that 80 out of 1000 students, on average, with a standard deviation of 5, exhibit ADD. A town near a nuclear plant has an average of 84 students per 1000 with ADD. Should the citizens be concerned? Why or why not?

5. A family had 22 children: 14 boys and 8 girls. Calculate the Chi Square value for such a cross and determine whether the results are due to a random mating or an environmental or genetic disorder.

crossbreeding (cross•breed•ing)
the pollination between plants of differing phenotypes, or varieties

10.5 Asexual Plant Propagation

Imagine a plant that grows 50% faster than similar plants. Imagine a different plant of the same species that has particularly deep-red-colored flowers. If one wants faster growing plants with redder flowers, one needs offspring that have traits from both plants.

Using breeding techniques, a plant biotechnologist can produce variety in the offspring of selected parental plants. Pollen is taken from one of the plants (maybe the one that grows faster) and is used to fertilize the flowers of the other plant (the red-flowered one). This is called **crossbreeding** because pollen is transferred between plants. The resulting offspring would have a variety of phenotypes and, by chance, one or more might have the desired characteristics, perhaps a deep red color and a fast growth rate. Crossbreeding results in a considerable amount of variation within a species, since two parents contribute genetic information to the offspring. The variety seen in domesticated roses (American Beauty, Yellow Gold, White Dog, Countryman, etc) as well as in many fruits and vegetables is the result of selective breeding (see Figures 10.32 and 10.33).

Figure 10.32. For centuries, roses have been bred to achieve changes in petal number, color, and size. The White Dog rose is considered a beautiful, delicate specimen. Note the single layer, or whorl, of petals and the abundance of stamen.
© Katharina Rau/iStock.

Figure 10.33. The Countryman rose has been bred to produce many, showy petals.
© Mark Bolton/Corbis Documentary/ Getty Images.

Figure 10.34. These stem cuttings are being grown hydroponically (in water). Hydroponically grown plants need less space and can even grow on the Space Shuttle.
© Patrick Johns/Corbis/VCG.

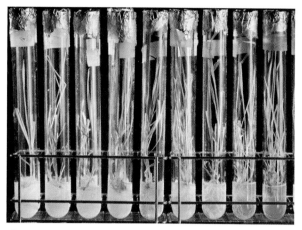

Figure 10.35. These plant-tissue-culture tubes contain clones of grass plants. The cloned plants are grown from tiny fragments or single cells of a "parent" plant. The cells reproduce into a mass that develops roots and shoots like normal seedlings. Plant tissue culture allows the production of genetically identical plants for large-scale agriculture applications.
© Premium Stock/Corbis.

asexual plant propagation (a•sex•u•al plant prop•a•ga•tion) a process by which identical offspring are produced by a single parent; methods include the cuttings of leaves and stems, and plant tissue culture, etc

runners (run•ners) the long, vinelike stems that grow along the soil surface

plant tissue culture (PTC) (plant tis•sue cul•ture) the process of growing small pieces of plants into small plantlets in or on sterile plant tissue culture medium; plant tissue culture medium has all of the required nutrients, chemicals, and hormones to promote cell division and specialization

Often, plant growers do not want variety in their plant crops. Imagine an African violet producer who needs thousands of identical plants to ship to the local nurseries throughout North America. Exact duplicates in the offspring are desired, with no variation. In other words, they want clones. **Asexual plant propagation** is the method by which identical offspring, or clones, are produced by a single parent. Asexual plant propagation has been practiced for hundreds of years. Cuttings of stems and leaves, for example, are relatively easy to grow into entire plants (see Figure 10.34). Some plants, such as strawberries, produce "**runners**." Runners are long, vine-like stems that grow along the soil surface. It is easy to separate runners from the parent and grow them into new plants.

Asexual plant propagation includes many traditional botanical methods, as well as some highly technical ones. The most challenging method of asexual plant propagation is **plant tissue culture (PTC)**. In PTC, small pieces of plants, down to even a few cells, are grown in sterile conditions into small plantlets. Plant tissue culture requires a special agar media that provides all of the nutrients that the isolated plant cells need (see Figure 10.35). Plant hormones or other ingredients are added to regulate the type of growth desired.

To produce a new plant through asexual propagation, meristematic tissue must be present. The meristematic regions of a plant contain cells that continue to actively divide, or retain the potential to divide. Plants grow because meristematic cells divide repeatedly to produce new cells (see Figure 10.36). As with other plant tissues, actively growing and dividing meristematic cells are found only in certain parts of the plant, such as the tips of roots, shoots, and branches. Meristematic tissue is also found within the perimeter of the stem and root in the vascular cambium. These cells are responsible for lateral growth, which adds girth (width) and bark to the plant. New cells add to the length and width of the plant. As cells are produced, they differentiate into tissues specialized for such functions as storage (parenchyma), water (xylem) or food transport (phloem), and protection (epidermis).

Asexual plant propagation can be as easy as taking a small piece of a parent plant, including some meristematic tissue, and placing it in the right media. For example, young branches are a common source of stem pieces for propagation. A young branch is cut from the plant, and two or more nodes are cleared of leaves. The cutting is placed in a medium, such as water, soil, or soil substitute (eg, perlite or vermiculite), so that the cleared nodes are submerged. Sometimes, commercial artificial plant

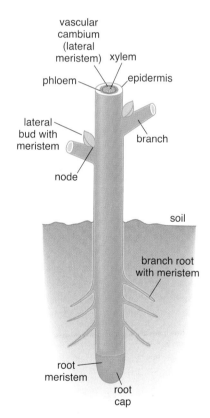

Figure 10.36. Meristematic Tissues. Each plant's growing region (meristem) contains meristematic cells, which are necessary for asexual reproduction. Apical meristem is not shown.

Figure 10.37. Workers in Ecuador prepare cacao stem cuttings for rooting. Each stem cutting has a few leaves attached. The cuttings are dipped in a synthetic plant hormone rooting compound and then planted. Through this type of stem cloning, all the plants on this plantation will have the same genetic qualities. Cacao plants produce cacao seeds used in chocolate production.
© Owen Franken/Corbis Documentary/ Getty Images.

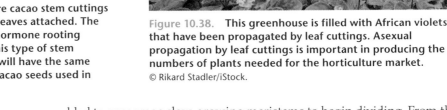

Figure 10.38. This greenhouse is filled with African violets that have been propagated by leaf cuttings. Asexual propagation by leaf cuttings is important in producing the numbers of plants needed for the horticulture market.
© Rikard Stadler/iStock.

hormones are added to encourage slow-growing meristems to begin dividing. From the nodes, new roots, stems, and leaves may grow. Some common plants grown through stem cuttings include coleus, geranium, and hydrangea (see Figure 10.37).

Leaf cuttings can also be used to propagate some plants. Since there is meristematic tissue at the base of the leaf and around each leaf vein, a leaf has the potential to develop roots and shoots. Leaf cuttings can be propagated in the same conditions as stem cuttings, but they are often a little more difficult to get started. The African violet is an example of a plant propagated through leaf cuttings (see Figure 10.38).

plant growth regulators (plant growth reg•u•la•tors) another name for plant hormones

Cloning by Plant Tissue Culture

Plant tissue culture (PTC) is the most challenging of the asexual propagation methods because it requires sterile technique. In PTC, a sterilized sample of a few cells or a piece of plant are grown in a sterile media, such as agar. The cells of the sample must be given everything they need to survive, including sugar, vitamins, and the right concentration of hormones (see Figure 10.39). Under optimum conditions, the cells of a PTC may differentiate into shoots, leaves, and roots of a new plant in several weeks to a month.

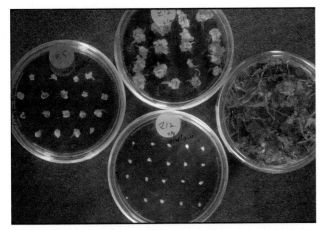

Figure 10.39. In a sterile dish, sterilized plant pieces are cut into sections (explants), as shown in the bottom plate, for culture in sterile PTC media. The pieces develop into callus tissue (top plate) and then into new roots and leaves (right plate). Until roots and stems are present, the explants will have to be given everything they need.
© Lowell Georgia/Getty Images.

In plants, like in animals, growth is regulated by the presence and concentration of molecules called hormones. Plant hormones (also called **plant growth regulators**) interact with other molecules in the nucleus, cytoplasm, or on membranes of specific cells and control an assortment of activities. Plant hormones regulate seed germination, as well as the growth of shoots, leaves, roots, and flowers. Although plant hormones are made in several plant organs, most hormones are produced at the root and shoot meristems (growing tips) and diffuse to other areas to modify growth. Depending on their relative concentrations, plant hormones cause root, shoot, or branch growth and must be present for propagation by plant tissue culture.

Two important groups of plant hormones are **auxin** and **cytokinin**. Auxin is produced primarily in shoot tips, where it causes cell elongation and leaf development. As auxin diffuses down to the root tip, it causes root tip elongation, but only when it is present in very low concentrations. Indoleacetic acid (IAA) is a naturally occurring auxin that regulates many processes, including lateral (side) branch growth. When the IAA concentration is high, as it is at a plant's apical meristem, lateral branch growth is inhibited. The impact of decreasing amounts of IAA on stem growth is seen in the conical shape of most evergreen trees.

Several companies produce synthetic auxins for commercial use. Used by both residential and commercial growers, a synthetic auxin, 1-naphthaleneacetamide, is the active ingredient in many commercial rooting compounds. Another well-known synthetic auxin is the herbicide 2,4-D. At one time, 2,4-D was the most widely used weed killer in the United States. Because 2,4-D causes cell division at such a rapid rate, some plants literally grow themselves to death. In the 1960s, concerns about the environmental impact and toxicity of 2,4-D led to studies whose results caused 2,4-D to be banned for most commercial purposes.

Cytokinins are hormones that cause cell division. A commercial version of cytokinin, called kinetin, is used to increase cell division in **explants**, the sterile plant tissue sections grown in PTC. Plants are very sensitive to individual hormone concentrations and to the concentration of one hormone in relation to another. The ratio of auxin to cytokinin is important in both stem and root elongation and in cell differentiation.

Several other plant hormones regulate processes in plants and may be used by scientists and farmers. **Ethylene** controls fruit ripening and leaf development. **Gibberellin** (gibberellic acid) promotes seed and leaf bud germination, stem elongation, and leaf development. **Abscisic acid** (ABA) regulates bud development and seed dormancy (see Figure 10.40). **Phytochrome**, which is actually a pigment that acts like a hormone, controls flowering.

Starting a Tissue Culture   The medium for a PTC contains a mixture of hormones at some optimal concentration, plus agar and other nutrients. The media is poured into test tubes and sterilized. The explants are placed on the media surface. Nutrients from the media diffuse into the plant tissue and trigger cellular processes. The first evidence of activity is usually the swelling of the explant, which is a result of a large number of cell divisions. The mass that results is called callus (see Figure 10.41).

auxin (aux•in) a plant hormone produced primarily in shoot tips that regulates cell elongation and leaf development

cytokinin (cy•to•ki•nin) a class of hormones that regulates plant cell division

explants (ex•plants) the sections or pieces of a plant that are grown in or on sterile plant tissue culture media

ethylene (eth•yl•ene) a plant hormone that regulates fruit ripening and leaf development

gibberellin (gib•ber•el•lin) a plant hormone that regulates seed germination, leaf bud germination, stem elongation, and leaf development; also known as gibberellic acid

abscisic acid (ABA) (ab•sci•sic ac•id) a plant hormone that regulates bud development and seed dormancy

phytochrome (phy•to•chrome) a pigment that acts like a hormone to control flowering

Figure 10.40.   Elizabeth Kohl, a Research Associate at Mendel Biological Solutions, measures the amount of the hormone, ABA, in plant extracts from thousands of samples from various transgenic plant lines. The amount of ABA in each extract indicates the degree to which the plants respond to stress compared with control plants. The instrument used is a gas chromatograph (GC)/mass spectrometer (MS). The GC portion heats the chemicals to a gaseous state. The rate at which the chemicals become a gas is characteristic of the compound. That information and the mass of the compound (determined by the MS) are used to identify and quantify such compounds as ABA.
Photo by author.

Figure 10.41.   Callus is a large mass of undifferentiated cells (white and brown mass) that results from the activation (increased cell division) of meristematic cells by plant hormones. If the hormone concentrations are correct, stems, leaves, and roots will develop shortly after the appearance of callus.
© Science Photo Library/Photo Researchers.

Figure 10.42. Shoots (stems) develop from callus if the hormone concentrations are correct. These plant clones, developed at Native Plants, Inc. in Salt Lake City, Utah, have been transformed to produce petroleum.
© Jonathan Blair/Getty Images.

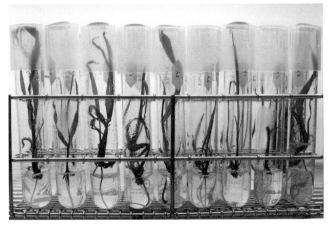

Figure 10.43. These young corn plant clones, with shoots and roots, are ready to be transplanted into sterile soil and covered with plastic to maintain high humidity.
© Lowell Georgia.

Figure 10.44. Once roots and shoots are present (top right), clones can be transplanted into a soil or soil-like medium (bottom right). The clones are genetic duplicates of the parents. These plants were started as explants rotating in liquid media (top left) to encourage callus development before shoot and root development.
© Jim Sugar.

By manipulating the concentrations of auxin and cytokinins, callus tissue can be induced to produce shoots, roots, or both. It has been found that if a high cytokinin-to-auxin ratio is maintained, certain cells will differentiate in the callus, giving rise to buds, stems, and leaves (see Figure 10.42). However, if the cytokinin-to-auxin ratio is lowered, root formation is favored. By selecting the proper ratio, the tissue may develop into all of the organs of a new plant.

Once shoots and roots are growing, it takes a few months for the clone to mature sufficiently with enough roots to be transplanted into soil or another medium. Until they are ready to be transplanted, PTC clones are kept in their culture tubes under fluorescent lights and at an appropriate temperature (see Figure 10.43). When transplanting the PTC clones, the humidity of the transplantation environment must be high enough to ensure that the clones do not dehydrate and die.

Advantages of Plant Tissue Culture Propagation   There are several advantages to PTC over traditional reproductive methods. Often, PTC produces more plantlets more quickly than other reproductive methods (see Figure 10.44). In addition, PTC produces clones of the parent with no unwanted variant offspring. Finally, some plants are difficult to grow by seed germination or stem or leaf cuttings, while PTC may easily propagate them.

Orchid propagation, for example, involves many challenges, including the fact that it takes 5 years to grow orchids from seed to flowers. One of the biggest obstacles in orchid propagation is that orchid seeds are so tiny (one million fit in a teaspoon) that they are difficult to germinate and manage. Like many other monocots (eg, irises, lilies, and grasses), orchids do not readily reproduce from leaf or stem cuttings. Therefore, advances in PTC have significantly improved the availability of orchids and other monocots.

Factors to Consider in Plant Tissue Culture Propagation

Several factors must be considered when conducting PTC:

- the species and variety of plant material
- the media and media ingredients
- the preparation of plant samples, media, and equipment (sterility and temperatures, etc)

The most important factor in PTC is sterility because bacteria and mold can easily overtake a culture if sterile technique is not followed. PTC must be done in a sterile environment, preferably a laminar flow hood (see Figure 10.45).

Figure 10.45. **This technician is handling an animal cell culture inside a sterile laminar flow hood. Animal cells in culture, like plant cells in culture, must be kept free of microorganism contamination.** Photo by author.

BIOTECH ONLINE

HEPA, a Heap of Filtering Power

High efficiency particulate air (HEPA) filters are found in the laminar flow hoods used for sterile cell and tissue culture.

Go to biotech.emcp.net/globalrph to learn how laminar flow hoods work and how HEPA filters are used in a laminar flow hood. In a short paragraph, summarize how laminar flow hoods work.

Find an additional website that answers the following two questions. Record the Web address you used as a reference and the answers to the questions.

1. What does the acronym HEPA stand for?
2. What are at least two other applications of HEPA filters?

Section 10.5   Review Questions

1. Which of the following are examples of asexual plant propagation: PTC, crossbreeding, stem cuttings, leaf cuttings, or runners?
2. Leaf or stem cuttings must include at least some of what kind of tissue to form new roots?
3. Auxin is responsible for what kind of plant growth regulation? Cytokinin is responsible for what kind of plant growth regulation?
4. How can a plant tissue culturist know that an explant is beginning to respond to the hormones in the PTC media?

Chapter Review

Speaking Biotech

Page numbers indicate where terms are first cited and defined.

abscisic acid (ABA), 331
alleles, 320
asexual plant propagation, 329
auxin, 331
average, 324
breeding, 308
callus, 311
Chi Square, 326
crossbreeding, 328
cuttings, 311
cytokinin, 331
degrees of freedom, 327
differentiation, 317
dihybrid cross, 323
diploid, 319
dominant, 320
embryo, 308

ethylene, 331
explants, 331
gametes, 309
germination, 316
gibberellin, 331
haploid, 319
herbaceous plants, 317
heterozygous, 320
homozygous dominant, 320
homozygous, 320
homozygous recessive, 320
mean, 324
meiosis, 309
meristematic tissue, 314
meristems, 315
mitosis, 316
monohybrid cross, 322

null hypothesis, 327
phytochrome, 331
plant growth regulators, 330
plant hormones, 317
plant tissue culture (PTC), 329
pollination, 307
polygenic, 320
Punnett Square, 322
radicle, 317
recessive, 320
runners, 329
selective breeding, 310
sexual reproduction, 308
standard deviation, 325
tissue culture, 311
woody plants, 317
zygote, 308

Summary

Concepts

- Plants propagate through sexual and asexual reproductive means. Sexual reproduction requires two parents (diploid) and the fusion of their sex cells (haploid) to produce a zygote (diploid), which will grow into the new plant (diploid). Asexual plant propagation requires only one parent (diploid), and it produces identical (diploid) offspring (clones). Some methods of asexual plant propagation include cuttings and tissue culture.
- Sexual reproduction controlled by scientists or farmers is called breeding. Selective breeding occurs when specific parent plants are chosen and bred. The breeder then transfers the pollen (pollination) from one plant to another.
- Predictions of the results of a cross can be made using a Punnett Square. If the parents' genotype can be discovered, possible gametes (sex cells) and the various ways they can fuse are predicted.
- Plants have organs and tissues. Recognizing them is important for agriculture, breeding, and asexual propagation. The presence of meristematic tissue in a sample is required for asexual propagation.
- Plant growth occurs due to mitosis within meristematic tissue. Meristems are found in all growing tips and in the vascular cambium. Mitosis creates cells with identical chromosome numbers; thus, 2N cells create 2N daughter cells. Cells created in meristems will specialize into other tissues. Plant hormones regulate mitosis and other aspects of growth.
- Plants go through an "alteration of generations," in a normal life cycle: from a diploid parent plant to haploid eggs and sperm, to a diploid zygote, which grows into a diploid embryo, and finally into a diploid parent plant. Each generation repeats this cycle.
- During a life cycle, the chromosomes carry genetic information as they pass from parent to gametes to zygote, embryo, and parent. The genes carried are the genotype, and the expression of the versions (alleles) of a gene in offspring is the phenotype.
- Many traits are represented by two different alleles of a gene. Because offspring inherit a chromosome each from their parents, they have two alleles for each gene.

- Some versions of alleles may be dominant, or expressed to a greater degree over another (recessive) allele. When expressing a genotype in writing, use capital letters to represent dominant alleles, for example, "H," and recessive alleles by small letters, such as "h." If offspring inherit the same alleles (eg, HH or hh), they are called homozygous. If they inherit different alleles, they are called heterozygous (Hh).
- Geneticists frequently analyze monohybrid (one trait) and dihybrid (two traits together) crosses by writing out a genetic-cross problem. When setting up a genetic-cross problem to predict the expected genotypes and phenotypes of offspring, the geneticist must determine the parents' and gametes' genotypes. Then, they use the genotypes to complete a Punnett Square and predict the likelihood of producing a particular type of phenotype in the offspring.
- The results of crosses and other experiments can be analyzed for accuracy and validity through averages and standard deviation, the 10% rule, and by Chi Square analysis.
- The standard deviation is a value that describes the range on either side of the mean (average) where data are considered valid. It describes how tightly the data are clustered. Smaller average deviations are expected around averages based on a large sample size. In most experiments, values outside two standard deviations are suspect.
- An χ^2 value is calculated to determine whether data that deviate from expected results are still within an acceptable range. During an χ^2 analysis, the sum of each experimental group's deviation is added together, and the χ^2 value is looked up on an χ^2 table. If the χ^2 value is less than the value at a probability of 0.05 ($P < 0.05$), then the value has a high probability of being the same as expected. The data are accepted as valid.
- Breeding introduces variety into offspring, and asexual plant propagation produces identical offspring, which are clones of the parent plant.
- Several methods of asexual plant propagation are used for agricultural and horticultural purposes, including cuttings, runners, and tissue culture. All require the inclusion of some meristematic tissue growing at a meristem, at a bud, or around the veins. Added plant hormones can encourage hard-to-propagate tissues to respond.
- Several hormones, or growth regulators, are important in plant growth and development: auxin regulates cell elongation, cytokinin regulates cell division, ethylene is important in fruit ripening and leaf development, gibberellins promote seed and bud germination as well as stem elongation, and phytochrome regulates flowering.
- In plant tissue culture, a small piece of plant tissue (an explant) is sterilized and placed on sterile PTC media containing all the vitamins, minerals, hormones, and organic nutrients needed to encourage cell division and elongation, until the plantlets can photosynthesize.
- Explant cells are sensitive to even small changes in hormone concentration. Individual hormone concentration and the relative concentrations of one hormone to another impact the promotion of stem, leaf, and root growth.

Lab Practices

- Flowers are the reproductive structures of most agricultural plants. Understanding flower structure helps explain their reproduction and evolution.
- The characteristics of flowers are used in the classification and identification of plants.
- Selective breeders choose pollen from plants with specific characteristics. They transfer the pollen to other plants with specific characteristics in the hopes of producing offspring with specific characteristics.
- Seeds are formed when sperm (pollen) fertilize eggs. Different seeds are specialized for different types of environments and dispersal.
- Seeds have cotyledons that contain endosperm, the seed's food source. These parts of the seed ensure its survival until the baby plant (embryo) inside the seed can photosynthesize.
- Seed germination can be studied using seed germination chambers, which allow observation of sprouting seeds and measurements of their growth. Seedling growth rates can be determined in centimeters per day.
- *Brassica rapa* (also called Wisconsin Fast Plants) seeds have been bred for fast germination and growth. They have a short generation time, with plants going from seed to flower to seed within about 8 weeks. These characteristics make it an ideal model organism for breeding experiments.

- Wisconsin Fast Plants (WFPs) have specific growth requirements, including a 24-hour light source and a need for constant water, fertilizer, and staking. Bee sticks are used for pollen transfer in many selective breeding experiments.
- Plants of known genotype can be crossed with other plants of known genotype, and predictions can be made as to the genotypes and phenotypes of the offspring. This is valuable for selective breeders who are attempting to produce specific offspring for a population of parents.
- Before conducting a cross, breeders show the predictions of the expected results of the cross, using a Punnett Square.
- The results of a breeding experiment may be analyzed in several ways. A common method is an χ^2 analysis, using the equation $\chi^2 = \Sigma[(O\text{-}E)^2/E]$, which examines the deviation from the expected for all of the actual phenotypic groups. Once the χ^2 value has been calculated, it can be checked on a chart to see if it falls within the acceptable probability (P) values for the amount of deviation expected from the type of cross being conducted.
- Some plants can be easily cloned through asexual propagation techniques, such as stem and leaf cuttings or runners. Not all species propagate equally well with each technique and should be evaluated independently.
- Cuttings require some exposed meristematic tissue. Maintaining appropriate watering is important so that cuttings do not dehydrate or rot (called fungal damping off).
- Asexual propagation yields can be improved by the addition of plant growth regulators (hormones) at the right concentration. Concentration assays can be developed to determine the optimum concentration for root growth in stem cuttings.
- A plant tissue culture (PTC) can be used to propagate small pieces of plants or seeds. A PTC requires perfect sterile technique and the use of laminar flow hoods. Plant tissue media must be prepared with the right concentration of hormones and nutrients. The PTC media must be autoclaved for just the right amount of time for sterilization but not so long as to degrade the temperature-sensitive hormones.
- Samples for PTC (explants) must contain at least some meristematic tissue and the right concentrations of hormones to produce callus (the undifferentiated tissue that grows new plantlets). It takes approximately 4 weeks to see callus formation on explants.
- Shoots develop first from African violet explant calluses. Then, upon transferring to new media, roots are encouraged to grow. Plantlets with substantial roots and stems may then be transferred to potting media in high-humidity growth chambers. Dozens of plantlets can arise from a single explant.

Thinking Like a Biotechnician

1. It takes approximately 5 years for an orchid to breed, produce seeds, grow into the offspring plants, and flower. It takes about 2 years for an orchid tissue culture to grow from a protocorm (an undifferentiated mass of cells) until it matures and flowers. Why would plant scientists at an orchid company breed plants instead of propagating through stem cuttings?
2. Why is meristematic tissue needed for tissue culture and other asexual propagations?
3. What are the changes that one can expect to observe as a plant tissue culture develops?
4. Why must meiosis occur to produce plant sex cells? Use the terms "diploid" and "haploid" in your answer.
5. A horticulturist has a supply of several different plant growth regulators (hormones), including zeatin (a cytokinin), IAA (an auxin), ethylene, gibberellin, and ABA. Which of the growth regulators would be a logical choice to test for its potential to do the following?
   a. Produce an ivy plant with extra large leaves.
   b. Produce corn seeds that germinate after only a 4-week dormancy period.
   c. Produce an apple tree bearing fruit that ripens in half the normal time.
6. A sample of cellulase is used to release juice from apple cells. After 50 trials, the average amount of juice produced by a 1000-µL volume of cellulase in a 50-mL volume of crushed apples is 7.8 mL. One sample has a juice extraction value of 7.2 mL. Using the 10% rule, is the sample's value valid and acceptable?

7. An experiment is conducted to determine how the amount of the plant hormone, indoleacetic acid (IAA), affects bean-plant stem growth. Untreated bean stems grow to an average height of 100 cm. When 1 mL of 20 ng/mL of IAA was sprayed on bean plants, 120 plants grew an average of 115 cm with a SD of 17 cm. Two plant samples gave readings that looked questionable to the technician. One had a value of 125 cm and the other had a value of 93 cm. Should these two plants' growth data be considered valid? Why or why not?

8. Consider a cross between two parent plants carrying genes for purple stems. Purple (P) is dominant to green (p) stems. One parent is heterozygous for the trait and one is homozygous dominant. Show the cross (problem), including the chances of this breeding giving green-stemmed plants.

9. Consider a cross between two parent plants carrying genes for purple stems and yellow flowers. Purple (P) is dominant to green (p) stems. Yellow (Y) is dominant to white (y) flowers. One parent is heterozygous for purple stems but homozygous recessive for yellow flowers. The other parent is homozygous recessive for purple stems but heterozygous for flower color. Show the cross (problem), including the chances of this breeding producing green-stemmed, white-flowered plants.

10. A cross of plants with the genotypes given in problem No. 9 results in 200 offspring plants as follows:

> 51 have purple stems and yellow flowers
> 53 have purple stems and white flowers
> 41 have green stems and yellow flowers
> 55 have green stems and white flowers

Conduct a Chi Square analysis of the cross and determine whether the results show that a valid dihybrid cross was conducted.

Biotech Live

Plant Biotech Companies: Who are they? What do they do?

Activity 10.1

Plant biotechnology companies represent a large segment of the biotechnology industry. According to a report by Dr. C. Ford Runge, "The Economic Status and Performance of Plant Biotechnology", six large plant biotechnology companies spent over $2.7 billion dollars on R&D in 2002.

Some plant biotech companies continue to practice the traditional methods used by plant breeders, food and seed producers, florists, and foresters. Several large, well established companies, though, such as Monsanto Co., as well as some small companies, such as Mendel Biotechnology, Inc. are applying the advances developed through rDNA technology and pharmaceutical production to plants. The work of these companies has resulted in a large number of new agricultural and pharmaceutical plant biotechnology products.

Several states that have had a agriculture-based economy have become an incubator for plant biotechnology, supporting companies that want to focus on agricultural biotechnology. North Carolina is an good example with over 2 dozen plant biotechnology companies.

A lab technician monitors the growth of a sunflower plant, raised from an embryo in controlled conditions, as a genetic engineering project at Sungene Technologies, a BASF Plant Science Company.
© Lowell Georgia/ Getty Images.

Find a plant biotechnology company in your state by using the "By State" lists of companies at thelabrat.com/jobs/companies/BiotechPharmaUnitedStates.shtml. If your state does not have a plant biotechnology company, visit the NC Biotech Crop and Plant Biotechnology page at directory.ncbiotech.org/taxonomy/term/393 to find a list of companies in North Carolina with a focus in agricultural biotech. Pick a company and learn enough about it to produce a Microsoft® PowerPoint® presentation. Record the answers and Web-site references to the following questions, and create a five-slide presentation to share your answers. Make sure there is at least one graphic on each slide.

- What are the main objectives or goals of the company?
- Where is the company located? Is it a relatively large company (more than 1000 employees) or a smaller company? Briefly describe the company. Include information in which a potential employee might have interest.
- What are some of the company's marketed plant products (or those in production) and their applications or uses? Identify two or three important products.
- Using **biotech.emcp.net/finance**, check to see if the company is publicly traded. Find the company stock trading symbol by clicking on "Symbol Lookup." Then, using the trading symbol, find the price per share of the company's stock. Print a 3-month chart of the company's stock performance. If you cannot find a company or its symbol, it may be privately owned, and shares may not be available to the public. Report that in your presentation.
- Click on the "News" link and find a recent article about the company or its products. If the company is not publicly traded, use Google.com to search for a news article. Summarize the article in one paragraph. Also, explain how the article might affect stock prices or the company's ability to conduct business.

The Economics of Plant Biotechnology

Show the economic impact of the traditional plant biotechnology industry.

Create five concise, easy-to-read Microsoft® PowerPoint® slides (or one poster) addressing the following items:

1. Using pictures and text, explain what plant biotechnology is and the industry's products and target markets.
2. Using Internet resources, find the annual revenues for plant-based industries such as florists, nurseries, food crops, livestock feed, timber, fiber, etc, for some year in the past decade. Create a graphic or chart that shows the economic impact of these plant biotechnology sectors.
3. Create a graph, chart, or table that illustrates the job opportunities in one of the above plant-based industries.
4. Search Monsanto's website to find the company's annual revenues in their annual report. Construct a graph to show the percentage of the annual revenue for each of the different divisions of the company.
5. On the fifth slide, list what references (URLs) were used for each of the other slides.

Activity 10.3 Agricultural Biotech: How far has it come?

"Modern" biotechnology began in the 1970s with the creation of the first human-made rDNA molecule; however, because of the complexity of plant and animal cells, agricultural genetic engineering has taken many years to be implemented. Learn how far agricultural biotechnology has come by accessing the following websites: **biotech.emcp.net/USDAnews** and **www.fb.org**.

Make a timeline to show the major events in recent agricultural biotechnology history.

1. Using an 80-cm length of adding-machine tape, make an "Agricultural Biotechnology Timeline" by marking every centimeter along the bottom of the tape. Each centimeter should represent 6 months. Start at the left end and mark the first centimeter, 1970 AD. Label every 10 cm. The most current years should be at the right-hand side of the timeline.
2. Add to your timeline all of the major events on the website. Insert other events to the timeline as you or your group (if assigned) "discover" them.
3. All events should be clearly labeled, and artwork should be added to make your timeline a more interesting, effective teaching tool. Add a minimum of three other agricultural biotechnology events.
4. Record your name and the date on the back of the timeline, along with the Web-site

Wisconsin Fast Plants... How Fast Is Fast?

Use the Internet to gather background information about the life cycle of a "model organism," *Brassica rapa*.

Find an article on the Internet that describes the major events in the life cycle of *Brassica rapa*. A circular, life-cycle diagram is preferred. Print the first page of the Internet article, any diagrams of interest, and the website's URL. Use the article to fill in the timeline chart below.

The Role of Plant Hormones in Plant Growth and Regulation

Time (days)	Description of Wisconsin Fast Plant's Life-Cycle Events
0	
2	
5	
7	
10	
13	
15	
20	
23	
36	
56	

By definition, a hormone is produced in one area and then moves to another target area to become active. Different plant hormones (also called growth regulators) are made in different parts of the plant and are responsible for different activities, depending on their location and concentration. The action of each plant hormone is due to its absolute concentration and its relative concentration to other hormones.

Demonstrate how and where plant hormones regulate plant physiology.

1. Using a diagram of a plant found on the Internet as a guide, draw the body of a "typical" flowering plant on 11 in × 17 in paper. Include all of the plant organs and tissues listed below, and leave room for labeling and explanations.
    a. shoot (stem) with apical meristem, nodes and internodes, branches, leaves, and lateral buds
    b. primary root (taproot), root meristem, root hairs, lateral (branch) roots, and root nodules
    c. flowers, flower buds, and fruits with seeds
2. Show the expected location of the following plant hormones in your plant by adding them, in different colors, to your drawing. Show the hormones as dots. Show the relative concentrations of each hormone at each location by varying the density (number of dots per unit area) of dots. Use Internet resources to find the location and action of these plant growth regulators.

    auxin = brown   cytokinin = blue
    ethylene = purple   abscisic acid = yellow
    phytochrome (a flowering pigment) = orange   gibberellins = pink

Activity 10.6 Methods of Commercial Plant Propagation

Although many methods exist for propagating plants, some are more practical than others. For example, orchids can be propagated using seeds, but orchid growers prefer to clone orchids using tissue culture. This is because orchid seeds are so tiny (1 million of them will fit into a teaspoon) that it is difficult to work with them. In addition, seeds are the result of sexual reproduction, and the offspring vary from the parents. Using tissue culture, growers can reproduce exact clones of valuable orchids.

Another example is the raspberry. Raspberry growers prefer to grow new plants from stem cuttings. Where it is difficult to get raspberry seeds to germinate, stem cuttings, with a small amount of rooting hormones added, will root to produce new plantlets in just a few weeks. It is faster and more economical to propagate raspberries in this way.

Learn the primary methods of commercial propagation of some common agricultural plants.

1. Use Internet access, as well as gardening and botany books, to discover the methods of propagation for the following agricultural products and related varieties:

navel oranges	broccoflower
cabbage	daffodils
strawberries	seedless watermelon
roses, such as the American Beauty Rose	grapes, such as wine grapes
apple trees, such as Fuji or MacIntosh	tangelos
tomatoes, such as Sweet 100 or Early Girl	bananas
garlic	asparagus
lettuce, such as romaine, oak leaf, or red leaf	berries, such as blackberries
conifer trees, such as Douglas fir	onions, such as Walla Walla sweets
potatoes, such as russet or Yukon Gold	

2. Produce a file folder of reference materials gathered while researching. Include a bibliographical entry at the top of each page. Include a table of contents at the front of the folder.

3. Create a poster that illustrates the principal methods by which each plant is produced commercially. Next to each entry, include the bibliographical references you used for your information. Make sure that what you write is complete, makes sense, and is in *your own words*. Label anything that is not completely obvious. On the poster include the following:

   - Common name of the plant (eg, baby blue eyes)
   - Scientific name of the plant (eg, *Nemopila insignis*)
   - Three or more labeled photos (citing source) or diagrams of the plant
   - *Detailed* descriptions (including diagrams/drawings of the plant's propagation methods, both sexual and asexual reproduction)
   - Interesting facts about the plant or the plant's propagation
   - General references/bibliography

   Make sure the poster is interesting to look at and effective as a teaching tool.

4. Be prepared to present a 5-minute explanation to your colleagues (the class) covering the types of propagation used to commercially produce "your" plant.

Alternative: Microsoft® PowerPoint® Presentation (seven to 10 slides)

Slide 1	Title page
Slide 2	Table of contents
Slides 3 and 4	General information on the agricultural product with photos/diagram
Slides 5 and 6	Description of propagation methods
Slide 7	References

Note: Evaluation includes the presenter's presentation style. (Do not read from posters or computers.)

Seed germination
Photo by author.

Bioethics

Alien DNA in Your Food?

Many food crops contain DNA that has been genetically engineered into them from other species. Some varieties of corn, for example, have had genes from certain bacteria, fungi, or insects inserted into them. Some new varieties of corn contain foreign DNA that codes for proteins that act as pesticides. When attacked by certain insects, the new corn proteins kill the corn pests. Do you know whether or not you are eating corn that produces these proteins? Do you care?

Many people *do* care. There has been considerable outcry against GMOs. The GMO controversy regularly appears in the news as factions of the public demand public debate.

Europe has seen many public demonstrations against genetically engineered food crops. The politicians in several European nations have passed laws requiring labeling on food products containing GMOs. In the United States, there has been publicity on the anti-GMO movement, but the public as a whole has not shown much concern about GMOs and foods containing them.

Do you have a pro-GMO, anti-GMO, or neutral position?

Before beginning the following activity, write a paragraph that 1) defines the term "genetically modified food"; and 2) describes your "gut reaction" to eating genetically engineered food. Does it bother you? Yes or no? Why or why not?

1. Using Internet resources, find 10 food crops (plant, animal, fungi, or bacteria) that have been genetically modified through the addition of DNA.
2. For each of the 10 food crops, list three food products that could be manufactured using each of these GMOs. Think about products that you can buy in a grocery store.
3. Decide which foods you would not eat if you knew that they were genetically engineered. Are there any foods that you are willing to give up? Are there any that you are not willing to give up?
4. Organize your information and decisions on a chart.
5. Discuss why genetic enhancement of food crops is so controversial.

Biotech Careers

Photo provided by Bob Creelman, PhD.

Plant Biologist

**Bob Creelman, PhD, Senior Scientist
Mendel Biological Solutions,
Koch Agronomic Services, LLC,
Hayward, CA**

Dr. Creelman is interested in understanding how plants respond and adapt to changes in their environment, such as drought or cold. Many studies have demonstrated that changes in gene expression as well as the levels of several metabolites (small molecules) are associated with the adaption to various stressful conditions. For example, several metabolites accumulate during drought conditions and may help protect the plant from wilting. In the photo, Dr. Creelman is preparing to use a Gilson 215 Liquid Handler robot to help purify some metabolites from plants. The use of robots makes sample preparation more efficient, improves precision and accuracy, and reduces the potential for errors. Scientists at Mendel Biological Solutions study physiological mechanisms in several species of plants. One of the most popular, due to its rapid growth rate and short life cycle, is the model organism ***Arabidopsis thaliana*** (**Arabidopsis**). Arabidopsis is often the organism of choice for breeding, cloning and genetic experiments because its entire genetic code is known.

***Arabidopsis thaliana
(A. thaliana) (a•rab•i•dop•sis
thal•i•an•a)*** an herbaceous plant, related to radishes, that serves as a model organism for many plant genetic engineering studies

11 Agricultural Biotechnology

Learning Outcomes

- Define and contrast the terms agriculture and agricultural biotechnology
- Give specific examples of agricultural and horticultural biotechnology applications, including genetically modified organism (GMO) crops, hydroponics, and plant-made pharmaceuticals
- Explain how genomic and plasmid DNA can be isolated from cells, including the additional steps required for plant cell DNA isolation
- Discuss how proteins of interest may be purified from plant samples and how DNA or protein samples may be assayed for their concentration and purity
- Describe the role that *Agrobacterium tumefaciens* plays in producing genetically modified plant crops
- Summarize the methods used to produce transgenic plants, and explain the selection processes for identifying transformed plant cells
- Describe the role of biotechnologies in food production, food processing, and food security

11.1 Applications of Biotechnology in Agriculture and Horticulture

Agriculture is the practice of growing and harvesting animal (livestock) or plant crops for food, fuel, fibers, and other useful products.

The word "agriculture" makes one think of farms. In the past few decades, the term "farm" has come to mean different things to different people. Some farms are small and run by a single family. Imagine a few acres, with a barn, a tractor, and a dozen chickens. During the past few decades, however, many farms have been bought by large agribusinesses. Many of these farms are hundreds or even thousands of acres in area and are operated by hundreds of employees. The term "farm" is also applied to other commercial operations that produce crops. Tree farms are acres of forests where timber trees are grown, cut down, and replanted. Wood from timber is processed into building materials and all kinds of paper. Fisheries, or fish farms, grow and harvest fish in a regulated environment. Salmon and trout farms are common now. Many people also consider ranches of cattle, sheep, goats, and pigs to be farms.

Agriculture includes all the practices of growing and harvesting crops, including soil management, water management, plant and animal breeding and hybridization, asexual plant propagation, seed production and improvement, and crop management. It also includes the use of fertilizers, herbicides, insecticides, and pesticides, as well as the use and improvement of farming tools and equipment. In the past few decades, new biotechnology techniques have been applied to improve the quantity and quality of agricultural products. These techniques include genetic testing, plant tissue culture, DNA manipulation and gene transfer, protein manipulations, genetic engineering, and plant and animal cloning.

BIOTECH ONLINE

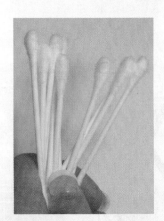

Cotton Swabs.
Photo by author.

A Picture of Crop Production in the US

In the United States, corn, soybean, wheat, and cotton are grown more than any other crops. This may not surprise you when you consider their importance in the human food, livestock feed, and textile industries. But just how much land is used to grow these crops?

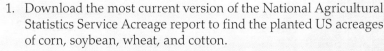

Go to biotech.emcp.net/mannlib.

TO DO

1. Download the most current version of the National Agricultural Statistics Service Acreage report to find the planted US acreages of corn, soybean, wheat, and cotton.

2. Create a pie graph that shows the amount of each crop planted (in millions of acres) reported in the document. Report the values as a percent of the total as well.

3. Within each section of the graph, list a state (in green) where production of each crop has increased. Specify the time period covered.

4. Within each section of the graph, list a state (in red) where production of each crop has decreased. Specify the time period covered.

5. Lastly, go to biotech.emcp.net/allaboutfeed and list three trends in crop production that are expected in this decade.

Figure 11.1. Genetic eye disorders are common in King Charles Cavalier spaniels. These may include microphthalmia, an inherited defect that is particularly common in the King Charles Cavalier. In the case of microphthalmia, one or both of the dog's eyes is smaller than normal, resulting in restricted vision and possible blindness.
Photo by author.

Farmers have a long history of breeding plants and animals with certain desired characteristics in an attempt to produce offspring with very specific traits. In the breeding process, one parent contributes tens of thousands of genes to another parent, who also contributes tens of thousands of genes. Sometimes the offspring get the desired genes and express them to give the desired phenotypes; in the case of agriculture, maybe the result is a new variety of tomato that ripens faster.

More often, though, selective breeding results in offspring that do not possess the desired traits, or offspring receive additional genes that produce undesirable traits. This is due to the fact that during sexual reproduction, random crossing-over of chromosomes introduces new combinations of genes that are passed to the offspring.

It is difficult, though, to predict the negative consequences of this kind of genetic mixing. During meiosis, new gene combinations result, so, often, "undesirable genes" are carried along with desirable ones to the next generation. If this occurs, the offspring may express undesirable phenotypes along with the desired ones. This is especially common if the bred organisms are closely related (**inbreeding**) since a certain undesirable gene may be more frequent among relatives. An example of the chronic disorders or problems that may result from inbreeding is seen in dogs (see Figure 11.1). German shepherds, for instance, are known to have hip problems (hip dysplasia) because of selective inbreeding for a particular desired posture. The randomness of recombination during breeding limits how quickly and efficiently new gene combinations result in new varieties with specific and desired characteristics.

According to the USDA, agricultural biotechnology is a range of tools, including traditional breeding techniques, which alter living organisms or parts of organisms to make or modify products, improve plants or animals, or develop microorganisms for specific agricultural uses. Agricultural biotechnology includes the molecular biology tools of genetic engineering, gene manipulation, and protein chemistry. Using gene transfer technologies, biotechnologists can add or modify specific genes to produce both different proteins and offspring with predicable, desired traits with less risk of unwanted characteristics from a wide range of donor organisms.

Using gene modification, agriculturists have developed plants that can be grown with fewer applications of chemicals or pesticides. Since overuse of chemicals and pesticides is a growing concern, plants that can grow well with little or no pesticide treatment, have considerable economic and ecological interest. Examples include Bt corn, Bt soybeans, and Bt potatoes (see Figure 11.2). Bt crops are resistant to several pests since they produce a compound toxic to insects (an insecticide). The compound (a protein called Bt Cry1Ab delta endotoxin) was originally made in a specific species of bacteria, ***Bacillus thuringiensis (B. thuringiensis,* or Bt)**. By inserting the Bt gene from bacteria into these plants, the plants become protected, to some degree, from insect damage. This kind of genetic engineering of food crops decreases the amount of pesticide sprayed on large acreages of crops. Decreased pesticide spraying results in less runoff into forests and streams.

The use of Bt products and other genetically engineered crops though is not without controversy. Several public interest groups are actively opposed to the growing and use of agricultural GMO products. They claim that adding these types of genes to our food crops puts helpful insects at risk and may produce proteins that are allergenic to humans. Still, according to Genetically Engineered Crops (USDA 2014), in 2013 about one half of all crop land in the U.S. was planted in genetically engineered corn, cotton, and soybeans. Genetically engineered crops are attractive to farmers because they reduce insecticide use, enable the use of less toxic herbicides, and save time and money.

Other agricultural biotechnologies have improved our food supply. Specific selective animal breeding, along with DNA fingerprinting and gene introduction, has produced livestock with improved nutritional value (eg. pork with less fat), fewer feed additives (eg. cattle), or increased growth rate (eg. AquAdvantage® Salmon). The same techniques are used to produce plants. For example, some are resistant to selected viruses (eg. papaya and banana), some have a higher nutrient content (eg. "Golden Rice" with higher levels of a precursor of vitamin A), and some require less fertilizer (eg. nitrogen-fixing plants) or herbicide (eg. Roundup Ready®). Visit the United Nations website at **www.fao.org/english/newsroom/focus/2003/gmo7.htm** to consider the benefits of using GMOs in agriculture.

An area of agriculture that has a large economic impact is **horticulture**. Horticulture is the practice of growing plants for ornamental purposes and includes plants grown for indoor decoration, landscaping, and fruit, vegetable, and flower gardens. Horticulturists use the same technologies as other agriculturists, including breeding, asexual plant propagation, and gene manipulation. The field of horticulture has employment opportunities in landscape design and maintenance of yards, parks, public gardens, and golf and sports turf. Horticultural scientists may work in applied plant research at nurseries, farms, or wineries focusing on increasing yield or variety or on combating plant disease (plant pathology). Greenhouses and nurseries have employment in management, sales, and marketing.

Figure 11.2. **Here is an example of corn borer larvae damage to non-Bt corn. Although Bt products decrease the amount of insecticide that has to be sprayed on corn in the field, Bt products have raised concerns among some health advocates and environmentalists.**
© Custom Medical Stock Photo.

inbreeding (in•breed•ing) the breeding of closely related organisms

***Bacillus thuringiensis (B. thuringiensis,* or Bt)** (Ba•cill•us thur•in•gi•en•sis) the bacterium from which the Bt gene was originally isolated; the Bt gene codes for the production of a compound that is toxic to insects

horticulture (hor•ti•cul•ture) the practice of growing plants for ornamental purposes

Figure 11.3. Here, basil plants are being grown hydroponically in water with no soil. The plants must be supported so that they do not fall into the water. The solution has to be aerated because roots need air or they will rot.
© Joseph Sohm; ChromoSohm Inc.

Hydroponics

Growing plants in a soilless, water-based medium is known as **hydroponics** (see Figure 11.3). Although hydroponics has been practiced for several thousands of years, it is gaining in popularity since in many areas of the world space is limited or soil is unavailable or unsuitable to support plant agriculture. In countries or regions with little or poor soil, hydroponics offers a good alternative method of growing several agricultural or horticultural crops. Several countries are investing in hydroponic crop production, including Saudi Arabia, Israel, Australia, and Japan.

At its simplest, hydroponics is growing a plant in water with the minerals needed to support plant growth. In practice, plants are suspended in a hydroponic tank with their roots submerged in water containing very specific concentrations of an assortment of **macronutrients** (nitrogen, phosphorus, potassium, calcium, sulfur, and magnesium) and **micronutrients** (iron, zinc, boron, molybdenum, etc) necessary for plant growth. Air is pumped into the hydroponic solution to prevent rotting. By manipulating the hydroponic environment, agriculturists can customize crop growth conditions.

BIOTECH ONLINE

Figure 11.4. Hydroponic Growth Chamber. Plant physiologist checks radishes (*Raphanus sativus*) grown using hydroponic techniques in a plant growth chamber at the Space Life Sciences Lab, Kennedy Space Center, USA. Scientists are researching plant growth under different conditions including types of light, different carbon dioxide concentrations and different temperatures. Such research and technology will be used to develop methods for growing fresh vegetables in space, to add more variety and nutrition to the heavily processed heat stabilized diets astronauts receive during space missions. Such technology would be especially important for the multi-year journeys necessary to take people to Mars.
© Science Photo Library/Photo Researchers.

Hydroponics "Out of This World"

Since some plants can be grown hydroponically, suspended over a nutrient solution without soil, hydroponics is perfect for growing plants in space (See Figure 11.4). The National Aeronautics and Space Administration (NASA) has supported several experiments to improve hydroponics, both on earth, and in orbit.

Go to biotech.emcp.net/CELSS and click on the link "Growing Crops in a CELSS."

In your notebook, describe what CELSS is, discuss the issues that must be addressed to grow crops in a CELSS, and list the advantages and disadvantages of hydroponics in a space CELSS.

hydroponics (hy•dro•pon•ics) the practice of growing plants in a soilless, water-based medium

macronutrients (ma•cro•nu•tri•ents) the minerals required by plants in high concentrations

micronutrients (mi•cro•nu•tri•ents) the minerals required by plants in low concentrations

Section 11.1   Review Questions

1. What is it called when very closely related animals are bred? Why is breeding closely related animals discouraged?
2. List several employment opportunities for people interested in the science or business of horticulture.
3. Name two advantages of growing plants hydroponically.

11.2 Advances in Agriculture through DNA Technology

Once bacteria cells had been genetically modified to produce nonnative proteins in the 1970s, plant scientists attempted similar techniques (recombinant DNA production and transformation) to introduce foreign DNA into plants to modify their phenotypes. As in medicine and industry, agricultural scientists began developing the processes necessary to genetically engineer agricultural products on the molecular level. Plant crop scientists, in particular, have produced several new and unique crops using DNA technology (see Table 11.1).

Table 11.1. **Just a Few of the Recombinant Plant Crops on the Market**

Crop	DNA Added to Produce a Particular Trait	Source of Transferred Genes
cotton	resists damage to sprayed herbicide	tobacco, bacteria
papaya	resists papaya ringspot virus	bacteria, virus
potato	expresses the Bt toxin to inhibit insect pests	bacteria, virus
tomato	alters ripening to enhance market appeal	bacteria, virus

Source: Nova Online at: http://biotech.emcp.net/novaharvest

As a first step in plant DNA modification, scientists had to develop methods of isolating plant **genomic DNA (gDNA)**. The gDNA is the chromosomal DNA of a cell. Each chromosome is composed of one very long DNA molecule. Therefore, if a cell contains several chromosomes, it contains several DNA molecules. All the chromosomes together from a cell are its genome.

There are 46 chromosomes in each human cell, therefore, 46 DNA molecules per cell, and 46 chromosomes in the human genome. On the other hand, a bacterial cell has only one long DNA molecule in its genome. Onion cells contain 16 chromosomes per cell; so, 16 DNA molecules comprise an onion's genome. The amount of a single cell's gDNA varies, depending on the organism, from 30,000 bp (bacterial genome) to over 3 billion bp in some animal or plant genomes.

genomic DNA (gDNA) (ge•no•mic DNA) the chromosomal DNA of a cell

Isolating Genomic DNA

Genomic DNA can be isolated from most cells in a straightforward fashion. First, the cells must be burst open. A cell lysis buffer is usually used to burst open the source cells. As the cells explode, all of the molecules in the cell are released into the sample tube. Since the majority of molecules in a cell are proteins, the next step is to use a protein-degrading buffer to denature the protein contaminants. By spinning the sample, the protein fragments are precipitated and removed from solution. Finally, the ribonucleic acid (RNA) is destroyed using RNase, and the remaining DNA is precipitated out of solution with ethanol and/or isopropanol, often using a DNA binding column and one or more a centrifugation steps. Since gDNA is very long, if handled properly, it may even be spooled onto glass rods. The purity and concentration of extracted DNA in a sample may be determined using UV spectrophotometry.

It was quickly learned, though, that extracting DNA from plants is significantly more challenging than working with bacteria or animal cells. This is largely due to pectin-impregnated cellulose fibers in the cell wall surrounding plant cells. The age and degree of cell wall development determines the amount of cellulose and pectin present and affects the ability to extract DNA from many plant cells. To improve DNA extraction yield, several protocols for plant cell DNA extraction have additional steps to remove the pectin and cellulose barrier. For example, freezing the plant tissue with liquid nitrogen

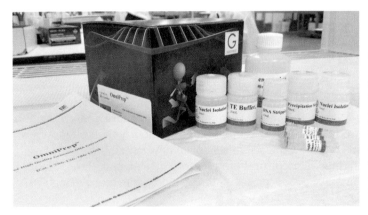

Figure 11.5. A genomic DNA isolation kit, such as this one from G-Biosciences, Inc., contains premade buffers for gDNA isolation. The buffers could be made from scratch, but it is often easier and more cost efficient to purchase a kit.

Photo by Colin Heath.

Agrobacterium tumefaciens (A. tumefaciens) (ag•ro•bac•ter•i•um tum•e•fac•i•ens) a bacterium that transfers the "Ti plasmid" to certain plant species, resulting in a plant disease called crown gall; used in plant genetic engineering

Ti plasmid (Ti plas•mid) a plasmid found in *Agrobacterium tumefaciens* that is used to carry genes into plants, with the goal that the recipient plants will gain new phenotypes

Figure 11.6. The tumor-inducing gene carried on the Ti plasmid causes crown gall tumors, such as the growths on the stem of this plant. The Ti plasmid may be transferred to a plant by the bacterium, *A. tumefaciens*. Ti plasmids may be isolated from *A. tumefaciens*, spliced open, and one or more genes of interest, from other sources, inserted (rDNA). For rDNA and transformation purposes, the tumor-inducing gene must first be removed.

© Geoff Kidd/Science Photo Library.

apparently ruptures the cell wall, releasing the cell's contents. A chloroform/isoamyl alcohol mixture is also commonly used to remove the cell wall, carbohydrates, and other impurities from the plant DNA. Several companies have developed gDNA extraction kits to improve the speed and yield of gDNA isolation (see Figure 11.5).

Inserting Foreign DNA Into Plant Cells To modify the genetic information in a plant to give it new characteristics, small sections of foreign DNA must be added to one or more cells, and then those cells must be grown or cultured into an entire plant.

For bacterial transformations, plasmids are used as vectors to carry genes of interest into host cells. Isolated plasmids up to many thousand base pairs in length are easily inserted into competent bacteria cells. Unlike bacteria cells, plant cells will not take up naked plasmids. Fortunately, plant scientists have discovered a particular species of bacteria that lives in plants and is able to transfer plasmids from its own bacterial cells into the plant cells. In essence, these bacteria conduct plant transformations.

The bacterium, **Agrobacterium tumefaciens** (A. tumefaciens), contains a plasmid (called the **Ti plasmid**) responsible for the transformation process. In nature, *A. tumefaciens* and the Ti plasmid cause crown gall, a plant disease (see Figure 11.6). Recently, scientists have learned how to grow *A. tumefaciens,* and how to isolate and manipulate the Ti pDNA for use in plant gene transfers. The Ti plasmid, like other plasmids, can be cut open and genes of interest inserted. The recombinant Ti plasmid can be returned to *A. tumefaciens,* which will then transfer the recombinant plasmid, or pieces of it, into target plant cells. If the target plant begins expressing the transferred genes, it is considered genetically engineered. Biotechnicians interested in plant transformation must learn to purify and manipulate Ti plasmids using plasmid preps (See Chapter 8), as well as how to grow *A. tumefaciens.* The next section discusses *A. tumefaciens* and the Ti plasmid in more detail.

Another technique to get pieces of DNA into plant cells is the use of a "gene gun." A gene gun is an apparatus that takes a plastic bullet covered with DNA-coated particles of gold or tungsten and blasts them into plant tissue, such as a leaf disk. As the bullet penetrates the tissue, some of the DNA-coated particles penetrate some cells although many cells don't survive. Some of the surviving cells incorporate the DNA sections into the plant genome and start expressing the genes. These cells can be grown by tissue culture into transformed plants. Particle gene gun transformation has successfully transformed some grains, including varieties of wheat and rice.

Several recombinant plant crops (genetically modified organisms or GMOs) have been produced with *A. tumefaciens*/Ti plasmid technology or using a gene gun (see Table 11.1).

How Much Do You KNOW about GMOs?

Go to **biotech.emcp.net/GMOfoods** and take the quiz to test your knowledge about foods made from GMOs. Record your answers on paper.

Do you know enough about GMO foods to take a pro or con position on the growth and distribution of these products in your county, state, country, or worldwide?

Section 11.2   Review Questions

1. Which is larger, gDNA or pDNA, and by how much?
2. Plant DNA is difficult to get out of plant cells. List a few "tricks" used by biotechnicians to isolate plant DNA.
3. Why is the bacterium, *A. tumefaciens,* of interest to biotechnologists?
4. Why is Ti plasmid of interest to biotechnologists?

11.3 Plant Proteins as Agricultural Products

Plant proteins are interesting for many pharmaceutical and agricultural reasons. Many plants contain proteins of commercial value, and recently scientists have learned how to engineer plants to make new proteins that confer desired qualities in crops. Scientists have even learned how to engineer plants to make human proteins for pharmaceutical use.

Plant Characteristics

Each plant has specific characteristics. These define its phenotype, the observable expression of the genes it received from its parents' genes, or genotype. A soybean plant, for example, produces seeds in pods. The seeds and pods are usually green. The leaves of the soybean plant are also green and smooth. The soybean leaves alternate up the stem. Soybeans grow best in warm climates and are sensitive to temperature and salt levels in the soil. Each of these phenotypic characteristics is the result of one or more proteins working together. Figure 11.7 shows examples of the phenotypes of another green plant, the pea.

The phenotype of a plant, tissue, or cell is directly related to the proteins it produces. Of course, it is the DNA sequence in the chromosomes that determines whether a certain protein will be made. Scientists can extract and use native proteins, or they may alter the genes or genetic information in plants through genetic engineering to make new or modified proteins. The engineered plants, or genetically modified organisms (GMOs), express new characteristics (new phenotypes). Many of these GMOs are of commercial value, such as a disease resistant corn plant.

If a particular protein is lacking in a plant, then the characteristic associated with that protein also would be lacking. Conversely, if a new protein is produced, then a plant may exhibit a new phenotype. Such is the case with herbicide-sensitive soybean plants. The herbicide Roundup® is used to destroy weeds crowding the growth of cash crops such as soybeans. Traditionally, Roundup® could not be used in soybean fields because soybeans die when exposed to it. However, by modifying the genetic

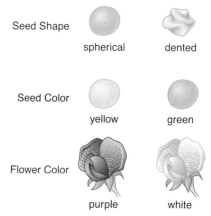

Seed Shape — spherical, dented

Seed Color — yellow, green

Flower Color — purple, white

Figure 11.7. **Phenotypes are the observable expressions of genes. Spherical is a phenotype. Dented is another phenotype. Phenotypes result from the presence or lack of specific proteins.**

code and adding a new gene, scientists have been able to produce Roundup®-resistant soybeans. Roundup®-resistant soybeans produce a new protein that allows them to survive Roundup® exposure by interrupting a biochemical pathway in the plant. The result is Roundup Ready® soybeans, which came to market in 1998.

Plant-Made (Plant-Based) Pharmaceuticals

Since the 1970s, companies have been making recombinant human pharmaceutical proteins in transformed bacteria, fungi, or mammalian cells at an incredible rate. Now, several companies are focusing on producing human pharmaceuticals in plants. These are called **plant-based pharmaceuticals (PBPs)** or plant-made pharmaceuticals (PMPs). Most PMPs are proteins or other compounds that require certain regulatory proteins to be made. Having plants produce human proteins for medicinal purposes is a fairly new application of plant biotechnology.

There are several advantages to engineering plants to make human proteins. First, plants are easy to grow in large numbers, and they require less manpower and specialized equipment. It is not much different growing 1000 acres of plants instead of 10 acres of plants. On the other hand, the scaling-up of bacteria cultures from 10 L to 1000 L to produce a desired protein is considerably more difficult, and it requires specialized equipment and staff. It is also more costly to produce human proteins in bacterial fermentation. Another reason to make human proteins in plants instead of using bacterial fermentation is that plants are eukaryotic and, as such, plant cells are equipped to assemble complicated eukaryotic proteins, such as antibodies and some enzymes.

To produce a PMP, a plant is genetically engineered to make a protein normally not made in the plant. The transformed plant is cloned into several new plants and planted into fields. When mature, the plants are harvested and the recombinant human proteins are extracted and purified from the plants. The goal of PMP scientists is to produce a wide range of therapeutic antibodies, enzymes, or hormones in plants. Currently, several companies and universities are making human pharmaceuticals in corn and tobacco plants, including antibodies to the West Nile virus, human immunodeficiency virus (HIV), and hepatitis B (see Figure 11.8). Some companies are producing PMPs in plant tissue culture. Like all recombinant protein products destined for pharmaceutical use, these PMPs must be purified from the production plants and then undergo clinical trials. The current generation of PMPs is not expected to be on the market for several years.

plant-based pharmaceutical (PBP) (plant-based phar•ma•ceu•ti•cal) a human pharmaceutical produced in plants; also called plant-made pharmaceutical (PMP)

Figure 11.8. Several companies, including Devgen NV, are developing human therapeutic proteins in corn. Devgen's products include a human antibody that is used to control inflammation. Another is a diabetes treatment. Plant-based pharmaceuticals appear to be an effective way to make large amounts of recombinant product. However, some people have great concerns about the development of plant-based pharmaceuticals. Can you think of any reason to have concern over human proteins made in plants?

© Jaroslaw Tomczak/iStock.

Extracting Protein Molecules from Plant Cells

A company may be interested in a native or a genetically engineered protein as a potential product. Some companies send botanists to tropical rainforests to find plants with medicinal value. Some of the molecules in which they are interested are proteins. No matter what the source of the protein, a scientist isolates it from all other molecules in the cell, including thousands of other proteins.

Extracting molecules such as DNA, RNA, or proteins from cells is required for virtually all biotechnology research, but extracting these molecules from plant cells presents special challenges. Plants are dense and sometimes very woody. Grating or mechanically breaking a plant sample increases the extraction yields (see Figure 11.9). Grinding samples in liquid nitrogen (N_2) or on dry ice may also increase yields. Plant cells, unlike animal or bacterial cells, have thick, sticky cell walls that must be removed or weakened to burst the cells open. When the cell walls have been removed, molecular isolation is similar to protocols for molecular isolation from other samples.

Figure 11.9. Grating a plant sample breaks up the plant tissue into smaller pieces in order to increase the surface area where solvents can work.

BIOTECH ONLINE

Not all Tobacco Plants are Bad for You

Several crops are currently being engineered to manufacture recombinant proteins for use as pharmaceuticals. These plant-based pharmaceuticals can often be made faster and cheaper in plants than in other production organisms such as bacteria or fungi.

Read the article at www.reuters.com/article/2014/10/01/us-flu-vaccine-analysis-idU SKCN0HQ2YO20141001#gQGodhBGI5Cz7MsV.97 to learn more about how tobacco plants may be engineered to make human flu vaccines.

- Discuss how companies such as GlaxoSmithKline are currently manufacturing vaccines.
- Give reasons why vaccine producers might want to produce vaccine proteins in plants instead of by current methods.
- Discuss the pros and cons of producing human pharmaceutical proteins such as vaccines in plants.

Removing or Weakening Cell Walls

Tough cellulose fibers impregnated with pectin molecules make plant cell walls strong. Several approaches are used to weaken or remove these structural molecules. Sometimes enzymes are used to degrade or remove cellulose and pectin from the walls. Cellulase and pectinase enzymes may be purchased from supply houses for this purpose. These enzymes weaken the cell walls by degrading cellulose and pectin, respectively. With the cell walls removed, the cells change shape into round protoplasts with only a thin plasma membrane separating the cell from its environment (see Figure 11.10). Protoplasts are valuable in many ways. Since they have no cell walls, it is easy to get DNA into protoplasts for genetic engineering purposes. Scientists actually can use a needle or "gene gun" to introduce foreign DNA into protoplasts. In addition, protoplasts are easy to burst open to retrieve a cell's DNA or proteins.

Cells can be also be opened by crushing their cell walls. Plant samples in buffer can be homogenized using a Tissuemizer® (IKA® Works, Inc.), which acts like a blender. Used at a low speed, the Tissuemizer® grinds the tissue without causing the sample to foam. Foaming may denature the extracted proteins in solution, which could result in changes in the structure or activity of the proteins of interest.

Another method to burst open plant cells is the "freeze fracture" technique. Plant samples are placed in a mortar and pestle, flash-frozen by pouring liquid nitrogen over them, and ground to a fine powder. The powdered plant material is transferred to a buffered solution. When liquid nitrogen is not available, samples can be frozen and ground with dry ice. It is essential that the cells are frozen solid for them to be completely macerated.

Plant cell contents released by these methods are treated differently, depending on the type of molecules desired. For proteins, one approach is to place the cell contents into a neutral solution, such as a potassium phosphate buffer. The majority of the plant cell proteins are extracted into the buffer. Another approach is to conduct an acetone extraction or an ammonium sulfate precipitation. These methods cause the thousands of proteins in a cell extract to fall out of solution. Depending on the conditions, fractions of proteins of different solubility can be separated from each other. Protein samples may be run and visualized on acrylamide gels using a vertical gel box (see Figure 11.11). The technique is similar to the one used to separate and visualize animal proteins that was discussed in earlier chapters.

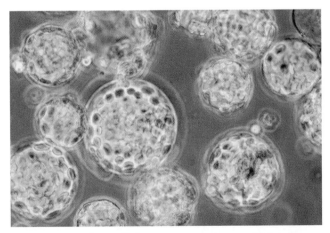

Figure 11.10. Plant cells have turgor pressure, which is due to water pressing outward on the plasma membrane. Since cell walls are rigid, when turgor pressure increases, most plant cells will not burst. But protoplasts have no cell wall; it is simple to increase the turgor pressure in a plant cell protoplast, causing it to burst. This technique may be used to retrieve plant cell DNA and protein.
© Science Photo Library/Photo Researchers.

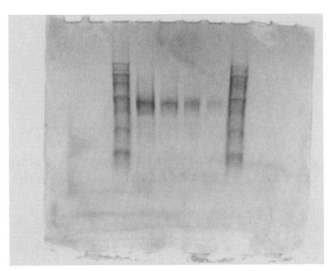

Figure 11.11. Sizing standards on each side of the lanes of a purified soybean peroxidase extraction.
Photo by author.

Visualizing Protein Samples

If a gel shows evidence of some protein of interest, further studies are conducted on the cell extract. Extracts may be tested to reveal proteins with a particular activity (see Figure 11.12). For example, a particular plant may be thought to contain a certain protein that acts as a neurotoxin to insect pests. Once the insecticidal protein is identified, it may be used directly to inhibit insect damage, or the gene for the protein can be isolated and used to engineer plants to become insect resistant. This method was used to create genetically modified Bollgard® cotton (Monsanto Co). It contains a neurotoxin protein, originally found in another species, which kills insects. Since insects die if they eat the plant, they do not live long enough to do significant damage, resulting in an insect-resistant corn plant. The gene for the neurotoxin protein was transferred to cotton in the 1980s. The Bollgard® cotton expresses the gene and synthesizes the insecticidal protein in the plant. The molecular weight of the neurotoxin protein is known, and it is easy to identify and isolate the active protein for further study.

The next section discusses in greater detail the isolation and manipulation of plant DNA and plant proteins to produce biotechnology products. Some of these products are the most promising products of the near future, including new GMOs that are important as food crops or as producers of plant-based pharmaceuticals, including human vaccines and human therapeutics made in plants.

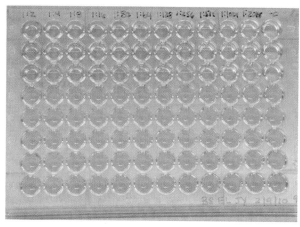

Figure 11.12. An activity assay is conducted on decreasing concentrations of a plant peroxidase. Peroxidase oxidizes a clear compound, 3,3',5,5' tetramethyl benzidine (TMB), to a blue compound. Adding hydrochloric acid (HCl) changes the TMB to a yellow color that is more stable and easier to evaluate. If available, a spectrophotometer can distinguish tiny differences in color and can reveal the concentration of protein in the sample.
© Photo by author.

Section 11.3   Review Questions

1. Distinguish between a plant's phenotype and genotype using examples.
2. What does GMO stand for? Explain how Monsanto's Roundup Ready® soybeans are an example of a GMO.
3. What is the most challenging part about trying to isolate plant DNA or plant proteins from cells?
4. How are PMPs related to genetically engineered organisms?

11.4 Plant Genetic Engineering

As discussed in Section 11.2, plant transformation involves moving genes into plant cells. When small pieces of plant tissue are incubated with certain bacterial cells (*Agrobacterium tumefaciens*) under specific conditions, the bacteria cells will inject their Ti pDNA into the plant tissue cells. The plant cells receive new genes and may start expressing these genes. Given appropriate media and a suitable PTC environment, the transformed plants cells may regenerate into new plantlets. These plants are known as **transgenic plants**, or GMOs. Transgenic plants are gaining importance in the agricultural, horticultural, and pharmaceutical industries. In addition, ecologists are interested in the applications of plant genetic engineering to help preserve several species in danger of depletion from disease and human encroachment (see Figure 11.13).

transgenic plants (trans•gen•ic plants) the plants that contain genes from another species; also called genetically engineered or genetically modified plants

Using *A. tumefaciens* to Genetically Engineer Plants

A. tumefaciens is unique in its ability to hold and pass the Ti (tumor-inducing) plasmid. This ability was discovered when a scientist determined that *A. tumefaciens* caused crown gall disease in plants. Crown gall disease is characterized by the formation of tumors in plants after infection by the bacterium. The bacteria carry the Ti plasmid containing the tumor-inducing gene into the wound site.

When plant tissue is infected, masses of undifferentiated cells are produced as a crown gall tumor. The tumor develops due to the expression of growth hormone genes that have been inserted into the mature plant cells by the bacterial plasmids. These growth hormones are produced in the transformed cells in quantities normally found only in embryonic plant cells.

The naturally occurring Ti crown gall-inducing plasmid is approximately 200 kb in length. The Ti plasmid is very large compared with a typical bacterial plasmid of 5 to 8 kb. The particular segment of the Ti plasmid responsible for tumor production is called tDNA (13 kb). Because tDNA carries genes for plant growth hormones (auxins and cytokinin), transferring tDNA to a plant via genetic engineering means the plant cells begin to produce growth hormones, causing tumors to develop.

Because of its ability to insert foreign DNA into plant cells, *A. tumefaciens* is used in plant genetic engineering. Ti plasmids can be isolated from *A. tumefaciens* cells in broth culture. Then, by using restriction enzymes and DNA ligase, technicians can insert genes of interest into the Ti plasmid and remove the growth hormone genes. Also, on the Ti plasmid is a region that controls transformation. It contains virulence (vir) genes that direct the transfer of the tDNA from the plasmid into the plant chromosomal DNA (see Figure 11.14).

One commonly used Ti plasmid is the genetically engineered Ti plasmid, pBI121. The pBI121 plasmid has two additional selection genes added to it (see Figure 11.15). One of the selection genes is the **neomycin phosphotransferase (NPT II) gene**. It codes for the production of the enzyme, neomycin phosphotransferase. This enzyme confers kanamycin resistance. Most plant cells are kanamycin-sensitive and will normally die in the presence of the antibiotic. If transformed with the NPT II gene, the plant cells will be able to survive in the presence of kanamycin, and they are called kanamycin-resistant. This allows scientists to differentiate transformed and nontransformed plant cells.

A second selection gene, the **GUS gene**, is present on the genetically engineered pBI121, Ti plasmids. This gene codes for an enzyme called β-glucuronidase. If the GUS gene is added to plant cells, the new enzyme may be produced. β-glucuronidase breaks down certain carbohydrates. When given a special carbohydrate, 5-bromo-4-chloro-3-indolyl-beta-D-glucuronic acid (X-gluc), which is normally colorless, the enzyme breaks it down to a blue product. The X-gluc carbohydrate is added to the dishes containing transformed cells. Transformed cells containing the enzyme, meaning they received and are expressing the GUS gene, will produce a blue precipitate.

The Ti plasmid can be spliced and other genes of interest inserted into the plasmid. When the plasmid is inserted into cells, gene transfer is recognized by the expression of the selection of genes. During *A. tumefaciens*-mediated plant transformation, the bacterium does not actually enter the target cell (see Figure 11.16). Instead, it attaches to the outside and inserts part of, or the entire, Ti plasmid into the cell. The plasmid or plasmid pieces are then incorporated into one of the plant cell chromosomes. Transcription of the newly placed genes results in new protein production and new phenotypes.

Figure 11.13. Eucalyptus trees are the focus of genetic engineering for the purpose of increasing their genetic diversity. Populations without enough genetic diversity (variations in the DNA sequence) are susceptible to threats in nature, such as diseases and pests. If a population's individuals are too similar then they will react in the same way to viruses and disease putting the whole population at risk.
© W. Wayne Lockwood, M.D./Corbis/VCG/Getty Images.

NPT II (neomycin phosphotransferase) gene (NPT II gene) a gene that codes for the production of the enzyme, neomycin phosphotransferase, which gives a cell resistance to the antibiotic kanamycin

GUS gene (GUS gene) a gene that codes for an enzyme called beta-glucuronidase, an enzyme that breaks down the carbohydrate, X-Gluc, into a blue product

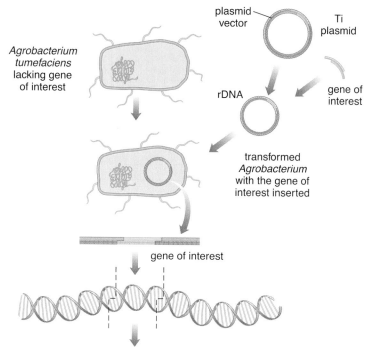

Figure 11.14. **Transforming *Agrobacterium*.** Before *A. tumefaciens* can be used to transform a plant, its Ti plasmid must be transformed with the gene(s) of interest.

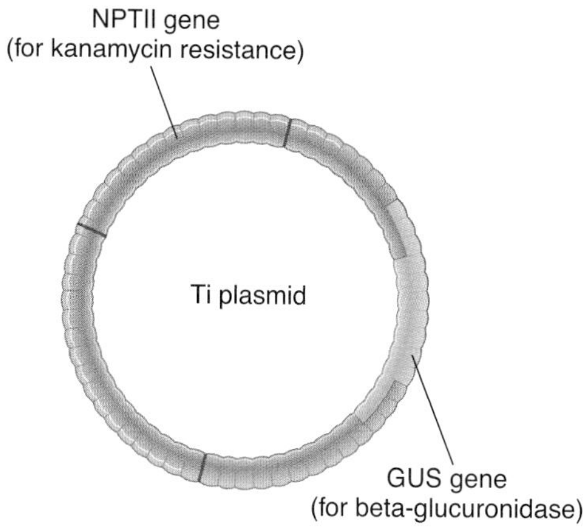

Agrobacterium produces a new "foreign" protein from its newly acquired foreign gene. This results in a new trait.

gene of interest

transformed *Agrobacterium* with the gene of interest inserted

Agrobacterium tumefaciens lacking gene of interest

plasmid vector

Ti plasmid

rDNA

gene of interest

Figure 11.15. **Ti Plasmid.** This Ti plasmid has two selection genes on it, NPT II and beta-D-glucuronidase (GUS), so that when it gets into plant cells, the plasmid transfer can be recognized. Cells receiving this plasmid will be able to survive on kanamycin-containing agar (from NPT II expression). They will also be able to convert a white carbohydrate in the medium to a blue color (due to GUS expression), which makes the entire colony blue, allowing the researcher to ascertain successful DNA transfer.

NPTII gene (for kanamycin resistance)

Ti plasmid

GUS gene (for beta-glucuronidase)

Arabidopsis thaliana, a Model Organism for Plant Genetic Engineering

Arabidopsis thaliana (Arabidopsis) has been the target of many plant genetic-engineering studies (see Figure 11.17). This is because, as a model organism, *A. thaliana* biology and chemistry is well understood. Once mechanisms of genetic manipulation and expression are understood in *A. thaliana,* the technology can be transferred to other plants of economic or ecological importance (see Figure 11.18).

Arabidopsis plants are relatively easy to transform. With a healthy overnight culture of transformed *A. tumefaciens* cells, a technician is able to paint, spray, dip, or dunk plants into a culture. The *A. tumefaciens* cells will move into the plant, attach to the cell walls of a few cells, and inject part or all of the Ti plasmid carrying the genes of interest. Not all cells are transformed, but many, including egg and seed cells, have a good chance of getting the new DNA. Suspected transformed cells are grown up in PTC selection media. Selection methods, such as inserting the green fluorescent protein (GFP) gene into the Ti plasmid, make it easier to see which cells are transformed.

Following transformation, it is a good idea to make sure that the correct DNA (gene of interest) actually was inserted into the cells. This can be accomplished through protein assays to confirm gene expression, polymerase chain reaction (PCR) to identify a section of the gene, or DNA sequencing (see Figure 11.19).

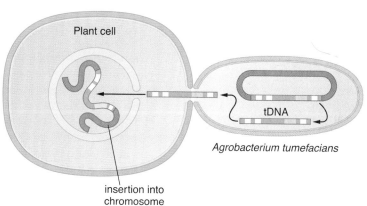

Plant cell

insertion into chromosome

tDNA

Agrobacterium tumefacians

Figure 11.16. **Ti Transfer.** During *A. tumefaciens*-mediated plant transformation, the bacterium attaches to the outside of the cell and inserts part of, or the entire, Ti plasmid into the cell.

Figure 11.18. In her work at Mendel Biotechnology, Inc. Dr. Katherine Krolikowski uses a LI-COR® Biosciences system to measure the amount of carbon dioxide (CO_2) and oxygen (O_2) being released from transgenic *Arabidopsis thaliana* leaves. The leaves are exposed to CO_2, O_2, and nitrogen (N_2) of known concentrations in an airtight compartment at the tip of the LI-COR® wand. The amounts of CO_2 and water released from the leaves reveal the rate of photosynthesis.
Photo by author.

Figure 11.17. Left: A flat of hundreds of *A. thaliana* plants, about 6 weeks old. *Arabidopsis* is the "*E. coli*" of the plant world. An "organism of choice" of geneticists, the entire genome of *Arabidopsis* has been determined. Right: A close-up of *Arabidopsis* shows its tiny flowers.
Photos by author.

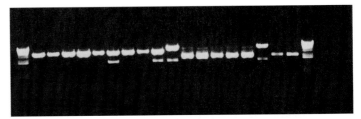

Figure 11.19. DNA that has been isolated from transformed cells may be further analyzed through PCR or DNA sequencing. These PCR product bands represent slightly variant DNA sequences. Chapter 13 presents more information on PCR.
© 2005 Custom Medical Stock Photo.

Section 11.4   Review Questions

1. What is the name of the naturally occurring bacterium and the plasmid that can infect plants and transfer DNA molecules?
2. Name at least two selection genes that are used to confirm that Ti plasmid transformation has occurred.
3. How does GUS act as a selection gene?
4. Why are so many plant genetic-engineering experiments conducted with *Arabidopsis*, even though it has little, if any, economic value?

BIOTECH ONLINE

Sustainable Agriculture and a Code of Business Conduct

Monsanto is a large company with thousands of employees producing hundreds of products. Some of the products are controversial for one reason or another. Genetically modified organisms (GMOs), for example, are in the news a lot. Many groups boycott agricultural GMOs because they say they are unsafe or harmful to the environment. Whether claims are substantiated or not, employees should learn as much as possible about each product and the controversy surrounding them. To help educate and prepare employees to work with agricultural products or practices that are sometimes controversial, Monsanto, Inc. has developed acode of conduct that is called "Our Pledge."

Go to the Monsanto, Inc. business conduct "Our Pledge" at: biotech.emcp.net/ monsantopledge. Review the eight items that are Monsanto's commitment to how they conduct agricultural business in a responsible way. From the links in the margin, summarize, in 3–4 sentences each how Monsanto promotes sustainable agriculture and works to alleviate global hunger.

11.5 Biotechnology in Food Production and Processing

Advances in agricultural biotechnology have impacted food production and processing mainly through improved selective breeding of plant crops and livestock, the genetic modification and cloning of food crops, and microbial, genetic, and protein testing of food products. The results are improved food supplies, increased nutritional content of some foods, and increased food safety and security.

Biotech and Food Production

A variety of foods have been produced through advances in biotechnology. Some of the biggest improvements in food yield and quality have come through recent advances in selective breeding of livestock and crop plants. Genetic testing and DNA fingerprinting allows breeders to better select parent plants and animals for breeding purposes to increase the chance of producing offspring with desirable characteristics, including cattle that grow faster, have more meat, or produce more milk. Another example is wheat bred to be more nutritious by containing a higher protein content than other wheat products.

Biotechnology advances have impacted the production of several fermented products. The manufacturing of different kinds of cheese, kim-chee, beer, and wine are each dependent on the type of microorganism used. Genetic testing and genetic modification are used to determine the best microorganism to use to give the desired product. A new genetically modified yeast has been used by some beer, wine, and bread manufacturers to produce these products at a fast rate. One version of the GM yeast is also thought to reduce the levels of a known carcinogen naturally found in fermented beverages.

Foods produced through genetic modification (genetic engineering) are increasingly reaching the marketplace. Each new GM crop has one or more desirable traits that could not be achieved by selective breeding. Corn containing the Bt gene is a good example. The Bt gene provides protection from a large number of insects so that fields may be sprayed with less insecticide. In another example, adding a gene to papaya protects the fruit from the ringspot virus. The genetic modification literally saved the Hawaiian papaya industry, which was being wiped out by a viral epidemic. Another example is the genetic modification of rice with a gene that increases levels of beta-carotene. The resulting "Golden Rice" is more nutritious and may be used in countries where poor nutrition is affecting vision (see Figure 11.20). As discussed in Section 11.3, it is expected that it will be common in the future to produce genetically modified food crops engineered, with one or more therapeutic proteins, as pharmaceutical-containing food.

Recently a new gene has been found and engineered into crops that that significantly increases protein content without adversely affecting yields (**www.futurity.org/crops-gene-protein-1051182-2**). This is a very important breakthrough since most of the world's people rely on plants as a major protein resource and protein that comes from animal sources requires more water, energy, and resources to produce plants. Food plants such as rice, corn, wheat, and soybean that have a higher protein content are better for individuals and better for the world.

In another biotechnology application, many foods are made with additives or by enzymes produced using genetic engineering. The beverages industry uses very large amounts of high-fructose corn syrup as a sweetener. Several recombinant enzymes, including amylase, are used to produce high-fructose corn syrup from corn mash. Another

Figure 11.20. The provitamin-A gene was inserted into a Ti plasmid and used to *A. tumefaciens*-transform rice plants. The result, called "Golden Rice," produces a high content of the precursor molecule to vitamin A. In many Third World countries, growing this nutrient-rich rice could improve the population's nutrition.
© AJ/IRRI/Getty Images.

Figure 11.21. Detecting pesticides in food (cereals) with ELISA (enzyme linked immunosorbent assay) technique.
© age fotostock/SuperStock.

■■■■■■■■■■■■■■■

feedstock (feed•stock) the raw materials needed for some process, such as corn for livestock feed or biofuel production

foodborne pathogens (food•born path•o•gens) disease-causing microorganisms found in food or food products

example of processed foods that have biotechnology ingredients are products that contain GM soybean oil that has a higher content of "healthy" fatty acids. Some foods have been processed using enzymes produced by genetic engineering, such as apple juice processed using recombinant cellulase or cheese produced using recombinant chymosin.

Biotech and Food Security

Obviously biotechnology is having an impact on food production, but advances in DNA and protein technologies are also increasing food security and are being used to protect food crops and food products.

In many parts of the world, pests and/or environmental change are decreasing the yield or quality of food crops. Food crop protection is needed from environment threats such as drought, unexpected or excessive heat or cold, increases in salinity, and leaching of soils. Each of these environmental threats to food crops puts the local population at risk of famine. In an effort to reduce the threat, plants are bred or genetically modified for tolerance to one or more of the environmental stresses. For example, drought-tolerant varieties of oilseed rape and maize tolerate water shortages significantly better than non-GM rape and maize. Protecting against drought ensures greater yields of these foods and **feedstock**. These particular GM products are expected on the market in the next few years. Related articles and information can be found at: biotech.emcp.net/guardian.

Equally important threats to food crops are pests, including viruses, fungi, insects, and worms. Protection of a host of plants from a myriad of biological pests has been a goal of selective breeding and genetic modification efforts. In addition to the example of the ringspot virus protection in papaya, several plants have been genetically modified to be resistant to biological pests including transgenic rice that is resistant to rice yellow mottle virus. This GM food product is being field-tested in drought-ravaged areas of North Africa with the hope that it may be used to decrease famine (*Transgenic Plants and World Agriculture*, National Academy of Sciences, 2000).

Food security also includes food product safety, which must be monitored during production and processing. Throughout production and processing, foods and food ingredients are tested for pathogens, toxins, pesticide residues, allergens, and ingredient contamination.

Testing for microbial contamination is standard in all plant and animal food products. Using cell culture, DNA testing (using DNA probes, DNA sequencing, PCR, and microarrays), and protein testing (using ELISA and other antibody test systems), deadly **foodborne pathogens** such as *Staphylococcus aureus*, *Escherichia coli*, *Pseudomonas aeruginosa*, *Salmonella*, *Candida* (a yeast), *and Aspergillus* (a mold) may be detected in food products. Several companies produce test kits specifically for use by inspectors and quality control departments. Neogen's GeneQuence® DNA probe technology (biotech.emcp.net/neogen) allows for the rapid and extremely accurate processing of food samples for the bacteria *Salmonella*, *Listeria*, and the pathogenic *E. coli* O157:H7.

Naturally occurring and environmental toxins can also contaminate food products. Several chemical and immunological tests have been developed to assay for the presence of these dangerous compounds. One group of very dangerous, naturally occurring toxins are the aflatoxins. These toxic compounds are made by some species of mold that commonly grow on grain. If not treated properly, the molds and their toxins sometimes get into grain products. Several companies have developed ELISA test kits for the detection of aflatoxins in grain and grain products. Grain and food inspectors use the detection kits to detect aflatoxins before they are accidentally ingested. Other kits are available for detection of many different fertilizer and pesticide residues (see Figure 11.21).

Allergen testing is another important food safety issue. There are countless compounds that cause severe allergic reactions in large numbers of people. Although it is not possible to design testing for every compound, biotechnology companies have designed genetic tests and ELISAs or ELISA test kits for some of the most common and some of the most severe allergens. One important allergen test is for gluten. Patients with celiac disease (CD) have a hypersensitivity to gluten, a compound naturally occurring in some grain products (wheat and related grains). Romer Lab® Inc. (**biotech.emcp.net/romerlabs**) developed an antibody assay, basically an ELISA on a strip, to use for screening gluten in food samples. Detecting gluten in food before it is eaten can prevent damage to the digestive system of CD patients and can be lifesaving.

Another application of biotechnology to food testing is for use in species identification in food products. It is not uncommon to find contaminant food products due to food processing. For example, slaughterhouses butcher and package several different kinds of meat livestock. Equipment must be cleaned and free of one species before switching to the processing of a different species. ELISA technology is used to test that species contamination has not occurred due to poor processing or cleaning practices. In one test, an ELISA can detect pig-specific proteins in beef products to ensure that pork does not contaminate more expensive beef products. In another example, fish samples may be DNA-tested to ensure that endangered or protected fish are not used or sold in specific products.

Risk Assessment and Biotech Food Products

The use of biotechnology in food crops, food production, or food processing raises several concerns by different individuals and groups. Some that question the general safety of GM food products point to concerns with transferring GM traits to wild populations and the concern that it is not possible to predict the negative consequences. Others raise the concern that genetic modification introduces new proteins in the food supply that could be toxic or allergenic. The National Academy of Sciences takes the position that that the potential risks associated with GMOs are the same as those created by traditionally bred organisms and thus should be regulated in the same fashion.

In the US, GMO and non-GMO food crops and products are regulated by several agencies, including the United States Department of Agriculture (USDA), (see Figure 11.22) the Food and Drug Administration (FDA), and the Environmental Protection Agency (EPA).

The Animal and Plant Health Inspection Service (APHIS), one department of the USDA, is responsible for protecting agricultural products from disease and pests and is responsible for field-testing GM crops. Another department, Food Safety and Inspection Service (FSIS), ensures that all meat and poultry products meet high safety standards.

The Food and Drug Administration (FDA) regulates the labeling of GM food if the food is determined to be significantly different than a conventional food product. In the US though, special labeling for genetically modified foods is not required unless a significant

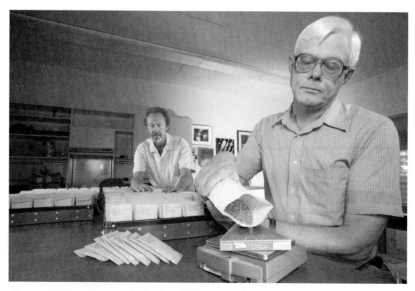

Figure 11.22. **Plant pathologist Blair Goates (left) and agronomist Harold Bockelman prepare seed samples from the National Small Grains Collection to be sent to east Africa for testing against new races of the stem rust pathogen.** Photo by USDA by Peggy Greb.

potential for food allergy or a substantial change in nutrient composition or product identity is shown.

The Environmental Protection Agency (EPA) regulates the use of pesticides and herbicides and determines their safe use. The EPA regulates biotechnology products that have been modified to produce pesticides or herbicides.

The Food and Agriculture Organization of the United Nations (FAO) "supports an ongoing science-based evaluation system that objectively determines the benefits and risks of each individual GMO. This calls for a cautious case-by-case approach to address legitimate concerns for the biosafety of each product or process prior to its release." This and other information from FAO can be found at the following website **biotech.emcp.net/fao_biotech_safety**.

Food biotechnology employs the same molecular biology tools and techniques used to study and manipulate biological molecules as other biotechnology disciplines. In food biotechnology, though, the goal is to improve food or food products and to monitor the process or quality of food products. Food and agricultural biotechnologists, scientists in other food science sectors, or employees in food production, processing, or regulation have a promising future with many career opportunities. Visit the United States Department of Labor's Bureau of Labor Statistics at **biotech.emcp.net/bls_agsci** to learn more about job descriptions and employment outlook for agricultural and food scientists.

BIOTECH ONLINE

Not in My Backyard!

Go to the website at: biotech.emcp.net/GM_wheat and read about Monsanto's difficulty in bringing genetically modified wheat to market. Then write responses to the following questions:

1. Describe the biotechnology wheat product involved in the controversy between Canada and Monsanto Canada, Inc.. What distinguishes it from other kinds of wheat? What new phenotype does it have?

2. What were the reasons for opposition to Monsanto's distribution of genetically modified wheat?

3. What alternatives were proposed to herbicide-resistant, genetically modified wheat?

Section 11.5   Review Questions

1. Give an example of how genetic modification has led to disease resistance in a plant crop.
2. List some food-borne pathogens that may be detected by biotechnologies.
3. What type of biotechnologies can be used for allergen testing of food products?
4. Summarize the roles of the USDA, FDA, and EPA in biotechnology food products and processing.

Chapter Review

Summary

Concepts

- Agriculture is a broad field that includes growing and harvesting crops, soil management, water management, plant and animal breeding and hybridization, asexual plant propagation, seed production and improvement, crop management, and the use of fertilizers, herbicides, insecticides, pesticides, farming tools, and equipment.
- Agricultural biotechnology is the application of advancements in DNA and protein technologies to improve crop yield and quality.
- Horticulture is the area of agriculture focused on growing plants for ornamental purposes. Horticulturists use the same technologies as other agriculturists, including breeding, asexual plant propagation, and gene manipulation.
- Hydroponics is a method used to grow cuttings and entire plants in a nonsoil medium, usually water. Where space in limited or soil is of poor quality, hydroponics can be used to grow plants for agricultural or horticultural applications.
- Genomic or plasmid DNA can be isolated from cells by preparing extraction buffers and other reagents. Many companies produce DNA extraction kits for gDNA or pDNA extractions. The DNA extracted for the purpose of plant genetic engineering must be very pure. The concentration and purity of DNA extractions are determined on a UV spectrophotometer.
- During plant transformation, foreign DNA is transferred into cells using a Ti plasmid as a vector or it must be injected into plant cells using a "gene gun."
- New gene expression and protein production are the products of genetic modification in plants. New proteins give GMO crops their new traits.
- Protein isolation and purification from plants require several additional steps as compared to protein isolation from bacteria or animal samples, including methods to remove the cell wall and the large amounts of carbohydrate present in plant samples.
- Plant-based pharmaceuticals are a new area of plant biotechnology. These products are plants that have been genetically engineered to produce human proteins for human diagnostic and therapeutic purposes. Several companies are already making and testing these products, but the general public has many concerns about the safety and environmental impact of these products.
- Agriculture and horticulture have recently seen the introduction of GMOs. One of the first GMOs was Bt corn. Bt crops contain a gene from a bacterium, *Bacillus thuringiensis*, which produces an insect-killing toxin. Although Bt products decrease the amount of insecticide that has to be sprayed on corn in the field, Bt products have raised concerns with health advocates and environmentalists.

- Plant genetic engineering often utilizes the bacterium, *A. tumefaciens*, and its Ti plasmid. *A. tumefaciens* can infect plants, and may inject part or all of the Ti plasmid into cells. If the transferred DNA incorporates in a chromosome and is expressed, the plant is considered transformed (a GMO).
- Foreign genes can be inserted into the Ti plasmid (recombinant Ti plasmid). The Ti plasmid can be reinserted into *A. tumefaciens* cells and used to deliver the recombinant plasmid, with its foreign genes, into plant cells.
- The Ti plasmids used for most plant genetic engineering contain selection genes that can be used to confirm transformation. These may include the NPT II, GUS, and GFP genes.
- Several important food crops, including corn, soybeans, wheat, potatoes, and papaya, and the model organism, *Arabidopsis*, have been genetically engineered using the *Agrobacterium*/Ti plasmid technology.
- Biotechnology advances have improved the production of several food crops and food products and play an important role in protecting the food supply by monitoring microbial, allergenic, and toxic contaminants, as well as providing some crops the ability to tolerant extreme environmental conditions.

Lab Practices

- Hydroponic tanks can be set up using inexpensive materials. The tanks can be used to conduct macronutrient (eg, calcium nitrate) studies on plant growth.
- Several protocols for the extraction of genomic DNA are available online. Using process development strategies, a technique, such as DNA extraction, can be improved through valid testing and manipulation of single variables.
- Commercial kits are available for isolating gDNA and pDNA. Each must be evaluated for its ease of use, yield, and cost. The DNA extraction yields are measured using UV spectrophotometry and gel assays.
- The concentration of DNA in a sample can be determined by measuring the absorbance of a sample at 260 nm on a UV spectrophotometer. The absorbance is compared with the ratio, 50 µg/mL = 1 au.
- The purity of DNA in a sample can be determined by measuring the absorbance of a sample at 260 nm and at 280 nm on a UV spectrophotometer. By dividing the 260 nm value by the 280 nm value, the ratio of DNA to protein may be determined (A260/A280). An A260/A280 value of 1.8 represents very pure DNA.
- Proteins may be isolated from plant samples using similar methods to other protein extractions with one or more additional steps, including the use of liquid nitrogen, nonpolar solvents, acetone precipitation, and/or ammonium sulfate precipitation.
- *Arabidopsis* ovules can by transformed with Ti plasmid carrying the NPT II and GUS genes by dipping flowers in *A. tumefaciens* cultures containing the desired recombinant plasmid. Suspected transformed seeds are grown on selection media that have both kanamycin and X-gluc in the agar. Nontransformed seeds die due to the presence of kanamycin. Transformed seeds survive the kanamycin and convert X-gluc to a blue compound (a result of their new GUS gene).
- Transformed plant DNA can be extracted, and specific sequences can be recognized and amplified using PCR technology with specific primers. The inserted GUS gene can be recognized in DNA from suspected transformants. If a plant's DNA contains a newly acquired GUS gene, then the PCR reaction will produce a single band visible on an agarose gel.

Thinking Like a Biotechnician

1. Name at least five transgenic plants of economic value.
2. Alcohol is used in virtually every DNA isolation. What is the purpose of alcohol in these protocols?
3. If a cell extract is thought to contain a specific protein, how might a technician check to see if the protein is present?
4. Besides using the Ti plasmid, name another way that foreign genes might be transferred to plant cells.
5. A researcher wants to insert a gene into strawberry cells to prevent strawberries from freezing. Propose how that could be done using protoplast technology.
6. A sample of genomic DNA is isolated from soybeans known to produce protein in higher-than-normal concentrations. It is suspected that a gene regulating transcription is responsible. What procedure may be used to recognize a specific gene sequence in the gDNA?
7. Suggest a method for creating a recombinant soybean plant that contains and expresses a human insulin gene.
8. Suggest a method for screening soybean tissue that has been inoculated with *A. tumefaciens* for the purpose of the insertion of a recombinant Ti plasmid.
9. What might be the value and application of plants such as the one discussed in question No. 7?
10. How might insulin be purified from the recombinant insulin soybean plant discussed in question No. 7?

Biotech Live

Specialty Areas in Agriculture

Activity 11.1

Agriculture is a very large field, and employees in agriculture specialize in one of many areas. College programs with a focus in one or more areas of agriculture are common. Depending on your skills and interest, there are a multitude of options for agricultural scientists.

 Create a poster describing the job opportunities and academic training programs that lead to an agricultural specialty.

1. Select an agricultural specialty from the list below:

agronomy	food science	silviculture
aquaculture	horticulture	soil science
arboriculture	hydrology	sustainable agriculture
biofuel feedstock	pest management	viticulture
floriculture	plant pathology	

2. Create a poster with the following features:
   a. In the center, the definition of the agricultural specialty or field
   b. Surrounding the central definition, five to six colorful images and text presenting the following:

   * Examples of what is studied in the specialty or field
   * Examples of job titles of workers in the specialty or field
   * Colleges or universities with degree programs or certificates in the specialty or field
   * Other interesting facts or statistics about the specialty or field
   * A list of websites to find more information about the specialty or field

Plants Can Get Sick Too!

A pathogen is a disease-causing agent. For example, a soil-borne fungus can cause brown rot and death in lettuce, as shown in the photo. Plant pathogens are a big problem for farmers and horticulturalists. Bacteria, fungi, and viruses all cause hundreds of millions of dollars in damage to food crops and ornamental plants. There are many jobs for scientists interested in determining how plant pathogens infect plants and how to combat them.

Butter lettuce afflicted with Sclerotinia drop.
© George D. Lepp/Getty Images.

Go to www.dpvweb.net/dpv/dpvnameidx.php to learn more about plant viruses. Take a moment to notice the enormous number of viruses known to infect plants.

Click a letter from the alphabetical list (maybe the first letter in your name). It will take you to a list of plant viruses starting with that letter. Pick one, and create a short information sheet about the virus and its host (the organism it infects). Include the following information:

- virus name
- host name (both the scientific name and common name)
- geographical region where it is a problem
- how it is transmitted from host to host
- symptoms of infection
- website URL that was used

Go to gmoanswers.com/ask/can-genetic-engineering-protect-plants-disease to learn more about how gene modification and genetic engineering technology is being used to combat plant disease. Give 3 examples.

Livestock Biotechnology

In a team of four, create a seven-slide Microsoft® PowerPoint® presentation with annotated photos and graphics that demonstrates how biotechnology has impacted the livestock industry. Include the following slides:

1. Title page
2. Breeding animals
3. Cloned animals
4. Use of recombinant and nonrecombinant hormones, feed, antibiotics
5. Genetic engineering of animals
6. Production of organs and tissues
7. References/bibliography

Each slide should have a minimum of text, but still include this information:

- definition/explanation
- one or two specific examples of how the technology is used and on which types of animals
- photos
- products expected in future
- concerns about the technology

Divide the work so that each person in the team is responsible for creating one or two slides. Then, have a team leader create a master slide and title page. The members of the group must provide the team leader with the slide files on which they worked so that they can be imported into the group's final presentation.

Arabidopsis thaliana: A Model Organism for Plant Genetics

Plant geneticists and genetic engineers prefer to use *Arabidopsis* for experiments because it is easy to grow, it grows fast, and, based on experience, conditions can be created to ensure the best results.

Use the Internet to learn more about *Arabidopsis* by finding the information listed below. List the website URL that you used next to the answers you record. Create a one-page fact sheet to distribute to your colleagues.

A wild-type *Arabidopsis* plant (right) and a dwarf mutant (left). The wild type shows the low leaves and flower spike of a typical plant.
Photo by author.

1. List the following classification *groups* for *Arabidopsis.*

   kingdom plantae
   division magnoliophyta
   order
   family
   genus *Arabidopsis*
   species *thaliana*

2. Copy and paste a photo of *Arabidopsis* into your fact sheet.
3. What other common plants are related to Arabidopsis?
4. Give an example of how *Arabidopsis* is used in plant research.
5. Go to the TAIR website at: **biotech.emcp.net/arabidopis**. Go to the "About TAIR" link, and describe the purpose of TAIR.
6. Go to the Portal pull down menu, and then to the Education and Outreach menu and click on the "About *Arabidopsis*" link, and list the seven reasons why *A. thaliana* is such a good model organism for plant research.

The Only Thing to Fear...

Examine the benefits and risks, and take a personal position on whether or not the United States should grow GMOs.

1. Go to the Public Broadcasting System website at www.pbs.org/wgbh/harvest/exist.
2. You will be asked the following question seven times: "Based on what you now know, do you think we should raise genetically modified (GM) crops?"
3. On a sheet of paper, write "yes" or "no" to the question, with at least one sentence explaining the reason for your answer.
4. Depending on your answer, you will be presented with a new argument meant to challenge your position. Your position may change from pro to con, or con to pro, depending on how convinced you are by the information presented.
5. After you are presented with the sixth argument and you have taken your seventh position, you may review all 12 arguments, pro and con.
6. At this point, take a final stand, and write three to five sentences that summarize your position. You may choose one of the following positions:

   * Yes, we should raise GMOs.
   * No, we should not raise GMOs.
   * I am undecided about whether we should raise GMOs.

   If you are still undecided, check out other viewpoints at
   www.pbs.org/wgbh/harvest/viewpoints

Activity **11.6** **"Well, Hello Dolly, Dolly, and Dolly?"**

Dolly (the large sheep in the photo) was one of the very first cloned animals. Her creation and early death have raised several questions about the future of cloning animals.

© Reuters.

TO DO

Go to novaonline.nvcc.edu/eli/evans/ HIS135/events/dolly96/Dolly_Module.html to learn about how Dolly was created and what was learned when she was cloned. Write answers to these questions:

1. What is "nuclear transfer," and what does it have to do with Dolly?
2. How did Dolly die, and what do scientists believe to be the cause of her death?
3. According to this website, what attempts have been made in the field of human cloning?
4. Do you think human cloning through nuclear transfer should be allowed? Under which circumstances?
5. Go to novaonline.nvcc.edu/eli/evans/HIS135/events/dolly96/Dolly_Module.html and read through the "22 Reasons for Human Cloning." List three that are most convincing. Come up with a reason you would not support human cloning.

Activity **11.7** **Is Biotech Safe??**

As of 2004, the popular cereal, Kix®, by General Mills, Inc., may have had "no preservatives, additives, or artificial colors," but it did have extra genes that have been inserted into the corn plants used to make the cereal. Many such biotechnology products come to market every year. Several are GMOs containing DNA that has been added or changed in some way. Before, during, and after an agricultural or pharmaceutical biotechnology product is manufactured and marketed, it is tested and reviewed for user and environmental safety.

TO DO

Learn about the industry's regulatory mechanisms, and decide if there is enough consumer protection regarding existing biotechnology products and potential future products created by advances in biotechnology techniques.

1. Use the Internet to find out what happens and who is involved in safety and efficacy reviews at the agencies, listed below, when new genetically modified crops are manufactured.

   Safety of Biotech - Federal Regulatory Agencies
   US Food and Drug Administration
   US Department of Agriculture
   US Environmental Protection Agency

2. At www.ers.usda.gov/media/323484/aer786_1_.pdf, read through the article from the US Department of Agriculture (APHIS), titled "Genetically Engineered Crops for Pest Management." Read the front matter report, which presents the positive aspects of using GMOs. List four of the benefits given for using GMOs for pest management.

3. Now, find a different website that presents the dangers of using GMOs, such as those for pest management. List four of the potential risks of using these types of GMOs.

4. Finally, write two or three sentences describing your level of concern about producing or using all, or parts of, GMO products.

Bioethics

Monarchs: What's All the Fuss About?

Corn is an important agricultural product and its pollination occurs mainly by wind. But, corncobs have long, pollen-filled silks that are a food source for many insects. Some of the insects are beneficial to the corn or other crops, but some are harmful pests. Many insect pests produce larvae that destroy corncobs and, thus, have a huge financial impact on crop production and profits. Corn has recently been genetically engineered with "Bt" genes to help protect it from insect pests. The new Bt corn produces a toxic compound that kills insect larvae.

Monarch butterflies are not a "pest" to corn plants. Instead, they eat milkweed leaves in the vicinity of cornfields. When corn pollen lands on milkweed leaves, monarch butterflies can eat the corn pollen along with the leaves. Many people feel that there is a significant threat to Monarch butterfly populations because they may occupy the same ecosystem as the Bt corn.

Examine the controversy surrounding the use of "Bt" corn and its impact on the ecology of native insects, such as the Monarch butterfly.

1. Find an article(s) or website(s) that describes Bt corn, including how it is produced, how it differs from other corn, and why scientists produced it. Copy and paste this article (including the bibliographical information) into a Microsoft® Word® document. Highlight the sections that explain the topics given above.

2. Find an article(s) or website(s) that describes Monarch butterflies. Copy and paste this article (including the bibliographical information) into a Microsoft® Word® document. Highlight the sections that give the scientific name of the butterfly and its lifestyle, including where it lives, what it eats, and its behavior.

3. Find articles that discuss opposing positions to the question, "Does Bt corn present a significant threat to Monarch butterfly populations?" Copy and paste these articles (including the bibliographical information) into a Microsoft® Word® document.

   a. Locate two articles that support the position that there is little or no risk to Monarch populations because of Bt corn. Show the evidence that support this argument. Do you think the evidence is compelling?

   b. Locate two articles that argue that there is a significant risk to Monarch populations because of Bt corn. Show the evidence that supports this argument. Do you think the evidence is convincing?

4. Create a poster or a 5-minute Microsoft® PowerPoint® presentation (a minimum of eight slides) summarizing your research. Include photos and diagrams on every slide that help explain both sides of the controversy. On the second to the last slide, present your personal opinion of the level of risk that "Bt" corn production has on Monarch populations and what action you would or would not take. On the last slide, list a "bibliography" that leads the audience to at least eight websites for further information on the Monarch butterfly/Bt corn controversy.

   If you choose to create a poster, you must include all of the elements required for the PowerPoint presentation.

5. Discuss why genetic enhancement of food crops is so controversial.

Biotech Careers

Photo by author.

Genetic Nurse Practitioner

Kim VanYsseldyk, MN, NP-C, AOCNP
Comprehensive Cancer Center, Sutter Roseville Medical Center
Roseville, CA

Kim began her career as a Registered Nurse and worked in an acute care cancer unit. After earning a Masters in Nursing, Kim worked as a Family Nurse Practitioner for 5 years. Kim found an opportunity to work with oncology patients at Sutter Roseville Medical Center (SRMC). SRMC launched the Cancer Genetics Program in 2005 and with medical oncologists Kim developed their Cancer Risk Program. Kim received specialized training in the cancer genetics training program at The City of Hope Medical Center and ongoing cancer genetics education courses.

Kim spends the majority of her day seeing patients who have cancer or those with a strong family history of cancer. She evaluates the cancer genetic risk, provides genetic counseling and genetic testing. Kim sees a large number of breast cancer patients who have a strong family history of breast, ovarian and other cancers. In a 90-minute appointment, Kim discusses the patient's medical and family history of cancer and genetic mutations that may be affecting the cancer risk. Kim also discusses with her patient the pros and cons of receiving genetic test results and how new genetic information may affect a patient's future.

If a patient chooses to have cancer genetic testing, Kim collects a blood sample that is submitted to a specialized cancer genetics laboratory. In the lab, specific DNA (genetic) codes are analyzed to look for mutations associated with hereditary cancer syndromes. Kim then meets with the patient to review the results. Medical management recommendations are made based on the genetic test results. Patients who carry cancer genetic mutations are managed aggressively with either intensive screening or preventive surgeries to decrease cancer risk.

In the past few years, genetic testing has expanded from testing for a small number of genes to tests that screen for dozens of genes. These larger screens are called **panels**. Multiple gene panels are made possible because of advances in the law (rulings that genes can not be patented) and new instruments and technology (Next-Generation Sequencing). Genetic testing and Next-Generation Sequencing are discussed in more detail in this chapter and in Chapter 14.

12 Medical Biotechnologies

Learning Outcomes

- Discuss the scope and role of medical biotechnology in the healthcare industry
- Describe the function of drugs and how they may be created with combinatorial chemistry
- Explain how scientists test the effectiveness of antibiotics and antimicrobials and discuss the significance of antibiotic resistance
- Explain how high-throughput screening methods are used to discover potential drug activity
- Describe the methods for synthesizing peptides and oligonucleotides and discuss the uses of each
- Detail the multiple uses of antibodies and vaccines in medical biotechnology
- Describe how genetic testing can reveal information about a patient's risk of developing a disease or disorder
- List examples of recent advances in medical biotechnology and expected new applications

12.1 Drug Discovery

A **medicine** is something that treats or prevents disease or alleviates its symptoms. Thus, medical biotechnology includes all areas of research, development, and manufacturing that work toward that goal. **Medical biotechnology** uses the techniques of molecular biology to develop diagnostic tools and therapies to improve patients' health and quality of life.

Disease may be caused by pathogens, including some bacteria, fungi, protozoa, or viruses (see Figures 12.1 and 12.2). Other causes are environmental agents (including air pollution or toxic waste) and genetic disorders (such as chromosome damage, mutation, or inheritance). Medical biotechnologists develop the tests that identify pathogens or the cause of disease. They also develop the vaccines to protect against them.

medicine (med•i•cine)
something that prevents or treats disease or alleviates the symptoms of disease

medical biotechnology (med•i•cal bi•o•tech•nol•o•gy)
all the areas of research, development, and manufacturing of items that prevent or treat disease or alleviate the symptoms of disease

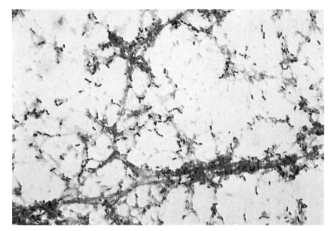

Figure 12.1. Various bacteria, fungi, and viruses cause meningitis, a life-threatening disease of the brain and spinal cord. The gram-positive bacterium *Streptococcus pneumoniae* (pairs or short chains of purple-blue spheres in the micrograph) is responsible for pneumococcal meningitis, one of the most common forms of bacterial meningitis. ~1000X
© CDC.

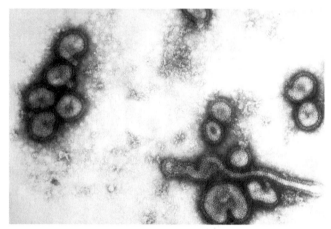

Figure 12.2. The *Haemophilus influenzae* bacterium (shown here) is the most common cause of meningitis in children under the age of 2. Older children most commonly get meningitis caused by viruses. ~10,000X
© Lester V. Bergman/Getty Images.

vaccine (vac•cine) an agent that stimulates the immune system to provide protection against a particular antigen or disease

Biotechnologists develop the medications or instruments used to treat or monitor patients if they are afflicted.

Medical biotechnology includes research and manufacturing in the following areas:
- causes of human diseases and disorders
- drugs, pharmaceuticals, and medicines derived from nature or through genetic engineering and/or combinatorial chemistry
- disease prevention, **vaccines**, and gene therapy
- diagnostics (testing) for medical applications
- devices or instruments to deliver drugs or monitor and assist patients

The medical biotechnology field is expansive, with many opportunities for interesting, meaningful careers working on understanding diseases and how to prevent or treat them.

BIOTECH ONLINE

New Medications: Hope for Suffering Patients

According to PhRMA, the Pharmaceutical Research and Manufacturers of America (**biotech.emcp.net/phrma_innovation**), it takes about 15 years for the average new medication to be developed, manufactured, and approved by the FDA, and only one in every 10,000 potential medicines makes it to market. Even so, for patients that are suffering, a new treatment gives them hope that they may live a more normal life.

Go to www.drugs.com/newdrugs-archive/January-2015.html and use the information given for an entire calendar year to create a chart of recent FDA drug approvals and then answer the following questions.

1. What disease or disorder had the most new drug approvals?
2. What company had the most drug approvals?
3. How many of the new drug approvals were for vaccines (a method of disease prevention)?
4. In your opinion, which drug approval may have the potential for relieving the most suffering? Give reasons for your selection.

Drug Development

The Food and Drug Administration (FDA) defines a **drug** (pharmaceutical) as "articles intended for use in the diagnosis, cure, mitigation, treatment, or prevention of disease in man or other animals; and (other than food) intended to affect the structure or any function of the body of man or other animal." In practice, drugs are chemicals that alter the effects of proteins or other molecules associated with a disease-causing mechanism. A drug is usually specific for a particular disease process. For example, a virus may recognize a protein on the surface of a cell that it infects (see Figure 12.3). An antiviral target could be the recognition molecule on the surface of the host cell or the virus. Scientists could attempt to design a "blocker" molecule (drug) that would bind on the cell-recognition molecule and block the virus from attaching. This is one of the mechanisms that scientists are currently using to design a drug treatment to block the human immunodeficiency virus (HIV) from infecting cells.

Drug discovery is one of the fastest growing areas of medical biotechnology. Drugs may be discovered in, and harvested from, nature. Drug molecules can be synthesized in a laboratory from simpler, preexisting molecules (**organic synthesis**) (see Figure 12.4). Genetic engineering has also led to several new drug molecules. In all of these drug discovery strategies, hundreds, thousands, or even millions of molecules are isolated and tested. This is called **screening**. In some ways, drug discovery is like a treasure hunt. The treasure is a molecule that shows activity on or against a target in an ocean of inert, ineffective, or harmful compounds.

To isolate or design a drug that will treat a disease or condition, scientists must understand the characteristics of the disease very well. They must understand how an organism contracts the disease and the course that the disease takes in its host. Scientists need to understand the biochemistry of the disease and which molecules the pathogen or host cells produce, and when. A disease's origin and development, including its evolution, hosts, and means of transmission, is called **pathogenesis**. With this information, scientists may find a "target" to aim for.

Antibiotics were some of the first drugs isolated to interfere with a disease-causing organism's pathogenesis. Antibiotics and antimicrobials are substances that kill or stunt bacteria and some other microorganisms. The first antibiotics discovered were produced in certain fungi or bacteria to interfere with other competing microorganisms. Antibiotics slow the growth or actually kill microbes by interfering with some cellular process. Ampicillin, for example, interferes with cell wall synthesis so that bacteria cells cannot grow and divide. Another antibiotic, streptomycin, interrupts protein synthesis at the prokaryotic ribosome, interfering with required cellular reactions. Many different antibiotics have been discovered in nature or have been developed from molecules originally found in nature.

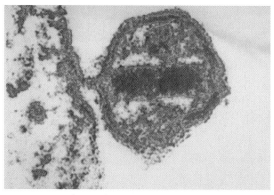

Figure 12.3. **In this scanning electron micrograph, HIV particles are bound to the surface of a human immune system cell (CD-8). Binding occurs due to specific interactions between the glycoprotein 120 (gp120) molecule on the HIV and a gp-120 protein receptor on the CD-8 cell. One approach to blocking HIV infection and, therefore, AIDS, is to block the gp120 receptor protein. 200,000X**
© Lester V. Bergman/Getty Images.

drug (drug) a substance that alters the effects of proteins or other molecules associated with a disease-causing mechanism; also called a pharmaceutical

drug discovery (drug dis•cov•er•y) the process of identifying molecules to treat a disease

organic synthesis (or•gan•ic syn•the•sis) the synthesis of organic molecules in a laboratory from simpler, preexisting molecules

screening (screen•ing) the assessment of hundreds, thousands, or even millions of molecules or samples

pathogenesis (path•o•gen•e•sis) the origin and development of a disease

Figure 12.4. **Dr. Willard Lew, Medical Chemist** As Associate Director of Medicinal Chemistry at Sunesis Pharmaceuticals, Inc., Dr. Lew directed a team that worked with small molecules that inhibit Aurora kinase, an enzyme involved in cell division. These molecules target cells in mitosis and inhibit Aurora kinase activity. The Aurora kinase inhibitor molecules may be a future cancer therapeutic. Currently, Dr Lew is a Senior Research Scientist II at Gilead Sciences, Inc. Gilead develops medications for patients with life threatening diseases, including HIV/AIDS, liver disease, and serious cardiovascular and respiratory conditions.
Photo by author.

Penicillin: One of the Best Finds in Nature

Penicillin was the first antibiotic to be isolated from nature and used therapeutically. The story of how it was discovered and used is an excellent example of the importance of an effective drug. Penicillin is also a great example of how growth in the drug discovery industry is not likely to slow.

TO DO

Learn more about antibiotics and the important role penicillin has played in medical advances. Go to biotech.emcp.net/penicillin.

Write answers to the following questions:

1. How was penicillin discovered? What organism produces it?
2. What does penicillin do? How does it work? On which cell parts?
3. What other penicillin-like drugs exist and why were they developed?

---

antiseptic (an•ti•sep•tic) an antimicrobial solution, such as alcohol or iodine, that is used to clean surfaces

A problem with antibiotic use is that some bacteria become resistant to the action of many types of antibiotics. Bacteria populations mutate so quickly that often, by chance, changes in DNA may allow previously antibiotic-sensitive bacteria to produce new proteins that can act against the antibiotic action. If the new mutant bacteria are able to live in the presence of the antibiotic, then they are considered "antibiotic resistant." If a patient takes a particular antibiotic to combat some disease-causing bacterium, and the bacteria are resistant to it, then the antibiotic will not work.

Recently, one particular antibiotic resistant bacterium is causing particular concern among healthcare professionals. It is methicillin-resistant *Staphylococcus aureus*, MRSA for short. MRSA is a type of staphylococcus that is resistant to several antibiotics. In the general community, MRSA can cause skin and other infections. In a healthcare setting, such as a hospital or nursing home, MRSA can cause severe problems such as bloodstream infections, pneumonia and surgical site infections. Although MRSA can lead to life-threatening infections, since 2005, MRSA deaths have decreased significantly due to the development of new antibiotics. To learn more about MRSA, visit the CDC's site at **www.cdc.gov/mrsa/pdf/SHEA-mrsa_tagged.pdf**.

Developing new antibiotics to replace older ones is an important focus of biotechnology. You may be familiar with several of the modifications to penicillin, such as amoxicillin, carbenicillin, and ampicillin. In addition, researchers look to species of bacteria and fungi for new compounds. Some *Bacillus* species, for example, have given rise to some new antibiotics, including cephalosporin.

Antimicrobials include antibiotics and other compounds that kill microbes. These compounds may include **antiseptics**, such as alcohol, Bactine® (Bayer Corp), iodine, astringents, and toxins. One way to test plant extracts for antimicrobial properties is to add extract-soaked filter-paper disks to bacteria cultures spread on Petri plates. Plant extracts containing compounds effective against bacteria leave clear halos, indicating bacterial death around the soaked disks in the bacteria lawns. The plant extracts demonstrating these clear areas on Petri plates are then further processed and screened for the specific ingredients causing bacterial death (see Figure 12.5).

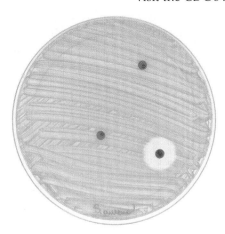

Figure 12.5. **Agar in Petri dishes is spread with bacteria. Then, antibiotic-soaked disks are laid on the agar. Each disk has a different antibiotic on it. Antibiotic No. 1 is at "one o'clock." Tetracycline is at "four o'clock." The clarity of the zone (halo) around the tetracycline disk, not the size, shows the effectiveness of inhibiting the bacteria. The other samples are not showing evidence of inhibition.**
© Custom Medical Stock Photography.

Sources of Potential Drugs The study of drugs and their composition, actions, and effects is called **pharmacology**. The pharmacology of a pharmaceutical must be understood to evaluate whether it has a disease-impacting effect. Understanding the pharmacology of a compound is one part of the drug discovery process. In drug discovery, the goal is to find a compound that shows activity against one or more molecules associated with a disease. Generally, this process is lengthy and often tedious, since hundreds, thousands, or millions of compounds may have to be screened as potential drug candidates. Even if a compound shows promise as a potential drug, it may have to be modified and tested repeatedly to meet FDA or other requirements.

Several medicinal compounds have come directly from plants or animals with very little modification. For example, until recently, the most common form of insulin used by diabetic patients was harvested from cow, pig, and sheep pancreases. These organs are ground up and the insulin protein is purified from the cell extracts for human use. Another example is digitalis, the compound produced by foxglove plants (see Figure 12.6). Digitalis is extracted from foxglove, and used to regulate and normalize irregular heartbeats. Efforts to isolate compounds from nature have been discussed in earlier chapters.

If a molecule cannot be found in nature in the exact form needed, scientists have the ability to modify naturally occurring compounds in an attempt to improve them. If a molecule of interest is found only in small quantities, it can be synthesized and mass-produced. These are some of the tasks of organic chemists employed by biotechnology companies.

Synthesis of simple organic molecules has been accomplished for more than 100 years. A simple two-atom molecule, such as salt, can be made in a laboratory by combining sodium and chlorine. Larger organic molecules can be synthesized by adding together smaller organic molecules. This process is known as **combinatorial chemistry**.

Aspirin (acetylsalicylic acid) is the most frequently used pharmaceutical drug and a good example of combinatorial chemistry.. Since aspirin is such a safe and effective pain reliever, with only a few side effects, it has a huge market and is in great demand (see Figure 12.7).

Figure 12.6. **Foxglove (*Digitalis purpurea*) in bloom. The foxglove plant produces compounds used as a heart stimulant to treat irregular heartbeats and heart failure.**
© Royalty-Free/Corbis.

pharmacology (pharm•a•col•ogy) the study of drugs, their composition, actions, and effects

combinatorial chemistry (com•bi•na•tor•i•al chem•is•try) the synthesis of larger organic molecules from smaller ones

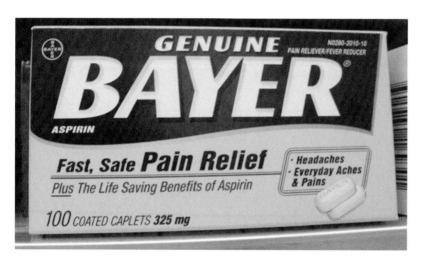

Figure 12.7. **Since natural sources of chemicals may be limited, organic chemists have learned to synthesize medium- to larger-sized molecules that were originally found in nature. A good example is commercial aspirin. Aspirin's main ingredient is acetylsalicylic acid, made by modifying the naturally occurring salicin in willow trees. The chemical name for salicylic acid is 2-hydroxybenzoic acid. The name reveals that the structure of the molecule is a benzene ring with some side groups. The white willow produces a significant amount of 2-hydroxybenzoic acid, which scientists can extract.**
Photo by author.

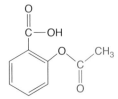

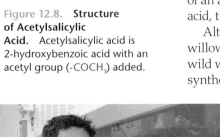

Figure 12.8. Structure of Acetylsalicylic Acid. Acetylsalicylic acid is 2-hydroxybenzoic acid with an acetyl group (-COCH₃) added.

Native people chewed on willow bark and leaves for pain relief. Eventually, the precursor to commercial aspirin was found in the leaves and bark of the white willow tree, *Salix alba*. The active ingredient in willow is actually salicin. Unfortunately, salicin damages soft tissues in the mouth and stomach. Chemists learned how to convert salicin to salicylic acid, which is less caustic than salicin, and more people can tolerate it. Several years after the discovery of salicin, salicylic acid was significantly improved by the addition of an acetyl group (-COCH₃), which produced an even gentler substance, acetylsalicylic acid, the active ingredient in modern-day aspirin.

Although aspirin is in high demand, white willow trees are in limited supply. If willows were the only source of salicin or salicylic acid, it would not be long before wild willows became short in supply. Thus, scientists have had to learn to synthesize a synthetic precursor to aspirin, 2-hydroxybenzoic acid (salicylic acid). They start with a phenol ring. Phenol is a by-product of burning coal. Carefully combining carbon dioxide with phenol, under the right conditions, produces 2-hydroxybenzoic acid. The 2-hydroxybenzoic acid was first synthesized in a lab in 1897 at the Bayer Corporation in Germany. The 2-hydroxybenzoic acid had serious side effects so it was modified by the addition of the acetyl group. The result was acetylsalicylic acid (2-ethanoylhydroxybenzoic acid) (see Figure 12.8). Today, it is more cost effective to synthesize the 2-ethanolhydroxybenzoic acid for aspirin production than to harvest it from willow trees.

Combinatorial chemistry companies, such as Infinity Pharmaceuticals, Inc., Sunesis Pharmaceuticals, Inc., and Pharmacopeia, Inc., mass-produce a myriad of small organics used in the R&D of all types of biotech products, including drugs (see Figure 12.9). Scientists also have learned how to synthesize more complicated molecules, such as peptides and oligonucleotides, which in nature are only produced inside cells.

Figure 12.9. Before becoming a senior investigator in External Discovery and Preclinical Sciences (XDPS) at Merck & Co., Inc., Kenneth Barr, PhD, was a Medicinal Chemist at Sunesis Pharmaceuticals, Inc. He is pictured here working at the chemical fume hood, preparing samples as part of the synthesis and isolation of novel organic compounds for possible drug candidates. Sunesis Pharmaceuticals focuses on R&D of small-molecule therapeutics for inflammation and cancer.
Photo by author.

Section 12.1   Review Questions

1. What kinds of organisms cause disease?
2. Where are drugs typically discovered?
3. How can a technician know if a certain type of bacteria is sensitive to an antimicrobial substance?
4. How is aspirin an example of combinatorial chemistry?

Figure 12.10. White willow leaves and bark produce salicin, a precursor to the active ingredient in aspirin. Since there is a great demand for aspirin, but only a limited supply of willow trees, scientists have learned to make aspirin and its precursors in the lab.
© Martin B. Withers/Corbis Documentary/Getty Images.

12.2 Creating Pharmaceuticals through Combinatorial Chemistry

Sometimes researchers find natural compounds that have potential as drugs, but they must be modified in some way to be more useful or effective. As described in the last section, the production of aspirin from simpler compounds was initially accomplished in the late 1890s (see Figure 12.10). Adding a simple acetyl group (-COCH₃) to the salicylic acid resulted in acetylsalicylic acid, the main ingredient in aspirin. Aspirin has been the most successful pharmaceutical in history.

The number of therapeutic products that are derived directly from plants and currently on the market is impressive. Both naturopathic and modern scientific medicine make use of plants or parts of plants that are thought to have therapeutic value. A large list of herbs used in the treatment of several conditions is found in Chapter 6, Table 6.2.

Techniques for Creating New Drugs

Until recently, it may have taken a chemist several months or years to isolate or create and test a single potential drug compound. Thus, finding a drug to use as a therapeutic could take many years, if it ever happened at all. Recently, though, chemists have begun to use combinatorial chemistry, along with **parallel synthesis**, high-throughput screening (making and testing, respectively, many batches of similar compounds simultaneously), and robotics to mass-produce and test thousands of potential drug candidates at the same time (see Figure 12.11).

Combinatorial chemistry is the process of creating new and varied organic compounds by linking chemical building blocks. The synthesis of acetylsalicylic acid, discussed in Section 12.1, is a simple example of combinatorial chemistry, in which a two-carbon acetyl group is combined with a ringed, salicylic-acid molecule. Today, combinatorial chemistry protocols tests thousands of different chemical building blocks with other organic compounds in all possible combinations, resulting in large numbers of different products. One or more of the possible products may show promise as a drug-like compound.

Advances in combinatorial chemistry methodology have sped the discovery of potential drug candidates. Using new technology, including robotics and miniaturization, minute volumes are placed in multi-well plates. In industry, it is common to use 96-well, 384-well, and even 1536-well plates (see Figure 12.12). In each tiny well, compounds may be added and the assayed or screened. Millions of unique compounds with potential therapeutic value can be created and screened in multiple replications. High-throughput technologies aid scientists in developing and evaluating (screening) new compounds more quickly and efficiently.

Several companies specialize in producing millions of compounds using combinatorial chemistry and miniaturization. These compounds have potential use in R&D or the discovery of new therapeutic products. Technicians must be able to keep track of these compounds along with their test results. The collection of compounds is called a "**library**." A compound library may include hundreds of plates with hundreds of wells of samples. Substantial documentation is necessary to track and manage each library. At some facilities, plates/samples are actually bar-coded so that a compound's location and test results can be easily documented.

An example of an important therapeutic product that is being improved using combinatorial chemistry is a group of molecules called protease inhibitors. Protease inhibitors are molecules that interfere with the activity of specific protease molecules to breakdown their protein substrate. Blocking the protease activity of a protease produced by the HIV/AIDS virus has been one of the most successful strategies in controlling the disease in infected individuals. Unfortunately, the HIV virus mutates quickly and the original protease inhibitors that worked well two decades ago, are not as effective now.

In an attempt to develop better HIV therapies, combinatorial chemists at Bayer Pharmaceuticals have taken one of the original protease inhibitors, hydroxyethylamine, and have combined it with thousands of other small organic molecules to create a library of potential new protease inhibitors that might work well to block the action of HIV proteases. They use high-throughput screening assays and robots to comb through the compound libraries, of maybe three million substances, for molecules that show promise as a protease inhibitor. Surveying new therapeutic molecules may take just a few weeks versus several years. The new compounds must interact with their target and not to other molecules. New therapeutic molecules

Figure 12.11. **In a chemical fume hood, a technician at a drug discovery company prepares a preliminary screen of several samples in parallel. Combining smaller organic molecules produces new larger compounds. Once tested, the samples are sent to high-throughput screening for additional assessment.** Photo by author.

parallel synthesis (par•al•lel syn•the•sis) the making large numbers of batches of similar compounds at the same time

library (li•brar•y) a collection of compounds, such as DNA molecules, RNA molecules, and proteins

Figure 12.12. **Combinatorial Chemistry Compound Screening Using Pipeting Robot** Thousands of samples may be tested (high throughput screening) and data collected using pipeting robots and spectrometry. Here a researcher screens samples for antibacterial properties. © Science Photo Library/ Photo Researchers.

must show some desired activity and also have the least amount of undesirable characteristics. Several new HIV protease inhibitors are currently in process development by Bayer and other biotechnology companies.

At AA ChemBio in San Diego, California, technicians create organic molecules for researchers at other biotech companies. This is called contract research and manufactuing. A client, for example, may contract AA ChemBio to create a library of compounds for use in a particular process. At most organic synthesis companies, a development department creates new compounds through combinatorial chemistry. An analytical department tests the characteristics of the newly produced compounds, including checking the molecular weight and purity of the new sample. Finally, a production department scales-up the manufacturing of the newly produced compounds for sale.

New Screening Technologies   Some drug screening and testing must take place in cells, either animal or human. Most organic compounds can be screened on sophisticated instruments in the lab to ensure their molecular structure, activity, purity, and/or concentration (see Figure 12.13).

High-throughput technologies are used to create new compounds, as well as to identify drug targets and screen reactions. In addition to the screens that are done in microtiter plates, scientists have developed ways to add minute amounts of samples to small fabricated disks called **biochips**. Biochips are, a type of **microarray** that can hold thousands of samples on a chip the size of a postage stamp (see Figure 12.14). Affymetrix, Inc. in Santa Clara, California, developed the first biochip, a GeneChip® array that contains millions of samples of synthesized fragments of DNA. The tiny pieces of DNA each have a specific sequence, to which scientists hope to assign a function. Each DNA sequence acts as a probe and may be bound with a labeled, complementary RNA or DNA strand (search strands) if it is distributed over the array. After the chip is flushed with labeled search strands, the chip is put into a chip reader and all the wells are scanned in search of the RNA-DNA hybrids. Using the chip in this way, researchers can locate and retrieve a single section of RNA or DNA that represents the expression of the DNA sequence of interest. With this technology, scientists can begin to discern the possible functions of an entire genome. Microarrays are discussed in more detail in Chapter 13.

In another application of microarray technology, thousands of therapeutic compounds may be screened for activity against a target molecule. If the substances being screened on the chip were proteins, we would call this a "protein chip." For instance, a blocker compound, such as an HIV blocker, may be needed to bind to a recognition protein on a cell surface and block the site where HIV attaches. Suspected compounds that may bind the recognition protein are attached at specific locations

biochip (bi•o•chip)   a special type of microarray that holds thousands of samples on a chip the size of a postage stamp

microarray (mi•cro•ar•ray)   a small glass slide or silicon chip with thousands of samples on it that can be used to assess the presence of a DNA or RNA sequence related to the expression of certain proteins

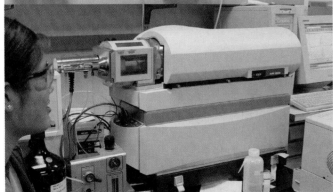

Figure 12.13.   **A mass spectrometer is used to verify the molecular weight of newly synthesized organic-compound libraries created through combinatorial chemistry. The graph on the computer monitor shows two peaks that represent two compounds in the sample, present in relatively high concentration.**
Photo by author.

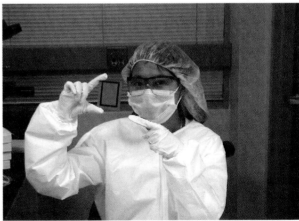

Figure 12.14.   **A technician tests the amount of DNA hybridization on a prototype microarray slide produced at Life Technologies, Corp. Microarrays allow hundreds of gene target samples to be screened in a short period of time.**
Photo by author.

or "wells" on the microarray chip. The recognition protein, with some sort of tag, such as a fluorescent dye, is added to the wells. If any of the compounds in any of the wells bind to the recognition protein, a chip reader will show fluorescence. The compound(s) that binds to the protein would be of interest as a potential therapeutic agent to block HIV from attaching to and infecting cells.

To assist researchers in designing compounds, scientists use molecular modeling and computer-assisted design technologies. Computer programs have the ability to create three-dimensional models of a molecule's structure from the data it has received. Then, other compounds can be designed that will bind or react to different areas of the targeted molecule. On the computer, these molecules can be altered and the results can be seen before wet-lab testing is performed. Eventually, though, all compounds are synthesized and tested in the lab (see Figure 12.15).

Figure 12.15. In the lab, a medical chemist concentrates and purifies samples using a "rotovap." The rotovap removes low boiling organic chemicals, usually solvents, from a mixture of compounds. Samples are then analyzed by mass spectrometry (see Figure 12.13). Photo by author.

BIOTECH ONLINE

Gail T. Colbern, DVM, MS, DACT is a Lab Animal Veterinarian and Director of Biopharmacology at a biopharmaceutical company. Dr. Colbern is responsible for ensuring that the animals used to test new medicines for safety and effectiveness are treated with the best possible care. She is also responsible for making certain that the animals do not suffer pain or distress during the studies. With a team of technicians, she makes sure that the animals have plenty of good food and water, and are comfortable in their environment. Monitoring the animals requires a lot of documentation.
Photo courtesy of Dr. Gail Colbern.

Animal Testing of Pharmaceuticals

Most people have a deep fondness for all kinds of animals. Dogs, cat, rabbits, hamsters, and birds are pets to millions of people around the world. It is difficult to imagine them being caged in laboratories for the purpose of testing drugs. But what if your husband or wife needed a cancer drug? You would probably want to be sure it was proven safe before it was available for human use. It is sometimes hard to accept that any kind of animal testing is necessary, but by knowing the contributions that animals have made to medical advances, it is easier to understand why animal testing is conducted.

Use Internet resources to learn more about animals in biomedical research and how you feel about such practices.

1. Use the CSBR website at biotech.emcp.net/CSBR to learn about the contributions of animals to medical research. List five contributions made by research rabbits, dogs, or livestock.

2. Go to biotech.emcp.net/animalresearch and view the video clips on the benefits of animal research. Record three contributions of animals to medical research presented in the video clips.

3. Find two websites that you think give the best arguments against using animals in biomedical research. List four reasons that are presented on these sites as arguments against using research animals.

4. Create a one-page document that describes your own personal position on using animals for biomedical research. List the kind of testing that you believe is acceptable and the types of animals that you are comfortable with being used as test subjects.

Section 12.2 Review Questions

1. What is the value of combinatorial chemistry?
2. How are chemical compound libraries related to high-throughput screening?
3. What is the value of microarray technology? gene therapy? Cite an example of how it can be used.

12.3 Creating Pharmaceuticals through Peptide and DNA Synthesis

Proteins are long chains of amino acids that are folded into a functional unit. **Peptides**, on the other hand, are amino acid chains, up to a few dozen amino acids in length, which are too short to be significantly folded (see Figure 12.16). Even though they are rather small molecules, peptides have become important in medical R&D in several ways.

- A peptide can be used in an attempt to identify regulator molecules. In medical research, regulatory molecules may control the manifestation of a disease. Often, a peptide can be synthesized that will bind to a regulatory molecule and directly interfere with disease expression. If scientists understand the molecule they are trying to block, they can design and synthesize a peptide to try to block it.
- A peptide may be used as a vaccine antigen to initiate an antibody response. For example, the HIV gp120 molecule is very large, and it can only be made inside mammalian cells. Scientists have found that only a small section of gp120, and not the whole molecule, may be sufficient to cause the human body to produce antibodies that will recognize and destroy HIV. It is relatively easy to synthesize a small part of the HIV gp120 molecule in the lab. This strategy has been used with several different vaccines. Unfortunately, early trials have not yet shown strong efficacy of the gp120 vaccine.
- Peptides are often synthesized for use in the purification of other proteins. In affinity chromatography, peptides may be attached to chromatographic resin beads, and used to bind and separate other proteins as a mixture flows through a column.

Several companies synthesize peptides for a variety of customers. Most utilize column technology to make peptides, in which resin beads bind to a "charged" amino acid. The first type of amino acid is added in excess to a column. The column is washed and the excess amino acids that do not bind to the resin beads are washed off. The second amino acid is flushed through the column, and a reaction binds it to the amino acid already on the beads. The column is washed again and excess amino acids that did not bind to the resin beads are washed off again. The process is repeated over and over until the peptide is the desired length with the desired amino acids in the desired order. The peptide is eluted off the column; then, a mass spectrometer and high-performance liquid chromatography (HPLC) are used to check the peptide's molecular weight and purity.

The development of **peptide synthesizers** automated the synthesis of peptides. Peptide synthesizers can be programmed to run a column, adding the correct amino acids in the correct sequence to make a desired peptide (see Figure 12.17). A synthesized peptide has a maximum length of about 50 amino acids. Many companies have developed and marketed synthesizers, including Protein Technologies, Inc., of Tuscon, AZ, and C S Bio Company, Inc., of Menlo Park, California.

Peptides are being tested as therapies for several disorders, including prostate cancer, type 1 diabetes, multiple sclerosis, and some other autoimmune diseases.

peptides (pep•tides) the short amino acid chains that are not folded into a functional protein

peptide synthesizer (pep•tide syn•the•siz•er) an instrument that is used to make peptides, up to a maximum of a few dozen amino acids in length

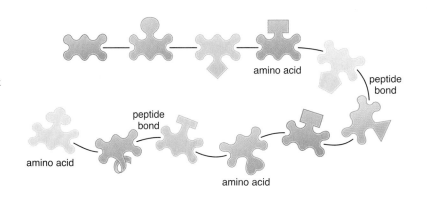

Figure 12.16. Structure of Peptides. Peptides are important in drug discovery and as potential pharmaceuticals. Using peptide synthesizers, peptides are made outside of cells by binding one amino acid at a time to another. Each amino-acid pair is connected by a peptide bond.

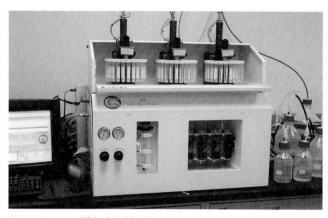

Figure 12.17. This C S Bio Company, Inc. peptide synthesizer automates the formation of peptide chains. The instrument is programmed to bind specified amino acids in a specific order. Along the top are 20 tubes filled with the 20 amino acids.
Photo by author.

Figure 12.18. A DNA synthesizer by Applied Biosystems™ (Life Technologies Corp.) can be programmed to produce small DNA lengths (up to about 50 bases) for use as primers or probes. The tiny bottles contain the A, T, G, and C nucleotides. The large bottles contain other reagents and buffers.
Photo by author.

BIOTECH ONLINE

Small but Powerful

Peptide therapies are a new research focus in the effort to combat several disorders including multiple sclerosis, Alzheimer's disease, Parkinson's Disease and several cancers. Learn how peptides composed of short chains of specific amino acids are being studied as therapeutics to some important human diseases.

1. **Read the article about cancer and peptide therapeutics at www.ncbi.nlm.nih.gov/pmc/articles/PMC3539351. Summarize, in** 6–7 **sentences, the different ways peptides are being tested for therapeutic purposes**.

2. **Read the article at cancer.osu.edu/news-and-media/news/Combination-Peptide-Therapies-Might-Offer-More-Effective-Less-Toxic-Cancer-Treatment to learn how combining multiple peptide therapies may impact the onset of an aggressive breast cancer. Give an example of combination peptide therapy.**

Oligonucleotide Synthesis

In addition to the synthesis of peptides, scientists have learned how to make small pieces of DNA, called **oligonucleotides** (oligos, for short). Oligos are segments of nucleic acid, usually 50 nucleotides or less in length. Oligos may be used as primers, probes, and recognition or blocker molecules. Oligonucleotides are critical in research for the identification of genes involved in disease, or for use in disease prevention.

Primers are produced by oligonucleotide synthesis for polymerase chain reaction (PCR) and DNA sequencing applications (See Chapter 13). Primers are segments of about 25 nucleotides in length that can specifically bind to a single-stranded DNA molecule. Primers are used to start synthesis and sequencing reactions.

DNA synthesis is done on a **DNA synthesizer**. The technology used to synthesize oligos is similar to the technology used in peptide synthesis (see Figure 12.18). In oligonucleotide synthesis, a column is set up with resin beads with a starter nucleotide. A second nucleotide is added to the column, and it binds to the first nucleotide. The column is washed, and a third oligonucleotide is added, and so on. Several companies, such as BioAutomation Corp. from Plano, Texas, market automated DNA synthesizers to other biotech companies for on-site oligonucleotide production. Oligos are so important to basic research that some companies have their own "oligo factories."

oligonucleotides (ol•i•go•nu•cle•o•tides) the segments of nucleic acid that are 50 nucleotides or less in length

DNA synthesizer (DNA syn•the•si•zer) an instrument that produces short sections of DNA, up to a few hundred base pairs in length

Creating antisense strands of nucleotides (strands that match the noncoding complement of a gene) also gives researchers a way to block or interfere with gene expression. This is currently being studied for use in some future therapeutics, including one to treat a certain type of lung cancer and another that may bind to HIV nucleic acids to block protein expression and disease symptoms. Find other examples of antisense therapeutics at **kcancer.com/node/96**.

Many companies, such as Integrated DNA Technologies, Inc., produce oligos for outside clients (contract manufacturing). A client submits an order form with a desired sequence. The company produces the volume and concentration of the requested oligo and ships it to the customer.

Section 12.3 Review Questions

1. What are three uses of peptides in medicinal biotechnology?
2. How does a peptide synthesizer make peptides?
3. How does a DNA synthesizer make oligonucleotides?
4. Of what value are oligonucleotides?

12.4 Creating Pharmaceuticals by Protein/Antibody Engineering

In the 1970s, biotechnologists learned how to manipulate cells in culture to take up foreign genes (rDNA) and use them to produce pharmaceutical proteins. Now, dozens of recombinant protein pharmaceuticals are produced in marketable volumes when transformed cells are grown in large volume bioreactors. Many pharmaceutical proteins, including rInsulin (for diabetes), rReteplase (for heart attacks), and rFactor XIII (for hemophilia), plus many other antibodies, enzymes, hormones, and structural molecules are produced using recombinant DNA and genetic engineering technologies.

Several examples of recombinant protein pharmaceuticals have been given through this text. A list of the top 25 independent pharmaceutical biotechnology companies and links to the recombinant therapeutic products they make, can be found at **en.wikipedia.org/wiki/List_of_largest_biotechnology_companies** (see Figure 12.19).

One of the most important groups of pharmaceutical biotechnology products are antibodies. The structure and function of antibodies was covered in Chapter 5. In this section, the role of antibodies and associated molecules in medical applications is discussed in more detail.

Antibody Technologies

Engineering cells to make antibodies is an expanding focus in medical biotechnology as scientists look for ways to improve a patient's immune system to prevent or fight disease. According to IBC Life Sciences, engineering antibodies to recognize specific molecules is expected to produce more than $3 billion in therapeutics annually.

Figure 12.19. **Genentech, Inc. was the first, and is one of the largest, recombinant protein pharmaceutical companies in the world. Genentech scientists engineer *E. coli* and Chinese hamster ovary (CHO) cells to make human proteins, including several antibodies, for therapeutic drugs and vaccines.**
Photo by author.

As discussed in Chapter 5, antibodies are the most complicated of all proteins. They are large, with four peptide chains bound together in the shape of a "Y." The tips of the Y vary from one type of antibody to another and are called variable regions. The tips of the antibody recognize and bind to other molecules called antigens (see Figure 12.20). Antibodies are usually very specific, recognizing only one or a few different types of antigens. The body produces millions of different antibodies that very specifically recognize only one or a few specific molecules.

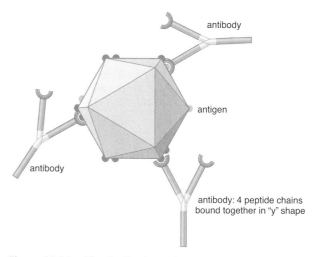

Figure 12.20. The Antibody-Antigen Reaction. Antibodies recognize and attach to antigens. If several antigens are present, several antibodies may attach to them. Different antibodies recognize different antigens.

Antibodies in Medical Research and Pharmaceutical Production

Antibodies are used to recognize molecules in medical research applications. For example, fluorescent dye molecules can be attached to antibodies, and then the fluorescently tagged antibody can recognize surface proteins on cells, such as those on cancer cells. Any type of cell can be targeted, since antibodies can be made to recognize them. In **flow cytometry**, fluorescent antibodies attach to surface proteins on certain cells. The flow cytometer can recognize the cells by their fluorescence and sort them into separate vessels (see Figure 12.21). This technology is valuable for research and diagnostics.

Antibodies are also used in pharmaceutical manufacturing in the final steps of purifying a pharmaceutical protein product from cell culture. Antibodies may be bound to the resin beads used in affinity chromatography columns (see Section 9.2), and when a mixture of proteins is passed through the column, the antibodies bind to the protein product, pulling them out of the solution.

Using Antibodies and Antigens in Vaccines

Antibody-antigen reactions are important in disease prevention, and several biotechnology companies produce antibodies or antigens for use in vaccines or vaccine research (see Figure 12.22). A vaccine is something (with an antigenic region) that, when given to an individual, increases the production of antibodies (in the body) against the compound. The compound (which is an antigen to the antibody) may be a virus particle, a protein, a carbohydrate, or even part of a nucleic acid.

flow cytometry (flow cy•tom•e•try) a process by which cells are sorted by an instrument, a cytometer, that recognizes fluorescent antibodies attached to surface proteins on certain cells

B cells (B cells) specialized cells of the immune system that are used to generate and release antibodies

memory cell (mem•or•y cell) a specialized type of B cell that remains in the body for long periods of time with the ability to make antibodies to a specific antigen

immunity (im•mu•ni•ty) protection against any foreign disease-causing agent

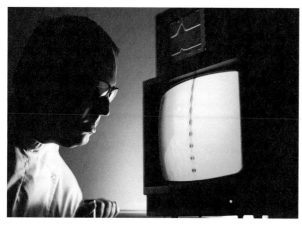

Figure 12.21. A researcher studies a stream of cells being counted and sorted by a flow cytometer. As cells pass through a laser beam, the scattering of the beam by antibody-tagged proteins provides a way to identify specific cells. The cells are disrupted into drops that are then sorted. Up to 50,000 cells can be counted and sorted per second.
© Science Photo Library/Photo Researchers.

Figure 12.22. In the 1990s, Cell Genesys, Inc. was an up-and-coming vaccine-producing company. In 2008, Cell Genesys reduced its staff to 25% of its former size because its lead product, a prostate cancer vaccine (GVAX) could not pass its Phase III clinical trial. Fortunately, other biotech companies, such as Dendreon Corp. in Seattle, Washington, has had success producing proteins that function as a prostate cancer vaccine.
Photo by author.

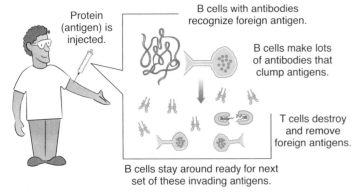

Protein (antigen) is injected.

B cells with antibodies recognize foreign antigen.

B cells make lots of antibodies that clump antigens.

T cells destroy and remove foreign antigens.

B cells stay around ready for next set of these invading antigens.

Patient is injected with an antigenic agent (in this case a protein). Patient's body increases B cell production and production of the specific antibody to the antigen. Even after the antigen is removed by the antibodies and white blood cells, some B cells live for a long time, ready to make these antibodies if exposed to the antigen in the future (immunity).

Figure 12.23. Vaccine Immunity. A vaccine injection of antigen initiates an immune response.

When antibody production increases in the body, the **B cells** that make the antibodies may remain in the body for a long period of time. They are called **memory cells**. Some memory cells stay in your body all of your life (eg, antibodies to smallpox). Some remain for several years, but eventually decrease in numbers (eg, antibodies to tetanus). The vaccine provides long-term protection (**immunity**) from invading particles because memory cells recognize invaders, and they quickly produce the specific antibodies that start an immune response that eliminates the intruders from the body before one becomes ill (see Figure 12.23).

BIOTECH ONLINE

© Bureau L.A. Collections/ Rhythm & Hues.

Getting Sick is No Laughing Matter

TO DO

Learn about how the human immune system recognizes and disables foreign invaders defending the body from disease. Report this information in a cartoon format.

1. Find one or more websites that describe the actions of the following immune system components. Write a description of the function or action of each component and record the URL of the website(s) you used to gather the information.

   • lymph system, including the thymus and spleen
   • bone marrow
   • white blood cells, including:
      • leukocytes (granulocytes, lymphocytes, monocytes)
      • B cells and the antibodies they produce
      • T cells (helper T-cells, killer T-cells, suppressor T-cells)
      • phagocytes
      • macrophages
   • complement system

2. Using one of the comic strip generating websites on the Internet such as **stripgenerator.com**, create a three-panel comic strip or one-panel cartoon that explains, teaches, or informs readers about some aspect of how the immune system works. Comic strips that are humorous as well as scientifically accurate will be garner the most glory.

Chapter 12

Developing vaccines that allow the body to recognize and destroy proteins on the surface of disease-causing organisms is a strategy for many biotechnology companies. Some disease-causing viruses and bacteria have several strains or mutate very quickly. Often a vaccine that works one year does not work another year. Such is the case with the flu virus and the annual flu vaccine. Each year, scientists try to predict what the new flu virus will be, and then they try to develop the correct vaccine in time to make many millions of doses available to the public (see Figure 12.24).

Antibodies for medical research and treatment are usually produced through genetic engineering of mammalian cells. Recently though, medical biotechnologists have learned how to grow human pharmaceuticals in plants and animals.

Section 11.3 contains a discussion of plant-based pharmaceuticals (PBPs) and the production of human medications, including human antibodies, in genetically modified edible crops like corn and bananas.

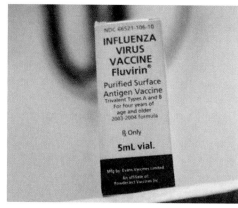

Figure 12.24. Chiron Corporation was the leading flu vaccine producer in the United States. Fluvirin® was Chiron's version of the 2004 flu vaccine. Somehow, it was contaminated during manufacturing, which resulted in a significant flu shot shortage and created a huge healthcare crisis.
© Shannon Stapleton/Reuters/Corbis.

BIOTECH ONLINE

Why Eating Your Vegetables Could Be Even More Important Than You Thought

Scientists are working on ways to genetically engineer common fruits and vegetables with DNA to produce antibodies to fight some horrible diseases.

Go to www.journalcra.com/sites/default/files/Download%20632.pdf. Read and summarize the article *"Edible" Vaccine - Vegetables As Alternative To Needles.* Would the added protection convince you to eat more broccoli and brussels sprouts?

Recently, medical biotechnologists have looked to livestock as production vessels for human pharmaceuticals. Animals that are genetically modified to make human pharmaceutical proteins are technically **transgenic** animals but are fondly called "pharm" animals. Some proponents of the new technology say that if cattle, sheep, goats, or chicken can be engineered to produce human pharmaceutical proteins, some drugs could be made in larger quantities and at lower cost. An additional benefit is that pharmaceutical proteins in the meat or milk of pharm animals may be easier to administer. Maybe, you could eat your medication instead of receiving a shot.

In 2009, the FDA approved the first therapeutic from transgenic livestock. It is the pharmaceutical ATryn®, a human blood-clotting agent, recombinant antithrombin, made in GM-goats. ATryn is produced by GTC Biotherapeutics, Inc. a Massachusetts company, which claims that in one year one of its goats can produce the equivalent of 90,000 blood donations' worth of antithrombin (biotech.emcp.net/goatdrug).

Transgenic animals are being used for pharmaceutical production, but they are also important for medical research. Some transgenic animals are used to study medical disorders. For example, transgenic mice with the human gene for gamma-secretase modulator (GSM) have shown an ability to decrease the amyloid plaque buildup associated with Alzheimer's disease (biotech.emcp.net/envivopharma). Now, knowing that GSM may help treat Alzheimer's disease increases the possibility of making GSM as a therapeutic or targeting the GSM gene for regulation.

There are some ethical concerns about using transgenic plants and animals as pharmaceutical factories. However, the potential of using livestock and plant crops to produce large volumes of a variety of new and improved disease therapies is exciting to medical scientists and practitioners with patients that are waiting for lifesaving therapies.

transgenic (trans•gen•ic) the transfer of genes from one species to another, as in genetic engineering

Monoclonal Antibodies

Monoclonal antibody technology is having a large impact on medical biotechnology and shows promise to increase treatment options for patients with a variety of diseases. Monoclonal antibodies (mAb, for short) are antibodies produced in special cells called hybridomas. Produced by fusing immortal tumor cells with specific antibody-producing WBCs (B cells), hybridomas grow rapidly, making large amounts of the specific antibodies that were coded for in the original B cells. The advantage of monoclonal antibody technology is that many identical antibodies to specific antigens are produced in large quantities. These can be used in genetic testing, medical diagnostics, and in disease treatment (see Figure 12.25)

One of the hottest fields in medical research and treatment is conjugated (or coupled) monoclonal antibodies. A conjugated antibody is one that has attached to it another molecule such as a drug or a radioactive material. Together the coupled pair can locate and bind specifically to a target in or on a cell. In this way, a cancer drug could be bound to an antibody and it could be delivered specifically to the target cancer cells.

Conjugated antibodies deliver drugs directly to specific cells and not others. Conjugated mAbs may be more effective in treating cells and may reduce the amount of drug circulating in the body causing unwanted side effects. An example is Brentuximab vedotin (Adcetris®) produced by Seattle Genetics, Inc. Brentuximab vedotin is a monoclonal antibody that targets the CD30 antigen found on certain cancerous lymphocytes (lymphomas). The mAb has a drug called MMAE, an antimitotic, conjugated to it. When the conjugated antibody finds its targeted lymphocyte it blocks cell division in those cells and not others.

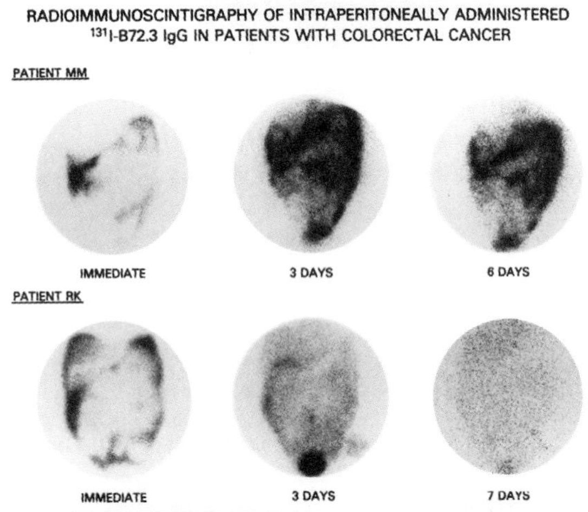

Antibody Recruiting Molecules

Several pathogens (ie, the HIV) are not well recognized by antibodies. You will remember that in an immune response, antibodies bind to antigens and to each other. Then white blood cells recognize the antigen-bound antibodies, engulf them, and destroy them. One theory explaining poor antibody activity against pathogens such as the HIV is that the antigenic molecules are too widely space on the pathogen for the antibodies to bind to them and each other.

One of the newest therapeutic technologies is the development of antibody recruiting molecules (ARM). These are molecules that have a great affinity for a specific antigen and the antibodies that recognize them. The ARM assists the antibody binding by creating a bridge between the antigen and the antibody. It is thought that ARM technology may be applicable to several pathogenic therapies, including assisting antibodies in the recognition of cancer cells. Read more about a specific ARM for HIV at **biotech.emcp.net/HIV_ARM**.

RADIOIMMUNOSCINTIGRAPHY OF INTRAPERITONEALLY ADMINISTERED
^{131}I-B72.3 IgG IN PATIENTS WITH COLORECTAL CANCER

PATIENT MM

IMMEDIATE   3 DAYS   6 DAYS

PATIENT RK

IMMEDIATE   3 DAYS   7 DAYS

Figure 12.25. **Monoclonal antibodies used to detect colon cancer.** Gamma camera scans of the abdomens (colons) of two patients. Each was injected with a monoclonal antibody that recognizes colorectal, ovarian, and breast carcinoma. Patient MM shows high concentration of the colon cancer-recognizing antibody.
National Cancer Institute.

Section 12.4 Review Questions

1. How are antibodies used in flow cytometry?
2. How does a vaccine provide immunity?
3. Give an example of a human pharmaceutical produced in a transgenic animal.
4. How are monoclonal antibodies different from naturally occurring antibodies?

Medical biotechnology is much more than just the engineering of organisms to make novel pharmaceuticals. In the broadest sense, medical biotechnology is the application of biology, chemistry, physics, and engineering to develop the methods, tools, and products that improve health care. In this section, some of the most promising recent discoveries and inventions in the field of medical biotechnology are presented.

Genetic Testing in Medicine

Genetic testing is the analysis of DNA sequences in order to learn if an individual has specific "genes of interest." The term, "genetic testing" is usually used when referring to the analysis of human DNA to identify genes thought to cause inheritable disease, but can also be used when identifying genes of interest in animals for veterinary or breeding purposes. The genes may have been inherited or resulted from a mutation during the individual's life. Genetic tests are used to diagnose disease, to identify increased risks of potential health problems, and to choose appropriate medical treatments.

According to the National Institutes of Health (NIH), there are over 2000 genetic tests currently available to detect human diseases and disorders. The NIH maintains an extensive website with medical and research information on genetic and rare diseases (rarediseases.info.nih.gov/gard).

It is known that several disease-causing genes are more common in people of certain ancestry. These diseases have a genetic predisposition. For many of these diseases, there are genetic tests for potential parents to predict the likelihood of passing on a genetic disorder. Tay-Sachs disease, for example, is a painful nerve cell disorder that usually kills afflicted children by the age of four. People of Eastern European Ashkenazi Jewish ancestry are more likely to carry the Tay-Sachs disease mutation. Although a baby would have to inherit a "bad" Tay-Sachs gene from each parent, knowing whether both parents carry the mutant gene and the likelihood of passing it on to their unborn children is important.

Similarly, two inheritable blood disorders are routinely screened through genetic testing. Thalassemia is an inherited mutation that results in faulty hemoglobin. It is known to be more common in Mediterranean and Far East Asian populations. Sickle cell disease is also an inherited hemoglobin disorder and is known to be more common in people of African ancestry. If parents or children are screened for these genetic disorders, they will be able to plan appropriate treatments to help prolong the life of afflicted individuals.

How are genetic disorders diagnosed? It depends on the disease, disorder or condition. A patient may visit a doctor for a physical examination: Physical characteristics are checked and measured. In the case of breast lumps or cysts, breast cancer may be a possibility. A doctor will do a breast exam by hand to feel for the presence and size of any lumps. A patient is then sent for a mammogram (X-ray image) and/or an ultra-sound (high frequency sound waves imaging) examination. For cancers and other conditions, doctors may order magnetic resonance imaging (MRI) or computerized tomography (CT) scans to visualize structures in the body.

Along with these tests, a patient's personal and familial medical history is taken. Genetic disorders are usually inherited and by mapping out the occurrence of a disorder through generations doctors can better determine which genetic tests should be run. A diagram that shows the members of a family that have or are carriers of the genes of an inherited condition is called a pedigree (see Figure 12.26). Pedigree charts often elucidate the likelihood of a disorder and the chances of passing it on, and may suggest that genetic testing is needed to help confirm a diagnosis.

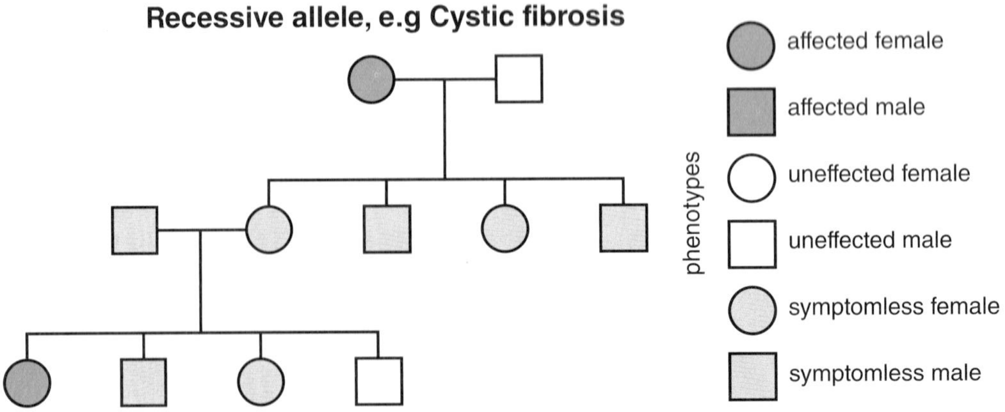

Recessive allele, e.g Cystic fibrosis

Figure 12.26. Genetic explanation. Since the condition is not shown in any of the offspring in the first generation but it reappears in the second generation, it must be caused by a recessive allele. The recessive allele is denoted by a lower case letter (for example, an "f") as distinct from the upper case version of the same letter (F) for the normal allele (dominant, in this case). In this case the appearance of the condition is independent of the sex of the individual.

phenotypes
- affected female
- affected male
- uneffected female
- uneffected male
- symptomless female
- symptomless male

genetic counselor (ge•ne•tic coun•sel•or) a clinician that reviews genetic histories and using data collected helps patients understand their risk of having a particular disorder

Genetic testing is done using a small sample of blood, body fluid (such as saliva or amniotic fluid) or a sample of cells. The sample is sent to a clinical laboratory that specializes in genetic testing. Depending on the disorders being tested, there may be one or many tests done on the sample. Genetic tests may include analysis of DNA sequences using DNA sequencing or PCR, whole chromosomal studies using microscopy, cell studies using biological markers in or on the surface of cells, or protein or enzyme testing.

Making the decision to undergo a genetic test is not always an easy decision. People have different reasons for being tested or not being tested. For some, not knowing that they have a genetic predisposition to a genetic disorder may give them peace of mind. Maybe not knowing that you have the gene for an old-age illness, such as Huntington's disease, allows you to live your life with less fear of the future. In some cases, there is no treatment for a positive genetic test result. For some though, it is important to know whether they are carriers of "bad" genes that could be passed to their children, or if a disease due to faulty genes could be prevented or treated. Test results might help a person make life decisions, such as family planning or insurance coverage. A **genetic counselor** can assess the risk of a genetic disorder by researching a family's history and by evaluating medical records. They can provide information about the pros and cons of testing and help a patient knows what their options are for the future.

BIOTECH ONLINE

Diagnosis this Genetic Disorder

Several famous people in history have had well-known genetic disorders. Use the Internet to diagnose the genetic disorder of the following historical figures.

Historical Figure	Symptoms	Genetic Disorder	Cause	website(s) Used
Abraham Lincoln (16th President of The United States)	Elongated skeletal features including very tall, slender, spindly fingers and toes, dislocated lenses of eye(s), weaken aorta,			
King George III (King of Great Britain and Ireland 1760 -1801)	Overly sensitive to sunlight, red or brown urine, increased hair growth			
Woody Gutherie (American folk musician, 1930s—1940s)	Uncontrolled, involuntary jerky movement, Difficulty with speech, posture, and balance, Progressive dementia			
Miles Davis (20th Century Jazz Musician)	Painful swelling of the hands and feet, fatigue, yellowish color of the skin, (jaundice) or whites of the eyes,			

A Case Study

A case study is an in-depth review of the scientific information available for an individual or group with a particular condition. There may be several different reasons for doing a case study but one may be to better understand the condition in general and apply it to other cases. Case studies often start with a patient's history, risk factors, and test results.

Consider a 50 year-old patient; let's call her Jayne, a mother of 3 children, learning after a biopsy of a lump in her breast, that she has an early-stage breast cancer tumor. What kind of breast cancer is it? There are several types requiring different treatment strategies. Is it a genetic cancer due to a mutant regulatory gene? Often these are more aggressive cancers. If Jayne has the mutant gene, which of Jayne's known family members also carry the cancer gene? What are the chances of Jayne passing the cancer-related genes to her children? What are the chances of Jayne's brothers or sisters having the cancer-related genes, getting a related cancer, and/or passing the genes on to their children? If Jayne's is not a known genetic cancer then is it a different known cancer with treatments that are shown to be more effective than others? What should Jayne expect in the short-term or long-term?

When Jayne's ancestry is studied it is learned that both sides of her family are from Poland and are Ashkenazi Jewish. There is breast cancer, prostate cancer, melanoma, and blood cancer on both her mother and father's sides of the family. This information makes Jayne at much higher risk of having one of the "more aggressive" breast cancers. If she carries the (mutant) gene for one of the genetic breast cancers then her children, both male and female, have a 50% chance of inheriting that gene, putting them at high risk of developing that or a related cancer.

Jayne meets with a genetic counselor that creates a pedigree chart showing where cancer is found in Jayne's ancestry (see Figure 12.27). The genetic counselor describes risks, in numerical values, of cancer developing in family members. With her ancestry and the presence of breast and prostate cancer in her family, Jayne is 5X more likely than other breast cancer patients to have one of the genetic breast cancers. Jayne will have a genetic test panel that will study her DNA from a blood sample and look for the presence of 49 genes that are known to have some connection to breast cancer as well as some other cancers in her family.

Jayne's Family Pedigree Chart

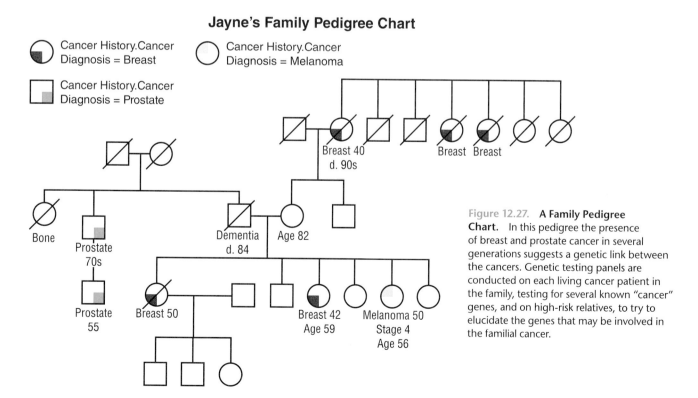

Figure 12.27. **A Family Pedigree Chart.** In this pedigree the presence of breast and prostate cancer in several generations suggests a genetic link between the cancers. Genetic testing panels are conducted on each living cancer patient in the family, testing for several known "cancer" genes, and on high-risk relatives, to try to elucidate the genes that may be involved in the familial cancer.

During Jayne's tumor biopsy, tissue samples are sent to a pathologist to determine the type of cancer cells present in the tumor. Many tests are done to see what proteins are present on the inside and on the surface of the cells to see if any known breast cancer marker proteins are present. When the pathology results of Jayne's tumor cells came back, it showed that her tumor cells were estrogen-positive (ER+). This means that Jayne's cancer cells have an elevated number of ER proteins on the surface of the cells where estrogen binds and stimulates abnormal growth. Fortunately, ER+ breast cancer responds well to estrogen inhibitors, which will be one strategy that will be used to treat Jayne after her tumor is removed.

Jayne has a decision to make though. Should she have the cancerous lump removed from her breast? This is called a lumpectomy and it will save her breast. If Jayne decides on the lumpectomy, it will be a 4–6 week recovery period followed by a month of daily high-dose radiation treatments and then long-term use of an oral estrogen inhibitor medication. Or, should she have a single or double breast removal, called a unilateral or bilateral mastectomy? Having both breasts removed reduces her chance to near 3% of having a reoccurring breast cancer in the future versus a 5% reoccurrence with the lumpectomy treatment. A bilateral mastectomy may give Jayne more peace of mind, however it is a significantly major surgery with 6–8 weeks of recovery and possibly follow-up surgical procedures. Of course, mastectomy significantly changes the outward appearance of the body. Jayne may be a candidate for breast reconstructive surgery. Reconstruction can add weeks to the recovery process and could have complications that lead to follow-up surgical procedures.

Jayne's oncologist shares with her the most recent scientific findings that one procedure over the other shows no long-term increased survival. What surgery should Jayne decide to have? It is a difficult decision. With the results of her genetic screen, Jayne is fortunate to learn that she does not have any of the 49 tested breast cancer genes. This helps Jayne decide to choose the lumpectomy treatment. Can you see why she made this decision? Do you agree with her decision? Jayne's negative genetic breast cancer testing does not mean her siblings (brothers and sisters) don't have any of those high-risk genes. Can you explain why? All of Jayne's siblings will undergo genetic testing. Can you explain why? Would you make the same choice?

Pharmacogenetics and Personalized Medicines

Since no two people's DNA have the same sequence, the response to different medications may differ between individuals or groups of people. With knowledge of the human genome, variations in the genome sequence, and variations in the resulting proteins, scientists and doctors will be able to design and modify drugs to better meet an individual's needs (see Figure 12.28). Medical scientists will be able to prepare a unique pharmaceutical molecule or pharmaceutical regimen personalized for each patient.

Pharmacogenetics is a branch of biotechnology that involves utilizing genetic and protein codes to design or improve medications. Pharmacogeneticists apply advances

pharmacogenetics (phar•ma•co•gen•e•tics) a branch of biotechnology that involves utilizing genetic and protein codes to design or improve medications

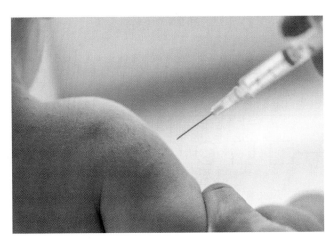

Figure 12.28. **Vaccinations protect children from some deadly bacterial and viral diseases. However, a small percentage of the patients have adverse reactions to vaccinations. Personalized vaccines, based on a person's genetic makeup, could be developed that would minimize adverse or allergic reactions.**
© Gajus/iStock.

in PCR, sequencing, microarrays, proteomics, and other molecular biology technologies (discussed further in chapters 13 and 14) to create new, personalized therapies (personalized medicine). Personalized medicines give an opportunity to tailor-make pharmaceuticals that are more effective, and may have less undesirable side effects, for different ethnicities, age groups, families, and individuals.

Pharmacogenetics is being used to fight certain kinds of cancers, such as those with a genetic basis. Human epidermal growth factor receptor 2 (HER2) antibodies recognize HER2 proteins on the surface of cancer cells. The overexpression of the HER2 gene overproduces the HER2 protein. By screening women (genetic testing) for the HER2 gene, doctors know the likely effectiveness of the anti-HER2 antibody treatment before they try it.

Using DNA sequencing, protein sequencing, or other assays, the identity of an individual's specific genetic and protein variations, compared with the majority of the population, will allow for individual drug design that could reduce the risk for allergic reactions and side effects and increase drug effectiveness. Since some mutations are common, variations of medications could be designed for groups of people.

Some companies are beginning to focus their research and manufacturing in pharmacogenetics. One of the first is Physician's Choice Laboratory Systems (PCLS) in South Carolina, which is developing personalized testing including molecular and genetic testing to detect variants of prostate and bladder cancers in different individuals. Other companies are working in cancer, cardiovascular, and mental health arenas. For the latest trends in pharmacogenetics, visit the American Medical Association's website at www.ama-assn.org/ama/pub/physician-resources/medical-science/genetics-molecular-medicine/current-topics/pharmacogenomics.page.

The promise of personalized treatments is leading to genetic screening panels that test for hundreds of genes at a time. These large screening panels are now available from several biotechnology companies, including Courtagen Life Sciences, Inc in Woburn, MA. As of 2015, using saliva samples, Courtagen had 10 genetic panels available that could test for a variety of neurological and mitchondrial disorders. For example, the Courtagen panel, epiSEEK®, test for 471 genes associated with Epilepsy and other seizure disorders. Learn more about this and other Courtagen gene panels at www.courtagen.com/test-menu-genetic-testing.htm.

BIOTECH ONLINE

A Medicine Just for YOU

Learn how having a certain mutation can make a certain personalized medicine work for you.

Go to biotech.emcp.net/lungcancerdrug. Read the article and summarize in a few sentences how this disease, mutation, and treatment are an example of personalized medicine. Then, click on the link and find the genetic answer to why this drug works well but only in some people.

© M. Miele/Corbis.

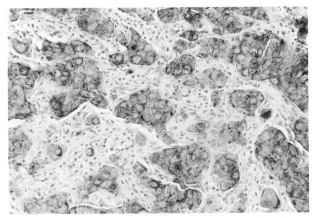

Figure 12.29. Human breast tumor cells are seen invading breast tissue. The cytoplasm of the tumor cells is stained brown with an antibody, which recognizes a specific antigen, called CEA, found in the cancer cells. 313X
National Cancer Institute.

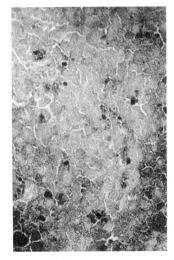

Figure 12.30. Gene therapy to treat melanoma. A metastatic melanoma (skin cancer) patient treated with tumor-infiltrating white blood cells carrying the gene for tumor necrosis factor (TNF). The TNF gene-carrying cells were injected into tumors (left) and these became necrotic (shriveled and died), leaving healthy tissue (right).
Dr. Steven Rosenberg.

■ ■ ■ ■ ■ ■ ■ ■ ■ ■ ■ ■

biomarker (bio•mar•ker)
a substance, often a protein or section of a nucleic acid, used to indicate, identify, or measure the presence or activity of another biological substance or process

Biomarkers and Diagnostics

One of the tools needed for the design of personalized medicines and genetic diagnosis is the ability to map individual patient's genes and protein expression and look at the differences between people who are sick and people who are not sick. These differences are markers, called **biomarkers**. Biomarkers may be sections of DNA or proteins in a certain cell or tissues, or they may be an organic or inorganic compound that is used to measure some aspect of health or treatment. Biomarkers may be used to diagnosis and treat disease, but they may also be used to monitor the effectiveness of a treatment.

Some examples of biomarkers are:

- protein that is monitored in blood or urine samples
- gene sequence that is present only in patients afflicted with a given disease
- protein on the surface of cancer cells (see Figure 12.29)
- radioactive dye used in CT scans or MRIs that binds to tumors
- unique antibody that is overexpressed in autoimmune diseases such as arthritis or lupus
- protein that is found in high concentration after a heart attack
- molecule that is found in higher concentrations after a medical treatment
- gene sequence that is only found in members of a specific ethnic group

One new application of the technology is the use of a biomarker dye to detect proteins in the saliva of breast cancer patients. The biomarker dye–cancer protein complex emits a detectable fluorescence. With this biomarker test, patients may be diagnosed earlier and possibly without the invasiveness of a breast tumor biopsy. Learn more about this "cancer spit test" and its potential application to other medical detection at: www.sciencedaily.com/releases/2011/08/110831155326.htm

Another exciting application of biomarker technology is the use of a biomarker to detect a gene that is associated with an adult onset disease such as maturity onset diabetes (type 2 diabetes). Maturity onset diabetes is found in middle age and elderly patients. About 5% of all diabetics carry a genetic marker for this form of diabetes. Adult diabetic patients with this marker do not respond well to insulin injections and respond better to an alternate treatment, sulfonylurea pills. Imagine learning that you can take pills instead of daily injections! Or if you are a parent, testing your child and learning that he/she carries the gene gives you an opportunity to try to control the progression of the disease through diet and exercise.

Gene Therapy

Gene therapy, briefly discussed in Section 4.3, is an area of medical biotechnology that has the potential to treat and even cure diseases that are inherited or genetic in origin. In gene therapy, functional genes are used to replace or improve the function of a defective gene, or new genes with new functions are used to correct a medical condition (see Figure 12.30). Research and development is in progress for several new gene therapies to correct or improve several common conditions and diseases, including hypertension (high blood pressure), hypercholesterolemia, muscular dystrophy, cystic fibrosis, and hemophilia.

Other New Medical Biotechnologies

Some other exciting new developments and applications of medical biotechnology include:

Artificial Organs   In the 1990s, biomedical technologists began to develop artificial tissues and organs that could replace or augment the function of damaged ones. Artificial skin for burn patients, corneas for patients with eye trauma, and artificial voice boxes (the larynx) are some of the results of early work. Now the field is expanding as the technology has expanded. Stanford University scientists have developed an artificial, portable kidney that can clean continuous clean blood and eliminate the need for lengthy dialysis three times a week. In a new technology, 3-D bioprinting, discussed in Chapter 14, is allowing the construction of biological structures including parts of the ear and several bones using a method similar to a copy machine.

In another advancement, scientists from Japan's RIKEN Institute have developed artificial versions of lymph nodes. Lymph nodes produce immune cells that recognize and fight pathogens. Artificial lymph nodes replace damaged or excised nodes, and they could be filled with disease-fight customized or engineered white blood cells and antibodies.

Biomedical Instrumentation   From tiny pill-sized cameras that may be swallowed and transmit images and data to physicians to new inhalers that better deliver asthma medications or other inhalable drugs to mechanical arms that use the same technology as in the robotic arms in the space program, the design and manufacturing of new medical devices is a limitless frontier. One exciting innovation is a walking simulator that tricks stroke patients into thinking that they are walking slower than they are and encourages them to walk farther, faster.

Regenerative Medicine   Regenerative medicine is a term that describes efforts to restore the function of diseased or damaged tissues or organs. Regenerative medicine uses any or all of the advance medical technologies described in this chapter, including genetic engineering, gene therapy, artificial tissues and organs, and biomedical devices. Regenerative medicine and the use of stem cells is discussed in greater detail in Chapter 14.

Section 12.5   Review Questions

1. How are genetic disorders diagnosed?
2. Give an example of a personalized medicine.
3. What is a biomarker and how can it be used to identify cancer cells?
4. How does pharmacogenetics allow for the development of personalized medicines?

Chapter Review

Summary

Concepts

- A medicine is a compound that treats, prevents, or alleviates a disease or its symptoms. Medical biotechnology includes all the areas of research, development, and manufacture of medicines, as well as disease prevention and treatment.
- Drugs are chemicals that alter proteins or other compounds responsible for diseases and their symptoms. Drugs are found and isolated from natural sources, or they are synthesized in the lab through chemical combination or genetic engineering.
- Aspirin was the first marketed therapeutic product made through combinatorial chemistry. It is the result of binding an acetyl group to the precursor of aspirin, salicylic acid.
- The number of potential drugs and products has increased as result of parallel synthesis and high-throughput production and screening. Hundreds of samples, in libraries, are processed and screened by technicians and robots. Some instruments and techniques for screening compounds include UV spectrophotometers, mass spectrometers, and microarrays.
- Peptides are important therapeutics and research tools. Peptide synthesis is conducted using peptide synthesizers. Peptide synthesizers are programmed to bind certain amino acids in a certain order. Peptide synthesizers have a resin column upon which the synthesis occurs.
- Oligonucleotides (oligos) are important therapeutics and research tools. Oligonucleotide synthesis is performed at several companies using DNA synthesizers. DNA synthesizers are programmed to bind certain nucleotides in a certain order. DNA synthesizers have a resin column upon which the synthesis occurs.
- Many companies produce antibodies for research or medicinal value. Antibodies are large molecules that specifically bind particular antigens. This characteristic is used in affinity chromatography when antibodies are used to pull antigens out of solution.
- Antigens in a vaccine trigger the body to produce large numbers of B cells and their antibodies. Some B cells remain in the body for years, remembering their antigen. They are called memory cells, and they eliminate their antigen before you become sick. This is the basis for immunity.
- Viruses and bacteria mutate rapidly, so new versions of vaccines are always in development.
- Pharmacogenetics, the study and use of genetic information to develop targeted therapies, has the potential to lead to personalized medicines for individuals or groups of people.
- Recent advances in medical biotechnology include the development of personalized medicine, the use of biomarkers and monoclonal antibody technology, new gene therapies, the construction of the first artificial organs, and new biomedical instrumentation for diagnostics and therapeutics.

Lab Practices

- Caffeine molecules can be extracted from plants. They are colorless and absorb light.
- Once lambda$_{max}$ is determined for the caffeine molecule, caffeine can be identified in other solutions by creating a standard curve of known caffeine concentrations.
- Materials safety data sheets (MSDS) include an extensive amount of important information besides safety guidelines. Physical characteristics necessary to identify compounds are on the MSDS.
- To test for antimicrobial activity of extracts, bacterial cultures are exposed to extract-treated filter-paper disks, so that zones of inhibition can be observed.
- The antibiotic ampicillin can be used as a positive control for the antimicrobial activity assay since *E. coli* is sensitive to it. Methanol is a known antiseptic agent and also is a good positive control.
- Aspirin (acetylsalicylic acid) can be produced by organic synthesis when salicylic acid is mixed with acetic anhydride. An acetyl group is transferred to the salicylic acid in the presence of sulfuric acid. Confirmation of the creation of acetylsalicylic acid may be made using a ferric nitrate test and a melting point determination.
- Melting point determinations are one of the methods used to identity a substance. The melting point of a substance is a defining characteristic. The melting point of most compounds is above 100°C, so the melting point is determined in oil rather than water.

Thinking Like a Biotechnician

1. What is the approximate maximum length of peptides made on peptide synthesizers?
2. Why are the peptides created on a peptide synthesizer not called proteins?
3. If scientists are screening thousands of samples daily or weekly, how can they process so much data?
4. How might a genetic disorder be detected on a microarray?
5. Using a DNA synthesizer, suppose you want to make a primer that will recognize the following sequence: 3'-TAC CCG GGC AAT TCC AGT-5'. What will the sequence on the primer have to be?
6. Allergies are caused by an overreaction of an antibody to an antigen. Suppose you are allergic to peanuts. Suggest an antibody therapeutic that might help you.
7. Explain how the antibody technology works on a pregnancy test strip.
8. How do scientists acquire enough antibodies to purify antigenic proteins for vaccine trials?
9. How might the active ingredient from the foxglove plant (*Digitalis*) be isolated from the plant? How would a technician know if it was isolated?
10. If a vaccine is needed for a certain cancer, for instance, breast cancer, what type of vaccine antigen might cause an immune response and be a possible therapeutic candidate?

Biotech Live

Antibiotic Resistance

Activity **12.1**

Antibiotics became widely available in the 1940s and improved the quality of human life in a big way. Before their introduction, it was not uncommon for people to die from bacterial infections (in wounds or from illness) that now are routinely cured with antibiotics. However, antibiotics have been overused, and many bacteria species or strains are now resistant to the very same antibiotics that once killed them. Today, doctors are reluctant to give antibiotic prescriptions because they fear that bacteria are becoming resistant to existing antibiotics before scientists can develop new ones.

As a starting place, use the FDA Antibiotic Resistance Web page at biotech.emcp.net/antibioticresistance to learn about the growing threat of antibiotic resistance.

Using the links on the website, create a four-slide Microsoft® PowerPoint® presentation, with at least one graphic image per slide, that includes the following:

- What is an antibiotic, and what is antibiotic resistance?
- What causes antibiotic resistance and why is it a major health concern?
- Give an example of antibiotic resistance that has significant health implications.
- Offer recommendations and actions to combat antibiotic resistance.

Activity 12.2   A Lot of Work for Medical Biotechnologists

It seems like every time you turn around, someone you know or care about has a serious disease. When you consider the number of people in the United States who have a serious disease, you can understand why so many people are motivated to work in medical biotechnology.

 Using data from the Internet, learn about the number of Americans diagnosed with life-threatening diseases each year and efforts to combat them.

Study the chart and graph shown at right. Then answer the following questions:

1. What was the estimated total number of newly diagnosed cancer cases in the United States for 2010?
2. From the US Patient graph, which diseases do you consider preventable, and what percentage of the total number are they?
3. If you were on the board of directors for a "start-up" biotechnology company, upon which of these diseases would you recommend that the company focus its therapeutic efforts? Why?
4. Pick a disease in which you have some interest, perhaps one that a friend or loved one suffers from. Find at least two companies that have a biotechnology drug on the market to treat the disease. Describe the type of molecule that is the main ingredient in each product. Record the website addresses you used as references.

Estimated Number of New Cancer Cases, 2010

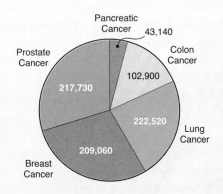

Total Number of Patients in the United States

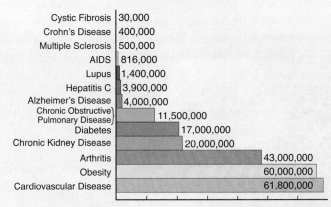

Source: Biotechnology Industry Organization, 2005

Cancer: A Cure in Our Lifetime?

Activity 12.3

Using the resources at the American Cancer Society website (**biotech.emcp.net/cancersoc**) or the Association of Cancer Online Resources (**biotech.emcp.net/acor**), find information about one type of cancer (assigned by your instructor) for the purpose of creating a poster and making a 5-minute oral presentation. Use other Web references as necessary to obtain the information required. On the poster include bulleted points and some diagrams, photos, or figures addressing the following topics:

1. The name of the cancer and the type of tissue and cells it affects.
2. The symptoms and effects of the cancer.
3. The treatments for the cancer and any side effects of each treatment.
4. A list of companies and products they are developing as therapeutics or vaccines for the cancer.
5. A list of references used in producing the poster.

What's the Risk in a Pedigree?

Activity 12.4

A pedigree analysis can shed light on the inheritance of a specific trait, genetic disorder, or disease. In a pedigree analysis, information is collected regarding what is known about the presence of a trait in a family's history. A pedigree chart is then drawn and used to make predictions about the risks of a family member inheriting the trait or disorder. The person that is the focus of the pedigree analysis is called the "Proband."

The information used in a pedigree may be phenotypic (the expression of genetic information), such as a child known to have cystic fibrosis. Or, the information may be genotypic (the presence of a specific gene sequence), such as a positive genetic test for the BRCA1 breast cancer gene.

Information recorded on a pedigree chart is recorded in a very specific way. On a pedigree chart, a male is represented by a square and a circle represents a female. Horizontal lines on the chart represent relationships such as marriages and siblings. Vertical lines represent children of a couple.

Filling the circle or squares of individuals denote that they possess the trait or disorder. A circle or square that is half-filled or contains a dot represents an individual that carries the gene responsible for the trait or disorder. See example of a pedigree chart below:

Pedigree Example:
Hitch-hiker's thumb

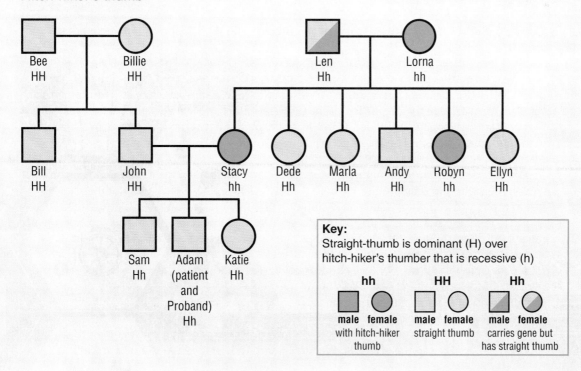

In this activity, a patient (Ed, born in 1949) diagnosed with hereditary melanoma weighs the risk of his siblings, children, and grandchildren inheriting melanoma. Melanoma is the least common skin cancer but it is the most deadly. Go to **www.myriad.com/patients-families/disease-info/melanoma/#unique-identifier** to learn more about the risk factors for melanoma.

Approximately 10% of melanoma cases are known to be genetic and several genes have been associated with hereditary melanoma, including CDKN2A (Cyclin-Dependent Kinase Inhibitor 2A). CDKN2A, a gene on chromosome 9, codes for several proteins including p16 and p14arf. Both of these proteins act as tumor suppressors and when mutations occur in the CDKN2A gene, the proteins may be dysfunctional and cancerous tumors may result.

A pedigree chart is drawn using information collected during an interview with Ed after receiving genetic test results that show he has a CDKN2A gene mutation on his chromosomes. Ed also had a biopsy of an irregular mole on his back that was confirmed as melanoma. Not only does he have melanoma and the CDKN2A gene mutation, Ed has relatives that have had skin cancer. Ed's maternal (mother's side) grandfather had skin cancer and is thought to have died of the disease. His mother was diagnosed with early stage melanoma but died tragically in an accident at 46 years old. Ed doesn't believe that anyone on his father's side has had melanoma.

Ed has 4 siblings, three sisters and a brother. His older, unmarried sister has had a skin cancer removed that was identified as melanoma. Ed's middle sister, her husband and their children (a boy and a girl) have had no skin cancers. Ed's youngest sister has been diagnosed with melanoma and also had a positive CDKN2A gene mutation test. Ed's brother is cancer-free.

Since Ed's diagnosis, he has requested that each of his three children, who are currently cancer-free, be tested for the CDKN2A gene mutation. His oldest son, Roy, was tested and does not have the mutation. Ed's middle son (Jared) is currently cancer-free but was tested and found to have the CDKN2A gene mutation. Jared's two daughters have not been tested yet. Ed's daughter, Camille, who is 35 years old, does not want to be tested at this time because she has 2 children (a son and a daughter) and she doesn't want to frighten them or be afraid that they may have the CDKN2A gene mutation.

With this information, a pedigree chart may be constructed and used to make recommendations for further testing and managing the risks of Ed's family members getting melanoma.

TO DO

*This activity draws a pedigree chart on a web-based tool. If that is not available, then the pedigree can be drawn by hand.

1. Go to **invitae.com/en/familyhistory** and sign up for the Web application if you have not already done so.
2. After signing up, go to **familyhistory.invitae.com/welcome** and "Start New Pedigree."
3. Use the information given above, about Ed, to start a pedigree chart. Use the current date as the MRN (Medical Record Number), for example, 080615 for August 8, 2015. Enlarge the screen so you can see the beginning of the pedigree chart and see Ed's placement on it.
4. Notice that the webpage is divided into 3 sections: on the left is the Risk Panel, in the center is the pedigree chart, and on the right is where to enter information for each person on the pedigree.
5. Start with Ed. Click on his square on the pedigree and fill in as much information as you can on the right side (Person information). Click "+ Add a New Field" and it will open a box to type in "melanoma." Click on Melanoma and hit "return" and it will give you a choice to label Ed' square as Affected, Carrier or Unknown. Choose "Affected" since Ed has melanoma. Since Ed had a positive CDKN2A genetic test, choose "Confirmed Molecularly" and add that notation. See the information displayed on the pedigree chart. Click on Ed's box to add the symbols for his siblings, children and grandchildren.
6. For each of the other members of Ed's family described above, add a box or square and the information to show if they have melanoma, have been tested for CDKN2A, and other information.
   * If you have problems knowing what to do next or why something shows up on the pedigree, pull down the HELP menu and check through the User Guide for hints.

7. When the pedigree is as complete as possible, check yours against another class member's pedigree for Ed and see if they have different information on the pedigree than you have. Next, click "ON" in the Risk Panel and calculate the risk of the cancers listed for Ed. What is his risk of his getting cancers other than melanoma?

8. In the "Actions" menu, use the "Change Proband" option, and for each of his children and grandchildren, determine their risk of getting melanoma?

9. Print up a copy of the pedigree chart you prepared. Record the melanoma risk value for each family member as determined on the pedigree. Based on this information, what would you say to Camille about being tested, having her children tested and how she might decrease their melanoma risk?

Who Passes the *BRCA1* Test?

This activity was developed using Bio-ITEST Genetic Testing materials by Chowning, J., Kovarik, D., Griswold, J., Porter, S., Spitze, J., Farris, C., and K. Petersen. Using Bioinformatics: Genetic Testing. Published Online April 2012. figshare. **dx.doi.org/10.6084/m9.figshare.936566** The development of these materials was funded in part by an ITEST grant from the National Science Foundation DRL-0833779

Activity 12.5

Approximately 8% of the female general population has breast cancer at sometime in their life. Although it is rare (about 0.05% of the population), men can also get breast cancer. Most breast cancers are not known to be linked to a genetic mutation. However, if an individual is tested and found to have one of the known mutations in the BRCA genes, his/her risk of developing cancer is increased to 87% for women and 8% for men.

BRCA1 is a tumor suppressor gene and when a patient has a mutation in the gene, the tumor suppression may be compromised and the cell may not able to repair damaged DNA. This may lead to uncontrolled or irregular growth of breast cells (and ovarian cells) and a rather aggressive form of breast cancer. The BRCA1 mutation is inherited like an autosomal, dominant mutation, which means that a child only needs to inherit one mutant gene from either parent to be BRCA1+.

A close friend of yours, Deborah (Deb) Lawler, is worried that she has a genetic predisposition for breast cancer. Deborah, who is 30 years old, has had several close relatives with cancer and some have died from the disease. Deborah's mother fought breast cancer when Deb was in high school, and Deb's maternal grandmother died from the disease before Deborah was born. Deborah was surprised when her Uncle Bob, her mother's only brother, was diagnosed with breast cancer. She really never thought about her male relative being at risk. In addition, Deborah's first cousin, Katherine, was diagnosed with breast cancer at the age of 33. It is obvious to Deborah that her chance of developing breast cancer is higher than the general population.

Deborah has asked for your help to understand the genetic testing for the mutations associated with breast cancer. Her goal is to decide if she wants the genetic test to learn if she carries the same BRCA1 mutation as her uncle. Deborah is not sure since if she does get a positive BRCA1 test result she will have to make some medical decisions that she may not be prepared to make. Also, a positive test result may provide information that her siblings may not want to know.

You tell Deborah that you think she would benefit from learning more about the BRCA1 mutation and the how that mutation affects breast cells. Using the NCBI nucleotide and protein databases as well as a molecular modeling tool, she can understand what happens when the normal tumor suppressor gene mutates. After Deborah learns more about the mutation, Deborah's family cancer history may be used to create a pedigree (inheritance chart) to help Deborah understand what the chances are of her having the BRCA1 mutation and maybe passing it on to her future children.

1. **Gather general information about the BRCA1 mutation.** Go to the National Cancer Institute (NCI) website at **www.cancer.gov/about-cancer/causes-prevention/genetics/brca-fact-sheet.** Scan through the links under BRCA1 and BRCA2: Cancer Risk and Genetic Testing. List 5 interesting things about the BRCA1 mutation and brevwast cancer that you would not have known if you didn't have access to this NCI information.

2. **In your notebook, construct and fill in a pedigree chart to assess Deborah's family members' risk of inheriting the BRCA1 mutation.** Creating a Lawler Family Pedigree diagram begins with known phenotypic data, in this case whether an individual is known to have breast cancer. Deborah's known (Lawler) family information is in the background above. Use the Pedigree Legend below the pedigree diagram to see how to fill in each square or circle (See Figure 12.31).

Figure 12.31. Lawler Family Pedigree A pedigree is first drawn to shown who in a family has a specific condition. If genetic testing information is obtained it can add information to a pedigree. For Deborah, what are the chances that she or one of her siblings has inherited a breast cancer gene? When Deborah has children, what are their chances of inheriting a breast cancer gene?

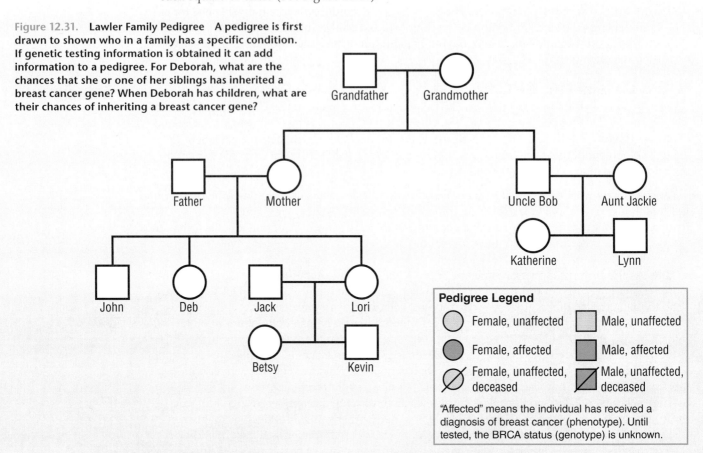

Pedigree Legend

○ Female, unaffected   □ Male, unaffected

● Female, affected   ■ Male, affected

⌀ Female, unaffected, deceased   ⊠ Male, unaffected, deceased

"Affected" means the individual has received a diagnosis of breast cancer (phenotype). Until tested, the BRCA status (genotype) is unknown.

3. **Use the NCBI Nucleotide database to analyze family members' DNA data for the BRCA1 mutation.** After learning more about the *BRCA1* mutation and her risk of inheriting the mutant *BRCA1* gene, several Lawler family members agree to have blood samples drawn and the genetic (DNA) test for the *BRCA1* mutation. The DNA sequences must be compared to known (reference) sequences for the mutant *BRCA1* gene versus the normal functioning gene. A consensus of research has determined that the *BRCA1* gene sequence is on chromosome number 17 and it is over 5700 nucleotides long. The sequence at the end of the gene is where the *BRCA1* mutation is found (See Figure 12.32). From the patient's DNA, only approximately 600 nucleotides of the gene are analyzed. (**research.nhgri.nih.gov/projects/bic/Member/brca1_mutation_database.shtml**).

The *BRCA1* reference sequence and many other DNA sequences, is stored in biological databases such as those at the National Center for Biotechnology Information (NCBI). The NCBI houses biological information in over 30 databases related to genetics and molecular biology. All of these databases can be searched using one search engine (Entrez). Of all the databases, NCBI's Nucleotide and Protein databases are most commonly used.

To use the NCBI Nucleotide database to compare Deborah's (or her family's) *BRCA1* genetic test sequence to the reference BRCA1 sequence, a bioinformatics tool called BLAST (short for Basic Local Alignment Search Tool) is used. BLAST can be used to compare the sequences of two or more proteins or nucleic acid molecules. A nucleotide BLAST search can compare Deborah's family *BRCA1* DNA sequences to the *BRCA1* DNA reference sequence.

A protein BLAST search can compare her family's *BRCA1* protein sequences to a *BRCA1* protein reference sequence. The sequences being compared are lined up (aligned) to identify sequence differences and thus mutations in the *BRCA1* sequences.

*Note: The amount of information at the NCBI is overwhelming. While you use NCBI for this limited application, you may want to explore to see the amazing depth and breadth of information and bioinformatics tools available at the NCBI.

Figure 12.32. **The BRCA1 gene is located on human chromosome #17. The mutation that is associated with the increased risk of breast cancer is found at the end of the gene, circled in red.**

```
ATGGATTTATCTGCTCTTCGCGTTGAAGAAGTACAAAATGTCATTAATGCTATGCAGAAAATCTTAGAGTGTCCCATCTGTCTGGAGTTGATCAAGGAACCTGTCTCCA
CAAAGTGTGACCACATATTTTGCAAATTTTGCATGCTGAAACTTCTCAACCAGAAGAAAGGGCCTTCACAGTGTCCTTTATGTAAGAATGATATAACCAAAAGGAGCCT
ACAAGAAAGTACGAGATTTAGTCAACTTGTTGAAGAGCTATTGAAAATCATTTGTGCTTTTCAGCTTGACACAGGTTTGGAGTATGCAAACAGCTATAATTTTGCAAAA
AAGGAAAATAACTCTCCTGAACATCTAAAAGATGAAGTTTCTATCATCCAAAGTATGGGCTACAGAAACCGTGCCAAAAGACTTCTACAGAGTGAACCCGAAAATCCTT
CCTTGCAGGAAACCAGTCTCAGTGTCCAACTCTCTAACCTTGGAACTGTGAGAACTCTGAGGACAAAGCAGCGGATACAACCTCAAAAGACGTCTGTCTACATTGAATT
GGGATCTGATTCTTCTGAAGATACCGTTAATAAGGCAACTTATTGCAGTGTGGGAGATCAAGAATTGTTACAAATCACCCCTCAAGGAACCAGGGATGAAATCAGTTTG
GATTCTGCAAAAAAGGCTGCTTGTGAATTTTCTGAGACGGATGTAACAAATACTGAACATCATCAACCCAGTAATAATGATTTGAACACCACTGAGAAGCGTGCAGCTG
AGAGGCATCCAGAAAAGTATCAGGGTAGTTCTGTTTCAAACTTGCATGTGGAGCCATGTGGCACAAATACTCATGCCAGCTCATTACAGCATGAGAACAGCAGTTTATT
ACTCACTAAAGACAGAATGAATGTAGAAAAGGCTGAATTCTGTAATAAAAGCAAACAGCCTGGCTTAGCAAGGAGCCAACATAACAGATGGGCTGGAAGTAAGGAAACA
TGTAATGATAGGCGGACTCCCAGCACAGAAAAAAAGGTAGATCTGAATGCTGATCTCCCTGTGTGAGAGAAAAGAATGGAATAAGCAGAAACTGCCATGCTCAGAGAATC
CTAGAGATACTGAAGATGTTCCTTGGATAACACTAAATAGCAGCATTCAGAAAGTTAATGAGTGGTTTTCCAGAAGTGATGAACTGTTAGGTTCTGATGACTCACATGA
TGGGGAGTCTGAATCAAATGCCAAAGTAGCTGATGTATTGGACGTTCTAAATGAGGTAGATGAATATTCTGGTTCTTCAGAGAAAATAGACTTACTGGCCAGTGATCCT
CATGAGGCTTTAATATGTAAAAGTGAAAGAGTTCACTCCAAATCAGTAGAGAGTAATATTGAAGACAAAATATTTGGGAAAACCTATCGGAAGAAGGCAAGCCTCCCCA
ACTTAAGCCATGTAACTGAAAATCTAATTATAGGAGCATTTGTTACTGAGCCACAGATAATACAAGAGCGTCCCCTCACAAATAAATTAAAGCGTAAAAGGAGACCTAC
ATCAGGCCTTCATCCTGAGGATTTTATCAAGAAAGCAGATTTGGCAGTTCAAAAGACTCCTGAAATGATAAATCAGGGAACTAACCAAACGGAGCAGAATGGTCAAGTG
ATGAATATTACTAATAGTGGTCATGAGAATAAAACAAAAGGTGATTCTATTCAGAATGAGAAAAATCCTAACCCAATAGAATCACTCGAAAAAGAATCTGCTTTCAAAA
CGAAAGCTGAACCTATAAGCAGCAGTATAAGCAATATGGAACTCGAATTAAATATCCACAATTCAAAAGCACCTAAAAAGAATAGGCTGAGGAGGAAGTCTTCTACCAG
GCATATTCATGCGCTTGAACTAGTAGTCAGTAGAAATCTAAGCCCACCTAATTGTACTGAATTGCAAATTGATAGTTGTTCTAGCAGTGAAGAGATAAAGAAAAAAAAG
TACAACCAAATGCCAGTCAGGCACAGCAGAAACCTACAACTCATGGAAGGTAAAGAACCTAACAAGCTGGAGCCAAGAAGGTAACAAGCCAAATGAACAGACAAGTAAAA
GACATGACAGCGATACTTTCCCAGAGCTGAAGTTAACAAATGCACCTGGTTCTTTTACTAAGTGTTCAAATACCAGTGAACTTAAAGAATTTGTCAATCCTAGCCTTCC
AAGAGAAGAAAAAGAAGAGAAACTAGAAACAGTTAAAGTGTCTAATAATGCTGAAGACCCCAAAGATCTCATGTTAAGTGGAGAAAGGGTTTTGCAAACTGAAAGATCT
GTAGAGAGTAGCAGTATTTCATTGGTACCTGGTACTGATTATGGCACTCAGGAAAGTATCTCGTTACTGGAAGTTAGCACTCTAGGGAAGGCAAAAACAGAACCAAATA
AATGTGTGAGTCAGTGTGCAGCATTTGAAAACCCCAAGGGACTAATTCATGGTTGTTCCAAAGATAATAGAAATGACACAGAAGGCTTTAAGTATCCATTGGGACATGA
AGTTAACCACAGTCGGGAAACAAGCATAGAAATGGAAGAAAGTGAACTTGATGCTCAGTATTTGCAGAATACATTCAAGGTTTCAAAGCGCCAGTCATTTGCTCCGTTT
TCAAATCCAGGAAATGCAGAAGAGGAATGTGCAACATTCTCTGCCCACTCTGGGTCCTTAAAGAAACAAAGTCCAAAAGTCACTTTTGAATGTGAACAAAAGGAAGAAA
```

```
ATCAAGGAAAGAATGAGTCTAATATCAAGCCTGTACAGACAGTTAATATCACTGCAGGCTTTCCTGTGGTTGGTCAGAAAGATAAGCCAGTTGATAATGCCAAATGTAG
TATCAAAGGAGGCTCTAGGTTTTGTCTATCATCTCAGTTCAGAGGCAACGAAACTGGACTCATTACTCCAAATAAACATGGACTTTTACAAAACCCATATCGTATACCA
CCACTTTTTCCCATCAAGTCATTTGTTAAAACTAAATGTAAGAAAAATCTGCTAGAGGAAAACTTTGAGGAACATTCAATGTCACCTGAAAGAGAAATGGGAAATGAGA
ACATTCCAAGTACAGTGAGCACAATTAGCCGTAATAACATTAGAGAAAATGTTTTTAAAGAAGCCAGCTCAAGCAATATTAATGAAGTAGGTTCCAGTACTAATGAAGT
GGGCTCCAGTATTAATGAATAGGTTCCAGTGATGAAAACATTCAAGCAGAACTAGGTAGAAACAGAGGGCCAAAATTGAATGCTATGCTTAGATTAGGGGTTTTGCAA
```

You will need the DNA sequence data for Deborah and the rest of the Lawler family who agreed to be tested. This is available at: digitalworldbiology.com/dwb/bio-itest-genetic-testing
Download: Genetic-Testing-Lawler-Family-Sequences-Lesson4.doc
The instructions for how to use the NCBI databases to do the BCRA1 analysis can be found starting on page 18 of the pdf, Genetic_Testing_Lesson 4 pdf at **digitalworldbiology.com/dwb/sites/default/files/itest/Genetic_Testing_Lesson4_Bio-ITEST-2014.pdf**
Start at PART I: Aligning DNA Sequences to a Reference Sequence. As you work through the analysis, record the data and answers to questions in the lesson in your notebook.
Record the names of the Lawler family members who have a *BRCA1* mutation.
Do all those that have a mutation in the *BRCA1* gene have the same mutation?

4. **Use the NCBI Protein database to analyze family members' protein data for the *BRCA1* mutation.** Is the mutation found in the Lawler family one that is likely to cause cancer or is the mutation one that makes no significant change in the tumor suppressor protein's structure and function? To answer this question, the protein's amino acid sequence must be analyzed.

You will need the protein sequence data for Deborah and the rest of the Lawler family who agreed to be tested. This is also included in the document: Genetic-Testing-Lawler-Family-Sequences-Lesson4.doc
The instructions for how to use the NCBI databases to do the *BRCA1* protein analysis can be found starting on page 23 the Genetic_Testing_Lesson 4 pdf at **digitalworldbiology.com/dwb/sites/default/files/itest/Genetic_Testing_Lesson4_Bio-ITEST-2014.pdf**
Start at PART II: Aligning Protein Sequences to a Reference Sequence. As you work through the analysis, record the data and answers to questions in the lesson in your notebook.
What differences or changes to the amino acid sequence are due to the mutation in the *BRCA1* gene?

Where is the location of the change? Count carefully using the reference numbers at the end of the sequence row.

The *BRCA1* mutation that affects Lawler family occurs at amino acid number?

_____

The mutation is an amino acid replacement of a methionine with an arginine. This is abbreviated as a M1775R mutation.

Which individuals in the Lawler family have the change (mutation) in tumor suppressor protein amino acid sequence?

Given the nucleotide and protein test results, fill out as much additional information as you can on the Lawler family pedigree.

5. Use 3D modelitng to learn how a mutation can change the shape and function of a protein. Knowing the position of the BRCA1 mutation, you can see which part of the linear protein sequence is most likely affected by the mutation. However, how will that mutation affect the shape and function of the tumor suppressor protein? Looking at the tumor suppressor protein in three dimensions, with and without the amino acid mutation, sheds light on the impact of changing a single amino acid in this protein.

The instructions for how to use the Molecule World iPad app to view three-dimensional molecular models including that of the *BRCA1* tumor suppressor protein can be found starting in Lesson 5 Using Molecular Models Student Instruction.pdf available at **digitalworldbiology.com/dwb/sites/default/files/itest/Lesson_5_student_insructions_Apr2015.pdf**

Write the answers to the questions in the lesson in your notebook.

*Note: if an iPad tablet is not available, a computer can be used to access a NCBI molecular modeling tool called Cn3D. The lesson for this version (Genetic Testing Lesson 5 (pdf) is available at **digitalworldbiology.com/dwb/gt/lesson-five-learning-to-use-cn3d**.

Activity 12.6 Getting Better: 200 Years of Medicine

The New England Journal of Medicine (NEJM) is dedicated to bringing medical professionals the most important, scientifically accurate medical research information in an understandable and clinically useful way.

NEJM is the oldest continuously published medical publication and is currently available in hardcopy or electronic form at **www.nejm.org**. Since 1811, the NEJM has published reports from the frontiers of medical science and practice since its early days, including the first public demonstration of ether anesthesia in 1846, the first full description of a spinal-disk rupture in 1934, and the first successes in the treatment of early childhood leukemia in 1948. More recent articles include some of the earliest descriptions of AIDS and then of its treatment, of aspirin and cholesterol-lowering agents in the prevention of heart disease, and of new molecular advances in the treatment of chronic leukemia and lung cancer. More than 600,000 people in 177 countries read it each week. NEJM is cited more often in scientific literature than any other medical journal.

NEJM is the most widely read, cited, and influential general medical periodical in the world. As it evolves to meet the changing needs of its readers in the 21st century, it is committed to maintaining that reputation and integrity, while using innovative formats and technologies for new features and faster delivery and access.

Watch this YouTube video produced by the NEJM to see just how far medical practices have come in the past 2 centuries.

In your notebook, list 2 ways in which surgical practices have improved in the last 200 years.

List 2 ways that early research in leukemia advanced cancer therapies.

Describe 2 breakthroughs in HIV/AIDS that have allowed some patients to fight the retrovirus infection more successfully then others.

Go to the New England Journal of Medicine's website at **www.nejm.org/page/about-nejm/200th-anniversary-articles** and find an additional article that was published as a celebration of NEJM"s 200 years of continuous publication. Read through it and summarize it in a few sentences.

Bioethics

How Do YOU Decide Who Lives and Who Dies?

The genetic disorders a baby can be born with vary greatly in their severity, from simple cosmetic or minor structural differences (see polydactyly) to entirely hopeless problems (see acephaly). We now have the products and technology (DNA, RNA, and protein tests) to reveal, often in advance, a wide range of genetic disorders.

Through this exercise, you will learn about several human genetic disorders, including their causes, symptoms, and treatments. Most of these disorders can now be detected by prenatal tests, such as ultrasound and amniocentesis, allowing parents and doctors to know about the disorders a baby may have before it is born.

This knowledge raises difficult ethical questions that have not been solved to the satisfaction of our society. Many parents choose to terminate a pregnancy if the child will have a severe disorder when it is born. The main ethical question that arises is the following: When is it right to terminate a pregnancy because of a genetic disorder identified through a test?

Problem: If a test reveals a genetic defect, should a pregnancy be terminated?

Part I: How Serious Are the Disorders?

(Do this part with a partner.)

Using the Internet, go to the March of Dimes website (**biotech.emcp.net/modimes**). Find the list of birth defects/genetic disorders, and select 10 to study. Create a chart that lists each disorder and its symptoms. Rank the genetic disorders from least severe (1) to most severe (10). You should consider many things in making your list, including the medical, personal, social, and economic impacts of each disorder.

Part II: Where Do You Draw Your Line?

(Do this part individually.)

A. Think of yourself as a pregnant woman, the husband of a pregnant woman, a friend of a pregnant woman, or a doctor. Ultrasound tests have shown mostly normal features on the fetus, but have also created some suspicion of a genetic disorder. Amniocentesis (genetic examination of fetal cells in the amnion) is being performed to test for genetic disorders. You are waiting for the results and thinking about what you might do.

B. Examine your ranked list. Somewhere on this list is an imaginary line:

   1. Above this line are the disorders that are mild enough that you would personally support continuing the pregnancy.
   2. Below this line are the disorders that are so severe that you would not feel personally obligated to support the pregnancy.

C. For you, this line could be any of the following:

   - anywhere on the list
   - above the first disorder on the list (meaning you would *not* support continuation of the pregnancy with *any* of the disorders)
   - below the last disorder on the list (meaning you would support continuation of the pregnancy with *any* of them)

D. Give the line some deep thought. Then, draw this line on the ranking where you think you would place it. As with many decisions concerning ethics, drawing this line can be very uncomfortable. With increasingly sophisticated technology, it is a decision that has to be made with increasing frequency. Be able to give reasons for your placement of the line. Consider quality of life for parents and offspring, costs, treatment difficulties, and feelings about abortion.

Biotech
Careers

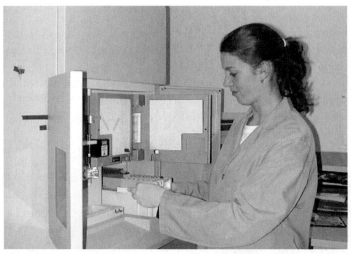

Photo courtesy of Ellie Salmon.

Forensic Scientist/DNA Analyst

Ellie Salmon, BS, MS
Forensic Analytical Laboratories Inc.
Hayward, CA

Ellie Salmon works on a wide variety of forensic cases involving biological material. She performs DNA testing on evidence collected by police departments or investigators, communicates her findings to the police agencies or attorneys, and, ultimately, testifies to the results in court. Her typical day involves analyzing clothing or swabs from a crime scene, and performing biological screening to determine whether blood or saliva or others biological materials are present. A DNA test is then performed on the item, and the DNA profile is compared with a known sample from a suspect or victim.

Ms. Salmon is an expert in the area of DNA analysis, including the isolation of biological materials and subsequent preparation, analysis, and interpretation by polymerase chain reaction (PCR), DNA sequencing, bioinformatics, and traditional serological techniques. Here, Ellie works on a capillary DNA sequencer to analyze the genetic fingerprint of biological samples left at a crime scene. In the past, she processed backlogged forensic casework from various police agencies using the FBI-designated 13-CODIS-STR loci and a high-throughput, convicted-offender DNA database.

The 46 chromosomes in human cells are actually 22 **homologous pairs**, plus two sex chromosomes. The pairs of chromosomes are copies of those contributed to the offspring from its parents. One copy of each chromosome was contributed by a sperm cell and one copy of each chromosome was contributed by the egg cell. The chromosomes are visible in cells when they shorten and thicken during mitosis (cell division) (see Figure 13.2).

The chromosome pairs are homologous because each chromosome is matched to another containing the same genes in the same order, although the specific information within a gene may be different. So, the two No. 1 chromosomes each contain genetic information for the same traits, for example, the gene for nerve growth factor production (or not) and the gene for coagulation factor V production (or not). Chromosome pair No. 2 contains different genes than chromosome pair No. 1 and, thus, codes for different proteins such as the gene for the production of alsin, a motor neuron protein (or not) (see Figure 13.3). Information about which genes are known on each chromosome is easily found with a quick search on the Internet.

Not all of the genetic code found on the chromosomes is decoded into proteins at the same time. Although each cell contains all the genetic information, only certain genes, coding for certain proteins, are transcribed (expressed) at any given time. Thus, different cells make different proteins at different times. For example, ovarian cells make estrogen, but only at certain times of a woman's life. Estrogen production is regulated at the gene level. Regulated gene expression is also the reason pancreas cells are the only human cells that synthesize insulin. The insulin gene is "turned off" in other cells.

Cells are specialized, producing only certain molecules, depending on which genes are "turned on" and which are "turned off." For example, pituitary cells synthesize human growth hormone, but not insulin. Salivary glands produce amylase, but not insulin or growth hormone.

DNA Replication   A typical human body is estimated to have over 20 trillion cells. These cells all originate from a single fertilized egg cell by means of **DNA replication** and subsequent cell division (mitosis). It is easy to see how important accurate DNA replication is. When new cells are made during growth and development, or to replace damaged cells, they each receive complete copies of all the DNA molecules and the complete genetic code (genome) for the entire organism.

During DNA replication, each DNA molecule in the cell is "unzipped" and copied (see Figure 13.4). In cells, DNA replication requires several enzymes working collaboratively to complete the replication in a timely manner. Once the DNA (chromosomes) in a cell has been replicated, one copy of each chromosome, called chromatids, moves to each of the new cells during cell division. In this way, each cell of the organism has the same genetic information as every other cell.

homologous pairs (ho•mol•o•gous pairs)   the two "matching" chromosomes having the same genes in the same order

DNA replication (DNA rep•li•ca•tion)   the process by which DNA molecules are duplicated

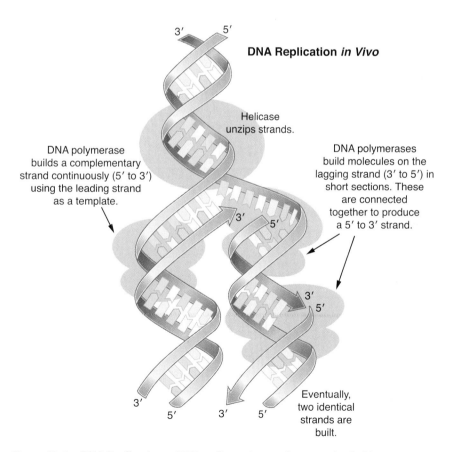

DNA Replication in Vivo

Helicase unzips strands.

DNA polymerase builds a complementary strand continuously (5' to 3') using the leading strand as a template.

DNA polymerases build molecules on the lagging strand (3' to 5') in short sections. These are connected together to produce a 5' to 3' strand.

Eventually, two identical strands are built.

Figure 13.4.   **DNA Replication.**   DNA replicates in a semiconservative fashion; one strand unzips and each side is copied. It is considered semiconservative because one copy of each parent strand is conserved in the next generation of DNA molecules. Six different types of enzymes are involved in the replication process.

in vivo (in vi•vo) an experiment conducted in a living organism or cell; literally, "in living"

helicase (he•li•case) an enzyme that functions to unwind and unzip complementary DNA strands during *in vivo* DNA replication

topoisomerase (to•po•is•om•er•ase) an enzyme that acts to relieve tension in DNA strands as they unwind during *in vivo* DNA replication

RNA primase (RNA prim•ase) an **enzyme that adds primers to template strands during** *in vivo* **DNA replication**

primer (prim•er) a short piece of DNA or RNA (15–35 bases) that is complementary to a section of template strand and acts as an attachment and starting point for the synthesis strand during DNA replication

DNA polymerase (DNA po•ly•mer•ase) an enzyme that, during DNA replication, creates a new strand of DNA nucleotides complementary to a template strand

RNase H (RNase H) an enzyme that functions to degrade RNA primers, during *in vivo* replication, that are bound to DNA template strands

in vitro synthesis (in vi•tro syn•the•sis) any synthesis that is done wholly or partly outside of a living organism (eg, PCR); literally, "in glass"

probes (probes) the labeled DNA or RNA sequences (oligonucleotides) that are used for gene identification

In the cell (**in vivo**), replication occurs simultaneously at many sites along a chromosome. At each site, replication commences when two DNA **helicase** molecules bind to the DNA and begin to unwind and unzip it (initiation). The strands separate along the hydrogen bonds holding the two sides together and form a replication bubble. At the same time, another enzyme, **topoisomerase**, relieves tension along the untwisting strand. Each of the DNA strands acts as a template to build a new second strand (see Figure 13.5). A small series of complementary RNA nucleotides are assembled on each DNA template strand by a **RNA primase** (polymerase) molecule. The RNA nucleotides act as a **primer** strand upon which the new DNA strand can be constructed. Another enzyme, **DNA polymerase**, adds DNA nucleotides onto the end of the RNA primer, complementary to the DNA template (elongation). The new strand is assembled until it runs into the next section being synthesized. At this point, the **RNase H** enzyme edits out the RNA primer, and DNA polymerase fills in the gap. Another enzyme, DNA ligase, connects the two replicated sections together (termination).

Biotechnicians often need small pieces of DNA for research and development (R&D). Small pieces of DNA can be synthesized in the lab, outside of cells. This procedure is called **in vitro synthesis**, and it produces small lengths of DNA called oligonucleotides (oligo means "few"). Synthesized DNA fragments may be used as genes of interest for transformation or for other applications.

Oligonucleotides are also used as **probes** for gene identification (as in microarrays) or as primers for both sequencing and PCRs. For these purposes, the oligos are tagged with some other molecule for recognition. Enzyme tags, radioactive labels, and fluorescent labels are commonly linked to primers and probes. These synthesis products, and their use in research and industry, are discussed in later sections.

One method of synthesizing DNA *in vitro* is by using an existing strand of DNA as a template to make a new strand. This can be performed in a test tube. To make DNA outside of a cell, in a test tube, all the necessary enzymes, cofactors, and buffers usually used for DNA synthesis in a cell must be added. If given the proper conditions and ingredients, a DNA strand will act as a model or template to make another strand.

Making DNA in a test tube (by hand) is rather labor-intensive, so companies have designed DNA synthesizers to make oligonucleotides in small or large quantities. Driven by the need to sequence the human genome quickly, scientists developed DNA synthesizers that could produce large numbers of varied primers (see Figure 13.6).

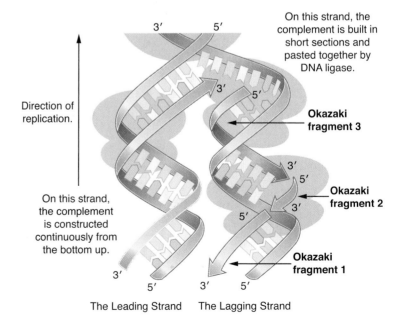

Figure 13.5. **DNA Okazaki Fragments.** DNA molecules have directionality. There is an up-to-down side and a down-to-up side. Due to this antiparallel nature (3' → 5' and 5' → 3'), the replication of copies on each template happens somewhat differently. On one side, the leading strand, the DNA is replicated continuously, building an entire complement. On the other side, the DNA is replicated in short pieces called Okazaki fragments.

Figure 13.6. **Inside this ABI 3900 High-Throughput DNA/Oligo Synthesizer® (Applied Biosystems by Life Technologies Corp.) are 48 columns of synthesis resin. The 3900 DNA synthesizer can produce 48 different types of oligos on the 48 columns simultaneously. The instrument can be programmed to produce single-stranded pieces of DNA of up to approximately 288 bases long. The technician adjusts the columns at the top of the instrument.**

Photo by author, by permission of Applied Biosystems by Life Technologies Corp.

BIOTECH ONLINE

© Veronique Beranger/zefa.

Tell Me About Telomeres

Telomeres are hot stuff in biotechnology these days. At the Telomere Information Center website listed below, several links lead to the latest research on telomere and telomerase activity.

biotech.emcp.net/telomereIC

Using the Telomere Information Center website as a starting point, create a one-half page fact sheet explaining what telomeres are and their relationship is to DNA replication, aging, cancer, and other age-related diseases. Include 2–3 additional references and diagrams or photos.

The reagents needed to complete an *in vitro* DNA synthesis include a DNA template, primer, DNA polymerase, nucleotides, reaction buffer containing magnesium ions, and **dithiothreitol (DTT)**. The function of each in the synthesis is described below.

DNA Template The **template** DNA is the strand from which a new strand is synthesized. As in a cell during DNA replication, the enzyme DNA polymerase reads the A, C, G, T code on the template. The DNA polymerase then assembles a new strand of nucleotides complementary to the template. Polymerase reads the template in a 3' to 5' direction, building a complementary molecule in a 5' to 3' direction.

Primer As you have learned earlier, a primer is a short piece of DNA or RNA, usually 15 to 35 nucleotides long, running 5' → 3', that has a nitrogenous base (A, C, G, and T or U) sequence complementary to the 3' → 5' template strand. When mixed together under the proper conditions, the primers will anneal, or bond, to each template strand. During synthesis, DNA polymerase uses the primer's 3' end as an attachment starting point to which the next nucleotide in the synthesis strand can be added. Sometimes an additional molecule is attached (or tagged) to the 5' end of a primer strand for visualization purposes.

DTT (DTT) the abbreviation for dithiothreitol, a reducing agent that helps to stabilize the DNA polymerase in DNA synthesis, PCR, and DNA sequencing reactions

template (tem•plate) the strand of DNA from which a new complementary strand is synthesized

dNTP (dNTP) the abbreviation for nucleotide triphosphates, which are the reactants (dATP, dCTP, dGTP, and dTTP) used as the sources of A, C, G, and Ts for a new strand of DNA

dATP (dATP) the abbreviation for deoxyadenosine triphosphate, the cell's source of adenine (A) for DNA molecules

dCTP (dCTP) the abbreviation for deoxycytidine triphosphate, the cell's source of cytosine (C) for DNA molecules

dGTP (dGTP) the abbreviation for deoxyguanosine triphosphate, the cell's source of guanine (G) for DNA molecules

dTTP (dTTP) the abbreviation for deoxythymidine triphosphate, the cell's source of thymine (T) for DNA molecules

reaction buffer (re•ac•tion buff•er) a buffer in PCR or some other reaction that is used to maintain the pH of the reaction

Nucleotides Nucleotide triphosphates, or **dNTPs (dATP, dCTP, dGTP, and dTTP)** are the reactants used as the sources of A, C, G, and T for the new strand. When DNA polymerase catalyzes the building of the new strand, it reads the template and binds the complementary nucleotide.

DNA Polymerase Enzymes called polymerases build large molecules (polymers) from smaller molecules (monomers). In DNA synthesis, a DNA polymerase builds a complementary DNA strand while moving down the template strand. The polymerase uses the four deoxynucleotides, (dNTPs), dATP, dCTP, dGTP, and dTTP, and cleaves off two phosphate groups of each, allowing the mononucleotides (dAMP, dCMP, dGMP, and dTMP) to be added to the synthesis strand. The cleavage of the phosphate groups from the dNTP actually provides the energy for the polymerization.

Reaction Buffer **Reaction buffer** is used to maintain the pH of the synthesis reaction. The DNA polymerase is pH sensitive, and works best at a constant pH of approximately 8.0. The reaction buffer contains salts and water at the optimum pH. DTT is a reducing agent that helps to stabilize the DNA polymerase during synthesis, PCR, and sequencing reactions.

The reagents for a synthesis are mixed in a tube and, within a few minutes, the polymerase catalyzes the synthesis of thousands of copies of a template strand. The synthesis is confirmed by running synthesis products on a polyacrylamide gel or sequencing the sample (see Figure 3.7).

The development of DNA synthesizers has automated DNA synthesis. Multiple samples of oligonucleotides of varying lengths can be made in just a few hours. A synthesizer has tiny columns of resin onto which an oligonucleotide strand is built. Bottles hold samples of primers, dNTPs, buffers, and other reagents. A technician types in the oligonucleotide sequence to be synthesized, and a computer regulates the order in which the reagents are added. At the end of the synthesis, the oligonucleotides are cleaved from the column, collected, and tested. The oligonucleotides may be used as primers or probes. Some companies have entire oligonucleotide factories to meet their DNA synthesis needs. Other companies order their oligos from oligo supply companies. In fact, you can even order and purchase oligos online.

Figure 13.7. **After mixing synthesis reagents, the technician runs samples on a polyacrylamide (PAGE) gel to confirm the production of the desired fragments. Although the DNA fragments are colorless, the loading dye in the samples, which moves at the same rate as some of the smallest DNA pieces, is visible in the middle of the running gel. When the dye has traveled about two-thirds of the way down, the gel is stopped and blotted onto a membrane for DNA staining and visualization.** Photo by author.

Snip Snip Here, Snip Snip There

TO DO

Go to biotech.emcp.net/oxfordcarcin and read the scientific article about how an enzyme that normally repairs damaged DNA increases the risk for lung cancer. Discuss the risk for smokers, nonsmokers, and drinkers of alcohol. Summarize how scientists determined the relationship between the DNA repair enzyme and the increased lung cancer risk.

Section 13.1   Review Questions

1. How many DNA strands does an *E. coli* cell contain? How many chromosomes does a human body cell contain?
2. What are homologous pairs, and where do they come from?
3. Name six enzymes involved in *in vivo* DNA replication.
4. How is *in vitro* DNA synthesis in a test tube different than *in vitro* DNA synthesis on an automated synthesizer?

13.2 DNA Synthesis Products

DNA molecules of various sizes are produced, used, and studied in academic and commercial laboratories. DNA is commonly synthesized for use in the following applications:

- probes for gene or DNA identification, as in DNA fingerprinting and microarrays
- primers for use in PCR and DNA sequencing
- PCR **amplification** products for genetic identification and genetic studies

amplification
(am•pli•fi•ca•tion)   an increase in the number of copies of a particular segment of DNA, usually as a result of PCR

Probes

Probes are relatively short pieces of DNA (or RNA) with a nucleotide sequence complementary to another sequence being searched for. Probes are used when scientists are trying to find a specific piece of DNA (see Figure 13.8). For example, probes were used to locate the human insulin gene. Since the pancreas produces large quantities of insulin, the messenger RNA (mRNA) for insulin should be in abundance in these cells. The insulin mRNA can be isolated and marked with a fluorescent or radioactive label. When mixed with single-stranded DNA from a human cell, the labeled insulin mRNA will bounce around until it hits the sequence for insulin on one of the DNA strands. Since the insulin mRNA

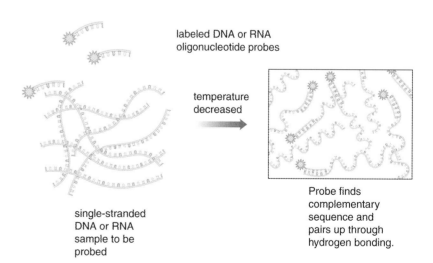

labeled DNA or RNA oligonucleotide probes

temperature decreased

single-stranded DNA or RNA sample to be probed

Probe finds complementary sequence and pairs up through hydrogen bonding.

Figure 13.8. Hybridization.   Primers and probes have sequences complementary to a DNA or RNA sequence of interest. To use a primer or a probe, the sample must be a single strand. If it is double-stranded DNA, it must be denatured by heat to single strands. The primer or probe is added and it binds to the complementary segment. If equipped with a colorimetric, fluorescent, or radioactive label, the hybridization may be visualized.

is complementary to the insulin gene, the two strands will bind to each other. The scientist will be able to identify the insulin gene because of its color, fluoresence, or radioactivity. The gene can then be cut out of the genome, identified, and used for recombinant DNA (rDNA), gene therapy, or other application.

The insulin probe described above was a naturally occurring RNA molecule, but more often, probes are synthesized pieces of DNA complementary to a target sequence. Most often, automated synthesizers, such as those discussed in the previous section, are used to produce probes.

Blotting Probes may be used after DNA samples have been run on a gel, as in DNA fingerprinting, DNA sequencing, or genetic testing. After running samples on a gel, the samples are transferred from the gel to a membrane (usually nylon) or specially treated paper (nitrocellulose). The transfer process is called Southern blotting (see Figure 13.9). A Southern blot transfers DNA molecules and in some respects is similar to the Western blot of proteins (see chapter 6). Blotting is performed because the membrane is easier to handle, sturdier, and longer lasting than a gel. After the samples are transferred to the membrane, the membrane can be "probed."

To conduct a blot, DNA fragments are run on an agarose or acrylamide gel. The gel is laid on a flat surface, and a specially treated membrane or nitrocellulose paper is laid on top of the gel. Absorbent paper or toweling is laid on top of the blotting membrane and, through capillary action, the buffer and DNA samples are drawn up from the gel onto the membrane. The blotted membrane is exposed to UV light (cross-linking) in a **cross-linker** to ensure that the negatively charged DNA molecules bond to the

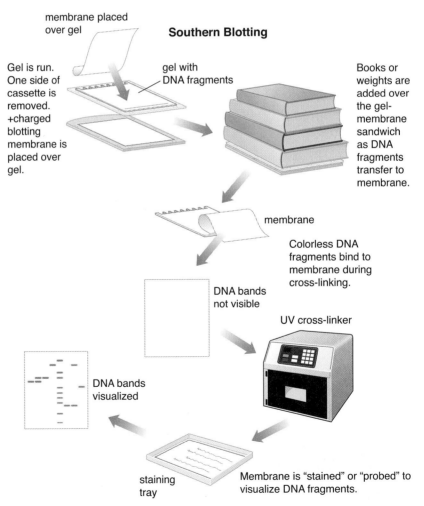

cross-linker (cross•link•er) an instrument that uses UV light to irreversibly bind DNA or RNA to membrane or paper

Figure 13.9. Southern Blot. A Southern blot is a transfer of DNA to a membrane or paper for visualization. When RNA is transferred, it is called a Northern blot. When proteins are transferred, it is called a Western blot.

positively charged membrane or paper. Cross-linking prevents samples from washing away during visualization. Then, the blot can be treated with the fluorescent, colored, or radioactive labels or probes to make the DNA fragments visible for study.

Microarrays   Probe technology has advanced recently with the development of microarrays. Microarrays are assemblies of large numbers of samples of DNA, or even RNA or proteins. The purpose of using a microarray is to be able to search through a large amount of genetic information, even an entire genome, to understand gene expression. Microarrays allow the probing, screening, and evaluation of many samples, up to a million, at one time.

DNA microarrays were first developed at Affymetrix, Inc., in Santa Clara, CA. Their microarray, known as a GeneChip®, contains millions of DNA samples. The Affymetrix GeneChip® is a silicon wafer containing hundreds of sections or "features" onto which DNA sequences called "probe arrays," of approximately 25 nucleotides in length, are built. Each feature contains multiple copies of an array probe representing a sequence that would produce the RNA for one of thousands of different proteins.

The probe arrays on a GeneChip® microarray are assembled by a robot through a process similar to the manufacturing of computer microchips. Nucleotides are added onto a strand in a specific order found on each sequence in the feature. The assembled probes are exposed to fluorescently labeled RNA (or cDNA) thought to be complementary to one or more sequences found in the array. Where a labeled RNA complement to a probe array sequence is found, it binds and the RNA-probe array combination shows a fluorescent signal. On a GeneChip® microarray, a black or dark blue color is the negative control, and a red or white color is a strongly positive signal. A feature with white color means that a large amount of the array probes hybridized with the labeled RNA or cDNA fragments. Arrays in white features would likely be targets of interest for continued investigation (see Figures 13.10a and b).

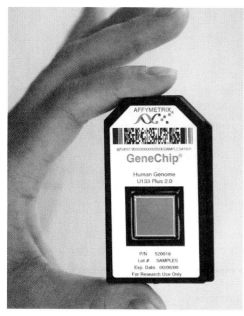

Figure 13.10a.   **A GeneChip® (Affymetrix, Inc.) is a silicon wafer, packaged in a cassette, containing thousands of gene segments (the probe array) to be studied. The DNA microarray is flushed with possible complements and the resulting hybrids of DNA and labeled sequences, or "output," can be analyzed using a computer. The arrays are also called "biochips" because they were developed in Silicon Valley using techniques similar to those used in semiconductor chip production.**
Photo courtesy of Affymetrix, Inc.

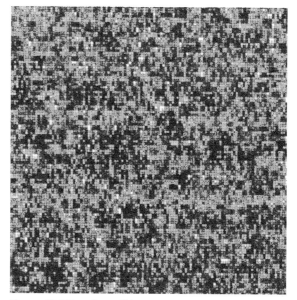

Figure 13.10b.   **GeneChip® Array Output (Affymetrix, Inc.): data from an experiment showing the expression of thousands of genes on a single GeneChip® probe array. Red or white squares indicate sequences complementary to the probe array feature.**
Photo courtesy of Affymetrix, Inc.

Figure 13.11. This microarray scanner was developed by Affymetrix, Inc. The GeneChip® probe array output is displayed on the computer monitor and analyzed using microarray software.
Photo courtesy of Affymetrix, Inc.

A **microarray scanner** is used to examine the features (see Figure 13.11). If a complementary RNA sequence to the probe array is found, the labeled complement binds to the DNA probe array in the feature. The scanner reads the degree of fluorescence, which is proportional to the probe array's degree of binding to a sequence of interest. In this way, hundreds of gene or RNA segments can be screened at the same time.

Microarray technology has revolutionized DNA screening, as well as applications in medicine and medical diagnostics, agriculture, food testing, environmental testing, evolutionary studies, industrial products, and other forms of research. With microarrays, scientists can study genetic variability and gene expression (or lack of it) between members of selected plant, animal, fungi, or microorganism species.

Different manufacturers and labs produce other types of microarrays. Some other common arrays have samples of DNA or RNA (eg, from tumors or plant extracts) spotted onto specially treated microscope slides. Like the GeneChip®, the samples on the array are probed with labeled DNA or RNA fragments. Samples that have a complement to the probe bind, and the amount of binding, produces varying amounts of a fluorescent signal.

Microarrays are an excellent way to screen RNA or cell samples when looking for a high amount of a particular RNA expression. High amounts of RNA usually indicate which proteins are being made in a cell and help scientists understand processes such as cancer. It also indicates which genes are being expressed and at what time. All this information leads to many targets (DNA, RNA, or protein) for therapeutics or other products.

microarray scanner (mi•cro•ar•ray scan•ner) an instrument that assesses the amount of fluorescence in a feature of a microarray

primer design (prim•er de•sign) the process by which a primer sequence is proposed and constructed

Constructing Primers

DNA synthesizers are also used to construct primer DNA, the pieces of DNA or RNA that can attach to a template strand and serve as a starting point for the DNA produced during PCR and DNA sequencing. Primers are required for the DNA replication in PCR and sequencing, since a DNA polymerase enzyme molecule can only add new nucleotides to an existing complementary nucleotide strand. The enzyme must have a short "starter" nucleotide strand already present, attached to the template, onto which additional nucleotides are added. The primer is a starter strand.

In cells, RNA primers are laid down on a template strand to start DNA replication. When scientists are trying to construct pieces of DNA in the laboratory, they must provide DNA primers for the DNA polymerase to work with. By knowing a little bit about the DNA needed, primers can be constructed to recognize a particular section. This is called **primer design**, and it is of critical importance in conducting a PCR or sequencing reaction. Some biotechnology companies' primary focus is the construction of primers for sale to R&D laboratories.

A "good" primer binds to one particular place in a genome and not at any other location. Therefore, when designing a primer, knowledge of the target sequence is mandatory. Some things to consider when designing a primer include the following:

- Select a primer sequence complementary to approximately 20 to 30 bases in the template target sequence. A primer that is too long or too short may have problems annealing to, or separating from, the template. Also, short primers are more likely to bind to some untargeted region of the template.

Figure 13.12. **A thermal cycler, like this GeneAmp PCR 9700® (Applied Biosystems by Life Technologies Corp.), can hold up to 96 PCR sample tubes (left). In PCR, two primers recognize and bind to a section of DNA, and allow for its replication millions of times. Primers for PCR are synthesized in "oligonucleotide factories." The instrument is called a "thermal cycler" because it cycles the samples through three temperatures, as seen in the display (right).**
Photos by author.

- Keep the primer composition at approximately 50% G-C, plus or minus (+ or –) 15%. Many G-C pairs means there are many spots where three H-bonds are holding the strands together. It is easier to split A-T pairs since only two H-bonds hold them together. The "proper" ratio of G–C and A–T ensures proper annealing and denaturing of the template and primer. Avoid Gs and Cs at the 3' end of the primer.
- Avoid sequences with long repeats of A, Cs, Gs and Ts; they are more likely to find each other and fold the primer over on itself.

PCR Amplification Primers are also used when trying to mark, identify, or amplify a piece of DNA. Primers for each end of a DNA section of interest can be synthesized. Once primers are bound to a section, the region can be copied millions of times, as in the PCR.

For PCR purposes, two sets of primers are used. One set binds to a forward section of targeted DNA, and one set binds to a reverse section hundreds of bases away from the forward primer. Care must be taken to design PCR primers so they will not anneal to each other. The next section discusses PCR, thermal cyclers (see Figure 13.12), and the use of primers in PCR.

Section 13.2   Review Questions

1. What is it called when DNA samples are transferred to a membrane for staining or probing?
2. How are probes used in microarrays?
3. Design a primer that would be good for recognizing the beginning of the following "sequence of interest." Describe why your primer is a good one.

   3'-ACACAGGATACGTGCTGCTCAATGCCATGATAGCCGGTCA-
   CAAGCTAATCCGATTATCGCGCAATTCCTAAATTCGCTAAAG
   CGAATCTTCAGGAAGGAACCCCGAAGGCCTTTT-5', and so on.

13.3 Polymerase Chain Reaction

Polymerase chain reaction (PCR) is a method by which millions of copies of a DNA segment can be synthesized in a test tube in just a few hours. In theory, PCR is simple. A section of DNA is unzipped and recognized by forward and reverse primers (see Figure 13.13). Then DNA polymerase molecules make copies of each template strand, starting at the end of each primer, to make two new fragments. The replication process is repeated 25 to 35 times with each generation of newly synthesized fragments (see Figure 13.14). Since each DNA fragment is doubled, the fragment number increases exponentially in each cycle. From one fragment amplified through 35 cycles of PCR, the result is more than 34 billion (2^{35}) fragments.

Since PCR technology can target a rather tiny section of DNA and replicate it over and over again (amplification), PCR technology has many advantages over other techniques. One advantage is that PCR can locate a targeted DNA fragment for a researcher, much like finding a DNA needle in a genomic haystack. This is very valuable since there is so much DNA in a cell. The PCR reaction accomplishes the same goal as a probe would in this situation, but it does so with much less lab work and less contamination.

Another advantage of PCR technology is that it can generate substantial amounts of DNA sample. Using PCR, researchers and technicians can produce large amounts of DNA sample, targeting the regions to be tested, and analyze for such purposes as research, medical testing, or criminal cases.

A third advantage is that PCR can be used to visualize the difference in size between similar DNA fragments. The tiny differences in size and sequence help to identify differences in individuals and explain evolutionary changes in populations and species. Examples of this technology are explained later when variable number of tandem repeats (VNTRs) is discussed.

Performing a PCR Reaction

To perform a PCR reaction (DNA synthesis in a test tube, not in a cell), a small quantity of the target DNA is added to a special thin-walled microtest tube (see Figure 13.15) that contains the following reagents with the indicated functions:

- reaction buffer: maintains pH
- forward primers: recognize one end of the fragment to be amplified.
- reverse primers: recognize the other end of the fragment to be amplified.
- *Taq* polymerase: special DNA polymerase that remains active at very high temperatures
- dNTPs: the four deoxynucleotides (A, C, G, T)
- magnesium chloride ($MgCl_2$): necessary cofactor for polymerase activity

The contents of the reaction tubes are thoroughly mixed, and the tubes are placed into the heat block of a thermal cycler. There are several brands of thermal cyclers, but all are basically just a heat block for which the temperature is controlled by a computer.

The tubes are placed in a retainer tray that sits on the thermal block in the thermal cycler (see Figure 13.16). The thermal block can be heated or cooled. The thermal cycler is programmed to cycle through three temperatures (hot, warm, and warmer) repeatedly. During each complete 3-cycle thermal cycle, each DNA fragment is replicated.

PCR 1st Cycle

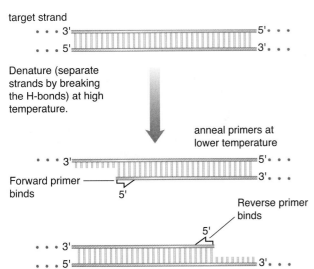

target strand

Denature (separate strands by breaking the H-bonds) at high temperature.

anneal primers at lower temperature

Forward primer binds

Reverse primer binds

Taq polymerase builds the rest of each complementary strand from the end of each primer as represented by "..."

Figure 13.13. PCR—1st Cycle.   DNA is unzipped during a denaturation stage, usually at about 95°C. By lowering the temperature, primers can bind (anneal) to each template at a target sequence. *Taq* DNA polymerase recognizes the end of the primer and builds the complementary strand to each template. Two strands result, ready to go through the cycle again.

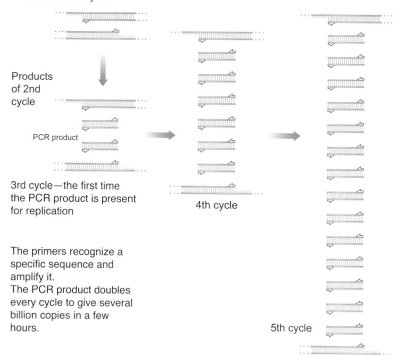

Products of 1st cycle

Products of 2nd cycle

PCR product

3rd cycle—the first time the PCR product is present for replication

4th cycle

The primers recognize a specific sequence and amplify it.
The PCR product doubles every cycle to give several billion copies in a few hours.

5th cycle

Figure 13.14. PCR Thermal Cycling.

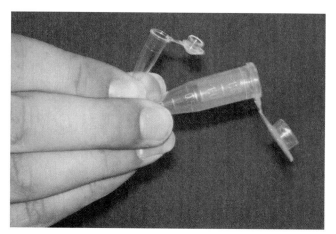

Figure 13.15. PCR samples are often prepared in thin-walled 0.2-mL tubes (left). The thin walls ensure rapid heating and cooling during the thermal cycling. A 1.5-mL tube (right) is shown for comparison.
Photo by author.

Figure 13.16. The heat block of this thermal cycler holds up to 96 samples. A thermal cycling program is typed into the computer, and the thermal cycler automatically changes between temperatures.
Photo by author.

During a PCR thermal cycle, a sample is taken through 20 or more cycles of three different temperatures (described below) to unzip and replicate the DNA fragments.

1. The first temperature is selected to provide complete **DNA denaturation** (unzipping of strands). It is common for denaturing temperatures to range between 94°C and 96°C and for a sample to be kept at that temperature for 30 to 60 seconds depending on the sample.
2. The second temperature is selected to provide optimal **primer annealing** to the template strands. Temperatures between 35°C and 65°C are generally used for a period of 30 to 60 seconds.

primer annealing (prim•er an•neal•ing) the phase in PCR during which a primer binds to a template strand

Figure 13.17. **PCR products are run on an agarose gel. Bands are sized by comparison to a set of sizing standards seen in the sixth lane from the right.**
© Custom Medical Stock Photo.

■■■■■■■■■■■■■■

extension (ex•ten•sion) the phase in PCR during which a complementary DNA strand is synthesized

optimization (op•ti•mi•za•tion) the process of analyzing all the variables to find the ideal conditions for a reaction or process

3. The third temperature selected is usually around 72°C, to provide the optimum temperature for *Taq* polymerase activity. During this part of the cycle, *Taq* polymerase is building new strands (**extension**) complementary to the template strands. *Taq* polymerase is an unusual enzyme that can withstand the high temperatures of each denaturing part of the cycle, but it works best at around 72°C.

A typical three-temperature cycle takes about 2 to 5 minutes. If a thermal cycling program requires 35 cycles, it would take approximately 3 hours to complete a PCR reaction.

Cycling Program The cycling program chosen depends on the type of sample to be amplified. An example of a thermal cycling program used to amplify a piece of mammalian DNA is shown in Table 13.1. A special form of *Taq* polymerase, AmpliTaq Gold® (Applied Biosystems by Life Technologies Corp.), is used. AmpliTaq Gold® requires a high-temperature activation step at the beginning of the thermal cycling. This helps reduce the amount of random, erroneous PCR product.

After a thermal cycling is completed, the PCR products are run on a gel and visualized. Usually a scientist knows what is expected and verifies the product by the band location on a gel. To verify PCR amplification products, 2% agarose gels or 10% polyacrylamide gels are commonly run with 100 bp standards (see Figure 13.17). Usually, the goal is to have a large band(s) of PCR product, of the amplified target DNA fragment(s). Southern blots can also be conducted with PCR gels if other detection methods are desired, such as using a fluorescent or radioactive probe.

Table 13.1. **Example of a Thermal Cycling Program**

Temperature	Time	Action
95°C	for 10 minutes	AmpliTaq® activation
35 cycles of: 95°C 54°C 72°C	 for 1 minute for 1 minute for 1 minute	denaturation primer annealing extension
72°C 4°C	for 10 minutes ∞	final extension hold/storage temperature

Challenges in PCR Technology PCR technology sounds straightforward; however, in practice, PCR can be a bit tricky for several reasons. One reason is that DNA samples are often compromised by contamination or mistreatment. For example, several coincidental PCR inhibitors are released (eg, DNase and RNase) when DNA is extracted from cells, or some inhibitors are ingredients in the extraction buffers (eg, EDTA). Also, DNA from past studies often contaminant lab workspaces. If recognized by a set of primers, the DNA may "show up" as erroneous PCR product on a gel.

Another problem in establishing a PCR regimen is that the concentration of the reagents, and the time and temperatures of the thermal cycling program may affect the quality of the results. A technician may spend many months testing one variable or another to optimize the conditions to give suitable PCR products at a high enough concentration.

Testing all the variables in a PCR protocol and determining the best conditions and concentrations is called "optimizing the reaction" or **optimization**. It is especially critical to optimize the concentration of $MgCl_2$ in a PCR amplification. A starting concentration of 25 mM $MgCl_2$ is often used. If 25 mM $MgCl_2$ does not produce the

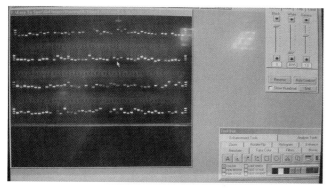

Figure 13.18. An agarose gel with 192 samples of PCR products. Several rows of wells allow a technician to screen many samples at one time. This is an example of high-throughput screening, in which hundreds of samples are analyzed at one time. The gel image is saved on a computer for further analysis.
Photo by author.

Figure 13.19. A Biomek® Laboratory Automation Workstation (Beckman Coulter, Inc.) is a robotic station that can pipet 96 samples at a time. This is valuable when preparing hundreds or thousands of similar samples, which is routinely done in high-throughput PCR screening.
Photo by author.

desired PCR products, a serial dilution is performed to prepare samples of decreasing $MgCl_2$ concentration. An amplification is conducted, testing each concentration of $MgCl_2$ until the concentration that gives the best product is determined. Each ingredient and reaction parameter in a PCR reaction must be optimized to ensure the best results.

Once a PCR reaction is optimized, the protocols can be scaled-up to larger volumes or a greater number of samples. This is what is done during high-throughput PCR and screening. When scientists survey hundreds or thousands of samples in search of some gene of interest, they prepare large batches of "master mix" that contain all the correct concentrations of enzyme, buffer, and cofactors. They prepare large batches of "primer mix" as well. Then they have enough to set up hundreds or even thousands of PCR reactions (see Figure 13.18). In some labs, with a large number of PCR samples or high-throughput screenings, robots do all the pipeting and sample prep in advance of the actual thermal cycling (see Figure 13.19).

BIOTECH ONLINE

PCR: Bigger is Not Always Better

PCR reactions are used in virtually all domains of biotechnology research and diagnostics. However, conducting a PCR reaction has been costly and time-consuming since thermal cyclers are expensive and not portable so samples must be brought to a lab that has the machine. The company, Amplyus, in Cambridge, MA is making it possible for PCR to be conducted by almost anyone, almost anywhere by designing a hand-held miniPCRTM thermal cycler.

Read the article, *PCR Heads into the Field*, in Nature Methods at **www.nature.com/nmeth/journal/v12/n5/full/nmeth.3369.html**. Free registration may be required.

Describe 4 other advances in PCR technology that makes it practical and affordable to do PCR reactions away from a typical laboratory, "in the field." Give examples of each technology is being used.

The new, portable devices and approaches let scientists look for specific genetic material in their samples as they hunt for pathogens affecting dynamic ecosystems, screen migrating people for infectious diseases or analyze rapidly changing water conditions. Some devices are prototypes; others are already being sold. Some designers use standard PCR instruments, others have developed new miniaturized hardware and still others avoid using instruments entirely.

What is mobile automated ddPCR instrument? Used for?

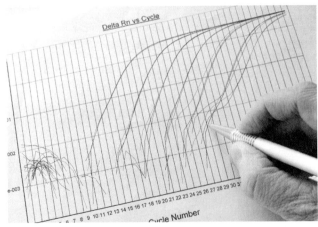

Figure 13.20. A scientist is analyzing the result of a qRTPCR run after a number of cycles. Real-time PCR is used to amplify and quantify a section of targeted DNA as it is being made, greatly speeding up the PCR process.

© Zmeel Photography/iStockphoto.

Advances in PCR Speed Research and Development

PCR makes it possible to isolate small sections of genetic information hidden among the hundreds or thousands of genes in a DNA sample. PCR is used for gene identification, disease detection, genetic fingerprinting, and gene synthesis.

Although "traditional PCR" is a powerful tool, until a few years ago, PCR was mainly limited by two factors: the sample itself (needed double-stranded DNA to run the "traditional PCR") and PCR product visualization technique (needed enough PCR product to see a sample on a gel).

Two new PCR processes allow scientists to overcome these limitations and use PCR to amplify nucleic acid sections from single-stranded samples (such as RNA) and to measure PCR product as it is made.

The first PCR advance is called reverse transcription PCR (RT-PCR). In RT-PCR, the enzyme, reverse transcriptase, found in retroviruses such as the HIV, may be used to produce a copy strand of DNA (cDNA) from a single-stranded RNA. The cDNA is the exact genetic code for the mRNA transcript and then may be amplified using traditional PCR. RT-PCR is used to detect and measure mRNA in samples as a way of understanding gene expression and the proteins that are present at specific times. Visit **biotech.emcp.net/RT-PCRanim** for an animation explaining RT-PCR.

The other PCR advance is called real-time PCR (qRT-PCR, for quantitative RT-PCR). The qRT-PCR process uses fluorescently-tagged probe technology to visualize and quantify the PCR product as it is made (see Figure 13.20). Thermocyclers that run qRT-PCR reactions use software to interpret the light produced from the fluorescent PCR reaction. A good discussion on real-time PCR may be found at **biotech.emcp.net/qRT-PCRdisc.**

Section 13.3   Review Questions

1. Why is *Taq* polymerase used in PCR instead of some other DNA polymerase?
2. What are the three parts to a thermal cycling reaction, and what is the difference in temperature between them?
3. What is it called when a PCR technician determines the best conditions for running a PCR protocol?

13.4 Applications of PCR Technology

Polymerase chain reaction has revolutionized the way DNA is studied and manipulated. Using PCR technology, scientists can gather and analyze smaller amounts of genetic information more easily and accurately than in the past. All of the following areas commonly use PCR:

- forensics/criminology
- missing children/soldiers
- paternity/maternity cases
- medical diagnostics
- therapeutic drug design
- phylogeny/evolutionary studies
- animal poaching/endangered species

Most applications of PCR technology involve some kind of genetic identification (see Figure 13.21). Genetic information can be studied on many levels. Before PCR, scientists could study nucleotide sequences on a chromosome by restriction fragment length polymorphism (RFLP) analysis (see Chapter 8) or DNA sequencing. Sequencing gives useful information about the genes that code for the amino acids found in proteins and for regulation of protein synthesis. However, the time and cost it takes to sequence the whole genome may be prohibitive. Also, the amount of information gathered during sequencing can be overwhelming, which in some genomes, like the human genome, totals over 3 billion base pairs. DNA sequencing is discussed further in the next secton.

Studies of whole chromosomes also give useful information. Chromosome studies using a microscope can produce a karyotype. As presented in the last section, a karyotype shows a cell's chromosomes magnified and matched with the homologous pair partners.

Karyotyping is useful for detecting certain genetic disorders caused by missing or extra chromosomes, or parts of chromosomes. These abnormalities are the causes of several severe genetic disorders, such as Turner syndrome, Cri-du-chat syndrome, and Down syndrome (see Figure 13.22). Scientists can also study gene expression using whole chromosomes and an assortment of regulatory molecules. However, studies of whole chromosomes do not usually elucidate genetic-code information.

The development of PCR techniques in the 1980s allowed scientists to more easily study DNA information on the gene level. For the first time, technicians could zero in on the fragments of DNA they actually wanted to study. As we have seen, sufficient DNA for genetic testing and DNA sequencing can easily be produced through PCR (remember, 35 cycles of PCR amplifications equal 2^{35} strands of DNA).

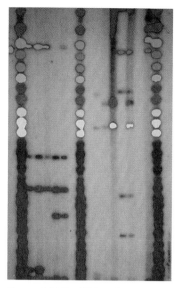

Figure 13.21. **This color-enhanced Southern blot of a genetic study is used to identify similarities and differences between parental DNA and that of a fetus.**
© Custom Medical Stock Photo.

karyotyping (kar•yo•typ•ing) the process of comparing an individual's karyotype with a normal, standard one to check for abnormalities

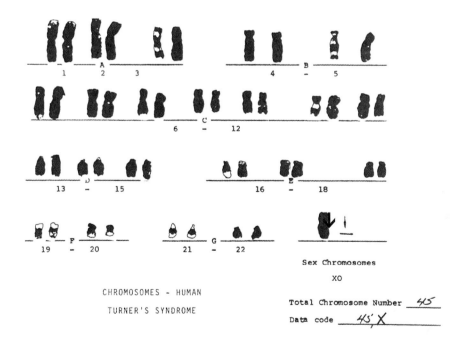

CHROMOSOMES - HUMAN

TURNER'S SYNDROME

Sex Chromosomes

XO

Total Chromosome Number __45__

Data code ____45, X____

Figure 13.22. **This is a karyotype of chromosomes from a patient with Turner syndrome, caused by a missing sex chromosome (arrow). The resulting genotype is shown as "XO" instead of "XX" for a "normal female" or "XY" for a "normal" male. Since the patient is lacking a sex chromosome, there are a total of 45 chromosomes per cell instead of the normal 46. Individuals with Turner syndrome appear to be females although they have underdeveloped secondary sex characteristics.**
© Lester V. Bergman/Getty Images.

DNA Fingerprinting

PCR technology came into the public spotlight during the 1992 O.J. Simpson murder trial. In the trial, Simpson was accused of murdering his ex-wife, Nicole, and her friend, Ron Goldman. Using the white blood cells obtained at the crime scene, prosecutors attempted to prove that Simpson's DNA was left at the crime scene and that the victims' blood was present in Simpson's car and in his home. Since each person's DNA is unique, a DNA fingerprint (DNA banding pattern) can be used to identify the blood from each person involved in the criminal investigation.

Prior to this trial, most DNA fingerprinting used restriction enzyme digestion (RFLPs) of genomic DNA (see Figure 13.23). The process works fairly well, but it gives data that are sometimes disputable or difficult to read. PCR technology has improved DNA fingerprinting by giving data that are more reputable. In addition, only a tiny amount of DNA is needed for PCR.

To obtain DNA for PCR from blood samples, white blood cells must be isolated. The white blood cells are burst open and the DNA is extracted. The PCR primers are designed to amplify several hundred regions of the human genome. The regions designated to amplify have been predetermined. Since so much of the DNA code is the same from individual to individual, specific regions of variability are chosen to amplify. These are sections that may contain regions with a variable number of tandem repeats, **VNTRs** of DNA sequence. (Note: VNTRs were first studied using RFLPs; however, PCR technology produces better data.)

VNTRs are regions on chromosomes that vary from person to person. They are sections of the chromosome where a sequence is repeated over and over again. Through evolutionary history, sections of the genome (9 to 80 nucleotides in length) are copied and inserted into adjacent positions along a chromosome. The numbers of repeats in a particular VNTR is characteristic of an individual and is inherited from one's parents. Anyone's VNTRs can be amplified using PCR as long as primers for the regions have been found. The differences in banding are due to the number of repeats a person has at a particular loci (see Figures 13.24 and 13.25).

A well known VNTR is D1S80, which is commonly used in forensics and other forms of human identification. It is a VNTR located on chromosome one and has a repeat section of 16 bp long. There are 29 known versions (alleles) of the D1S80 locus (location). The D1S80 locus can contain anywhere from 15 to 35 or more repeats. Since the PCR primers recognize DNA sequences on the outside end of the outer repeats, some of the D1S80 PCR products will be longer or shorter than others. A D1S80 with only one repeat would give a PCR product 161 bp in length, whereas a loci with two repeats would PCR products of 177 bp in length, 3 repeats would result in 193 bp, 17 repeats would produce 417 bp, 18 repeats would be 433 bp, and so on. These can be easily seen by running the PCR products on a gel with 50-bp DNA standards.

Individuals can have two copies of only one version of the alleles (homozygous) or two versions (heterozygous) of the D1S80 VNTR. They receive one from their father and one from their mother. Most people have two different D1S80 loci on their chromosomes. Sometimes, though, both parents have the same D1S80 loci on their chromosomes, in which case a person would inherit identical D1S80 loci. Since there are 29 different alleles and each person can have any two of them, there are 435 different genotypes possible for D1S80. By determining other VNTRs along with D1S80, technicians can elucidate a unique DNA fingerprint for an individual.

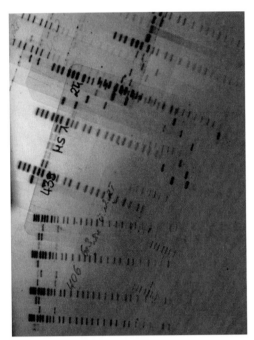

Figure 13.23. These autoradiograms show DNA fingerprinting using restriction fragment length polymorphism (RFLP) technology. RFLPs are unique DNA bands produced by restriction-enzyme digestion of DNA samples. DNA samples are treated by restriction enzymes, run on a gel, blotted onto a membrane, and treated with radioactive probes. Differences in banding are due to differences in the DNA sequence.
© Science Photo Library/Photo Researchers.

VNTRs (VNTRs) the abbreviation for variable number of tandem repeats, sections of repeated DNA sequences found at specific locations on certain chromosomes; the number of repeats in a particular VNTR can vary from person to person; used for DNA fingerprinting

Forensics

Forensics is the application of biology, chemistry, physics, mathematics, and sociology to solve legal problems, including crime scene analysis, analysis of accidents, cases of child support, and paternity. It involves collecting evidence to prove that a crime has occurred and who may have committed it.

Although many forensic investigations include DNA analysis, it is but one tool available to a forensic scientist. Blood, hair, glass, metal, plastic, fabric, soil, bone, dyes, paint, construction tools, tire skid marks, impact damage, burn damage, computer programs, bank accounts, and phone records are just some of the items that might be studied in forensic cases.

Forensic scientists use an assortment of tools to analyze data, including microscopes, DNA and protein gels, DNA sequencers, gas chromatographs, mass spectrometers, and spectrophotometers.

The field of forensics has many specialties, and many job opportunities exist. The main employers of forensic scientists are the state and federal governments, and local police departments (see Figures 13.26 and 13.27). With the increased concern related to bioterrorism and biodefense, the need for more biotechnologists, focused in forensics, increases. Table 13.2 lists examples of professionals who work in forensic science or in companies or departments that do forensic work.

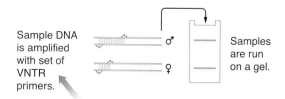

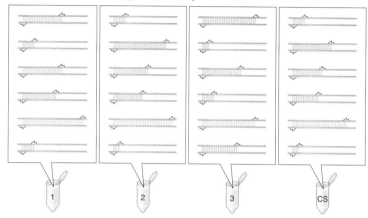

Samples from Suspects and Crime Scene

Figure 13.24. **VNTR PCR Reaction.** Samples from three suspects are taken, as well as DNA (from blood) left at the scene of the crime. The samples are prepared for PCR and amplified in a thermal cycler. Since small variations in DNA code space the VNTR primers at varying distances, different PCR products result in each individual. Many different VNTR analyses give the fingerprint (see Figure 13.25).

forensics (fo•ren•sics)
application of biology, chemistry, physics, mathematics, and sociology to solving legal problems including crime scene analysis, accident analysis, child support cases, and paternity

The banding in each lane represents the VNTR products that were created during PCR. The differences are due to differences in the DNA sequence.

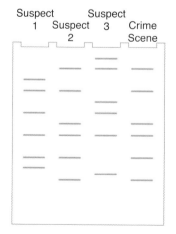

Figure 13.25. **VNTR PCR Gel.** The VNTR PCR products are run and visualized on a gel. Can you tell whose DNA matches the DNA left at the scene of the crime?

Figure 13.26. **Brooke Barloewen is a Criminalist in the Forensic Biology Section of the Crime Laboratory at the Santa Clara County Crime Laboratory in San Jose, CA. She examines evidence for the presence of biological stains, hairs, tissues, and skin cells. She then takes these samples through the multistep DNA process. Evidence samples are compared with known reference samples. Unknown samples are searched against a nationwide database. A report is generated and sent to the police agency. Sometimes, Brooke testifies about her results in court, explaining and interpreting her results to a jury. Sometimes, she is called out to a crime scene to collect the evidence herself.**
Photo courtesy of Brooke Barloewen.

DNA Technologies

Figure 13.27. Gary A. Stuart, MD, Forensic Pathologist. A forensic pathologist is a medical specialist who, working with a coroner's office or other institutions, determines the cause and manner of death and establishes major diagnoses of disease using the autopsy, microscopy of postmortem tissue specimens, and information from the CSI and toxicology studies. Dr. Stuart worked as a forensic pathologist in Bosnia and Kosovo with several international humanitarian organizations and with a legal arm of the United Nations. These organizations have been involved in an effort to recover and repatriate human remains of victims of atrocities that occurred during the Balkan conflicts of 1991–1999. Dr. Stuart also was involved in the collection of evidence to be used in the prosecution of war crimes, genocide, and crimes against humanity at The Hague. He is a member of Physicians for Human Rights, the United Nations International Criminal Tribunal for Former Yugoslavia (ICTY), and the International Commission of Missing Persons.
Photo by Mary Hansen.

Table 13.2. Career Titles in Forensics

Accountant	Criminalist	Linguist
Anthropologist	Dentist	Nurse
Artist	Document Examiner	Pathologist
Ballistic Expert, Firearm Examiner	Engineer	Photographer
Biodefense Specialist	Entomologist	Psychologist
Bioterrorism Specialist	Fingerprint Specialist	Serologist/Biologist
Chemist	Forensic Pathologist	Technician, Evidence Specialist
Computer Forensic Investigator	Forensic Scientist	Toxicologist
Crime Lab Technician	Geologist	Wildlife Forensic Scientist

BIOTECH ONLINE

CSI—Your Town

Advances in chemistry, biology, and physics have made the investigation of crimes vvery scientific. With the popularity of crime-solving shows such as *CSI*, many people are interested in forensics as a career. But, what do forensic scientists or criminalists actually do, and who employs them?

1. Go to **biotech.emcp.net/ccifaq** and answer the following questions:
   a. What is a forensic scientist?
   b. What is a criminalist?
   c. What are five areas of forensic science?
2. Using your Internet searching skills, find at least one job posting for either a criminalist or forensic scientist. Record the employer name, location, the job description, and one other interesting fact.
3. Record this information and share it with another person in the class.

Section 13.4   Review Questions

1. Restriction fragment length polymorphism technology was formerly used for DNA fingerprinting. What technology is currently used for DNA fingerprinting?
2. For a DNA fingerprint, many PCR targets are used. Each target is its own VNTR. What is a VNTR?
3. Why would looking for the persons responsible for sneaking endangered species (rare birds, for example) into the United States be considered a job for a forensic scientist?

DNA Sequencing

It is an understatement to say that advances in DNA science have changed the way biologists do their jobs. **DNA sequencing**, polymerase chain reaction (PCR), microarray, and bioinformatics have provided so much data that researchers must design, conduct, and report the results of their experiments in ways that are different from those that were standard just a generation ago.

DNA sequencing includes all the techniques used to determine the order of nucleotides (A, G, C, and T) in a DNA fragment. There are several reasons why scientists sequence DNA, including the following:

- to discover new genes
- to look for mutations (defects) or differences in genes or the genetic code (genotyping)
- to determine the exact structure of a gene, RNA, or protein
- to verify that an isolated gene is, indeed, the targeted gene of interest
- to determine or verify the nucleotide sequence of a synthetic nucleotide fragment
- to determine as much about the genome of an organism as possible, including the fine genetic structure, to better understand genetic expression (exome sequencing) and variations in gene expression (transcriptome sequencing)
- to describe populations and their characteristics, such as diseases, disorders, and evolution

Figure 13.28. Scientists from the National Animal Disease Center analyze DNA sequences to develop a vaccine made from a modified *Brucella abortus* bacterium that causes brucellosis disease in cattle.
Photo by Keith Weller. © ARS.

Until 2005, there really was only one method (Sanger) that produced the large amounts of data necessary to sequence genomes. However, over the past decade, several new methods of automated sequencing have revolutionized genomic sequencing providing a wealth of data in a fraction of the time and cost of Sanger sequencing. Scientists are now able to isolate genomic DNA from an organism and, within a day, recognize and sequence one or more genes or even entire genomes. No matter the method, sequencing DNA on gels (see Figure 13.28), in tubes, on chips, on beads, etc., large amounts of sequence data must be compiled, analyzed and referenced. In this section, a brief introduction to the current methods of DNA sequencing is presented.

Dideoxynucleotide Sequencing

In the first DNA sequencing performed, the bases were actually destroyed individually in a systematic way to determine their order. After a short period of time, the Sanger Method of **dideoxynucleotide sequencing** replaced this long, tedious procedure. The Sanger Method is simply a predictable, interrupted DNA synthesis.

The Sanger Method was the method by which The Human Genome Project scientists determined the entire consensus sequence of human DNA by the year 2000. Now the Sanger Method is primarily used to sequence relatively small pieces of DNA such as plasmids or PCR products.

In dideoxynucleotide sequencing, a tube containing millions of copies of the DNA of interest is given all the reagents needed to result in synthesis of all the DNA strands. However, in addition to the "normal" deoxynucleotides (dATP, dCTP, dGTP, and dTTP), some modified nucleotides are added that disrupt DNA synthesis. The modified nucleotides are **dideoxynucleotides** (ddNTPs), and their insertion into a synthesized strand stops further DNA synthesis in a predictable way.

DNA sequencing (DNA se•quenc•ing) pertaining to all the techniques that lead to determining the order of nucleotides (A, G, C, T) in a DNA fragment

dideoxynucleotide sequencing (di•de•ox•y•nu•cle•o•tide se•quenc•ing) a sequencing method that uses ddNTPs and dNTPs in a predictable way to produce synthesis fragments of varying length; also called the Sanger Method

dideoxynucleotides (di•de•ox•y•nu•cle•o•tides) the nucleotides that have an oxygen removed from carbon number 3, abbreviated ddNTPs

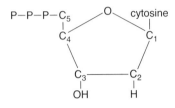

dCTP
deoxycytosine triphosphate

Dideoxynucleotides are identical to the "normal" nucleotides that are usually added to a DNA strand during replication, except that they have an oxygen molecule removed at carbon No. 3 (see Figure 13.29). This tiny change in the molecule has tremendous impact on DNA replication, since the oxygen at carbon No. 3 is where the next nucleotide in a strand normally attaches. Without that oxygen at carbon No. 3, an additional nucleotide cannot be added, and strand synthesis stops.

In setting up sequencing reactions, all the reagents necessary for DNA synthesis are required. Template, primer, dNTPs, and buffer with essential cofactors must be present for normal strand synthesis. In the original version of ddNTP synthesis, four test tubes were prepared, each representing the sequencing reactions needed to determine the location of each of the four nitrogen bases (A, C, G, and T) on the template (see Figure 13.30).

In the four tubes, scientists prepare the DNA synthesis reactions to include all of the dNTPs, plus just a few of only one of the ddNTPs. Synthesis proceeds and, occasionally, a ddNTP is inserted where a dNTP should be. This stops synthesis of the strand at that point. Since the ddNTP is in short supply, usually a "functional" dNTP is inserted. Because each tube contains millions of strands to copy, a "synthesis-stopping" ddNTP is eventually inserted at each position to stop synthesis and create synthesis fragments of differing lengths that all end in the ddNTP for that tube. The resulting fragments are run on a PAGE gel or in a capillary sequencer. Any band in a lane is known to end with the particular ddNTP for that sample tube. Since fragments are run on a gel proportional to their size, the order of nucleotides is determined by reading the bands in each lane

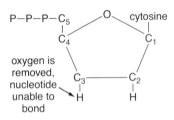

ddCTP
dideoxycytosine triphosphate

Figure 13.29. ddNTP structure. During DNA replication, if a dideoxynucleotide (ddNTP), such as ddCTP (bottom), is inserted instead of dNTP, such as dCTP (top), strand synthesis stops. In ddNTP sequencing, this principle is used to produce synthesis fragments of varying lengths, all ending in a known ddNTP.

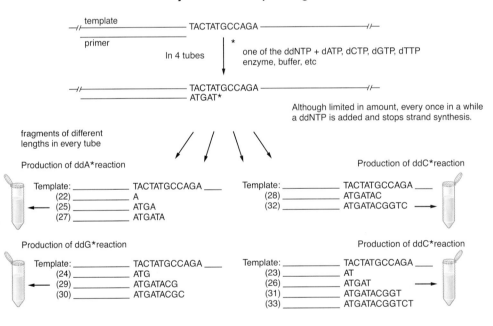

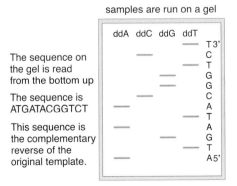

Figure 13.30. ddNTP Sequence Fragments. After the ddNTP synthesis reactions, partial synthesis products are run on a gel (in this case), or in a capillary system, to determine where synthesis was stopped. The synthesis stops on a DNA fragment because a ddNTP is inserted, and that information reveals which nucleotide is normally found in the sequence at that location on the strand.

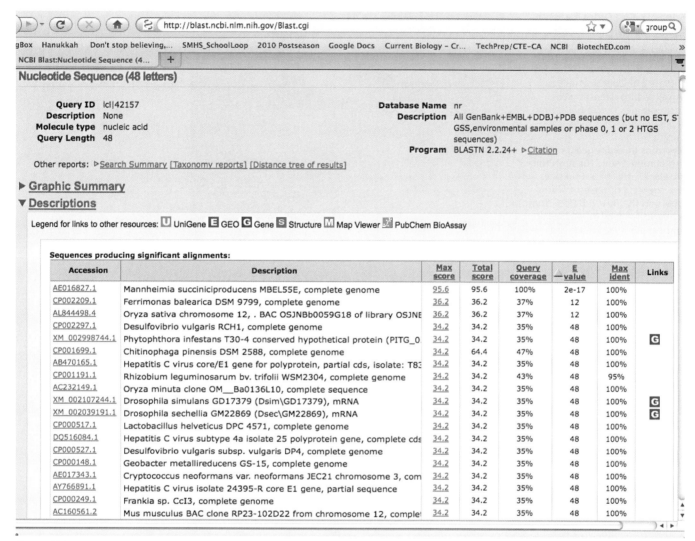

Nucleotide Sequence (48 letters)

Query ID	lcl|42157
Description	None
Molecule type	nucleic acid
Query Length	48

Database Name	nr
Description	All GenBank+EMBL+DDBJ+PDB sequences (but no EST, ST GSS,environmental samples or phase 0, 1 or 2 HTGS sequences)
Program	BLASTN 2.2.24+ ▷Citation

Other reports: ▷Search Summary [Taxonomy reports] [Distance tree of results]

▶ **Graphic Summary**

▼ **Descriptions**

Legend for links to other resources: U UniGene E GEO G Gene S Structure M Map Viewer ▨ PubChem BioAssay

Sequences producing significant alignments:

Accession	Description	Max score	Total score	Query coverage	E —value	Max ident	Links
AE016827.1	Mannheimia succiniciproducens MBEL55E, complete genome	95.6	95.6	100%	2e-17	100%	
CP002209.1	Ferrimonas balearica DSM 9799, complete genome	36.2	36.2	37%	12	100%	
AL844498.4	Oryza sativa chromosome 12, . BAC OSJNBb0059G18 of library OSJNE	36.2	36.2	37%	12	100%	
CP002297.1	Desulfovibrio vulgaris RCH1, complete genome	34.2	34.2	35%	48	100%	
XM_002998744.1	Phytophthora infestans T30-4 conserved hypothetical protein (PITG_0	34.2	34.2	35%	48	100%	G
CP001699.1	Chitinophaga pinensis DSM 2588, complete genome	34.2	64.4	47%	48	100%	
AB470165.1	Hepatitis C virus core/E1 gene for polyprotein, partial cds, isolate: T83	34.2	34.2	35%	48	100%	
CP001191.1	Rhizobium leguminosarum bv. trifolii WSM2304, complete genome	34.2	34.2	43%	48	95%	
AC232149.1	Oryza minuta clone OM__Ba0136L10, complete sequence	34.2	34.2	35%	48	100%	
XM_002107244.1	Drosophila simulans GD17379 (Dsim\GD17379), mRNA	34.2	34.2	35%	48	100%	G
XM_002039191.1	Drosophila sechellia GM22869 (Dsec\GM22869), mRNA	34.2	34.2	35%	48	100%	G
CP000517.1	Lactobacillus helveticus DPC 4571, complete genome	34.2	34.2	35%	48	100%	
DQ516084.1	Hepatitis C virus subtype 4a isolate 25 polyprotein gene, complete cds	34.2	34.2	35%	48	100%	
CP000527.1	Desulfovibrio vulgaris subsp. vulgaris DP4, complete genome	34.2	34.2	35%	48	100%	
CP000148.1	Geobacter metallireducens GS-15, complete genome	34.2	34.2	35%	48	100%	
AE017343.1	Cryptococcus neoformans var. neoformans JEC21 chromosome 3, com	34.2	34.2	35%	48	100%	
AY766891.1	Hepatitis C virus isolate 24395-R core E1 gene, partial sequence	34.2	34.2	35%	48	100%	
CP000249.1	Frankia sp. CcI3, complete genome	34.2	34.2	35%	48	100%	
AC160561.2	Mus musculus BAC clone RP23-102D22 from chromosome 12, complet	34.2	34.2	35%	48	100%	

Figure 13.31. **An example of a page showing BLAST search results.**

from the bottom of the gel up to the top. The ddNTPs can have tags attached to them, so instruments can read which fragment length has the particular ddNTP at the end.

Once a sequence is determined, it can be used to understand which RNA and protein molecules the DNA code of interest may produce. A scientist can also determine whether there are any known, similar sequences. One way to learn what is known about a sequence is to type the "determined" sequence into a program called "**BLAST**." By using various international databases, BLAST locates any similar sequences that researchers have previously found. If BLAST finds any close matches to a queried sequence, they are listed in the order of how good the match is. For example, say a sequencing experiment gives the sequence below. By going to www.ncbi.nlm.nih.gov/BLAST, and typing in:

ATTGGTCGCTTATTCTTGGCCGGTGTCGTTCCGGGTACAATGCTCTGT

a match with part of the DNA sequence from the bacteria, *Mannheimia succiniciproducens* (MBEL55E), is found (see Figure 13.31).

BLAST (blast) an acronym for Basic Local Alignment Search Tool, a program that allows researchers to compare biological sequences

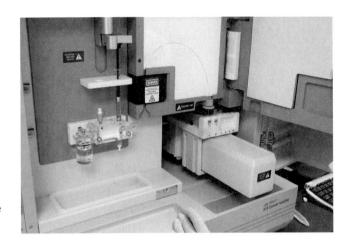

Figure 13.32. Older models of DNA sequencers have very large PAGE gels that run up to 96 lanes of 24 sets of four sample tubes. A newer model, the ABI PRISM® 310 Genetic Analyzer (Applied Biosystems by Life Technologies Corp.) has a tiny capillary tube that pulls up a sample with all four sequencing reactions mixed together. As the samples move through the capillary tube (based on size), a laser detects different-colored labels on the four ddNTPs. Some sequencers have 16, 48, 96, or more capillaries, which can run samples simultaneously (high-throughput sequencing).
Photo by author.

cycle sequencing (cy•cle se•quenc•ing) a technique developed in the late 1990s that allowed researchers to run synthesis reactions over and over on samples, increasing the amount of sequencing product and the speed of getting results

Human Genome Project (Hu•man Ge•nome Proj•ect) a collaborative 10-year project completed in 2000, which aimed to sequence the entire DNA code for the human organism

In the late 1990s, several events led to much faster DNA sequencing than did the process described above. One advance was the advent of automated DNA sequencers. These sequencers can run and interpret gels or tiny capillary tubes of sequencing product, handling dozens of samples at a time (see Figure 13.32). Using a laser, the large number of samples are read and the data stored and interpreted using computer programs. The second advance was a process called **cycle sequencing**, which utilized a thermal cycler to run synthesis reactions repeatedly on samples, increasing the amount of sequencing product as well as the speed of obtaining results.

Due to these technological advances, the sequencing of whole genomes rapidly began (see Figure 13.33). For example, the **Human Genome Project** (HGP), responsible for decoding the entire human DNA sequence (every nucleotide on all 23 chromosomes), was originally predicted to take up to 20 years to complete. Scientists, in fact, published the rough draft of the entire human genome in the year 2000, approximately 10 years earlier than expected. The final draft was completed in 2003. Geonomics is discussed in more detail in the next chapter.

Next-Generation Sequencing

In 2005, a revolution in micro- and nanotechnologies resulted in the development of what is called next-generation sequencing (NGS). Next-generation sequencing, also called massively parallel sequencing, allows simultaneous sequencing and analysis of millions of DNA samples (high-throughput sequencing) at virtually the same time. NGS is more accurate than Sanger sequencing and requires much less DNA sample per analysis (1 copy versus thousands). In addition, NGS has gotten significantly faster and cheaper to use in the past few years.

Figure 13.33. Craig Venter (left), President of the J. Craig Venter Institute and former President and Founder of Celera Genomics, a part of Applera Corporation; Leroy Hood (center), President of Institute for Systems Biology and formerly of the University of Washington; and Mike Hunkapiller (right), former CEO of Applied Biosystems, Inc. These men, together with Dr. Francis Collins, Director of the National Human Genome Research Institute (not shown), directed teams that designed the methods and instruments for rapid DNA sequencing. These advances led to the initial completion of the Human Genome Project in 2000.
Photo by Mike O'Neil, Applied Biosystems, Inc.

The first NGS system, Genome Analyzer® was created by Solexa in 2006 and was later purchased by Illumina, Inc. The Genome Analyzer®, gave scientists the ability to produce 1 gigabase (Gb) of DNA sequence data in a single run. Soon, Agencourt, Inc. (later purchased by Applied Biosystems, released a NGS platform called Sequencing by Oligo Ligation Detection (SOLiD®). Both these NGS use DNA template to create a library of DNA fragments. The library of template fragments are each amplified (by PCR) and the replicates PCR products are sequenced in parallel using fluorescent nucleotides. This generates millions or billions of sequencing "reads", each of which originated from a different template molecule.

Another next-generation sequencing platform, The Ion Torrent Personal Genome Machine (PGM), measured changes in pH as nucleotides are added during the sequencing reaction. Each of the A, C, G, T nucleotides are flooded over template one at a time and if that particular nucleotide that is added incorporates into a sequencing synthesis strand, the pH changes, and is then measured. From the pH changes, and knowing what nucleotide was added, the sequence is read.

These three NGS platforms are the most popular and widely used. By 2010, according to, "Comparison of Next-Generation Sequencing Systems" in the Journal of Biomedical Biotechnology (2012; 2012: 251364), using these next-gen platforms, NGS was sequencing DNA at the rate of 100 GB per 5 days; by 2013 the human genome could be sequenced in 15 minutes!

Simply, it could be said that to obtain nucleic acid sequence from the amplified template libraries, each NGS platform discussed above relies on sequencing by synthesis. The library fragments act as a template, off of which a new DNA fragment is synthesized. The sequencing occurs through a cycle of washing and flooding the fragments with the known nucleotides in a sequential order. As nucleotides incorporate into the growing DNA strand, they are digitally recorded as sequence. In actual practice each NGS platform is technically very different. An excellent presentation explaining the engineering technology used in these NGS platforms can be found at **users.ugent.be/~avierstr/nextgen/Next_generation_sequencing_web.pdf**.

Once sequencing is complete, raw sequence data must undergo several analysis steps. Although data production during NGS is much faster than the Sanger Method, huge amounts of data are produced and data analysis requires specially trained bioinformatics technicians to piece together the fragment sequences (align them) and analyze them for genetic variants including single nucleotide polymorphisms or SNPs, as well other insertions or deletions (see Figure 13.34). Analysis can also include the search for genetic variants, novel genes, or missing genes. Bioinformaticians use software programs to perform the needed alignment and analysis (see Figure 13.35). The data can reveal a wealth of genetic information including single nucleotide polymorphisms or SNPs, as well larger insertions or deletions, the identification of genetic variants, novel genes, or missing genes.

Figure 13.34. Next-Generation Sequencing at the Joint Genome Institute DNA sequencing determines the exact order of the four bases that make up a gene or genome. For processing by the NGS HiSeq2000, the DNA template is cut into fragments and each base is labeled with a different fluorescent dye so that a detector can distinguish them and their order in the sequence.
© Science Source/Photo Researchers.

Figure 13.35. **Garrett Lew, BS, Bioinformatics Research Associate, Five Prime Therapeutics, Inc. Garrett's interest in bioinformatics began as an undergraduate student at University of California at Davis. He worked with a PhD candidate and wrote Python scripts to manipulate and analyze sequence data used in studying the allele specific expression of infected poultry. With this experience in sequence analysis, Garrett was hired as a Bioinformatics Research Associate at Five Prime Therapeutics. At Five Prime, Garrett works closely with the Molecular Biology group who creates different DNA and protein constructs (libraries) that are needed by other departments. He analyzes both DNA sequences from NGS and amino acid sequences using chromatograms to check the quality of construct samples before they are given to other researchers downstream. By running different programming scripts, Garrett checks the sequence and alignment of different samples against references. He adds verified sequences to cloning and protein databases. Here, Garrett is reviewing chromatograms to check the nucleotide sequence quality, as well as the alignment of the sequence to a reference. For constructs that have not been fully verified, Garrett designs primers used to complete sequencing before updating the database. Another part of his genetic sequence analysis is exon splice site verification. Since Five Prime Therapeutics is a pharmaceutical biotechnology company, ultimately all of Garrett's work is to further the analysis of genes and proteins to determine their structure and function. This data may explain the effects that these molecules have and lead to finding treatments for human diseases or disorders.**
Photo courtesy of Garrett Lew.

Section 13.5 Review Questions

1. How is a ddNTP different from a "regular" dNTP?
2. When preparing sequencing-reaction tubes, each of the four dNTPs are added, but just one kind of ddNTP. Which are used in the highest concentrations, the dNTPs or the ddNTPs, and why?
3. Where on the Internet may one go to compare DNA sequence data?
4. What is the main difference between next-generation sequencing and dideoxynucleotide sequencing?

Chapter Review

Speaking Biotech

Summary

Concepts

- In a cell, DNA directs protein synthesis and its own replication. Each body cell in an organism has the same DNA and the same number of chromosomes as every other body cell in that organism. Sex cells have half the chromosomes (half the DNA) of body cells.
- Chromosomes in humans are arranged as 23 homologous pairs. Homologues have the same genes in the same order on the chromosome.
- Chromosomes are visible in cells when they thicken and shorten during cell division. Identification of chromosomes and matching them with their homologues is called karyotyping.
- As a result of mitosis, each daughter cell gets an exact copy of the chromosomes of the parent cell. This is possible because DNA is replicated prior to cell division.
- In cells, six enzymes control DNA replication. The enzymes unzip the complementary strands of the DNA double helix (helicase), release tension (topoisomerase), attach a primer molecule (RNA primase), synthesize a complementary strand to each template (DNA polymerase), remove primer (RNase H), and seal adjacent DNA replication sites (DNA ligase).
- DNA replication can be performed in the lab by technicians. *In vitro* DNA synthesis is done to produce primers, probes, and genes of interest for research. In a test tube, a template molecule is necessary from which a complementary strand is built. Then, all the other essential ingredients for DNA synthesis, including buffer, primer, dNTPs, DNA polymerase, DTT, and magnesium chloride, are added to the tube.
- Using an automated DNA synthesizer, technicians build DNA strands by coupling nucleotides in a specific order to resin beads in a column.
- DNA synthesis to make probes is an important application. Probe hybridization is used to screen samples of DNA or RNA for evidence of gene expression. Using microarray technology means hundreds of samples can be probed at the same time.
- For ease of use and higher-sensitivity assays, DNA fragments are often transferred to membranes or paper. This is called Southern blotting. To prevent DNA samples from being washed off blots, samples are cross-linked to the membrane by UV light exposure.
- DNA synthesizers also make primers for use in PCR and DNA sequencing. Primers have to recognize specific sequences, so appropriate primer design is critical. Primers should be of a certain length and nucleotide composition.
- PCR is a method used to recognize certain sequences of DNA, and replicate them enough times to have sufficient sample to test or use in research.

- PCR reactions are run in thermal cyclers, which control the temperature cycling of the PCR reaction automatically.
- During a PCR amplification, DNA is heated until the strands separate (denature), the sample is cooled and primer anneals or binds to each strand, Taq polymerase synthesizes the rest of each strand. Commonly, the denaturation step occurs at about 95°C, the annealing is at about 60°C, and the extension of the strands is at about 72°C. A complete cycle takes about 2 to 3 minutes.
- For each DNA strand in a sample, one strand becomes two, two become four, four become eight, and so on, exponentially, until there are a billion or more copies in the sample.
- A PCR product is run on a gel to confirm its presence, concentration, and purity.
- PCR protocols take months to optimize. To produce the best PCR product, the reactants' volume and concentration are tested, as well as the cycling time, temperature, and cycle number.
- Once a PCR reaction is optimized, it can be scaled-up for large sample numbers and high-throughput screening.
- Two new PCR processes, RT-PCR and qRT-PCR, allow scientists to amplify nucleic acid sections from single-stranded samples (such as RNA) and to measure PCR product as it is made.
- PCR is used in many applications, including forensics, missing persons cases, paternity cases, medical diagnostics, drug design, medical research, evolutionary studies, and endangered species study and protection.
- In recent years, VNTR PCR has become the main method of DNA fingerprinting. A VNTR is an allele that is present in a species in many forms due to mutations in the length of the region. Performing a PCR on VNTRs can determine a person's genotype for an allele. The pattern of VNTRs on a gel is called a DNA fingerprint.
- Forensics is the science of data collection for the purpose of solving a crime. Many methods of data collection and analysis are used in forensics, including DNA fingerprinting.
- Applications of DNA sequencing include gene identification, searching for mutants, confirmation of gene transfer, and comparative DNA studies.
- DNA sequencing is accomplished using dideoxynucleotides (ddNTPs). These nucleotides are different than the "normal" nucleotides, and they disrupt DNA synthesis and sequencing reactions.
- DNA sequencing technicians prepare sample tubes, each with a different ddNTP. Random insertions of different ddNTPs, instead of their "normal" dNTP counterparts, halt synthesis at different spots on the strand. Fragments of differing lengths result in a tube, but each ends with the ddNTP. The four sets of sequencing fragments can be loaded on a gel, and the distance each band travels in its respective lane reveals the sequence of the original DNA strand of interest.
- Dideoxynucleotide sequencing was formerly performed on large "slab" gels, but recently, automated sequencers with capillary tubes can run many more samples and report data more quickly. The advent of these instruments and computer programs to analyze the data has resulted in a genomics revolution.
- NGS can take a single DNA strand, amplify fragmented pieces of it using PCR, and sequence those strands in parallel creating sequenced data faster and cheaper than dideoxynucleotide sequencing.
- DNA sequencing data can be analyzed using many bioinformatics tools. A common one is BLAST, which allows comparisons of sequences within and between organisms. This information helps direct pharmaceutical and evolutionary studies, among other things.

Lab Practices

- Oligonucleotides can be made in a test tube using a template and a primer that recognizes the 3' end of the template. To get good annealing of the primer to the template, technicians mix the primer and template together, and heat them to a temperature that is high enough to separate them completely. Slow cooling allows accurate complementary annealing.
- DNA synthesis (in a test tube) proceeds to completion if all the required reagents are in the proper volume and concentration, including DNA polymerase, dNTPs, reaction buffer, $MgCl_2$, and DTT.
- DNA polymerase has maximum activity at approximately 37°C and is incubated with reactants for approximately 4 minutes at that temperature.
- DNA synthesis fragments are run on a boric acid/urea (TBU)-PAGE gel in boric acid/EDTA (TBE) buffer with DNA sizing standards. Before loading, they are heated. Heating the samples and the presence of urea helps ensure that the primer-synthesis product and template do not reanneal.

- For DNA samples in low concentration, Southern-blot visualization methods are more sensitive than staining a gel with ethidium bromide (EtBr).
- In a Southern blot, a moistened membrane or nitrocellulose paper is laid on a gel with DNA samples. Paper is laid on top, and a heavy weight is placed on top of the paper. Capillary action draws the samples from the gel to the membrane.
- The type of label on the DNA determines the type of visualization to be used on a blot. For a biotin-tagged primer, a series of chemicals is attached to the biotin. This results in alkaline phosphatase conversion of nitro blue tetrazolium (NBT) to a blue product that "stains" the DNA fragments on the blot. Sizes of the colored synthesis products can be determined by comparing them with the sizing standards on the blot.
- PCR is possible because of two scientific breakthroughs: the thermal cycler and Taq polymerase.
- During a thermal cycle, the DNA strands are separated, forward and reverse primers attach, and Taq polymerase replicates the strands. These reactions happen at certain temperatures, and the thermal cycler automates the process.
- The thermal cycler is set to repeat the three-part cycle of hot, warm and warmer temperatures for 25 to 35 times. Each cycle doubles the amount of the segment being replicated.
- PCR products are run on gels (PAGE or agarose) and commonly visualized with EtBr. The size of the PCR product is determined by comparing the fragments to sizing standards.
- PCR can be used to recognize a specific section for the purpose of genetic typing (genotyping or fingerprinting).
- Extraction protocols to prepare human DNA for PCR and sequencing analysis may include steps using salt and heat. To purify the DNA for study, proteins and ions must be removed.

Thinking Like a Biotechnician

1. People with Down syndrome have an extra chromosome number 21 in all of their cells (called trisomy 21). How can chromosome abnormalities such as these be identified in the lab?
2. If an *E. coli* DNA strand (genomic DNA) is approximately 4.6 million bp long with approximately 4400 genes, what is the average length of an *E. coli* gene?
3. DNA strands are antiparallel. Why is this important for PCR?
4. Six enzymes are needed for DNA replication in cells. How many are needed for DNA synthesis in a tube?
5. A microarray is set up to recognize a DNA sequence only found in men with a rare type of prostate cancer. The array reaction is run. Five wells with samples from five men out of 1000 glow 30% more than the negative controls. What do you conclude?
6. DNA synthesis reactions are run with the goal of producing strands between 60 and 100 bases long. How might these be visualized in a lab?
7. Southern blot membranes are positively charged. Considering what they are used for, why might that be important?
8. In PCR, special thin-walled microtest tubes are used. Why might it be valuable for PCR tubes to be thin-walled?
9. In a PCR reaction, every variable needs to be tested to determine the optimum condition to produce the maximum amount of PCR product. Design an experiment in which the primer-annealing temperature will be optimized.
10. Here is a DNA sequence that must be located in a genomic DNA sample. Below is a possible primer sequence. Evaluate the appropriateness of using this potential primer given the criteria in Section 13.2.

   Sequence of Interest:
   3'-ACATGCTGCTGCCGGTCACAAGGCAATTCCTAAAAAGGGAAGGAACCCC-GAAGGCCTTTT-5'

   Possible Primer to Sequence of Interest:
   5'-TGTACGACGACGGCCAGTGTTCCGTTAAGGAT-3'

Biotech Live

Activity **13.1** **Extreme Bacteria Produce Extreme Enzymes**

PCR requires a unique enzyme, *Taq* polymerase, that can withstand very high temperature that would denature most enzymes. *Taq* polymerase is an example of an extremozyme, an enzyme that functions in an extreme environment, such as a very high or low pH, a high or low temperature, or high or low salt concentrations.

Extremozymes are of extreme interest to biotechnologists who are looking for new and novel proteins to use in research, genetic engineering, or therapeutics.

Gather and share information about a known extremozyme.

1. Create a single Microsoft® PowerPoint® slide, to be printed and used as a fact sheet, that presents information about the extremozyme, *Taq* polymerase.

2. Find the source of the enzyme. Identify the organism from which it was originally found. Describe where the source organism lives. Include pictures of the environment or the organism, if possible.

3. Give examples of current or potential uses for the enzyme.

4. List the names of scientists, academic institutions, or companies involved in studying or producing the extremozyme.

5. Find one more example of another type of extremozyme and give a brief explanation of what it does.

6. Identify the main website(s) used to gather the information above.

Activity **13.2** **PCR Primer Design**

Background: PCR is a directed DNA synthesis reaction, using two primers to delineate and replicate a desired DNA sequence. In a PCR reaction, every ingredient and condition needs to be optimized so that the target sequence and only the target sequence is amplified. One of the most important factors in a successful PCR is designing the "best" primers. For each primer (forward and reverse), the primer should bind to only one target sequence in the sample and not at any other location.

Integrated DNA Technologies (IDT) is a company that produces and sells custom short strands of DNA called oligonucleotides. These custom oligonucleotides may be used as PCR primers, DNA sequencing primers or gene probes.

When designing PCR primers, a search of a piece of template DNA sequence is done to find primer sequences that will give a desired PCR product, or amplicon. Tools on the **www.idtdna.com** website will do the DNA sequence analysis and find the "best" forward and reverse PCR primers for the desired PCR product amplification. The goal of primer design is to find forward and reverse primers that have the following characteristics:

- *Primer sequence complementary to 20 to 30 bases in the template target sequence.* A primer that is too long or too short may have problems annealing to, or separating from, the desired location on the template. Short primers are more likely to bind to some undesired region of the template.

- *Primer composition that is approximately 50% G-C.* G-C pairs have three H-bonds holding the strands together. It is easier to split A-T pairs since only two H-bonds hold them together. The "proper" ratio of G–C and A–T ensures proper annealing and denaturing of the template and primer.

- *Avoid Gs and Cs at the 3' end of the primer because they impact end binding.* No more than 3 of the last 5 nucleotides should be G or C.

- *Forward and reverse primers that have Tm values within 1–2°C of each other.* The Tm is the temperature that 2 nucleotide strands will separate (denature). The Tm of the primers must be close to each other so that the thermal cycling can progress quickly and effectively at the denaturing step each cycle.

- Primer sequences that do not have long repeats of A, Cs, Gs and Ts (no more than 4 in a row) since they are more likely to find each other and fold the primer over on itself. These are called hairpin turns and "tie" up primers making them unavailable for the PCR reactions.

 Design PCR primers and use them to recognize Lambda phage genomic DNA during a PCR. The www.idtdna.com may be used to find a "good" primer set (forward and reverse primers) that will recognize specific sequences in the Lambda DNA Genome and produce a 500 bp PCR product found only once in the genome.

Procedure

1. Go to the National Center of Biotechnology Information at www.ncbi.nlm.nih.gov and chose "Nucleotide" from the Popular Resources list.
2. Do a search for the Lambda Phage DNA sequence by pasting "NC_001416.1" in the search box.
3. The search should result in the NCBI entry, "**Enterobacteria phage lambda, complete genome.**" Click on that link and scroll down the page to see all of the information provided. At the very bottom is the entire lambda viral DNA molecule's DNA sequence, but it is a format that is not useful for primer design.
4. At the top of the page you will find a NCBI Reference Accession ID for Enterobacteria phage lambda (**NC_001416.1**). With this ID number, primer design programs can pull up the DNA sequence in an appropriate format. You will need to copy and paste the Accession ID "**NC_001416.1**" into the IDT primer design program when instructed.
5. Go to www.idtdna.com and click on the "Tools" link.
6. Pick "PrimerQuest Tool" and paste the NCBI Reference Accession ID for Enterobacteria phage lambda (NC_001416.1) into the NCBI ID# box. Click "Get Sequence" and the entire ATGC code for Lambda phage will appear in a form that is useful for PCR primer design.
7. Next click on "Choose Your Design" and click the "General PCR (Primers Only)" option.
8. Look through the list of Custom Design Parameters on the page. In "Task Settings" make sure the "Results to Return" is 5. This will result in 5 pairs of PCR primers designed.

   Most of the other criteria will already be set to standard PCR parameters and should be left that way. However, you want Lambda bacteriophage DNA PCR product (amplicons) of 500 base pair lengths. Find the parameter "Amplicon Criteria" and set the minimum, optimum, and maximum lengths all to 500.

9. Then hit "Get Assays." This should give you a list of 5 "best" primer sets for the DNA sequence. There may be more primer sets for a 500 bp sequence but the IDT program has found that these five best meet the design parameters.
10. Study each primer set in light of the criteria set in the Background section. Use the link for each set labeled, "View Assay Details." How well does each forward and reverse primer in each set meet the criteria?

Determine the single primer set that you think is best for the Lambda PCR. Either print up the selected primer set results or record them in your notebook. For both the forward and reverse primer, include the:

- Primer sequence
- Primer length
- Primer start position in the lambda DNA sequence code
- Primer Tm
- Primer G-C%
- Number of repeated nucleotide sequences (4 or more of one of the A, T, G, C) in each primer strand

Activity 13.3 You Solve the Case

Forensic scientists use many of the tools of math, science and engineering to solve cases. Although it is interesting work, it is not exactly like the TV show, *CSI*.

Use the information at the Forensic Sciences Foundation (fsf.aafs.org/what-we-do) and the American Academy of Forensic Sciences (www.aafs.org) websites to learn more about what it is like to be a forensic scientist. What are:

1. The job duties of a forensic scientist. Summarize these in a paragraph
2. The credentials and qualities needed to be a forensic scientist. List these.
3. The discipline sections within the American Academy of Forensic Sciences (AAFS). Pick three that most interest you and list them.

Activity 13.4 Knock-Knock. Who's There? SNP. SNP Who?

Microarrays allow for the detection of nucleotide differences in the DNA sequence, even down to a difference in a single nucleotide. These single nucleotide differences are called SNPs, for single nucleotide polymorphisms. In several cases, microarrays and SNPs have been used to identify sections of DNA associated with a specific disease. This is an example of using microarrays for genetic testing. Microarrays and SNPs are also used to show sections of DNA that warrant further research for a connection with a particular disease. This application is important for research and development of new pharmaceuticals.

1. Go to **dnamicroarray.net** to learn how array oligonucleotide probes are produced and used to recognize diseased cells. Summarize what you find.
2. Next, go to the Howard Hughes Medical Institute (HHMI) website at: **biotech.emcp.net/HHMI** to learn more about arrays and specifically about SNP arrays. How are they different from the arrays used to recognize diseased cells (from step 1)? On the HHMI site, protein arrays are explained. How are they different, in form and function, from DNA or RNA arrays?
3. Finally, go to **biotech.emcp.net/23andMe** and click on "how it works" to learn how individuals can have their own DNA screened by a company that uses microarray and SNP technology. List three things you learned about what 23andMe can tell you about your own DNA.

Activity 13.5 Growing Up Too Fast

A small mutation in the DNA sequence of a gene can cause a child to mature and age so quickly that the child's body soon resembles the body of an old man or woman. Using advanced techniques, researchers at the National Human Genome Research Institute have identified the mutation that causes the premature aging syndrome known as Hutchinson-Gilford progeria syndrome (HGPS), the classic form of progeria.

Read the summary of an article in the scientific journal, *Nature* at biotech.emcp.net/hgps_disorder to learn more about the genetic cause or molecular mechanism of the HGPS disorder.

Sam, age 7, with friend John, age 15.
Photo courtesy of The Progeria Research Foundation.

First, write a summary of the differences in DNA and proteins that result in this premature aging syndrome. Describe how knowledge about the DNA sequence could lead to medical advances in diagnosing and/or treating the disorder.

Second, go to www.genome.gov/15515061 and learn about a new anti-cancer drug that may coincidentally hold promise for progeria patients. Include a synopsis of this information in your summary. To learn more about progeria, visit www.progeriaresearch.org.

Bioethics

Give Us Your DNA Sample, Like It or Not?

Does the end justify the means when it comes to solving a murder versus jeopardizing the privacy of an entire town of innocent men? In 2002, a woman was brutally murdered in a small Massachusetts town. More than 2 years later, unable to find a suspect, the police wanted DNA samples from every male in town. Should the police be allowed to force the entire male population to give a cheek-cell sample for DNA fingerprinting?

Weigh the advantages and disadvantages of DNA testing of an entire town's male population to find a murder suspect.

1. Go to **biotech.emcp.net/dnatesting** to learn more about the circumstances of the murder.
2. Find two other articles on the Internet that discuss the case and whether or not forced mass DNA testing should be allowed in this situation. In your notebook, summarize each article, and record the Internet website address.
3. Find another article that discusses a similar circumstance (mass DNA testing) in another murder in a different town. How was that resolved? Record the Internet website address.
4. Create a two-column chart, and list at least three reasons for requiring mass DNA testing and three reasons for not allowing forced, mass DNA testing.
5. Imagine that you are an innocent man who lives in the town. Would you support mandatory DNA testing of yourself and the rest of the male population to solve the murder? If you do support mandatory DNA testing, what punishment or penalty should there be for men who refuse to be tested? If you do not support mandatory testing, to what lengths are you willing to fight it? Are you willing to go to court or to jail to resist giving up your DNA privacy?

Biotech
Careers

Photo courtesy of Alshad S. Lalani.

Research Scientist

Alshad S. Lalani, PhD
Senior Director, Clinical Sciences
Puma Biotechnology, Inc.
San Francisco, CA

Dr. Alshad "Al" Lalani is a cancer research scientist. He currently serves as Senior Director for Clinical Studies at Puma Biotechnology. Puma is a biopharmaceutical company dedicated to the development of novel therapeutics for the treatment of cancer. As of 2016, Puma has nine therapeutic products (seven for breast cancer) in Clinical Trial Stage II and III.

Prior to moving to Puma, Dr. Lalani served as associate director of translational medicine at Regeneron Pharmaceuticals, Inc, a New York-based biopharmaceutical company that discovers, develops, and markets medicines for the treatment of serious medical conditions, including cancer.

Regeneron's first marketed therapeutic is ARCALYST® (rilonacept), an injected interleukin-1 blocker that is one of the recommended treatments for a rare, inherited, inflammatory condition called Cryopyrin-Associated Periodic Syndromes (CAPS). Regeneron has other therapeutic candidates in Phase III clinical trials for the potential treatment of gout, age-related macular degeneration, central retinal vein occlusion, and certain cancers.

In the past decade, Dr. Lalani's scientific interests in tumor biology, angiogenesis, virology, and inflammation have been applied in developing innovative therapies for blocking the growth and spread of tumors. He began his career in biotechnology as a staff research scientist at Cell Genesys, Inc, where he led a team developing immuno- and gene therapies. Dr. Lalani's research has been published in over 25 articles in peer-reviewed journals, and he has contributed to multiple published and pending patents.

14 Biotechnology Research and Applications: Looking Forward

The Biotechnology Industry Organization (BIO) (bio.org/articles/healing-fueling-feeding-how-biotechnology-enriching-your-life) uses the phrase "Biotechnology: Healing, Fueling, and Feeding the World" to describe the goal of its more than 1000 members worldwide. The industry is doing that and much more. The field of biotechnology includes academic, industrial, and regulatory organizations whose employees work on the research, development, and manufacture of innovative tools and products to improve health care, agricultural practices, and industrial products. Others in the field work on products and processes to protect the environment and remediate environmental damage. Still others work to protect society from those who would cause human suffering. In this chapter, some of the recent advances and future trends in the expansive area of biotechnology will be presented.

Learning Outcomes

- List some of the tools used in genomics and the advances made possible by them.
- Describe how bioinformatics and microarray technology are speeding genetic studies and the search for novel pharmaceuticals.
- Give examples of how RNA technologies impact research and development of new therapeutics.
- Explain how CRISPR and Cas9 technologies allow for precise and accurate editing of genetic code.
- Distinguish between different methods used for protein study and how they have led to advances in proteomic research.
- Explain how advances in stem cell research, regenerative medicine, 3D-bioprinting, and synthetic biology may lead to improved health care.
- Describe how advances in biofuels and other biotechnologies are being used to understand and protect the environment.
- Outline the important applications of the growing biotechnology fields of veterinary biotech, dental biotech, nanotechnology, bioterrorism, and biodefense.

14.1 Advances in Molecular Biology Lead to New Biotechnologies

Molecular biology is the study of biology on the molecular level. Molecular biologists study DNA, RNA, and/or proteins to understand how organisms, their cells and the cells' products work and how they might be manipulated. These macromolecules, their structure and function, and discoveries in molecular biology are the basis for most applications in biotechnology.

Scientists have been studying macromolecules for more than a century, but the most significant discovery, determining the structure of DNA in 1953 by Watson and Crick, led to what is now called the "biotechnology revolution." Since then, each decade has produced significant advances in molecular biology techniques including recombinant DNA technology in the 1970s, DNA sequencing/genomics in the 1980s, proteomics in the 1990s, and microarrays, RNA transcriptomes, and gene editing in the first decades of this century. Each of these big steps in molecular biology research techniques have led to imaginative applications in new products and processes.

Advances in Nucleic Acid Technologies

Genomics Simply stated, **genomics** is the study of an organism's genome, meaning all the genes and all the other noncoding sequences. Genomics includes determining the nucleotide sequence, how the sequence is read and regulated, and how the sequence and genes vary from organism to organism. Once a genome is sequenced, its genes, gene functions, and gene regulators (promoters, enhancers, transcription factors, etc.) are studied and compared. The ultimate goal of genome scientists is to understand the relationship between genomes and protein expression.

Since the development of DNA sequencing techniques has been rather recent, genomics is a relatively young field of science with its origins somewhere in the 1970s when Sanger sequencing (dideoxynucleotide sequencing) was used to sequence relatively long pieces of DNA (see Section 13.5). With the advent of more and more sophisticated automated sequencers and **bioinformatics** (computer programs and mathematical models that analyze and relate sequence data), genomics has mushroomed into a major discipline within biotechnology. It is estimated that over 1000 genomes have already been sequenced, including the first draft of the "human genome" in 2000.

Although Sanger sequencing is routinely used to produce data of several hundred to a few thousand base pairs in length, a genome has billions of base pairs. Such a large amount of DNA sequencing and analysis required new paradigms to make genome sequencing feasible. One advent was the use of shotgun cloning (shotgun sequencing). **Shotgun cloning** is a method that allows the entire DNA from an organism to be prepared and sequenced fairly rapidly. In shotgun cloning, restriction enzymes are used to chop the genomic DNA released from cells into manageable-sized pieces. The pieces are ligated into plasmids to create a library of genome fragments. The plasmids are used to transform cells where many copies of the recombinant DNA (rDNA) plasmids are made. Sequencing of the plasmids produce a section of genomic DNA sequence of about 500–1000 base pairs (bp). Computers analyze overlaps in the genomic fragments' sequence and determine how the pieces fit together. At less than 1000 bp per sequence, you can imagine how much time and work (hundreds of scientists working for approximately fifteen years at dozens of facilities) went into sequencing the entire 3 billion bp sequence of humans and completing The Human Genome Project (see Figure 14.1). From the Human Genome Project sequence data, it has been estimated that humans have between 20,000 and 30,000 functional genes.

genomics (ge•no•mics) the study of all the genes and DNA code of an organism

bioinformatics (bi•o•in•for•mat•ics) the use of computers and databases to analyze and relate large amounts of biological data

shotgun cloning (shot•gun clon•ing) a method of cloning commonly used during the sequencing of the human genome that involves digesting DNA into 500 bp pieces, generating libraries from those fragments, and eventually sequencing the libraries

The demand for faster, lower cost, and less labor-intensive sequencing has driven the development of high-throughput sequencing and has propelled the field of genomics. Next-generation sequencing [NGS]) technologies and parallel sequencing, discussed in Section 13.5, can produce millions of sequences in a short time (within a day or two). Assembly software programs can analyze the sequences and put them together to read entire genomes as a single, continuous sequence.

Once a genome sequence is determined the formidable task of understanding the sequence information begins. This is called **genome annotation** and includes all the bioinformatics research and testing to determine which sections of the genome sequence codes for proteins and then determining the structure and function of those proteins. Annotation also includes determining which sequences do not code for proteins and which code for regulatory elements.

There are many direct and indirect benefits of genomics in general and in completing the Human Genome Project specifically. Genome research has led to increased understanding of evolutionary relationships between organisms. By comparing genomes, we can determine how closely organisms are related (comparative genomics) and what mutations may lead to speciation. A recent genomic study of four strains of *Campylobacter* (only one of which was virulent) has led to an understanding of how the bacterium causes food poisoning. The hope is that these data will lead to a vaccine to prevent the virulent form of *Campylobacter* from making people sick. For further information, see the report at: **biotech.emcp.net/Campylobacter**. Genome data can also shed light on migrations of populations and species, and how species survive ecological pressures.

Figure 14.1. **In 1992 The Institute for Genomic Research (TIGR) was founded in Rockville, Maryland. TIGR was a not-for-profit center, with a large staff deciphering the genomes of many kinds of bacteria, protozoa, fungi, plants, and animals. In 1995, using shotgun sequencing, TIGR researchers were the first to sequence the entire genome of a free-living organism, the disease-causing bacterium, *Haemophilus influenzae*. TIGR also partnered with the FBI and the NSF to sequence the stain of anthrax used in terrorist attacks in 2001. In 2006, J. Craig Venter merged TIGR operations into the J. Craig Venter Institute (JCVI), a non-profit genomics research institute. JCVI has operations in Maryland and La Jolla CA where its research focus is medical, microbial, plant, and environmental genomics.**
© The Institute for Genomic Research.

genome annotation (ge*nome an*no*ta*tion) Using bioinformatics research and testing to determine the genome sequence that code for proteins, regulatory molecules, or are noncoding.

BIOTECH ONLINE

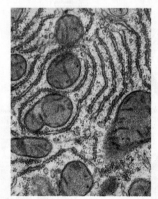

© Lester V. Bergman/Getty Images.

Mitochondria Have Sequences, Too

Mitochondria and chloroplasts are thought to have originated as free-living bacteria. One piece of evidence to support this is the fact that both have their own circular DNA strand. Scientists are just learning how mitochondrial DNA is involved in cellular processes and what can happen if mitochondrial DNA is modified.

Learn how sequencing can lead to understanding of disease mechanisms.

Go to **www.sciencedaily.com/releases/2005/01/050121100603.htm** to learn how mitochondrial DNA mutations may play a significant role in the study of prostate cancer. Read the article, and summarize in a few sentences which gene is affected and how it was determined.

Conduct an Internet search to find one reference that gives a patient profile for prostate cancer (who gets it, how common it is, and the survival rate). Write a summary of your findings.

Microarrays The amount of genetic information that is produced in the elucidation of an organism's genome is daunting. How does a scientist know what the genetic codes mean? Human genomics (determining human gene sequences) allows for identification of specific gene and protein targets for pharmaceuticals, gene therapies, and genetic testing. As you will remember from the introduction to microarrays in Chapter 13, at the same time as the human genome was being deciphered, microarray technology was developed and used to search for specific genes or gene products in the genome. This has led to hundreds of new and potential drugs and vaccines. The development of microarray technology has made the study of gene expression more manageable. Microarrays can show the most probable expression of a genetic sequence (the RNA and proteins produced by a (DNA sequence).

A microarray is a collection of oligonucleotide sequences (DNA, cDNA, RNA, or other synthesized nucleotides) that are linked at fixed locations on some type of solid support, either a glass slide, silicon chip, or nylon membrane. The array sequences are used as "probes" to recognize other nucleic acid sequences of interest. The sequences of interest are usually labeled with fluorescent tags for visualization. The array sequences are incubated with the labeled molecules. If any of the tagged molecules hybridize with the probe arrays, a color change can be detected and measured by a microarray reader.

For genomic studies, where the goal is to assign functions to gene sequences, the genomic arrays are incubated with specific labeled molecules that are related to the gene function, such as an mRNA strand that codes for a protein of interest. If any of the tagged molecules hybridize with the probe arrays, a color change can be detected and measured by a microarray reader. For example, if a scientist wants to know what genes are being expressed during a period of high protein production (let's say, insulin), cells producing a lot of insulin would be cultured and the mRNA for insulin extracted from the cells. The insulin mRNA would be tagged with a fluorescent label and washed over the genomic single-strand DNA array probe. The tagged insulin mRNA hybridize only at certain spots on the array depending on the genetic sequence, and bind with different intensities, depending on the match. This shows the researcher the DNA gene matches to them RNA insulin (see Figure 14.2). An animation of how labeled probes colorize a microarray, visit **www.youtube.com/watch?v=9U-9mlOzoZ8**.

Microarrays may be used for genetic screens and testing as well as for gene expression studies for any organism. Using microarray technology, scientists have begun to assign gene functions to the gene codes for all kinds of livestock and crops. Recently, pineapple scientists have used genomics to find the genes and genetic pathways that allow juicy pineapple plants to thrive in water-limited environments. Imagine if those genes could be identified and used to make other crops drought-tolerant.

Affymetrix Inc. was the developer of the first commercially available microarray, called the GeneChip® (see Figures 13.10a and b). Affymetrix now markets pre-designed arrays which contain known genomes "ready to probe" for molecular biologists hoping to better understand gene expression. As of 2016, Affymetrix had dozens of whole genome arrays available including those of dog, mouse, cow, buffalo, pig, chicken, salmon, rice, strawberry, corn, soybean, *Arabidopsis*, and humans. To learn more about these arrays, visit Affymetrix's array website at **www.affymetrix.com/estore/browse/level_three_category_and_children.jsp?category=cat390001&categoryIdClicked=cat390001&expand=true&parent=35923**.

Figure 14.2. **DNA microarray technology is a powerful research tool that allows scientists to assess the level of expression of a large subset of the 30,000 human genes in a cell or tissue. This technology can quickly show the genes that are active in a tumor cell, for example, narrowing the precise molecular causes of a cancer. Here, a scientist examines a microarray slide. A microarray analysis is displayed on his computer screen.**
© National Cancer Institute.

© Daniel Prudek/iStock.

The Buzz Is That the Bee Genome Is Done!

Why would anyone care about sequencing honeybee DNA? Scientists at the Baylor College of Medicine are excited because they have recently completed the first draft of the honeybee genome.

Learn how bee genomics may lead to important agricultural, ecological, and economic benefits.

1. Visit the Baylor College of Medicine Bee Genome website at **www.hgsc.bcm.edu/honey-bee-genome-project-0**. Scan through the information on the page and watch the bee genome video. List three facts about the honey bee genome project and why it is important.

2. Go to the NCBI bee genome portal at **biotech.emcp.net/bee**. List five links available at this website that you believe would be valuable if you were studying bee genes, bee proteins, or genes of organisms that are closely related to bees.

3. It is possible to obtain the entire honeybee genome data from oligonucleotide microarray assemblies. Search online and find a link to where this data may be located.

4. Use the Internet to find an article about the causes and concerns over the dwindling honeybee population. Summarize these. Explain how having the bee genome completed may help address some of these concerns. Describe how deciphering the honeybee genome may lead to improved management of this beneficial insect.

Bioinformatics   Imagine the enormous amount of genomic information that is generated by the thousands of scientists working on deciphering and interpreting the genetic information in genomes. How can a scientist manage all the data, study it, compare it to other data, and store it? High-powered computers, computer programs and databases of computerized information have been developed to manage the data see Figure 14.3). Bioinformatics is the use of computers and statistical analysis to understand biological data.

During experimentation, as large amounts of numerical and sequence data are collected, the data may be organized into a computer file called a biological database. A biological database may be as simple as a spreadsheet of sequences or may be files of sophisticated software programs. The database allows for searches or queries so that a user can find information quickly. The power in a database lies in the ability to program it to do more than just search. Many databases allow the comparison of large amounts of information to find similarities or differences in the data. You may have already used one or more of the databases available through the National Center for Biotechnology Information (NCBI). NCBI has several databases, including FASTA, VAST+, BLAST, GenBank, Nucleotide, Protein, Entrez, and PubMed. Visit **www.ncbi.nlm.nih.gov** and click "About NBCI" to learn what each one of these databases can do.

Figure 14.3.   HIVE Genomics Research Computer
At the FDA headquarters, Maryland, USA, a researcher shows a small subset of the available hard-drives that make up the storage cluster of the High-Performance Integrated Virtual Environment (HIVE). This powerful computer is used to help analyze data from Next Generation Sequencing (NGS) and parallel sequencing for use in genome analysis.
© Science Photo Library/Photo Researchers.

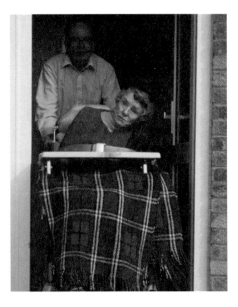

Figure 14.4. Huntington's disease (Huntington's chorea) is a genetic disorder caused by the inheritance of a single dominant allele. The expression of the allele results in the degeneration of certain parts of the brain, leading to jerky, involuntary movements (chorea), and eventually to dementia. The expression of the allele is often delayed until age 35 to 40. Researchers hope that interference RNA may be used to block the expression of the Huntington's allele.
© Science Photo Library/ Photo Researchers.

RNA and Genomics   Until recently, the majority of effort and funding to understand genes and gene expression went into DNA sequencing and genomics. However, since the mid-1990s, a significant amount of research is focused on how RNA is involved in gene regulation and protein expression.

It has been known since the 1950s that messenger RNA (mRNA) carries the code for proteins from genes on the DNA to ribosomes where proteins are assembled. However, it is just now becoming clear that there are several forms of RNA interacting with DNA molecules and other RNA molecules. This interaction turns some genes on and off, and interferes with some posttranscriptional processing and translation of RNA at ribosomes. In addition, useful forms of RNA can be constructed in the lab for the very purpose of gene recognition and gene blocking (see Figure 14.4).

Although several different types of regulatory RNA have been found, the following three are garnering the most interest as scientists try to learn how they regulate, interfere, or block gene expression:

- RNAi
- siRNA
- microRNA

RNAi stands for RNA interference. RNAi molecules are double-stranded RNA pieces, which, when they enter a cell, mimic viral DNA invasions. The cell responds by chopping the RNA into small pieces. The short, double-stranded RNA pieces unzip, and single strands then bind with proteins that, as a complex with the RNA, interfere with the cell's native RNA or DNA, blocking protein production. RNAi is an important tool that researchers use to identify genes and their protein products.

Short-interfering RNA **(siRNA)** is essentially the same as RNAi, except that scientists actually create the short-interfering single RNA strands themselves to learn which genes are turned off and which proteins are not synthesized due to siRNA activity (see Figure 14.5). The siRNA may also bind with native RNA, and the cell may mistakenly destroy the complex. Research into RNAi and siRNA is expected to lead to many new disease therapeutics.

MicroRNA molecules are also small pieces of RNA that are known to interrupt posttranscriptional RNA function. MicroRNA molecules bind to mRNA as soon as it is made, which prevents the RNA from being translated. It is thought that microRNAs may regulate the production of over 400 proteins in the cell and that some of the microRNA strands may have interference action on multiple mRNA transcripts.

Studying the presence and concentration of certain types of RNA in cells includes the use of **Northern blots**. Northern blots are similar to Southern blots in that samples (in this case RNA extracts) are run on a gel, transferred to a membrane, and then visualized. Using Northern blot techniques, researchers can learn how and where RNA is produced and active. Northern blots allow scientists to study a few RNA samples at a time, but with the advent of microarrays and following the lead of DNA genomic scientists, RNA researchers are now using microarrays to study the entire **transcriptome** (all the mRNA in a cell at a given point in time). The transcriptome tells a researcher exactly what genes are being expressed and what proteins are being made. Transcriptome microarrays are being used to compare samples from healthy cells to diseased cells at different time points. This will help identify what proteins are involved as a disease progresses.

RNAi (RNAi)   an abbreviation for RNA interference, a type of double-stranded RNA that is chopped into small pieces when engulfed by the cell, and binds to and interferes with the cell's native RNA or DNA, blocking protein production

siRNA (siRNA)   an abbreviation for short-interfering RNA, a type of single-stranded RNA oligo (fragment) that is created by scientists to target a gene for silencing

microRNA (mi•cro•RNA)   the small pieces of RNA that are known to interrupt posttranscriptional RNA function by binding to mRNA as soon as it is made

Northern blot (Nor•thern blot)   a process in which RNA fragments on a gel are transferred to a positively charged membrane (a blot) to be probed by labeled cDNA

transcriptome (trans•crip•tome)   all of the mRNA in a cell at a given point in time

RNA Interference-Simplified Mechanism

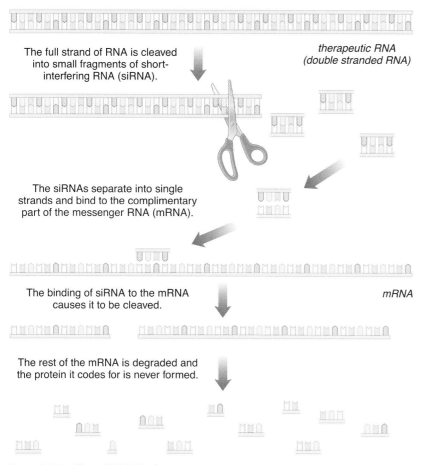

The full strand of RNA is cleaved into small fragments of short-interfering RNA (siRNA).

therapeutic RNA
(double stranded RNA)

The siRNAs separate into single strands and bind to the complimentary part of the messenger RNA (mRNA).

The binding of siRNA to the mRNA causes it to be cleaved.

mRNA

The rest of the mRNA is degraded and the protein it codes for is never formed.

Figure 14.5. **How siRNA Works.**
Reproduced by permission of HOPES, Department of Anthropological Sciences, Stanford University.

RNA studies using interference RNA, transcriptomes, RNA microarrays, etc, are helping scientists understand the mechanisms that lead to the overproduction of proteins that occurs in many diseases, including Huntington's disease and certain types of cancer.

BIOTECH ONLINE

Companies into Shutting Down

The first RNAi companies have just started making product. Their aim is to turn off gene expression, thereby turning off disease mechanisms.

Go online and find one company that is making at least one RNAi product.

Create a half-page fact sheet describing the company, its product, and its market. Include reference information and one or more images.

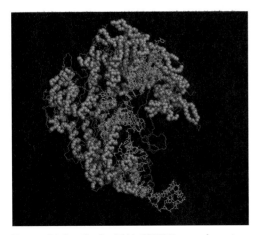

Figure 14.6. **Cas9-gRNA CRISPR complex** In this computer-generated model compiled in the Molecule World application, a Cas9-gRNA complex from the bacterium *Streptococcus pyogenes* (PDB: 4ZT9) is shown. The red and blue gRNA is shown embedded in the protein and adjacent its active site where the target DNA is bound. The secondary structure of the Cas9 enzyme is shown as alpha helices (green), beta-pleated sheets (brown) and the rest of the polypeptide in grey. Courtesy of *Molecule World*, Digital World Biology, LLC.

Gene Editing and CRISPR Technology   A new kind of RNA interference, called CRISPR, is allowing scientists to actually edit genes in cells. CRISPR technology is based on a discovery that bacteria have sections of repeated DNA sequence and associated genes for bacterial restriction enzymes that are used to disable invading viral DNA when the bacteria become infected.

CRISPR stands for "clustered regularly interspaced short palindromic repeats" and are short sequences of 5–10 nucleotides repeated throughout several bacterial genomes. When those CRISPR regions are linked to genes that code for a restriction enzyme (Cas9), scientists can use the combo to produce gRNA (guide RNA) that works together with Cas9. Guide RNA is made up of two strands, one that matches the template DNA and one that has a researcher-edited code that is desired. The gDNA and Cas9 enzyme complex can recognize and bind to specific DNA sequences and the edited RNA code is copied into the target DNA (see Figure 14.6). The CRISPR/Cas9/gDNA system can recognize and knockout or eliminate genes, block gene expression, or increase gene expression (**www.doublexscience.org/crisprcas-basic-research**).

For researchers it is relatively simple to produce and introduce guide RNA and the Cas9 protein to almost any cells giving CRISPR technology enormous potential to alter the genome of humans and other animals, plants and food crops, as well as helpful and harmful microorganisms. By delivering the Cas9 protein and appropriate guide RNAs into a cell, the organism's genome can be modified at virtually any desired location.

Genome editing using CRISPR technology has already been used to modify the genome in several model organisms including yeast, zebrafish, fruit flies, mice, monkeys, and even human embryos. CRISPR technology is already demonstrating some impressive results (see the MIT Technology Review article at **www.technologyreview.com/news/543941/everything-you-need-to-know-about-crispr-gene-editings-monster-year**) and is expected to usher in a new wave of gene therapies that can be used to correct faulty disease-causing genes.

Another genome editing technology being developed to turn off genes. It is called zinc finger nuclease technology (ZFN) and is similar in some ways to CRISPR/Cas9 technology. Zinc finger nucleases are proteins that recognize certain DNA sequences (similar to a restriction enzyme) and can cut at that location. Cellular repair enzymes attempting to repair the cut often introduction mutations, thus rendering a gene to be inactive. ZFN technology is in clinical trials as a HIV treatment. Learn more about ZFNs and HIV at **www.nature.com/news/gene-editing-method-tackles-hiv-in-first-clinical-test-1.14813**.

Advances in Protein Technologies

proteomics (pro•te•o•mics)
the study of how, when, and where proteins are used in cells

proteome (pro•te•ome) all of an organism's protein and protein-related material

Proteomics   Humans are thought to produce over 2 million different proteins from some 30,000 genes. Just by doing the math, it is obvious that some genes must be making multiple proteins. Scientists have shown that a single gene can encode for as many as 50 different protein forms. How is that possible? Obviously, there must be more to protein production than the dogma that "one gene codes for one mRNA that codes for one protein" in "The Central Dogma of Biology."

Proteomics is the study of how, when, and in what forms proteins are active. It includes the study of RNA splicing (exons—the expressed portion of a mRNA transcript, and introns—the sections of a mRNA transcript that are removed before translation) and other posttranscriptional modifications.

Since proteins are so numerous, diverse in structure, and sensitive to changes in their environment, the proteome is significantly challenging to study and understand. **Proteome** scientists use all of the tools discussed in the last section on genomics and more. Studies of DNA sequencing, genomics, microarrays, transcriptions, and RNA interference help protein scientists understand which sections of the DNA code for a protein, and help to shed light on some of the methods of gene regulation. Additional methods for studying the proteome include protein crystallography, mass spectrometry, nuclear magnetic resonance (NMR), molecular modeling, peptide sequencing, protein gels, liquid chromatography, protein arrays, protein databases, and protein diagnostic assays, such as enzyme-linked immunosorbent assays (ELISAs) and Western blots.

Some of the techniques and instruments of proteomics have been presented in other chapters (Chapters 2, 5, and 7). Here some of these technologies are reviewed in light of how they are used to better propel proteomics and protein product development.

X-ray Crystallography One of the most important proteomic research methods is **protein (xray) crystallography**. Protein crystallography visualizes the positions of atoms in a molecule based on x-ray wave diffraction, and the constructive and destructive interference that results off the atoms in the sample. Pure protein crystals are needed for x-ray crystallography. Once crystals are formed, they are mounted in the x-ray beam, and a detector measures the way the x-ray light bends (diffraction) and bounces back together (recombination), as shown in Figure 14.7.

The **x-ray diffraction pattern** of a protein crystal (see Figure 14.8) is difficult to interpret, but it gives the scientist data that can be loaded into computer programs to generate space-filling models of polypeptide chains (see Figure 14.9) and three-dimensional structural models of a protein (see Figure 14.10). These models provide

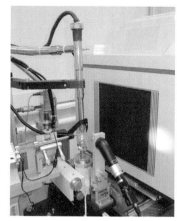

Figure 14.7. The glowing sample in the center is hit by the x-ray beam, which was generated and delivered by the horizontal metal tube on the left. The detector (black square on right) measures the amount of x-ray diffraction and interference. Photo by author.

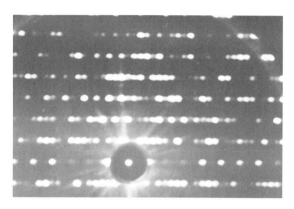

Figure 14.8. The x-ray diffraction pattern reveals the structure of a protein. The white spots that result from the diffraction of the x-ray light as it passes through the protein are interpreted by computer analysis. The data are used to create three-dimensional images of the protein.
© Science Photo Library/Photo Researchers.

protein (x-ray) crystallography (pro•tein crys•tal•log•ra•phy) a technique that uses x-ray wave diffraction patterns to visualize the positions of atoms in a protein molecule to reveal its three-dimensional structure

x-ray diffraction pattern (x-ray dif•frac•tion pat•tern) a pattern of light intensities that develops when an x-ray beam is passed through a mounted crystalline structure

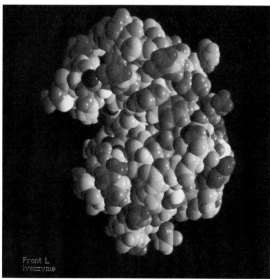

Figure 14.9. This space-filling model of the enzyme lysozyme shows how closely amino acids in the protein are associated with each other.
© Science Photo Library/Photo Researchers.

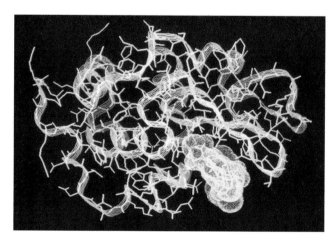

Figure 14.10. A computer-generated image of the three-dimensional shape of chicken egg-white lysozyme is shown here. The model is developed using x-ray diffraction data and shows the folding due to the amino acid sequence in the protein. A substrate (yellow) fits into lysozyme's active site.
© Science Photo Library/Photo Researchers.

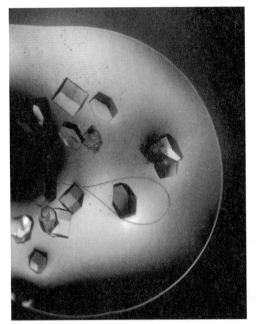

Figure 14.11. These are crystals of the protein lysozyme, isolated from chicken egg white. A wire loop is isolating a crystal to be used in x-ray crystallography. Only pure crystals, like these, which have sharp edges, may be used in x-ray crystallography.
© Science Photo Library/Photo Researchers.

■■■■■■■■■■■■■■

salting out (salt•ing out) a technique for crystallizing proteins that involves precipitating a sample of pure protein using a stringent salt gradient of sodium chloride, ammonium sulfate, or some other salt

scientists with an idea of what the protein looks like and how chains or parts of polypeptide chains interact so that they can better understand protein function.

Once x-ray crystallography data is obtain, the data can be shared in databases, such as the NCBI Structure database. The protein data can then be analyzed using bioinformatic tools such as CN3D and VAST+. The genes for the protein of interest can be studied using genomic databases and analysis tools.

Some proteins are easy to crystallize and some are not. The most common technique used to produce crystals is called "**salting out**," using a concentration gradient of sodium chloride, ammonium sulfate, or some other salt with a saturated solution of protein. The protein falls out of solution as the concentration of salt in a second solution slowly increases. If a protein crystal has formed perfectly, then all of the molecules in the crystal line up with the same orientation. This produces a clearer diffraction pattern, similar to the one in Figure 14.8. Comparing the x-ray diffraction data from similar proteins reveal differences in proteomes.

Salting out is usually performed by one of two methods: 1) hanging drops of protein solution over a concentrated salt solution, or 2) protein solution dialysis using dialysis tubing with a salt solution on the outside of the dialysis membrane. Both methods use the principles of diffusion (hypotonic—low solute concentration, and hypertonic—high solute concentration) to move water away from the protein molecules. This results in the protein "drying out" and falling out of solution. The slower salting out is performed (the more slowly the outer solution's concentration is increased), the better the chance is that the protein molecules line up and form clean, clear crystals (see Figure 14.11).

Mass Spectrometry The mass spectrometer (mass spec) is an instrument that measures the masses and relative concentrations of atoms and molecules. The mass spec ionizes (places a charge on) the proteins in a sample and moves them quickly toward a detector (see Figure 14.12). On its path, a magnetic force is applied on the moving, charged particles causing them to change direction proportional to their molecular mass. The change in direction of the molecules is measured as they move through the detector. Pure protein crystals are hydrated and used in aqueous solutions in the mass spec.

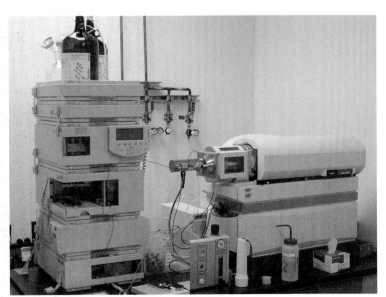

Figure 14.12. The mass spectrometer (right) is used to determine the molecular weight of a protein sample, and high-performance liquid chromatography (HPLC) measures the purity and concentration of protein in the sample (left). A sample must be pure, as in protein crystals, to obtain a good reading.
Photo by author.

Nuclear Magnetic Resonance Nuclear magnetic resonance measures the spin on nuclei (protons) of particular isotopes in a magnetic field (see Figure 14.13). This causes a magnetic or NMR signal. NMR produces data that are used to study physical, chemical, and biological properties of a variety of samples. NMR is used to study protein structure in aqueous (watery) solutions, which is important because all proteins actually function in the aqueous solution of cells, tissues, or organs. In the 1970s, NMR technology was applied to medical diagnoses and renamed magnetic resonance imaging (MRI) (see Figure 14.14).

Figure 14.14. A patient is placed in an MRI unit and exposed to a magnetic field. An image of soft tissue is displayed on the computer screen. MRI is a type of NMR.
© Peter Beck/Getty Images.

Figure 14.13. The NMR spectrometer contains large super-conducting magnets that produce the magnetic field for NMR. Computers gather and process data.
© Science Photo Library/Photo Researchers.

BIOTECH ONLINE

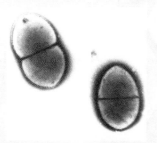

© Lester V. Bergman/Getty Images.

Protein Shape Is the Key

Learn how protein structural studies can lead to the understanding of disease mechanisms, preventions, and therapies.

Go to www.sciencedaily.com/releases/2005/01/050111182645.htm to learn how protein-shape studies may lead to a vaccine for pneumonia. Read the article and summarize in a few sentences what methods the scientists have used to understand the pneumonia bacterium (one species of pneumonia bacteria, *Streptococcus sp*, is pictured).

Conduct an Internet search to find another image of a different pneumonia bacterium or virus. Copy and paste the image into a Microsoft® Word® document. Compare and contrast the structure and function of each pneumonia-causing pathogen. Record the address of each website used.

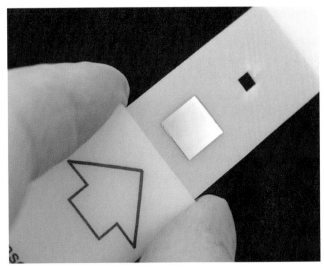

Figure 14.15. A scientist holds a protein microarray chip. Protein samples can be added to tiny sections on the chip and assayed. Protein arrays are critically important in proteomics research, in which the structure and function of all proteins in the body are studied.
© Science Photo Library/Photo Researchers.

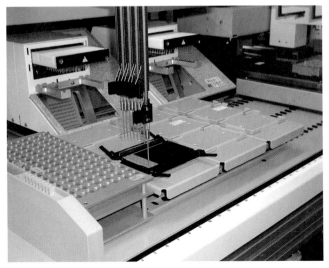

Figure 14.16. TECAN Genesis 2000 robot prepares Ciphergen SELDI-TOF protein array chips for proteomic analysis. Identifying the proteins of specific cancers on the protein array leads to specific and accurate cancer diagnoses.
© National Cancer Institute.

protein arrays (pro•tein ar•rays) a fusion of technologies, where protein samples bound on the glass slide (protein chip) are assessed using antibodies or other recognition material

Protein Arrays **Protein arrays** are one of the newest and most powerful of the proteomic technologies. Protein arrays may be thought of as a fusion of microarray and ELISA techniques, all processed on a "protein chip."

Protein arrays are set up with protein samples (antibody, antigen, or other protein) bound on a glass or silicon slide (protein chip) (see Figure 14.15); then a sample of interest is added (see Figure 14.16). Protein chip readers measure the amount of binding.

Protein array technology is exciting because it allows for the screening and comparison of large numbers of samples at one time for specific proteins. Companies are producing arrays of antibodies, antigens, and specific tissue samples, and these are readily available to serve as probes to blood, tumor, and other samples from patients.

Protein arrays are making great advances possible in the understanding of many kinds of protein-to-protein interactions, such as those between enzymes and substrates, and hormones and receptors. The applications of the protein-chip arrays will lead to the design of better drugs, better diagnostic tests, and better assays for proteins in the lab and in the field.

Section 14.1 Review Questions

1. How much of the human genome was sequenced utilizing shotgun cloning?
2. Name a plant and an animal whose entire genome has been put on a GeneChip® microarray for the purpose of studying gene function and expression.
3. MicroRNA, RNAi, and siRNA all otperate in a similar fashion. What do they do?
4. What is the name of the technique that uses an x-ray to study the structure of a protein crystal? What is the x-ray image called?
5. What causes proteins to crystallize out of solution?

14.2 New Biotechnologies to Address Some of Our Biggest Challenges

DNA, RNA, and protein biotechnologies have begun to address the problems that plague humanity, including disease, pollution, and famine. In previous chapters you learned about advances in agricultural and medicinal biotechnology that are increasing the yield and quality of food crops and targeting devastating conditions to improve the quality of life for vast numbers of patients. This section surveys four areas of biotechnology that hold great promise to improve human health and welfare, including:

- stem cells and regenerative medicine
- 3D-bioprinting
- biofuels
- biodefense

Stem Cells and Regenerative Medicine

Stem cells are unspecialized cells that have not yet differentiated into cells with a specific function. Recently, though stem cells have the been grown into specialized cells such as blood, bone, muscle, cartilage, and nerve cells that can be transfered into patients with the potential to regenerate tissue or organ function caused by disease or injury. Different types of stem cells are found in different places and at different times depending on the stage of development.

Embryonic stem cells are found in a developing embryo. These are the starter cells for all the specialized cells in tissues and organs. Since human embryos have the potential to become a functioning human, the use of embryonic stem cells is a controversial subject. The ethics of using these cells was considered in the Chapter 2 bioethics activity.

Adult stem cells are found in the tissues and organs of a person of any age, and they keep their ability to divide and specialize. Bone marrow, for example, contains adult stem cells that constantly replicate and differentiate into different kinds of blood cells (see Figure 14.17).

Some stem cells will only specialize into certain cells, such as bone or cartilage cells. Other stem cells are **pluripotent**, meaning that they can specialize into any type of specialized tissue (see Figure 14.18). Some stem cells that are not pluripotent may be manipulated into becoming pluripotent.

Pluripotent stem cells are the target of a significant amount of biotechnology efforts since these cells hold the promise of being used in the treatment of several disorders caused by dysfunctional cells, such as Parkinson's disease, Alzheimer's disease, or type 1 diabetes or replace cells damaged in spinal or brain injuries. Theoretically, pluripotent stem cells could be grown into any kind of tissue or organ. These healthy tissues or organs could be used to replace damaged or malfunctioning ones.

National Cancer Institute

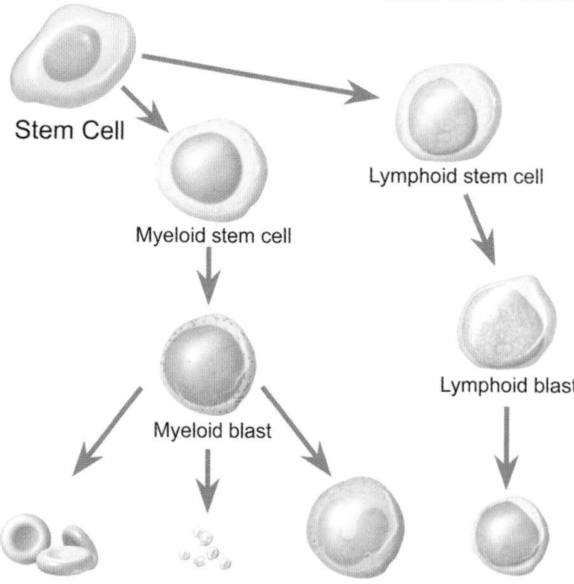

Figure 14.17. Adult stem cells located in bone marrow mature into one of three types of blood cells: red blood cells, platelets, and white blood cells. Precursor cells are also shown, including myeloid blasts, lymphoid stem cells, and lymphoid blasts.
© National Cancer Institute.

Stem Cell

Myeloid stem cell

Lymphoid stem cell

Myeloid blast

Lymphoid blast

Red blood cells Platelets White blood cells

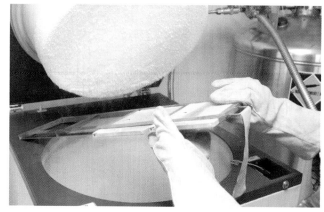

Figure 14.18. These stem cells are frozen in metal cassettes and stored in a liquid nitrogen (-196°C) tank (called cryostorage) for future use. As long as they are stored properly, they will remain pluripotent for many years.
© iStockphoto.

Getting to the Root of Stem Cell Therapies

Stem cell therapies have the potential to address some of the greatest challenges in medicine. This is because stem cells are pluripotent, meaning they can develop into almost any type of cell in the body. This characteristic offers endless opportunities to study cellular processes, develop and test new pharmaceuticals, and developing new ways to regenerate damaged tissue and repair diseased organs.

Learn more about how stem cell research is being used to develop new treatments for a variety of human diseases and disorders.

1. At www.cityofhope.org/blog/how-stem-cells-become-neurons, read the article describing research about how stem cells differentiate into nerve cells. Describe the role of cytosine modification in the process.

2. After reading the article, *Turning Stem Cells Into Sperm* at www.explorestemcells.co.uk/turning-stem-cells-into-sperm.html, give two reasons why the process has medical value and list two challenges that researchers have in perfecting the process.

3. The California Institute for Regenerative Medicine (CIRM) is committed to accelerating the development of new stem cell-based therapies for chronic, debilitating diseases. Visit their "Disease Programs" website (www.cirm.ca.gov/patients/disease-information) to learn what progress is being made in using stem cell therapies to cure several diseases. Pick three diseases/disorders from the list and summarize CIRM's stem cell research funding and progress toward cures.

Stem cell work is being conducted throughout the world with two main focuses. One is the use of either embryonic stem cells or adult stem cells in transplantation to augment or replace tissues or organs, as described above. The other stem cell research focus is to try to manipulate the body to make more stem cells that would function to replace or restore damaged tissues or organs. This is one area of a field called regenerative medicine.

Regenerative medicine includes all the practices (stem cell therapy, gene therapy, tissue engineering, and certain pharmaceutical therapies) that lead to repair or replacement of damaged tissues or organs. Angiogenesis, the growth of new blood vessels to stroke-damaged areas of the brain, is an example of regenerative medicine. Several companies are studying angiogenic factors that increase angiogenesis. To learn how regenerative therapies are being used to help wounded soldiers, visit the Armed Forces Institute of Regenerative Medicine (AFIRM) website at: **www.afirm.mil.**

3-D Bioprinting

One of the most incredible technical advances of the last decade is the development and use of three-dimensional printing to manufacture an assortment of different metal or plastic products. Using a computer image as a template, 3-D printing copies a three-dimensional model by laying down thin layers of plastic or metal molecules, on top of the other, until a replica is produced. Examples of products made by 3-D printing include the production of some plastic shoes, plastic phone cases, and metal machine parts.

Even more astounding than the use of 3-D printing to make innate objects is the application of three-dimensional printing to biological objects. Called **3-D bioprinting**, it uses a modification of ink-jet and stem cell technologies to lay down specific cells and tissues in pre-determined ways with high precision (see Figure 14.19).

3-D bioprinting
(three•D•bio•print•ing)
The use of computers, ink-jet technologies, and cells to produce three-dimensional biological materials such as tissues, and organs

Figure 14.19. **3-D Bioprinting Overview** 3-D printers such as this one from Zurich University of Applied Sciences (ZHAW) create precise, 3-D structures from living stem cells using computer-assisted design (CAD) and/or computer-assisted manufacturing (CAM) blueprints. The bioprinter syringes deposit a suspension containing thousands of cells per drop that coalesce to form layers of cells. Bioprinting allows 3D tissue production in a fast controlled manner by precise positioning of primary cells, biomaterials and growth factors. In the future, printed tissues could reduce animal experiments in drug development and foster regenerative medicine. Researchers foresee the use of bioprinting in almost all areas of medicine from facial reconstruction to artificial organs.
© BSIP/Photo Researchers.

Bioprinting is being developed to correct a variety of human conditions and is currently being used to build a variety of tissues (see Figures 14.20a–c) and some simple organ-like structures such as pieces of skin for burn patients and the outer ear structure for patients that have no outer ear due to a birth defect or an accident.

Although complex organs produced by bioprinting may be years away, in the future, organs such as a heart, pancreas, kidney, liver, or bone may be produced using bioprinting (see Figure 14.21). The advantage to bioprinting is that a patient's own stem cells may be cultured and used as the bioprinter "ink." This is a definite advantage as compared to the rejection risks involved in using transplant organs from another person.

Figures 14.20a–14.20b.
Bioprinting Muscle Tissue Researchers at Zurich University of Applied Sciences (ZHAW) are using a regenHU bioprinter to produce different human tissues. Bioprinting allows three-dimensional tissue production in a fast and controlled manner by the precise positioning of primary cells, biomaterials and growth factors. Tissues are produced in layers, alternating photopolymerized bioink and cells. The researcher here views printed muscle tissue under a microscope and it is displayed on the monitor. This printed muscle tissue is being developed for pharmaceutical testing. In the future, printed tissues could reduce animal experiments in drug development and foster tissue and organ use in regenerative medicine.
© BSIP/Photo Researchers.

Figure 14.21. **Bioprinting a Bone** A 3-D bioprinter lays down cultured cells in a specified fashion to create a cartilage-based spinal bone. Imagine the number of patients that might benefit from bioprinted vertebrae.
© ulldellebre/iStock.

I am All Ears for Bioprinting

Go to www.the-scientist.com/?articles.view/articleNo/37318/title/Printing-Ears and listen to the video about bioprinting in Dr. Bonassar's Cornell University laboratory.

In 5 steps, describe the process by which Dr. Bonassar's lab team has created a new outer ear using cartilage cells and a bioprinter.

Click on the link to "Organs on Demand" and read though the article. List 3 challenges to creating function organs by bioprinting.

Figures 14.22a and 14.22b.
The Joint BioEnergy Institute
Inside the laboratories of the JBEI Emeryville facility, researchers use the latest tools in molecular biology, chemical engineering, computational and robotic technologies, and pioneering work in synthetic biology to transform biomass sugars into energy-rich fuels. In doing so, our dependence on foreign oil and the amount of waste and pollution may be reduced.
© Roy Kaltschmidt, Lawrence Berkeley National Lab, The Regents of the University of California, 2010

bioalcohols (bi•o•al•co•hol)
alcohols produced by fermentation of plants, commonly used as a biofuel

cellulostic biomass biofuel (cell•u•los•tic bi•o•fuels)
biofuel derived from processing cellulose (cell wall material) of certain plants

Biofuels

Several biotechnology companies and research institutions are putting considerable efforts into developing alternate energy sources to coal and petroleum products (See Figure 14.22a and 14.22b). Biofuels are alternate energy compounds (petroleum alternatives) that are derived from living things or once-living things (collectively called biomass). With the goal of providing energy independence and decreasing greenhouse gas emissions, new biofuels are an enormous sector of the biotechnology industry and may soon have a great economic impact.

There are several kinds of biofuels, but the majority of research and development efforts are directed toward the ones discussed here. The categories below are somewhat artificial since some biofuels are included in several categories.

Bioalcohols are alcohols derived directly from plant fermentation and can be burned as fuel. An example of a bioalcohol is ethanol produced from corn. Ethanol won't work directly in automobiles and must be mixed with gasoline (eg, 15% ethanol: 85% gasoline). However, ethanol is not the best bioalcohol for fuel since it is a 2-carbon alcohol and thus has less energy stored in its bonds than alcohols such as butanol (a 4-carbon alcohol), isopentanol (a 5-carbon alcohol), and hexanol (a 6-carbon alcohol). Ethanol yields only about 70% of the energy per unit volume as does the hydrocarbon bonds of gasoline. On the other hand, butanol, isopentanol, and hexanol release more energy per unit volume and may be used as direct replacements for gasoline and other fuels such as jet fuel or diesel. Since they behave like petroleum products, butanol, isopentanol, and hexanol are called oil-like hydrocarbons. These oil-like hydrocarbons are the focus of considerable research and development (See Figure 14.23).

Cellulostic biomass biofuel are bioalcohols that result from the fermentation of plant materials. To produce these biofuels, the cellulostic biomass (cell wall material) in the crops or crop waste must be converted to six-carbon sugars (glucose and xylose).

The six-carbon sugars can then be used as inexpensive substrates for fermentation to produce bioalcohol fuels. Examples of biomass feedstock include corn stalks (corn stover), sugarcane leaves, some woody plants, some algae and some seaweed, and some grasses or grain crops that are left over from food crop production. Some of the feedstock biofuels are attractive alternative energy sources since they recycle crop waste products.

Currently, research efforts are focused on using non-food grasses such as *Miscanthus* or switchgrass and feedstock waste products instead of using agricultural crops for biofuel production. *Miscanthus* and switchgrass grow quickly and add significant cellulose biomass (impregnated with the polysaccharide, lignin) as compared to other crops see Figure 14.24). With the right technology including new breeds of plants and new and better chemical (ionic liquids) and enzymatic (i.e. cellulases and ligninases) treatments, the cellulosic biomass from these plants can be fermented to oil-like hydrocarbons and used as replacement fuels for gasoline, diesel and jet fuel in an economical and environmentally friendly way. Recently, even the lignin molecules from cellulose biomass are molecules of interest since ionic liquids are being developed that can convert lignin to biofuel precursors.

Biodiesels are basically vegetable oils and may be used alone or mixed with petroleum diesel (see Figure 14.25). Biodiesels burn more like petroleum, produce more energy than an equivalent amount of bioethanol, and may be more easily used than ethanol biofuels with less modification to automobiles. Traditionally bred plants may be the source of the biodiesel, or genetically engineered plants may be used. One GM-tobacco produces oils that are high-enough quality or quantity to use as biodiesel (biotech.emcp.net/tobacco_biofuel).

Algae-based or microbial biofuels are promising new replacement products for petroleum oil. New strains of algae have been created that produce significantly large amounts of oil as a byproduct of photosynthesis. Algae have the potential to be engineered to produce fifteen times more oil per acre than plants such as corn and switchgrass (biotech.emcp.net/algae_biofuel).

Figure 14.23. **Garima Goyal, Senior Research Associate in JBEI's Fuel Synthesis Division** Garima's current research focus is the engineering of the mevalonate pathway in *E. coli* to encourage greater conversion of glucose to isopentanol during bacterial fermentation. Isopentanol is a fuel bioalcohol that behaves like petroleum-based oils. Garima's day is full of molecular biology including PCR reactions, restriction digestion, transformation, cell culture, and metabolite analysis. Here, she prepares a bacterial cell culture. Garima uses computer software developed at JBEI to aid her research as well as robots for lab-automated, high throughput procedures. Photo by author.

biodiesels (bi•o•die•sel) oils made by plants that may be used as fuel

algae-based or microbial biofuels oils produced by algae or bacteria that may be used as biofuel

Figure 14.24. **Feedstocks Plant Growth and Testing Room** Researchers in the Feedstocks Division of the Joint BioEnergy Institute grow a variety of plants for use in biofuels research in a climate-controlled plant growth room.
Roy Kaltschmidt, Lawrence Berkeley National Lab, The Regents of the University of California, 2010.

Figure 14.25. **A US National Institute of Standards and Technology (NIST) biochemist prepares biodiesel fuel for injection into a gas chromatograph-mass spectrometer, an instrument that separates, identifies, and measures the purity of components in a mixture.**
© Ost/NIST.

Algae: Better for Oil than Dinosaurs?

Learn about the benefits and challenges of using algae to make biodiesels. Read the article at www.smithsonianmag.com/innovation/scientists-turn-algae-into-crude-oil-in-less-than-an-hour-180948282/?no-ist and list two advantages and two disadvantages of using algae to produce biofuel.

Biogas is the gaseous product of fermentation of plant materials. It is being studied and used in vehicles and to produce electricity.

For virtually all the biofuels, the biomass must be treated to break down cellulose and hemicellulose to sugars for fermentation, by microbes, to bioalcohols. Challenges for biofuel producers include determining the best chemical and enzymatic methods to use in cellulose breakdown as well as the best microorganisms (fungi and bacteria) to use for the fermentation processes. Another challenge is producing different kinds of biofuels for different applications, such as jet fuel, diesel, gasoline, and heating oil. In addition, the cost to produce and make biofuels available is dependent on the efficiency of all the processes involved. Currently, researchers are working to optimize existing procedures in biofuel production including how to grow fuel crops that have the most biomass and the most easily extracted biofuel molecules (the cellulose in plant cell walls and the lignin that cements cellulose fibers together). Scientists are engineering the plants, microbes, and enzymes used in biofuel production to grow plants with more cellulose (and less lignin) and to get better breakdown and fermentation of the sugars into the desired bioalcohols.

biogas (bi•o•gas) biofuel produced from gaseous waste product of plant fermentation

bioterrorism (bi•o•terr•or•ism) the use of biological agents to attack humans, plants, or animals

biodefense (bi•o•de•fense) all the methods used to protect a population from exposure to biological agents

biometric (bi•o•me•tric) the measurement and statistical analysis of biological specimens or processes

Biodefense: Protection Against Bioterrorism

It is fairly obvious why there is so much interest in biotechnological solutions to **bioterrorism**. Anthrax became a common word overnight in 2001 as this biological agent caused death and hardship for many people in several terrorist acts. There are many approaches to dealing with biological agents and bioterrorism.

Recently, scientists have developed an instrument that can detect biological agents. It is used to help identify potentially dangerous pathogens in public places, such as airports, stadiums, and governmental buildings. The instrument, developed by researchers at Lawrence Livermore National Laboratories in California, is called the Autonomous Pathogen Detection System. The hope is that it will help authorities limit the public's exposure to biohazards and allow them to start treating victims before they show symptoms of exposure. The instrument uses molecular and microbiological tests to detect familiar threats, such as anthrax, plague, and botulism bacteria toxin, in addition to several other bacteria, fungi, and viruses.

Biodefense is the term for all of the methods used to detect and protect the population from exposure, or the consequences of exposure, to biological agents. Scientists are developing **biometric** diagnostic testing and response systems. For example, scientists are developing vaccines to prevent illness in the event of exposure to a biological agent. A vaccine is an antigen that elicits a mild, immediate immune response and a strong, long-term immune response. A team at Cornell University reported the development of a fast-acting vaccine against anthrax. Using a new biotechnology technique, they were able to transfer a gene to a mouse, and in just 12 hours, the mouse showed immunity to the deadly bacteria. The mouse model shows that the technology could lead to a potential vaccine for humans, which could work faster and be more protective than any of the current anthrax vaccines.

BioBugs: Microscopic Terrorist Groups

The CDC (biotech.emcp.net/CDC_BTbugs) lists the following microbes as bioterrorism disease organisms or agents

Figure 14.26. *Bacillus anthracis* bacterial colonies from an overnight culture on sheep's blood agar. Note the ground-glass, non-pigmented texture with accompanying "comma" projections from some of the individual rough-edged colonies. A "tenacity test" causing the colony to 'stand up' like beaten egg white is a positive test for *B. anthracis*.

CDC/Megan Mathias and J. Todd Parker.

anthrax (*Bacillus anthracis*) (see Figure 14.26)
botulism (*Clostridium botulinum*)
brucellosis (*Brucella* species)
cholera (*Vibrio cholerae*)
cryptosporidiosis (*Cryptosporidium species*)
eastern equine encephalitis
Ebola hemorrhagic fever (Ebola virus)
Escherichia coli O157:H7 Infection (*E. coli* O157:H7)
Epsilon toxin poisoning (*Clostridium perfringens*)
glanders (*Burkholderia mallei*)
Hantavirus pulmonary syndrome-HPS (Hantavirus)

Lassa hemorrhagic fever (Lassa virus)
Marburg hemorrhagic fever (Marburg virus)
melioidosis (*Burkholderia pseudomallei*)
plague (*Yersinia pestis*)
psittacosis (*Chlamydia psittaci*)
Q fever (*Coxiella burnetii*)
ricin poisoning (ricin toxin)
salmonellosis (*Salmonella* species)
shigellosis (*Shigella* species)
smallpox (variola major)
tularemia (*Francisella tularensis*)
typhus fever (*Rickettsia prowazekii*)
Venezuelan equine encephalitis
western equine encephalitis

We can be exposed to these disease-causing agents by contact with contaminated air, water, or food. The National Bio and Agro-Defense Facility (NBAF) in Manhattan, Kansas, is part of the USDA. At NBAF, using diagnostic techniques and vaccine and therapeutic development, scientists work on protecting the nation's agricultural economy and food supply from natural or intentional introductions of diseases that could be transferred to humans.

Gather and share information about one of the microbes that pose a threat as a biological agent of mass destruction.

Create a one-page PowerPoint® presentation slide that can be used to warn the public about a biologic agent that could be used as a public threat. Include the names of the agent, an image of the agent or the disease it causes, the disease or disorder that results from exposure, how to prevent exposure to the agent, what to do in case of exposure, two other interesting facts about the agent, and two URLs where more information can be found.

Section 14.2   Review Questions

1. What does pluripotent mean and how does it apply to stem cells?
2. Give an example of a bioprinted product and describe how it is produced.
3. Why is biodiesel a preferred fuel as compared to ethanol?
4. What is the function of the Autonomous Pathogen Detection System?

14.3 Other Fields Impacted by Biotechnology

Many other applications of biotechnologies are being used in the fields of biology, medicine, and engineering to create new scientific subspecialties. In the future, these new biotechnology sectors are expected to expand with many new applications and marketed products.

Synthetic Biology

In May 2010, the journal Science reported the creation of the first synthetic cell, one that is regulated by a human-designed and computer-generated genome (www.nature.com/nature/journal/v465/n7297/full/465422a.html). These cells are an example of **synthetic biology**.

Synthetic biology includes all of the efforts to construct or redesign new biological molecules, cells, tissues, organs, or systems. At its simplest, the genetic engineering of cells to produce new recombinant proteins is an example of synthetic biology. The technology to create synthetic cells is in its infancy, but it is already being used to design and create new versions of cells that could perform as stem cells, conduct specific bioremediations, or produce specific biofuels.

Several new biosynthetic versions of economically and environmentally important molecules are being developed. One strategy in biofuel research and development is to use synthetic biology to identify better microbes as biofuel producers and then to modify them to generate higher yields of bioalcohols or other fuel molecules in less time for less cost. Instead of using the same bacterium, such as *E. coli*, to do all new biosynthetic research, some scientists are looking for new production organisms (See Figure 14.27). There are millions of bacteria species and strains. Logically, some may be better than others to culture (domesticate) and modify for production of selected biofuels.

One synthetic biology product being produced is a bio-based substitute for acrylic. Until now, acrylic has made from petroleum and because it is a long, strong molecule it is used in paper, paint, and diaper products to make them more durable. Unfortunately, making acrylic is expensive and produces greenhouse gases. Fortunately a company, OPX Biotechnologies, using synthetic biology strategies, redesigned a naturally occurring microbe to make a version of acrylic called BioAcrylic.

BioAcrylic has most of the desirable characteristics of petroleum-based acrylic but its production generates considerably less greenhouse gases.

Another example of synthetic biology is the creation of artificial skin. An artificial skin called Integra® is currently on the market. Artificial skin is composed of animal tissue and silicon and is used to cover large areas of damaged or missing skin as found in patients with severe burns. Over time, burn patients' own skin is triggered to replace the artificial skin with their own natural skin.

Those interested in synthetic biology should visit the website at: **biotech.emcp.net/syntheticbiology**. This is a member-edited website and has several resources available to a budding synthetic biotechnologist. In fact, there is actually a website, the iGem Registry of Standard Biological Parts (**biotech.emcp.net/partsregistry**), where a synthetic biologist may obtain many of the parts needed to construct cells, organelles, or macromolecules.

synthetic biology (syn•the•tic bi•o•lo•gy) the application of biotechnologies to design and construct new biological systems, such as macromolecules, metabolic pathways, cells, tissues, or organisms

Figure 14.27. New Strains through Synthetic Biology
Dr. Sarah Richardson is a Post-Doctoral Researcher in the Fuels Synthesis Department at the Joint BioEnergy Institute. Sarah's passion is Synthetic Biology and she created a synthetic yeast chromosome as part of her PhD thesis. At JBEI, Sarah works on building a pipeline to find and domesticate specific microbes for jobs that they might do better than traditional model organisms. Currently, her organism of interest is *Corynebacterium glutamicum*. Sarah thinks that *C. glutamicum* may be a better biofuel producer than engineered *E. coli*.
Photo by author.

Environmental Biotechnology

As you are probably well aware, humans and other species are at considerable risk due to natural and man-made environmental events. Pollution of the air, water, and soil jeopardizes the livelihood of many communities. Limited energy sources drive us to mine for energy products in fragile areas. Human encroachment and interference cause some species to become endangered or extinct. Scientists are using several biotechnologies to tackle some of these pressing ecological and environmental problems. **Environmental biotechnology** is a vast field with many applications for monitoring and correcting the health of entire species, populations, communities, and ecosystems. One of the most promising areas of environmental biotech is bioremediation, which focuses on correcting environmental problems to return an area to a "healthy" status. One example of bioremediation is when special bacteria are used to treat sewage water to reclaim it as usable water (see Figure 14.28). The need for bioremediation occurs when an environment is altered so that organisms cannot grow and reproduce in a "normal" way. Oil or chemical spills, runoff from sewage, fertilizer, or pesticides, or pollutants entering an environment by other means are factors that require bioremediation of an affected area. Several testing methods have been developed to measure the impact of each of these pollutants on the environment, and now some biotechnologies are being used to remediate the impacted areas.

Figure 14.28. **Bacteria are used in one of many steps used to turn sewage water into usable, clean water.** © iStockphoto.

environmental biotechnology (en•vi•ron•men•tal bi•o•tech•nol•o•gy) a field of biotechnology whose applications include monitoring and correcting the health of populations, communities, and ecosystems

Organisms have been found in nature that degrade or process the very pollutants that may harm other organisms in an ecosystem. Several species of bacteria and fungi, as well as a few plant species, can metabolize harmful compounds and convert them into harmless ones. An example is the bacterium, *Geobacter* sp, which has special metabolic pathways that allow it to use petroleum as a food source. It can take (harmful) petroleum from contaminated soil and break it down to (harmless) carbon dioxide. Some *Geobacter* species also are known to remove radioactive metals from soil. The hunt is on by scientists for more organisms that can break down, recycle, and render harmless many of the pollutants that humans produce.

In another example, scientists have recently completed the genome of the soil bacterium, *Dehalococcoides ethenogenes*. This bacterium has been used to clean up chlorinated solvents found in underground water supplies. The pollutants are the result of the release of solvents by dry-cleaning plants and other industries. Genomic studies may lead to the identification of other organisms that have similar unusual, but beneficial, mutant metabolic pathways.

Phytoremediation is a type of bioremediation in which plants are used to "clean up" an impacted area. Currently two species of the mustard family, *Brassica juncea* and *Brassica carinata*, are being studied to measure their effectiveness at removing large quantities of chromium, lead, copper, and nickel from contaminated soils. Read about these bioremediation efforts and more at: **biotech.emcp.net/bioremed**.

Another example of environmental plant biotechnology is the development of new "greener" plastics. Most bioplastics are made from renewable plant feedstock materials instead of oil and are easily biodegradable. To make the "green" plastics, biorefineries use fermentation and enzymes to break down plant materials to carbohydrates. The carbohydrates are used to create the bioplastic. Several national grocery chains, such as the Nugget Market, Inc., are already using the green bioplastic bags. Bioplastics can also be used to make other products including fabric for clothing and bedding, food containers, and packing materials.

Figure 14.29. Before endangered animals, such as this baby panda from the Smithsonian's National Zoo, are bred, their genealogy is determined using a variety of data that include genetic fingerprints. The genealogy of captive endangered species is all recorded in a database called the Species Survival Plan.

© Laurie Perry/Smithsonian National Zoo.

microbiology (mi•cro•bi•ol•o•gy) the study of organisms too small to be seen without a microscope

microbiome (mi•cro•bi•ome) all of the microorganisms living in or on another living thing

Environmental biotechnology also includes such practices as protecting endangered species. Scientists conduct genetic profiling of rare and endangered animals and plants in the hope of understanding how to maintain or restore their populations (see 14.29).

Microbial Biotechnology

Microbes (microorganisms) include bacteria, fungi, protozoa, microalgae, viruses and all other organisms too small to be seen without microscopy. Microbes live everywhere and their sheer numbers make them extremely important to humans and other organisms (see Figure 14.30). Microorganisms impact all areas of health, the environment, and the economy.

The study of microorganisms is called **microbiology**. Microbiologists may work in the field but most work in laboratory research. Using the tools of biotechnology, microbiologists study the structure and physiology of microbes, as well as their life cycles, genetics, and ecology. They develop methods to combat harmful microbes and techniques to use microorganisms to produce new products including foods, biofuels, and industrial products.

It has been estimated that 90% of cells in the human body are actually bacteria, fungi, or otherwise non-human microbes (www.ncbi.nlm.nih.gov/pmc/articles/PMC3709439). The microbes that are found on an individual are called a **microbiome**. According to a 2013 *Nature Biotechnology* article, "Medicines from Microbiota," altering the microbes in an individual's microbiome could be detrimental and may be associated with several diseases including obesity, type 1 and type 2 Diabetes, allergy, atherosclerosis, and nonalcoholic fatty liver disease (NAFLD). In the future, a person's microbiome composition might be used as a diagnostic or therapeutic tool.

Even in healthy individuals, without most of your microbiome, humans would not be able to survive. Microbes live in and on the body of plants and animals, aiding in digestion, nutrient absorption, and processing of waste. In nature, beneficial microbes act as decomposers, processing wastes and recycling them into beneficial substances.

Microbes have long been used to make a variety of food, beverages and pharmaceuticals. Biotechnologists have learned to use microbes in new beneficial ways. With the advent of genetic engineering, many microbes are utilized as production organisms to synthesis new or modified drugs, as well as new or improved agricultural, industrial and environmental products. However, many microbes are

Figure 14.30. **Biofilm on a Raspberry Leaf (SEM) ~10,000X** All organisms are covered to some extent, both inside and out, with a biofilm of microorganisms. On this raspberry (*Rubus idaeus*) leaf, the biofilm covers virtually the entire surface. The biofilm is a mixture of mutualistic and commensalistic fungi, bacteria, and other microscopic objects. Some biofilms are detrimental, such as the bacterial dental plaque on teeth. Some biofilms in nature are beneficial including those found in soils that act as decomposers and recyclers of different inorganic and organic substances. Biotechnologists have recognized that biofilms may place a role in improving the environment and have started research and development of biofilm products or processes.

© Science Photo Library/Photo Researchers.

the cause of some of our worst problems. Diseases caused by microorganisms are responsible for human suffering and agricultural losses.

Understanding the diversity and molecular genetics of microbes will lead to solutions to many of the problems caused by microorganisms. For example, it is expected that studies of the genomes of pathogenic microbes as compared to nonpathogenic ones will reveal the disease-causing genes and potential targets for therapies.

Microbial genomics is the study of microbial genes and gene functions. Microbial genomics has resulted in a huge amount of sequence data. Since the first microbial genome was sequenced in 1995, 1666 complete bacterial genome sequences have been determined, and more than 4900 are in progress (www.nejm.org/doi/full/10.1056/NEJMra1003071). NCBI's Microbial Genomes Resources (www.ncbi.nlm.nih.gov/genomes/MICROBES/microbial_taxtree.html) presents public data from prokaryotic genome sequencing projects. The Microbial Genomes Resources sequence collection contains data from finished genomes and partially assembled genomes. Microbial genomics is critical for advances in food safety, food security, agriculture, environmental remediation, medicine, and pharmaceutical production.

microbial genomics all efforts to study the entire genome of a microorganism

BIOTECH ONLINE

The Human Microbiome Project

An understanding of what microbes typically live in or on humans is needed to understand how microbes influence human health. Many researchers around the world are working together to catalog and characterize the human microbiome. They are part of The Human Microbiome Project described in detail at www.ncbi.nlm.nih.gov/pmc/articles/PMC3709439.

Learn more about The Human Microbiome Project using broadcasts presented on the National Public Radio websites at NPR.org.

1. Read the introduction and listen to the report titled, *From Birth, Our Microbes Become As Personal As A Fingerprint*, at www.npr.org/sections/health-shots/2013/09/09/219381741/from-birth-our-microbes-become-as-personal-as-a-fingerprint.

   • Describe similarities and differences in the microbiomes found in different parts of your body such as your nose, mouth, armpits, and your gut.

   • Explain how not understanding these microbiomes may harm your health.

2. Read the introduction and listen to the report titled, *Human Microbiome: Finally, A Map Of All The Microbes On Your Body* at www.npr.org/sections/health-shots/2012/06/13/154913334/finally-a-map-of-all-the-microbes-on-your-body.

   • Why is it important to sequence the genomes of microbes on or in the human body?

   • What were some of the first findings of the human microbiome genomic studies describe in the broadcast?

3. How may learning the sequence of your own personal microbiome impact your medical and personal privacy? Learn more about it by listening to the NPR broadcast, *Getting Your Microbes Analyzed Raises Big Privacy Issues* at www.npr.org/sections/health-shots/2013/11/04/240278593/getting-your-microbes-analyzed-raises-big-privacy-issues.

Marine Biotechnology

One of the most interesting areas in biotechnology is marine science and aquaculture. The potential benefits and/or the risks of ocean biotechnology are of great interest given the degree of human dependence on the ocean and the fact that so many species live in aquatic environments.

One example of a biotechnology product that could arise from marine sources comes from the mussel, a clam-like mollusk (see Figure 14.31). Mussels produce a glue-like substance so strong that it prevents them from being torn off rocks battered by powerful waves. The glue is of such high quality that it has industrial value. Biotechnologists have isolated the mussel glue genes and transferred them to tobacco plants that are then transformed into glue producers.

Since the ocean and the number of organisms in it is so vast, the applications of biotechnology to marine science is equally vast. One expansive area is the use of unique chemicals (toxins, pigments, and sensory or regulatory molecules) produced only in marine organisms as therapeutic or diagnostic compounds. For example, the sea squirt that grows on mangrove roots contains a powerful anticancer drug called ET743 (**biotech.emcp.net/ET743**). Scientists are currently examining how to commercially produce ET743 for therapeutic purposes.

Figure 14.31. The glue that holds mussels in place in their tidal environment has commercial value. The mussel's though, only produces a small amount of the glue. To produce large volumes of the glue, biotechnologists have engineered tobacco plants with mussel glue genes. Potentially, the transformed plants could produce commercial volumes of the mussel glue.
© Science Photo Library/Photo Researchers.

BIOTECH ONLINE

Marine Biotechnology: What Is All the Flap About?
Recently, there has been a widely publicized fight between groups in support of or against the development and possible release of transgenic salmon. In the following activity, learn more about this controversial marine biotechnology product.

TO DO

Learn about one product of marine biotechnology and take a position on its possible use.

1. View the video at: biotech.emcp.net/trans_salmon.

2. Using the Internet, find two articles, one in support of and one against, about transgenic salmon production. Make a chart with two columns (pros and cons), and make a complete list of the issues regarding the production and use of transgenic salmon.

3. Imagine you are a fisherman living on the coast of Alaska. Would you support the development and release of transgenic salmon? Why or why not?

4. Visit www.marinebiotech.eu to learn about specific examples of medical and environmental biotechnology research using other marine organisms.

Veterinary Biotechnology

The Humane Society of the United States reports that in the United States more than 77 million dogs and 85 million cats are owned as pets. Imagine the enormous market for products that improve the quality of life for these and other pets. In addition, consider the need for excellent health care for livestock, racehorses, or animals in zoos and aquaria. Pet owners, farmers, and zookeepers expect cutting-edge diagnosis and treatment for their animals. For these reasons, veterinary biotechnology is a rapidly growing field.

Veterinary biotechnology is already having an impact on how animals are kept healthy. Veterinarians use biotech products daily to treat diseases and disorders, such as heartworm and other parasites, arthritis, and allergies. Veterinarians prevent diseases using vaccines, such as those for rabies and feline HIV. They also use biotech-based diagnostic tests to look for confirmation of several diseases. Several pet genomes have been sequenced, accelerating the rate at which new diagnostics and therapeutics can be found and tested.

In 2010, according to a BIO report, sales of biotechnology-based, animal health products generate $2.8 billion (out of a total market for animal health products of $18 billion). The animal health industry invests more than $400 million a year in research and development of diagnostics and therapeutics, which by 2010 has resulted in 111 USDA-approved biotech-derived veterinary pharmaceuticals. Veterinary biotechnology is a large business sector with opportunities for all kinds of business and science employment.

The American Veterinary Medical Association (AVMA) supports the use of biotechnology for a variety of veterinary applications (**biotech.emcp.net/AVMA**), including:

- the benefit and protection of public (human, animal, and environmental) health and welfare,
- enhancing host resistance to infectious diseases and eliminating genetic-based diseases,
- increasing the efficiency of food and fiber production,
- improving the utility, nutritional value, and safety of human food and animal feeds,
- the production of improved animal medicinal products and diagnostic tools,
- the improvement and protection of the environment, and
- the mitigation of the environmental impact of crop and agricultural animal production.

The importance of veterinary biotech became clear in 2007 with a shocking case of melamine poisoning in dogs due to contaminated dog food. Melamine is a toxic compound that can cause kidney failure. When the kidneys fail, the dogs are unable to clear waste from the bloodstream. Melamine poisoning is life threatening. Fortunately, the use of ELISA technology has allowed all dog foods to be tested for melamine, protecting dogs and their owners from unnecessary suffering.

In another interesting veterinary biotechnology application, horse embryonic stem cells are being tested as a treatment for strain-induced tendon injury in horses. If the veterinary scientists are successful in regenerating functional tendon cells for use in horses, then that same technology might be applied to human injuries. Advances in veterinary biotechnology often have applications to human biotechnology.

Biotechnologies are also being used to ensure that zoo animals are not inbred. Zoo veterinarians must conduct genetic tests to ensure that animals that are allowed to breed in zoos are not closely related to each other, as that may lead to a high incidence of several genetic disorders (see Figure 14.32). The genetic information collected from captive endangered species is stored in an internal database called the Species Survival Plan.

Figure 14.32. In zoos, only certain animals are allowed to mate. To ensure that captive animals have a reduced risk of inherited disease, zoos conduct genetic testing and screening on all captive members of an endangered species that are of reproductive age, including gorillas.
© Photo courtesy of Steven Halford/Stock Xchng.

Nanotechnology

Nanotechnology is a broad field that includes the application of technological advances on a small scale (a billionth of a meter) to problems in biology, chemistry, and physics.

Nanotechnology uses nanoprocessors, nanocomponents, nanomicroscopes, and nanodelivery systems to create tiny instruments or systems. Examples of some nanodevices include new miniaturized telecommunication circuits, tiny electrical wires the width of a virus particle, and carbon tubes the diameter of a molecule. Several new nanotechnology advances are in the area of biotechnology, such as immunity nanoparticles (iron particles with antibodies on them) being used to fight ovarian cancer and an asthma nanosensor composed of carbon tubes for the detection of nitric oxide in the lungs of asthmatics (**biotech.emcp.net/popmech_nana**).

In medical applications, researchers are using nanotechnology to develop 3D vaccines in microscopic silica rods. The tiny rods can be injected under the skin . The rods have have nanopores, that contain vaccine antigens inside which attract white blood cells that turn on the body's immune recognition of the antigen in the future. 3D vaccine nanorods can be modified easily by changing the antigen inside. Learn more about this new nanotechnology at **www.ibtimes.com.au/3d-vaccines-cancer-other-infectious-diseases-1420060**.

To learn more about nanotechnologies already being used in medical research, disease detection, stem cell research, cancer therapeutics, bioinstrumentation, and more, visit **www.nanotechproject.org/inventories/medicine**.

BIOTECH ONLINE

Clearly, I See a Future in Nanotechnology

TO DO

Learn how nanotech is creating advances in an "older" technology, biotech.

The fields of biotechnology and nanotechnology overlap in many ways. Go to the following website to learn how advances in nanobiotechnology are being used to create a new kind of contact lens: **biotech.emcp.net/bioniceye**.

Read and summarize the article. Explain how these contact lenses have been modified using nanotechnology. Discuss the possible uses of the new contact lenses.

Section 14.3　Review Questions

1. How might synthetic biology be used to correct a disease such as diabetes?
2. What is the name of the process by which strategies are used to solve environmental problems, such as oil spills, soil erosion, or fertilizer pollution?
3. List a few examples of veterinary biotechnology products.
4. What is the approximate size of the instruments and products of nanotechnology?

14.4 Opportunities in Biotechnology: Living and Working in a Bioeconomy

The opportunities in biotechnology for scientific and business employment are great, and even with economic downturns, the public needs the products that biotechnology makes. With advances in genomics and proteomics, the outlook for employment in biotechnology is expected to be good for the rest of this century.

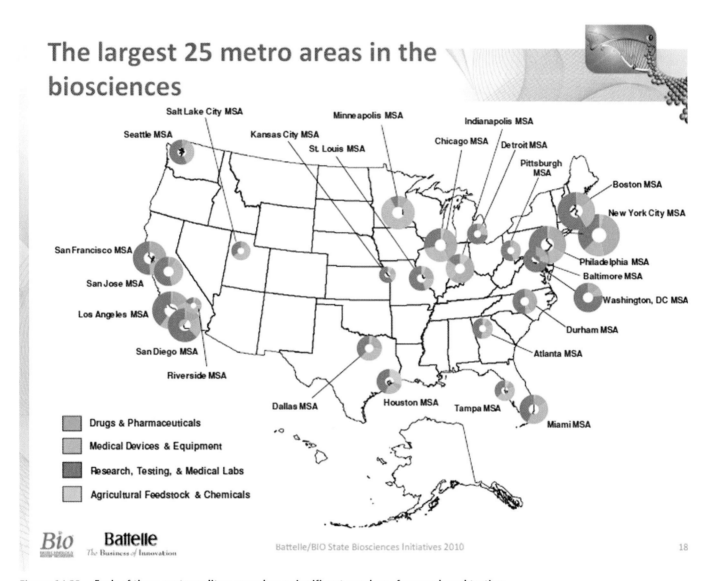

Figure 14.33.   Each of these metropolitan areas has a significant number of research and testing labs. Some of the regions have a focus in the development of pharmaceuticals, medical devices, or agricultural products as shown in the color-coded key. Visit www.genengnews.com/insight-and-intelligence/top-ten-biotech-jobs-most-in-demand-over-the-next-decade/77899666 for an idea of the careers that will be the most in demand over the the next decade.

© Source: The Battelle/BIO State Biosciences Initiatives 2010 (bio.org/battelle2010), used with permission.

Biotechnology Is a Global Endeavor

In 2012, SciDATA reported that there were 24,000 biotechnology companies in over 40 countries (**biotech.emcp.net/SciDATA**). In addition, many companies have research and development, manufacturing, or administrative facilities in several different countries. This gives a biotechnology employee many options for where to work and live, whom to work for, and in what type of role.

Scientific American Worldview 2009 cited the United States, Singapore, Denmark, Israel, and Sweden as the top five countries for biotechnology innovation. These countries lead in biotechnology innovation because they provide excellent protection for intellectual property, they spend a considerable amount on research and development of products, and they have excellent support for education and financing. In addition, in 2011, Biotech-Now.org cited that Brazil, Italy, Canada, India, and Malaysia are the companies to watch for significant growth in the industry. In these countries, robust biotechnology innovation comes with robust business development and outstanding employment opportunities. A quick search on the Internet can provide a lot of information about biotechnology opportunities in these or other countries.

Virtually every state in the United States has some kind of biotechnology industry and employment. However, several cities/regions in the US have a significantly large biotechnology-based economy (see Figure 14.33). These include:

New York City, NY	Chicago, IL
San Francisco, CA	Boston, MA
San Jose, CA	Philadelphia, PA
Los Angeles, CA	Seattle, WA
Houston, TX	Durham, NC
San Diego, CA	Salt Lake City, UT
Dallas, TX	Washington, DC
Minneapolis, MN	Indianapolis, IN
Miami, FL	Tampa, FL

These metropolitan areas have large concentrations of biotechnology employers, ranging from small start-up companies to large pharmaceutical, agricultural, and industrial complexes, as well as academic and government facilities. Each region also has extensive educational and training facilities to prepare a biotechnology-based workforce.

BIOTECH ONLINE

The Biotech Industry Where YOU Live

Learn about the biotech industry and opportunities for employment in the state in which you live.

1. Go to the BIO website at: www.bio.org/sites/default/files/Battelle-BIO-2014-Industry.pdf to find information about the state of the biotechnology industry across the country.

2. Use the report to learn about biotech in:
   - the state in which you currently live
   - a state that is known to have a large biotechnology industry such as CA, MA, or NC
   - a "small biotech state," such as NE or TN

3. Construct a chart that compares the opportunities for biotechnology employment and advancement in the three states. Use criteria that are important to you, such as industry focus, number of companies, cities that are biotech hubs, average salary, etc.

Biotechnology Is a Diverse Endeavor Requiring a Diverse Workforce

Biotechnologies are practiced in academic (university), regulatory (governmental), and corporate (industry) facilities. In each, many different employees are needed, some with scientific backgrounds, some with business backgrounds, and some with background and experience in both. According to the Batelle/BIO State Bioscience Initiatives, 2010 (**biotech.emcp.net/BIO_main**), 2014 BIO report, "Bioscience Economic Development in the States," U.S. bioscience companies employed 1.62 million employees. In the decade prior to the report, the biotech industry added nearly 111,000 new, high-paying jobs. The bioscience sector has significantly higher wages than other industries. In 2014, the average annual wage for a bioscience employee was $88,202, versus the national average of $49,130 for other employees in the private sector.

Rewarding careers with diverse job functions may be found in these areas:

- research and development, including basic research, discovery research, process development, product development, preclinical research, assay development, and clinical development
- technical training and education
- manufacturing, including fermentation, pilot plant, quality control, and operations regulatory, including quality assurance, regulatory affairs, medical affairs, and information management
- business, including administration/management, business development, finance, sales, marketing, corporate communications, IT, technical support, and legal
- human resources, including recruiting, compensation, benefits, and evaluation

How to Prepare for a Career in Biotechnology

Your training for a career in biotechnology may have started with this book. As you continue to prepare to enter the bioscience/biotechnology workforce, begin by going to the Bioscience Competency Model at **www.careeronestop.org/competencymodel/competency-models/bioscience.aspx**.

Each competency on the Bioscience Competency Model pyramid is important in your career preparation. By clicking on a competency, you will get more information that can guide you on your career preparation pathway, including descriptions of the technical and soft skills needed to meet a competency.

To meet these competencies, several educational and technical training programs (colleges/universities) exist. You may already be enrolled in one. To learn about the educational and training programs in your region and across the United States, visit the Bio-Link website at: **biotech.emcp.net/Biolink_home**.

Whatever training, degrees, certificates, or experience you have, you should possess certain qualities for any position that you apply for. These qualities should be reflected in your letters of reference, resume, and interview to increase the likelihood of being seriously considered for a position. These employee qualities include the following:

- working in a professional and confidential manner
- acting in an alert and safe manner
- maintaining accurate records
- working well as a team member, working well individually
- having a pleasant, positive attitude and being easy to get along with
- completing work in a reasonable amount of time
- being self-directed, recognizing tasks that need to be done
- reflecting on work done and the values and applications of the work
- maintaining a clean and orderly workspace
- contributing to the organization and cleanliness of common areas
- having excellent attendance and promptness
- dressing appropriately for the work environment
- communicating appropriately and in a timely manner
- staying current in required science and technology areas through professional development

Two excellent resources for prospective biotechnology employees is *Career Opportunities in Biotechnology and Drug Development* (Friedman, T.: Cold Spring Harbor Laboratory Press; 2008) and **biotechcareers.com**.

No matter where you work in the field, you should be proud that the biotechnology industry's goal is to improve the human condition. Whether you are a scientist, a research associate, or a technician, or you work in the business or regulatory side of biotechnology, the processes you study and the products you help produce or market could help sick people feel better or give them hope of a cure. They may help feed a hungry world or clean up a polluted planet. The biotechnology products that you or your team have worked on may make life easier for those that have substantial challenges living "normal" lives.

BIOTECH ONLINE

Resume for a Biotechnology Laboratory Position

After completing a program like the one presented in this curriculum, you will possess knowledge and skills that should open doors to you. Opportunities for workplace experiences (internships or externships), short- or long-term employment, or entry into advanced academic programs are out there for individuals who can clearly and effectively communicate their unique qualities. A well-constructed resume specific to the biotechnology workplace can help build your confidence and effectively communicate your goals and objectives.

Create a one-page, professional-looking resume that may be used to convince a potential supervisor that you have the best attributes to be selected for a laboratory position in a biotechnology facility.

1. Go to **biotech.emcp.net/resume** and review the features of an effective resume. Look over the examples of resumes on their site.

2. Create your own one-page biotechnology laboratory resume in Microsoft® Word®. Do not use a template to construct your resume since hidden formatting can make changing the resume problematic. Here are some general hints:

   a. The resume should fill one page but not be so overcrowded as to appear messy. A margin of 0.75" is sufficient. Too much empty space on your resume, around the edges and in the body, may imply that you don't have much to offer.

   b. Make sure that the resume is neat and easy to read. Use font sizes of 10 to 14 point. Avoid too many changes in font style. Use tabs (instead of the space bar) to align text.

   c. Use short phrase for descriptors instead of whole sentences. Use brief statements that are to the point. Make sure there are no spelling errors and that you have use appropriate terminology.

   d. Include your name and contact information at the top of the page in an easily read font. Include a "professional" email address.

   e. Move sections that demonstrate your science aptitude to the top half. This includes your science/math education, your lab experience, and your lab/computer skills.

   f. Don't forget the science skills you have developed in other classes or workplaces (ie microscopy, titration, distillation, statistical analysis, instrument maintenance, etc).

   g. Include a "Work Experience" and/or "Volunteer Experience" section (paid jobs, unpaid jobs, volunteer/community work, working in classrooms or at school, child care, etc.). Don't exaggerate. Put down what you have done and give some idea of the amount of time you have done it.

   h. If you have room, add other sections that demonstrate some of the other reasons someone would want to have you in the lab, such as "Personal Qualities" (hardworking, follows direction, keeps a clean lab station, attention to detail, good group member, works well individually) or "Other Interests" (sports, music, clubs, hobbies, etc).

   i. Don't lie or exaggerate and be ready to field questions about anything on your resume.

Chapter Review

Summary

Concepts

- Genomics is the study of all of the DNA in a cell.
- The Human Genome Project published a rough draft of the entire DNA code for the human organism. After hundreds of people worked on sequencing the code for more than 10 years, the project was completed, for the most part, in 2001.
- The use of shotgun cloning was one reason that the Human Genome Project was completed so quickly. Inserting fragments of the genome into plasmids and transforming cells with the plasmids provided many fragments to analyze quickly.
- The Human Genome sequence information has led to hundreds of studies in gene expression and protein production.
- Genome projects are being conducted for dozens of animals, plants, fungi, protozoa, and bacteria.
- Genome annotation documents which sections of the genome sequence codes for proteins, which sequences do not code for proteins, and which sequences code for regulatory elements.
- To study gene expression, the genomic arrays are incubated with the labeled RNA or cDNA molecules. If any of the tagged molecules hybridize with the probe arrays, a color change can be detected and measured by a microarray reader, indicating a gene section responsible for transcription and translation.
- Bioinformatics is the use of computers and statistical analysis to understand biological data. Computer programs and databases of computerized information have been developed to manage the enormous amount of genomic and proteomic data.
- RNA studies are being used to understand protein production and regulation. MicroRNA, siRNA, and RNAi are all small pieces of RNA that are used to block or modify genes or RNA transcripts. These types of RNA are thought to play a big role in turning off genes, gene products, and protein synthesis.
- RNA molecules can be identified and quantified on Northern blots.
- RNA researchers may now use microarrays to study an entire transcriptome (all of the mRNA in a cell at a given point in time). Transcriptome studies give scientists an idea of what proteins are actually being made at given time points, which helps explain processes in a cell.
- CRISPR/Cas9 (guide RNA and a restriction enzyme) technologies allow for precise and accurate editing of genetic code and the potential for gene therapy.
- Proteomics is the study of how, when, and where proteins are used in cells. Tools of proteomics include protein crystallography and x-ray diffraction, mass spectrometry, NMR, and several assays, including ELISAs, Western blots, and protein arrays.

- Protein crystals are needed for many of the proteomic techniques. The main way to produce protein crystals is to use a process called "salting out," in which protein molecules are forced out of solution by differences in salt concentration. Pure crystals develop only if the salting-out process is slow.
- Stem cells are unspecialized cells that have not yet differentiated into cells with a specific function.
- Embryonic stem cells are found only in a developing embryo and are naturally pluripotent. Adult stem cells are found in tissues and organs of a person from the time it is a fetus and through adulthood. Adult stems are programmed to make certain kinds of cells for replacement purposes.
- The goal of regenerative medicine is to replace or restore damaged tissues or organs. Stem cell technology is an example of regenerative medicine.
- 3-D bioprinting uses a modification of ink-jet technology, and with stem cell and computer technologies, lays down specific cells in pre-determined ways to produce new, replicate tissues and organs.
- Energy independence and decreasing greenhouse gas emissions are the driving forces in developing new biofuels, including bioethanol from corn, cellulostic biomass, or feedstocks and biodiesel from algae or genetically modified plants.
- There is much interest in biodefense as a response to possible bioterrorism attacks. The development of early biowarning biometric diagnostic systems and quick-acting vaccines are strategies currently being tested.
- Synthetic biologists use biotechnologies to construct or redesign new biological molecules, cells, tissues, organs, or systems. It is a new discipline but a promising one that could lead to novel, specialized cells that could be used to treat or cure disease or produce new industrial products.
- Environmental biotechnology uses biotechnology advances to improve the environment and/or the organisms in it. A principal practice in environmental biotechnology is bioremediation, which attempts to clean pollutants out of soil and water. Several naturally occurring organisms have been found that can metabolize pollutants, thus helping to restore environments.
- Understanding the diversity and molecular genetics of microbes, and documenting it in one of the microbiome projects, will lead to new uses of microbes and solutions to many of the problems caused by microorganisms.
- Using biotechnologies to study and use marine organisms is the focus of marine biotechnology. Some marine biotechnology products include transgenic fish, therapeutic compounds, and industrial compounds.
- The need for biotechnologies to address animal health is great due to the large number of pets and agricultural animals. Veterinary biotechnology has developed several diagnostics, vaccines, and therapeutics for pets and for farm and zoo animals.
- Nanotechnology is the study, use, and manipulation of things that are about a billionth of a meter in size. Nanotech has many applications in biotech, including medical nanoprocessors, nanocomponents, and nanodelivery systems.
- There are more than 24,000 biotechnology companies in 38 countries. This gives a biotechnology employee many employment options.
- Virtually every state has some kind of biotechnology industry and employment, although some large metropolitan areas have a significantly large biotechnology industry.

Lab Practices

- Bacteria may be used to remove compounds from contaminated soil, water, or air. This is called bioremediation. *Shewanella oneidensis* MR-1 broth cultures can metabolize chromium (chromate), essentially removing it from the solution. The decrease in potassium chromate is indicated using a diphenylcarbazide assay.
- Cellulase activity can be measured by measuring glucose concentration after samples are incubated with cellulase. Different cellulase samples can have different amounts of activity. Glucose concentration can be measured against glucose standards using a Benedict's solution aldose assay, using commercially available glucose test strips, or spectropotometric assays.
- D-limonene is a hydrocarbon found in citrus rinds. It has been shown to have insecticidal properties. It can be extracted and isolated using distillation.
- Suspected bioinsectides can be tested using wingless fruit flies. Fruit flies may be easily cultured in a biotechnology laboratory facility.

Thinking Like a Biotechnician

1. How is microarray technology accelerating the understanding of gene expression and how cells function?
2. If a genome has 4 billion bp in it, and a typical sequencing reaction can determine the code of about 500 bp, how many sequencing reactions will it take before a computer can start piecing together the entire genetic code?
3. DNA sequencing requires millions of exact copies of the same DNA strand. How does a technician get millions of copies of the same strand to sequence?
4. Zookeepers want to breed two koalas from different zoos. Another zookeeper is declining, saying that the koalas are two different subspecies. How could that be checked?
5. What would an x-ray diffraction pattern look like if the protein crystal used to make the x-ray pattern was not pure?
6. Which test could determine whether a fish, such as a salmon, was genetically modified?
7. How can advances in veterinary biotechnology benefit human as well as animal health care?
8. Propose an assay to test for oil-eating bacteria (a possible candidate for bioremediation).
9. A protein assay is needed to determine which bacterium produces cellulase. What should be bound on the protein chip? And what should be added to detect what is on the protein chip?
10. You are given a solution suspected to have a very high concentration of protein in it. How can you use a dialysis bag to crystallize the protein out of solution?

Biotech Live

Lambda (λ) Phage Genome: What's Up Lambda's Sleeve?

Activity 14.1

The lambda (λ) phage is one of the most important tools for scientists in the field of biotechnology. Since the lambda phage infects *E. coli* cells, it is simple to maintain and grow it in the lab. In the 1980s, scientists recognized that knowing the DNA sequence of the lambda phage would be important for several reasons, including cloning, genetic studies, and understanding how viruses work.

Since then, the entire genome of lambda phage has been sequenced, and its essential genes have been identified. It is interesting that the phage has very few genes, yet it is able to take over a cell and direct it to become a virus-producing machine.

Using Internet resources, find information about the lambda bacteriophage genome.

Make a model of λ DNA. Use the information to explain how the virus infects and reproduces in *E. coli* cells.

1. Find a website or a bioinformatics database that has a diagram of or information about the lambda bacteriophage's genome.
2. Find the total length of the λ DNA in base pairs. Cut a piece of adding-machine tape to a length that represents the total length of the λ DNA molecule. Let each centimeter represent 100 bp.
3. Most of the genes in the λ genome are involved in the reproduction of the virus. Using the Internet resources, label on the tape model the λ genes that scientists have found to be coded for on the λ DNA.

   - The gene that is responsible for recognizing the *E. coli* cells is the "tail" gene. Label it in orange, and put an asterisk (*) by the label.
   - Label the gene that is responsible for the insertion of λ DNA into the *E. coli* chromosome in blue.
   - Label genes that are responsible for the λ virus's critical structures (head, tail, and DNA) in green.
   - Label the gene responsible for exploding the infected cell, releasing new λ phages, in red.
   - Label the genes that are responsible for general control of the entire infection and reproduction process in black.

4. Use your knowledge of the λ genome to propose how the virus infects and controls an *E. coli* cell. Write your explanation on the back of the λ DNA paper model, along with the references you used to make your model.

Activity 14.2 What's the Difference? Alpha or Beta, It's All Amylase.

The bacterium *Bacillus cereus* produces two types of amylase (also called a glucosidase). One type, alpha-amylase, breaks down starch to maltose units by chopping a starch strand randomly anywhere along its peptide chain. The other type, beta-amylase, also breaks down starch to maltose but does so by specifically chopping maltose units off the ends of starch molecules.

As a molecular biologist at a biotechnology company, you are interested in finding the best amylase, and the gene for that amylase, for recombinant DNA and recombinant protein manufacturing. You need to find the similarities and differences between the alpha and beta forms of amylase and the genes that code for each so that each can be evaluated for further use.

Use the National Center for Biotechnology Information (NCBI) website to access databases and tools to learn more about the alpha and beta forms of amylase from *Bacillus sp.*

1. Create a chart to record information about the structure and genetics of an alpha-amylase as compared to a beta-amylase from the bacterium, *Bacillus cereus*. Have a place to record the name of each protein, the organism that was the source of the protein studied, and the anatomical or genetic information requested below. Include bibliographical information showing the URLs used to collect the information. The protein data should be able to be compared side by side.

2. Go to the NCBI website at: www.ncbi.nlm.nih.gov. Enter either "alpha-amylase Bacillus cereus" or "beta-amylase Bacillus cereus." Use the following databases to collect the following information:

   - Protein - The accession number, the number of amino acid residues, and the amino acid sequence (use FASTA link)
   - Nucleotide - The GenBank number, for the number of nucleotides in the sequence and the nucleotide sequence (use FASTA link)
   - Structure - The MMDB ID number and the PDB ID number (for each protein or a similar alpha or beta amylase). Use Cn3D and VAST+ to compare the structures. Make a copy of the images for the chart.

3. Search on NCBI for an article that describes a method to distinguish between alpha-amylase and beta-amylase using beta-limit dextrin azure. Explain how the test is conducted and how test results distinguish by the 2 forms of amylase. Record the bibliographic information for the article, citing your source.

Activity 14.3 Body Parts From 3-D Printing

TechRepublic.com reported that in 1992 a man from Wales (in the United Kingdom) was in a motorcycle accident resulting in serious damage to his head and face. The man broke his cheekbones, jaw, nose, and skull. In 2014, doctors at the Centre for Applied Reconstructive Technologies in Surgery used 3D printing to produce facial implants that successfully fixed injuries he sustained. CT scans were used to 3D print a model of his skull, and then a titanium implant to hold the new the bones in place. What other body parts could be repaired or replaced using 3-D bioprinting?

Learn about the successes achieved in 3-D bioprinting and companies leading the 3-D bioprinting movement.

1. Go to www.forbes.com/sites/robertszczerba/2015/06/17/no-donor-required-5-body-parts-you-can-make-with-3-d-printers-2 and use the article to learn more about five human body parts that are on the forefront of medical 3-D bioprinting. For each, describe the possible use of the bioprinted structure.

2. One of the first 3-D bioprinting companies is **Organovo**, Inc. in San Diego, CA. The company has already reached some impressive achievements including printing of human liver tissue. The printed liver tissue has use in therapeutics to combat several liver diseases including hepatitis. Watch the video at www.organovo.com/science-technology/bioprinting-process to learn how Organovo uses the bioprinting process. Another of Organovo's products is exVive3D™. Use Organovo's website to learn more about what exVive3D™ is used for.

3. Use the Internet to find three other companies that specialize in making 3-D bioprinters or making products using 3-D bioprinting. Make a chart the shows information about each company. Include the following information:

   - The name, address of the company, home website URL
   - The main focus or business goals of the company
   - An example of a 3-D product they are working on or have developed (name model # description)
   - One other interesting thing about the company

It Takes A Lot of Scientists to Fuel The World

To convert from a petroleum-based energy economy to one based on biofuels will require looking at fuel research and production differently and more collaboratively. The Joint BioEnergy Institute (JBEI) is an example of how scientists, government agencies, and industry can work together in new and novel ways.

JBEI is a U.S. Department of Energy (DOE) Bioenergy Research Center located in Emeryville, California. It is dedicated to developing advanced biofuels derived from plant biomass that can replace gasoline, diesel and jet fuels and to contribute to cleaner air, sustainable transportation and energy independence. JBEI's employees come from their partner institutions (see below) and conduct research activities under one roof. Being located in a biotech hub allows for JBEI science staff to access some of the world's best imaging, computational, and laboratory resources and conduct research and pilot scale-up in the most modern facilities.

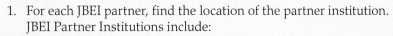

1. For each JBEI partner, find the location of the partner institution. JBEI Partner Institutions include:

   - Lawrence Berkeley National Laboratory
   - Sandia National Laboratory
   - Pacific Northwest Laboratory
   - Lawrence Livermore National Laboratory
   - University of California, Berkeley
   - University of California, Davis
   - Carnegie Institution for Science

2. JBEI's researchers work in four divisions that have separate, but often, mutualistic goals. Go to JBEI's research page at www.jbei.org/research to learn about each division:

   - Feedstocks
   - Deconstruction
   - Fuel Synthesis
   - Technology.

   Click through the links in each division to find the focus of the work being done there. Summarize the research focus in your notebook. Also, find one of the many projects being worked on in that division. Describe each in a few sentences.

3. JBEI's partner institutions hire employees to work labs at JBEI. Go to JBEI's Current Job Openings at www.jbei.org/about/jobs and pick 2 current job openings that might be of interest to you. For each, describe the position including some of the specific job duties. Then list 4 qualifications that you hope to have by the time you complete your college biotechnology training.

Activity 14.5 Party Bacteria... Green Party, That Is

Many bacteria species cause disease, but many more do much good both inside and outside the laboratory. Several species of bacteria are decomposers that recycle dead plants and animals, rebuilding the soil. Bacteria are now being engineered to do even more good for the environment. Scientists recently have transformed some species of bacteria to use petroleum oil as a food source. Now, these genetically engineered bacteria are released during an oil spill to help break down and remove the oil pollutants. The use of bacteria or other organisms to restore environmental conditions is called bioremediation. Many people hope that other bioremediation technologies can be developed to address environmental issues, such as the greenhouse effect and soil or water pollution.

 Learn about oil-eating bacteria—what they do and how they have ecological and economic value.

1. View the video at: biotech.emcp.net/oilbacteria. List three important points that the narrator makes. List three concerns you have about what the narrator presents.
2. Next, use the Internet to find information about the bacterium *Pseudomonas putida*, a naturally occurring bacterium that is known to digest compounds in petroleum. List several characteristics of these important bioremediation bacteria.
3. Go to biotech.emcp.net/NCBI_pubmed and do a search of the PubMed database to find article summaries that have to do with petroleum bioremediation or other bacteria that may be used for bioremediation. Find two articles you think others should read, list their URLs, and give a short statement about why each article is valuable.

Activity 14.6 The Evolution of the Science and Industry of Biotechnology

Inspired by an activity written by Ann Murphy and Judi Perrella of the Woodrow Wilson Biology Institute in 1993.

Many people believe biotechnology is a modern phenomenon that began in the 1970s with the creation of the first human-made rDNA molecule. However, humans have been manipulating organisms to produce products for hundreds of years, and that is what biotechnology is—manipulating organisms.

 Make an "Evolution of Biotechnology Timeline" to show the major biotechnological developments of the past 250 years.

1. Using a 125-cm length of adding-machine tape for your timeline, mark every centimeter along the bottom of the tape. Each centimeter represents 20 years. Start at the left end and label the first centimeter "1760 AD." Label alternate centimeters. The last centimeter on the right side should be the present year.
2. Major events will be added to the timeline as they are "discovered" by you or your group. Some are listed below, but some must be discovered through research at the library or online. A clue is given to help you search for the "event."
3. All events should be clearly labeled, and artwork should be added to make your timeline more interesting and more useful as a teaching tool.

Before 1750: Plants were used for food, and domesticated or selectively bred for desired characteristics. Show specific examples. Animals were used for food and work, or selectively bred for desired characteristics. Show specific examples. Microorganisms were used to make cheeses and beverages, and to ferment bread. Show specific examples.

1750–1850: The cultivation of leguminous crops increased, as well as crop rotation to increase yield and land use.

1797: Edward Jenner used living microorganisms to _____.

1820–1850s: Animal-drawn machines were first used. Horse-drawn harrows, seed drills, corn planters, horse hoes, two-row cultivators, hay mowers, and rakes came into use. Industrially processed animal feed and inorganic fertilizers were developed.

1864: Louis Pasteur proved the existence of microorganisms, such as _____. He showed that all living things are produced by other living things.

1865: Gregor Mendel investigated how traits are passed from generation to generation. He called them _____. We now know them as genes.

1869: Fredrick Meischer isolated DNA, which stands for _____, from the nuclei of _____.

1880: The steam engine was used to drive combine harvesters.

1890: Ammonia, naturally produced only in cells, was synthesized in the lab.

1893: Robert Koch published findings that microorganisms cause disease.

1893: Pasteur patented the fermentation process.

1893: German scientists from Lister Institutes isolated the diphtheria antitoxin, which led to the development of a vaccine.

1902: Walter Sutton coined the term "_____" and proposed that chromosomes carry _____.

1910: Thomas H. Morgan proved that genes are carried on chromosomes; the term "biotechnology" was coined.

1918: Germans isolated and used acetone produced by plants to make bombs. Yeast was grown in large quantities. Microorganism-activated sludge was used for the sewage-treatment process.

1927: Herman Mueller increased the mutation rate in fruit flies by exposing them to x-rays. The term "mutation" means _____.

1928: Frederick Griffiths noticed that a rough type of bacterium changed to a smooth type when some unknown "transforming principle" from the smooth type was present. He tested these virulent bacteria in little _____. This was the first example of a human-controlled transformation.

1928: Alexander Fleming discovered the antibiotic properties of certain molds. An antibiotic does what? _____

1920–1930: Plant hybridization was developed. Give specific examples.

1938: Proteins and DNA were studied by x-ray crystallography, which was used for _____ _____. The term "molecular biology" was coined.

1941: George Beadle/Edward Tatum proposed the "one gene produces one enzyme" hypothesis.

1943–1953: Linus Pauling described sickle-cell anemia, calling it a molecular disease. He studied the structure of DNA.

1944: Oswald Avery performed a transformation experiment with Griffith's bacterium. He determined that the transforming principle was actually _____.

1945: Max Delbruck organized a course to study a type of bacterial virus that consisted of a protein coat containing DNA.

1950: Erwin Chargaff determined that there is always a 1:1 ratio of adenine to thymine in the DNA of many organisms.

1952: Alfred Hershey/Margaret Chase used radioactive labeling to determine that it is the _____, not _____, that carries the instructions for assembling new phages. Thus, DNA was identified as the genetic material.

1953: _____ and _____ determined the double-helix structure of DNA.

1956: Fredrick Sanger sequenced _____, a protein, from pork.

1957: Francis Crick/George Gamov explained how DNA functions to make _____.

1958: Coenberg discovered DNA polymerase that functions to _____.

1960:	Messenger RNA (mRNA) was isolated.
1965:	Plasmids were classified. Plasmids are important because they are used to _____.
1966:	_____ and _____ determined that a sequence of just three nucleotide bases code for each of the 20 amino acids.
1970:	Reverse transcriptase was isolated. It functions to _____.
1971:	Restriction enzymes were discovered. They _____ DNA.
1972:	Paul Berg cut (spliced) sections of viral DNA and bacterial DNA with the same restriction enzyme. He pasted viral DNA to the bacterial DNA.
1973:	Stanley Cohen/Herbert Boyer produced the first rDNA organism that produced the human protein, _____. Genetic engineering was applied in industry.
1975:	There was a moratorium on rDNA techniques.
1976:	The National Institutes of Health guidelines were developed for the study of rDNA.
1977:	Bacterial cells produce human growth hormone (HGH). This was the first practical application of genetic engineering.
1978:	Genentech, Inc. used genetic engineering techniques to produce human insulin in the bacteria, _____. Genentech, Inc. was the first biotech company on the New York stock exchange.
	Stanford University scientists performed the first successful transplantation of a mammalian gene.
	The discoverers of restriction enzymes received the Nobel Prize in medicine.
1979:	Genentech, Inc. produced human growth hormone and two kinds of interferon.
	DNA from malignant cells transformed a strain of cultured mouse cells, resulting in a new tool for analyzing cancer genes.
1980:	The US Supreme Court decided that man-made microbes could be patented.
1983:	Genentech, Inc. licensed Eli Lilly to make insulin.
	Scientists accomplished the first transfer of a foreign gene into a plant.
1985:	The US Supreme Court decided that plants could be patented.
1986:	The first field trials were conducted of recombinant DNA plants that were resistant to insects, viruses, and bacteria.
1988:	The first living mammal was patented.
1990:	Chymosin became the first genetically engineered substance used in food (cheese).
1994:	FLAVR SAVR® (Calgene, Inc.) tomatoes were sold to the public.
1996:	"Dolly" the sheep became the first cloned animal.
2000:	The first complete plant genome, _____, was determined.
	"Golden Rice," a type of rice with a _____ gene, was first announced.
2001:	The rice genome was determined.
2004:	The first genetically engineered pet, the _____, was marketed.
2005:	The first successful cloning of a dog occurred in _____.
2006:	Illumina's next generation sequencer is first on the market.
2007:	Launch by the NIH of The Human Microbiome Project.
2008:	First 3-D bioprinting to create a biotube similar to a human _____.
2012:	Jennifer Doudna demonstrates the potential of CRISPR technology.
2015:	CRISPR/Cas9 technology first used to treat a mouse with _____.
2016:	(Add an event of your choice.)

(Continue adding years and events up to the present.)

Bioethics

NSF Funding Committee: Who Should Get Funded?

You have been asked to serve on this year's National Science Foundation's Grant Award Committee. The purpose of the committee is to review this year's grant applications, and decide how to split and award $30 million in funding. There are five proposals to be considered. Each proposal requests over $10 million in funding.

There are a total of 10 committee members. The committee must justify the allocation of funds, based on social, environmental, health, and/or political needs, as well as show evidence that a funding recipient can show significant progress along the way to meeting the ultimate objective.

 Decide who receives how much funding. Who receives none? How do you decide?

The proposals before the committee include the following:

- monarch butterfly breeding and conservation
- a genetically engineered protein chemotherapy/antinausea medication
- drought-resistant tomato development
- genetic-test development for presymptom identification of Alzheimer's disease
- HIV vaccine testing in Africa

1. A pair of committee members uses the Internet to research one of the funding proposals. Each team member collects and reviews at least five articles describing some aspect of the research focus of the funding proposal they are assigned to review.

2. Members make a 5-minute presentation to the rest of the committee on "their" funding proposal, giving the strongest arguments in favor of funding. The presenter should distribute a one-page information sheet to each member summarizing the research-proposal background information. Committee members may ask questions following the presentation.

3. Other committee members review each oral presentation and information sheet.

4. The committee discusses how funds should be allocated. Discussions should include the value of the proposal to both society and the advancement of science. Partial funding of a proposal is allowed as long as sufficient funds are awarded to support the research sufficiently.

5. The committee prepares a pie chart of funding awards to show how the funds were distributed.

6. Each committee member writes a one-page statement describing the committee's decision-making process and the member's opinion on why the committee awarded the funds the way they did.

Glossary

3-D bioprinting production of three-dimensional biologic products using living cells and computer technology

A

abscisic acid (ABA) a plant hormone that regulates bud development and seed dormancy

absorbance the amount of light absorbed by a sample (the amount of light that does not pass through or reflect off a sample)

absorbance spectrum a graph of a sample's absorbance at different wavelengths

absorbance units (au) a unit of light absorbance determined by the decrease in the amount of light in a light beam

acid a solution that has a pH less than 7

activity assay an experiment designed to show a molecule is conducting the reaction that is expected

adenosine triphosphate (ATP) a nucleotide that serves as an energy storage molecule

adult stem cells the unspecialized cells found in specialized tissues and organs that keep their ability to divide and differentiate into cells with specific functions

aerobic respiration utilizing oxygen to release the energy from sugar molecules

affinity chromatography a type of column chromatography that separates proteins based on their shape or attraction to certain types of chromatography resin

agar a solid media used for growing bacteria, fungi, plant, or other cells

agarose a carbohydrate from seaweed that is widely used as a medium for horizontal gel electrophoresis

agriculture the practice of growing and harvesting animal or plant crops for food, fuel, fibers, or other useful products

Agrobacterium tumefaciens (A. tumefaciens) a bacterium that transfers the "Ti plasmid" to certain plant species, resulting in a plant disease called crown gall; used in plant genetic engineering

alcoholic fermentation a process by which certain yeast and bacteria cells convert glucose to carbon dioxide and ethanol under anaerobic (low or no oxygen) conditions

alleles the alternative forms of a gene

amino acids the subunits of proteins; each contains a central carbon atom attached to an amino group ($-NH_2$), a carboxyl group ($-COOH$), and a distinctive "R" group

amplification an increase in the number of copies of a particular segment of DNA, usually as a result of PCR

amu abbreviation for atomic mass unit; the mass of a single hydrogen atom

amylase an enzyme that functions to break down the polysaccharide amylose (plant starch) to the disaccharide maltose

amylopectin a plant starch with branched glucose chains

amylose a plant starch with unbranched glucose chains

anaerobic respiration releasing the energy from sugar molecules in the absence of oxygen

anatomy the structure and organization of living things

anion exchange a form of ion-exchange chromatography in which negatively charged ions (anions) are removed by a positively charged resin

antibiotics molecular agents derived from fungi and/or bacteria that impede the growth and survival of some other microorganisms

antibodies proteins developed by the immune system that recognize specific molecules (antigens)

antimicrobial a substance that kills or slows the growth of one or more microorganisms

antiparallel a reference to the observation that strands on DNA double helix have their nucleotides oriented in the opposite direction to one another

antiseptic an antimicrobial solution, such as alcohol or iodine, that is used to clean surfaces

applied science the practice of utilizing scientific knowledge for practical purposes, including the manufacture of a product

aqueous describing a solution in which the solvent is water

Arabidopsis thaliana (A. thaliana) an herbaceous plant, related to radishes, that serves as a model organism for many plant genetic engineering studies

asexual plant propagation a process by which identical offspring are produced by a single parent; methods include the cuttings of leaves and stems, and plant tissue culture, etc

assay a test

autoclave an instrument that creates high temperature and high pressure to sterilize equipment and media

autoradiogram the image on an x-ray film that results from exposure to radioactive material

auxin a plant hormone produced primarily in shoot tips that regulates cell elongation and leaf development

average a statistical measure of the central tendency that is calculated by dividing the sum of the values collected by the number of values being considered

B

Bacillus thuringiensis (B. thuringiensis, or Bt) the bacterium from which the Bt gene was originally isolated; the Bt gene codes for the production of a compound that is toxic to insects

bacteriophages the viruses that infect bacteria

balance an instrument that measures mass

base a solution that has a pH greater than 7

base pair the two nitrogenous bases that are connected by a hydrogen bond; for example, an adenosine bonded to a thymine or a guanine bonded to a cytosine

B cells specialized cells of the immune system that are used to generate and release antibodies

beta-galactosidase an enzyme that catalyzes the conversion of lactose into monosaccharides

beta-galactosidase gene a gene that produces beta-galactosidase, an enzyme that converts the carbohydrate X-gal into a blue product

biochemistry the study of the chemical reactions occurring in living things

biochip a special type of microarray that holds thousands of samples on a chip the size of a postage stamp

biodefense all the methods used to protect a population from exposure to biological agents

bioethics the study of decision-making as it applies to moral decisions that have to be made because of advances in biology, medicine, and technology

bioinformatics the use of computers and databases to analyze and relate large amounts of biological data

biomanufacturing the industry focusing on the production of proteins and other products created by biotechnology

biomarker a substance, often a protein or section of a nucleic acid, used to indicate, identify, or measure the presence or activity of another biological substance or process

biometric the measurement and statistical analysis of biological specimens or processes

bioremediation the use of bacteria or other organisms to restore environmental conditions

biotechnology the study and manipulation of living things or their component molecules, cells, tissues, or organs

bioterrorism the use of biological agents to attack humans, plants, or animals

BLAST an acronym for Basic Local Alignment Search Tool, a program that allows researchers to compare biological sequences

blot a membrane that has proteins, DNA, or RNA bound to it

breeding the process of propagating plants or animals through sexual reproduction of specific parents

broth a liquid medium used for growing cells

buffer a solution that acts to resist a change in pH when the hydrogen ion concentration is changed

buffer standards the solutions, each of a specific pH, used to calibrate a pH meter

C

callus a mass of undifferentiated plant cells developed during plant tissue culture

carbohydrates one of the four classes of macromolecules; organic compounds consisting of carbon, hydrogen, and oxygen, generally in a 1:2:1 ratio

cation exchange a form of ion-exchange chromatography in which positively charged ions (cations) are removed by a negatively charged resin

CDC abbreviation for Centers for Disease Control and Prevention; the national research center for developing and applying disease prevention and control, environmental health, and health promotion and education activities to improve public health

cDNA abbreviation for "copy DNA," cDNA is DNA that has been synthesized from mRNA

cell the smallest unit of life that makes up all living organisms

cellular respiration the process by which cells break down glucose to create other energy molecules

cellulase an enzyme that weakens plant cell walls by degrading cellulose

cellulose a structural polysaccharide that is found in plant cell walls

cell wall a specialized organelle surrounding the cells of plants, bacteria, and some fungi; gives support around the outer boundary of the cell

cGMP abbreviation for current good manufacturing practices

Chinese hamster ovary (CHO) cells an animal cell line commonly used in biotechnology studies

Chi Square a statistical measure of how well a dataset supports the hypothesis or the expected results of an experiment

chlorophyll the green-pigmented molecules found in plant cells; used for photosynthesis (production of chemical energy from light energy)

chloroplast the specialized organelle in plants responsible for photosynthesis (production of chemical energy from light energy)

chromatin nuclear DNA and proteins

chromatograph   the medium used in chromatography (ie, paper, resin, etc)

chromosomes   the long strands of DNA intertwined with protein molecules

cleavage   process of splitting the polypeptide into two or more strands

clinical testing   another name for clinical trials

clinical trials   a strict series of tests that evaluates the effectiveness and safety of a medical treatment in humans

clones   the cells or organisms that are genetically identical to one another

cloning   a method of asexual reproduction that produces identical organisms

codon   a set of three nucleotides on a strand of mRNA that codes for a particular amino acid in a protein chain

cofactors   an atom or molecule that an enzyme requires to function

column chromatography   a separation technique in which a sample is passed through a column packed with a resin (beads); the resin beads are selected based on their ability to separate molecules based on size, shape, charge, or chemical nature

combinatorial chemistry   the synthesis of larger organic molecules from smaller ones

competent/competency   the ability of cells to take up DNA

concentration   the amount of a substance as a proportion of another substance; usually how much mass in some amount of volume

concentration assay   a test designed to show the amount of molecule present in a solution

control   an experimental trial added to an experiment to ensure that the experiment was run properly; *see* positive control and negative control

conversion factor   a number (a fraction) where the numerator and denominator are equal to the same amount; commonly used to convert from one unit to another

Coomassie® Blue   a dye that stains proteins blue and allows them to be visualized

CRISPR   clustered sections of repeating DNA code, used in combination with the Cas9 restriction enzyme to modify, block or eliminate sections of target DNA

crossbreeding   the pollination between plants of differing phenotypes, or varieties

cross-linker   an instrument that uses UV light to irreversibly bind DNA or RNA to membrane or paper

cuttings   the pieces of stems, leaves, or roots for use in asexual plant propagation

cycle sequencing   a technique developed in the late 1990s that allowed researchers to run synthesis reactions over and over on samples, increasing the amount of sequencing product and the speed of getting results

cystic fibrosis (CF)   a genetic disorder that clogs the respiratory and digestive systems with mucus

cytokinin   a class of hormones that regulates plant cell division

cytology   cell biology

cytoplasm   a gel-like liquid of thousands of molecules suspended in water, outside the nucleus

cytoskeleton   a protein network in the cytoplasm that gives the cell structural support

D

data   information gathered from experimentation

dATP   the abbreviation for deoxyadenosine triphosphate, the cell's source of adenine (A) for DNA molecules

dCTP   the abbreviation for deoxycytidine triphosphate, the cell's source of cytosine (C) for DNA molecules

degrees of freedom   a value used in Chi Square analysis that represents the number of independent observations (eg, phenotypic groups) minus one

denaturation   the process in which proteins lose their conformation or three-dimensional shape

deoxyribose   the 5-carbon sugar found in DNA molecules

deuterium lamp   a special lamp used for UV spectrophotometers that produces light in the ultraviolet (UV light) part of the spectrum (100–350 nm)

dGTP   the abbreviation for deoxyguanosine triphosphate, the cell's source of guanine (G) for DNA molecules

diabetes   a disorder affecting the uptake of sugar by cells, due to inadequate insulin production or ineffective use of insulin

diafiltration   a filtering process by which some molecules in a sample move out of a solution as it passes through a membrane

dialysis   a process in which a sample is placed in a membrane with pores of a specified diameter, and molecules, smaller in size than the pore size, move into and out of the membrane until they are at the same concentration on each side of the membrane; used for buffer exchange and as a purification technique

dideoxynucleotides   the nucleotides that have an oxygen removed from carbon number 3; abbreviated ddNTPs

dideoxynucleotide sequencing   a sequencing method that uses ddNTP and dNTPs in a predictable way to produce synthesis fragments of varying length; also called the Sanger Method

differentiation   the development of a cell toward a more defined or specialized function

dihybrid cross   a breeding experiment in which the inheritance of two traits is studied at the same time

dilution   the process in which solvent is added to make a solution less concentrated

diploid   having two sets (2N) of homologous (matching) chromosomes

direct ELISA   an ELISA where a primary antibody linked with an enzyme recognizes an antigen and indicates its presence with a colorimetric

disaccharide   a polymer that consists of two sugar molecules

DNA   abbreviation for deoxyribonucleic acid, a double-stranded helical molecule that stores genetic information for the production of all of an organisms's proteins

DNA fingerprinting   an experimental technique that is commonly used to identify individuals by distinguishing their unique DNA code

DNA ligase   an enzyme that binds together disconnected strands of a DNA molecule

DNA polymerase   an enzyme that, during DNA replication, creates a new strand of DNA nucleotides complementary to a template strand

DNA replication   the process by which DNA molecules are duplicated

DNA sequencing   pertaining to all the techniques that lead to determining the order of nucleotides (A, G, C, T) in a DNA fragment

DNA synthesizer   an instrument that produces short sections of DNA, up to a few hundred base pairs in length

dNTP   the abbreviation for nucleotide triphosphates, which are the reactants (dATP, dCTP, dGTP, and dTTP) used as the sources of A, C, G, and Ts for a new strand of DNA

dominant   to how an allele for a gene is more strongly expressed than an alternate form (allele) of the gene

double-blind test   a type of experiment, often used in clinical trials, in which both the experimenters and test subjects do not know which treatment the subjects receive

drug   a chemical that alters the effects of proteins or other molecules associated with a disease-causing mechanism

drug discovery   the process of identifying molecules to treat a disease

DTT   the abbreviation for dithiothreitol, a reducing agent that helps to stabilize the DNA polymerase in DNA synthesis, PCR, and DNA sequencing reactions

dTTP   the abbreviation for deoxythymidine triphosphate, the cell's source of thymine (T) for DNA molecules

E

efficacy   the ability to yield a desired result or demonstrate that a product does what it claims to do

ELISA   short for enzyme-linked immunospecific assay, a technique that measures the amount of protein or antibody in a solution

elution   when a protein or nucleic acid is released from column chromatography resin

elution buffer   a buffer used to detach a protein or nucleic acid from chromatography resin; generally contains either a high salt concentration or has a high or low pH

embryo   a plant or animal in its initial stage of development

embryonic stem cells   the unspecialized cells, found in developing embryos, that have the ability to differentiate into a wide range of specialized cells

endonucleases   the enzymes that cut RNA or DNA at specific sites; restriction enzymes are endonucleases that cut DNA

enhancer   a section of DNA that increases the expression of a gene

environmental biotechnology   a field of biotechnology whose applications include monitoring and correcting the health of populations, communities, and ecosystems

enzyme   a protein that functions to speed up chemical reactions

EPA   abbreviation for the Environmental Protection Agency; the federal agency that enforces environmental laws including the use and production of microorganisms, herbicides, pesticides, and genetically modified microorganisms

epitope   the specific region on a molecule that an antibody binds to

equilibration buffer   a buffer used in column chromatography to set the charges on the beads or to wash the column

Escherichia coli (E. coli)   a rod-shaped bacterium native to the intestines of mammals; commonly used in genetics research and by biotechnology companies for the development of products

ethics   the study of moral standards and how they affect conduct

ethidium bromide   a DNA stain (indicator); glows orange when it is mixed with DNA and exposed to UV light; abbreviated EtBr

ethylene   a plant hormone that regulates fruit ripening and leaf development

eukaryotic/eukaryote   a cell that contains membrane-bound organelles

exon the region of a gene that directly codes for a protein; it is the region of the gene that is expressed

explants the sections or pieces of a plant that are grown in or on sterile plant tissue culture media

exponential growth the growth rate that bacteria maintain when they double in population size every cell cycle

extension the phase in PCR during which a complementary DNA strand is synthesized

extracellular outside the cell

F

fast-performance liquid chromatography (FPLC) a type of column chromatography in which pumps push buffer and sample through the resin beads at a high rate; used mainly for isolating proteins (purification)

FDA abbreviation for the Food and Drug Administration; the federal agency that regulates the use and production of food, feed, food additives, veterinary drugs, human drugs, and medical devices

feedstock the raw materials needed for some process, such as corn for livestock feed or biofuel production

fermentation a process by which, in an oxygen-deprived environment, a cell converts sugar into lactic acid or ethanol to create energy

fermenters the automated containers used for fermentation, or growth of micro-organism cultures designed to be easily monitored and controlled

flow cytometry a process by which cells are sorted by an instrument, a cytometer, that recognizes fluorescent antibodies attached to surface proteins on certain cells

fluorometer an instrument that measures the amount or type of light emitted

foodborne pathogens disease-causing microorganisms found in food or food products

forensics application of biology, chemistry, physics, mathematics, and sociology to solving legal problems including crime scene analysis, accident analysis, child support cases, and paternity

formulation the form of a product, as in tablet, powder, injectable liquid, etc

fraction a sample collected as buffer flows over the resin beads of a column

frit the membrane at the base of a chromatographic column that holds the resin in place

fructose a 6-carbon sugar found in high concentration in fruits; also called fruit sugar

G

gametes the sex cells (ie, sperm, eggs)

gel electrophoresis a process that uses electricity to separate charged molecules, such as DNA fragments, RNA, and proteins, on a gel slab

gel-filtration chromatography a type of column chromatography that separates proteins based on their size using size-exclusion beads; also called size-exclusion chromatography

gene a section of DNA on a chromosome that contains the genetic code of a protein

gene therapy the process of treating a disease or disorder by replacing a dysfunctional gene with a functional one

genetically modified organism (GMO) an organism produced by genetic engineering that contains DNA from another organism and produces new proteins encoded on the acquired DNA

genetics the study of genes and how they are inherited and expressed

genome one entire set of an organisms's genetic material (from a single cell)

genomic DNA (gDNA) the chromosomal DNA of a cell

genomics the study of all the genes and DNA code of an organism

genotype the genetic makeup of an organism; the particular form of a gene present for a specific trait

germination the initial growth phase of a plant; also called sprouting

gibberellin a plant hormone that regulates seed germination, leaf bud germination, stem elongation, and leaf development; also known as gibberellic acid

glucose a 6-carbon sugar that is produced during photosynthetic reactions; usual form of carbohydrate used by animals, including humans

glycogen an animal starch with branched glucose chains

glycoprotein a protein which has had sugar groups added to it

glycosylated descriptive of molecules to which sugar groups have been added

graduated cylinder a plastic or glass tube with marks (or graduations) equally spaced to show volumes; measurements are made at the bottom of the meniscus, the lowest part of the concave surface of the liquid in the cylinder

gram (g) the standard unit of mass, approximately equal to the mass of a small paper clip

gravity-flow columns column chromatography that uses gravity to force a sample through the resin bed

green fluorescent protein a protein found in certain species of jellyfish that glows green when excited by certain wavelengths of light (fluorescence)

GUS gene a gene that codes for an enzyme called beta-glucuronidase, an enzyme that breaks down the carbohydrate, X-Gluc, into a blue product

H

haploid having only one set (1N) of chromosomes

harvest the method of extracting protein from a cell culture, also called recovery

HeLa cells human epithelial cells

helicase an enzyme that functions to unwind and unzip complementary DNA strands during in vivo DNA replication

herbaceous plants the plants that do not add woody tissues; most herbaceous plants have a short generation time of less than one year from seed to flower

herbal remedies the products developed from plants that exhibit or are thought to exhibit some medicinal property

heterozygous having two different forms or alleles of a particular gene (ie, Hh or Rr)

high-performance liquid chromatography (HPLC) a type of column chromatography that uses metal columns that can withstand high pressures; used mainly for identification or quantification of a molecule

high through-put screening the process of examining hundreds or thousands of samples for a particular activity

histones the nuclear proteins that bind to chromosomal DNA and condense it into highly packed coils

homologous pairs the two "matching" chromosomes having the same genes in the same order

homozygous having two identical forms or alleles of a particular gene (ie, hh or RR)

homozygous dominant having two of the same alleles for the dominant version of the gene (ie, HH or RR)

homozygous recessive having two of the same alleles for the recessive version of the gene (ie, hh or rr)

hormone a molecule that acts to regulate cellular functions

horticulture the practice of growing plants for ornamental purposes

Human Genome Project a collaborative international effort to sequence and map all the DNA on the 23 human chromosomes; completed in 2000

hybridization the binding of complementary nucleic acids

hybridoma a hybrid cell used to generate monoclonal antibodies that results from the fusion of immortal tumor cells with specific antibody-producing white blood cells (B cells)

hydrogen bond a type of weak bond that involves the "sandwiching" of a hydrogen atom between two fluorine, nitrogen, or oxygen atoms; especially important in the structure of nucleic acids and proteins

hydrogen ion a hydrogen atom which has lost an electron (H^+)

hydrophilic having an attraction for water

hydrophobic repelled by water

hydrophobic-interaction chromatography a type of column chromatography that separates molecules based on their hydrophobicity (aversion to water molecules)

hydroponics the practice of growing plants in a soilless, water-based medium

hypothesis an educated guess to answer a scientific question; should be testable

I

immunity protection against any foreign disease-causing agent

inbreeding the breeding of closely related organisms

indirect ELISA an ELISA where a primary antibody binds to an antigen and then a secondary antibody linked with an enzyme recognizes the primary antibody—the antigen presence and concentration is indicated by the degree of a colorimetric

induced fit model a model used to describe how enzymes function, in which a substrate squeezes into an active site and induces the enzyme's activity

insulin a protein that facilitates the uptake of sugar into cells from blood

intracellular within the cell

intron the region on a gene that is transcribed into an mRNA molecule but not expressed in a protein

Investigational New Drug (IND) application a document to the FDA to allow testing of a new drug or product in humans

in vitro synthesis any synthesis that is done wholly or partly outside of a living organism (eg, PCR); literally; "in glass"

in vivo an experiment conducted in a living organism or a cell; literally, "in living"

ion-exchange chromatography a separation technique that separates molecules based on their overall charge at a given pH

J

journals scientific periodicals or magazines in which scientists publish their experimental work, findings, or conclusions

K

karyotyping the process of comparing an individual's karyotype with a normal, standard one to check for abnormalities

L

lactic acid fermentation a process by which certain bacteria cells convert glucose to lactic acid under anaerobic (low or no oxygen) conditions

lactose a disaccharide composed of glucose and galactose; also called milk sugar

lag phase the initial period of growth for cells in culture after inoculation

lambda$_{max}$ the wavelength that gives the highest absorbance value for a sample

large-scale production the manufacturing of large volumes of a product

library a collection of compounds, such as DNA molecules, RNA molecules, and proteins

lipids one of the four classes of macromolecules; includes fats, waxes, steroids, and oils

liter abbreviated "L"; a unit of measurement for volume, approximately equal to a quart

load the initial sample loaded onto a column before it is separated via chromatography

lock and key model a model used to describe how enzymes function, in which the enzyme and substrate make an exact molecular fit at the active site, triggering catalysis

lysis the breakdown or rupture of cells

lysosome a membrane-bound organelle that is responsible for the breakdown of cellular waste

lysozyme an enzyme that degrades bacterial cell walls by decomposing the carbohydrate peptidoglycan

M

macerated crushed, ground up, or shredded

macromolecule large molecule

macronutrients the minerals required by plants in high concentrations

marine biotechnology the study and manipulation of marine organisms, their component molecules, cells, tissues, or organs

mass the amount of matter (atoms and molecules) an object contains

mass spectrometer an instrument that is used to determine the molecular weight of a compound

maxiprep a DNA preparation yielding approximately 1 mg/mL or more of plasmid DNA

mean the average value for a set of numbers

media preparation the process of combining and sterilizing ingredients (salts, sugars, growth factors, pH indicators, etc) of a particular medium

medical biotechnology all the areas of research, development, and manufacturing of items that prevent or treat disease or alleviate the symptoms of disease

medicine something that prevents or treats disease or alleviates the symptoms of disease

medium a suspension or gel that provides the nutrients (salts, sugars, growth factors, etc) and the environment needed for cells to survive

meiosis a special kind of cell division that results in four gametes (N) from a single diploid cell

memory cell a specialized type of B cell that remains in the body for long periods of time with the ability to make antibodies to a specific antigen

meristematic tissue the tissue found in shoot buds, leaf buds, and root tips that is actively dividing and responsible for growth

meristems regions of a plant where cell division occurs, generally found in the growing tips of plants

messenger RNA (mRNA) a class of RNA molecules responsible for transferring genetic information from the chromosomes to ribosomes where proteins are made; often abbreviated mRNA

methylene blue a staining dye indicator that interacts with nucleic acid molecules and proteins, turning them to a very dark blue color

metrics conversion table a chart that shows how one unit of measure relates to another (for example, how many milliliters are in a liter, etc)

microarray a small glass slide or silicon chip with thousands of samples on it that can be used to assess the presence of a DNA sequence related to the expression of certain proteins

microarray scanner an instrument that assesses the amount of fluorescence in a well of a microarray

microbial agents synonym for microorganisms; living things too small to be seen without the aid of a microscope; includes bacteria, most algae, and many fungi

microbiology the study of microscopic organisms such as bacteria, protists, and single cell fungi

microbiome the microbes that are found on an individual

microbial genomics the study of the structure and function of the genes of microorganisms

microliter abbreviated "µL"; a unit measure for volume; equivalent to one-thousandth of a milliliter or about the size of the tiniest teardrop

micronutrients the minerals required by plants in low concentrations

micropipet an instrument used to measure very tiny volumes, usually less than a milliliter

microRNA small pieces of RNA that are known to interrupt posttranscriptional RNA function by binding to mRNA as soon as it is made

midiprep a DNA preparation yielding approximately 800 micrograms/mL of plasmid DNA

milliliter abbreviated "mL"; a unit measure for volume; one one-thousandth of a liter (0.001 L) or about equal to one-half teaspoon

miniprep a small DNA preparation yielding approximately 20 micrograms/500 microliters of plasmid DNA

mitochondria membrane-bound organelles that are responsible for generating cellular energy

mitosis cell division; in mitosis the chromosome number is maintained from one generation to the next

molarity a measure of concentration that represents the number of moles of a solute in a liter of solution (or some fraction of that unit)

mole the mass, in grams, of 6×10^{23} atoms or molecules of a given substance; one mole is equivalent to the molecular weight of a given substance, reported as grams

molecular biology the study of molecules that are found in cells

molecular weight the sum of all of the atomic weights of the atoms in a given molecule

monoclonal antibody a type of antibody that is directed against a single epitope

monohybrid cross a breeding experiment in which the inheritance of only one trait is studied

monomers the repeating units that make up polymers

monosaccharide the monomer unit that cells use to build polysaccharides; also known as a "single sugar" or "simple sugar"

moral a conviction or justifiable position, having to do with whether something is considered right or wrong

multicellular composed of more than one cell

multichannel pipet a type of pipet that holds 4–16 tips from one plunger; allows several samples to be measured at the same time

N

nanometer 10–9 meters; the standard unit used for measuring light

nanotechnology all technologies that operate on a nanometer scale

negative control a group of data lacking what is being tested so as to give expected negative results

neutral uncharged

next-generation sequencing (NGS) a type of DNA sequencing that allows for simultaneous sequencing and analysis of millions of DNA samples at virtually the same time

NIH abbreviation for National Institutes of Health; the federal agency that funds and conducts biomedical research

nitrocellulose a modified cellulose molecule used to make paper membrane for blots of nucleic acids and proteins

nitrogenous base an important component of nucleic acids (DNA and RNA), composed of one or two nitrogen-containing rings; forms the critical hydrogen bonds between opposing strands of a double helix

NMR an abbreviation for nuclear magnetic resonance, a technique that measures the spin on nuclei (protons) of isotopes in a magnetic field to study physical, chemical, and biological properties of proteins including their structure in aqueous (watery) solutions

nonpathogenic not known to cause disease

normality a measurement of concentration generally used for acids and bases that is expressed in gram equivalent weights of solute per liter of solution; represents the amount of ionization of an acid or base

Northern blot a process in which RNA fragments on a gel are transferred to a positively charged membrane (a blot) to be probed by labeled cDNA

NPT II (neomycin phosphotransferase) gene a gene that codes for the production of the enzyme, neomycin phosphotransferase, which gives a cell resistance to the antibiotic kanamycin

nucleic acids a class of macromolecules that directs the synthesis of all other cellular molecules; often referred to as "information-carrying molecules"

nucleotides the monomer subunits of nucleic acids

nucleus a membrane-bound organelle that encloses the cell's DNA

null hypothesis a hypothesis that assumes there is no difference between the observed and expected results

nutraceutical a food or natural product that claims to have health or medicinal value

O

oligonucleotides the segments of nucleic acid that are 50 nucleotides or less in length

observation information or data collected when a subject is watched

open-column chromatography a form of column chromatography that operates by gravity flow

operator a region on the operon that can either turn on or off expression of a set of genes depending on the binding of a regulatory molecule

operon a section of prokaryotic DNA consisting of one or more genes and their controlling elements

optimization the process of analyzing all the variables to find the ideal conditions for a reaction or process

optimum pH the pH at which an enzyme achieves maximum activity

optimum temperature the temperature at which an enzyme achieves maximum activity

organ tissues that act together to form a specific function in an organism (eg, stomach that breaks down food in humans)

organelles specialized microscopic factories, each with specific jobs in the cell

organic molecules that contain carbon and are only produced in living things

organic synthesis the synthesis of drug molecules in a laboratory from simpler, preexisting molecules

organism a living thing

P

P-10 a micropipet that is used to pipet volumes from 0.5 to 10 μL

P-100 a micropipet that is used to pipet volumes from 10 to 100 μL

P-1000 a micropipet that is used to pipet volumes from 100 to 1000 μL

P-20 a micropipet that is used to pipet volumes from 2 to 20 μL

P-200 a micropipet that is used to pipet volumes from 20 to 200 μL

PAGE short for polyacrylamide gel electrophoresis, a process in which proteins and small DNA molecules are separated by electrophoresis on vertical gels made of the synthetic polymer, polyacrylamide

pancreas an organ that secretes digestive fluids, as well as insulin

paper chromatography a form of chromatography that uses filter paper as the solid phase, and allows molecules to separate based on size or solubility in a solvent

parallel synthesis the making large numbers of batches of similar compounds at the same time

patent protection the process of securing a patent or the legal rights to an idea or technology

pathogenesis the origin and development of a disease

pectinase an enzyme that weakens plant cell walls by degrading pectin

pepsin an enzyme, found in gastric juice, that works to break down food (protein) in the stomach

peptides the short amino acid chains that are not folded into a functional protein

peptide synthesizer an instrument that is used to make peptides, up to a maximum of a few dozen amino acids in length

peptidyl transferase an enzyme found in the ribosome that builds polypeptide chains by connecting amino acids into long chains through peptide bonds

percentage a proportion of something out of 100 parts, expressed as a whole number

% transmittance the manner in which a spectrophotometer reports the amount of light that passes through a sample

pharmaceutical relating to drugs developed for medical use

pharmacodynamic (PD) assay an experiment designed to show the biochemical effect of a drug on the body

pharmacokinetic (PK) assay an experiment designed to show how a drug is metabolized (processed) in the body

pharmacology the study of drugs, their composition, actions, and effects

phenotype the characteristics observed from the expression of the genes, or genotype

pH meter an instrument that uses an electrode to detect the pH of a solution

phosphodiester bond a bond that is responsible for the polymerization of nucleic acids by linking sugars and phosphates of adjacent nucleotides

phospholipids a class of lipids that are primarily found in membranes of the cell

phosphorylation adding phosphate groups

photosynthesis a process by which plants or algae use light energy to make chemical energy

pH paper a piece of paper that has one or more chemical indicators on it and that changes colors depending on the amount of H^+ ions in a solution

physiology the processes and functions of living things

phytochrome a pigment that acts like a hormone to control flowering

pI   the pH at which a compound has an overall neutral charge and will not move in an electric field; also called the isoelectric point

pigments   the molecules that are colored due to the reflection of light of specific wavelengths

pipet   an instrument usually used to measure volumes between 0.1 mL and 50 mL

pK$_a$ the pH at which 50% of a buffering molecule in aqueous solution is ionized to a weak acid and its conjugate base; the point at which there are an equal number of neutral and ionized units.

placebo   an inactive substance that is often used as a negative control in clinical trials

plant-based pharmaceutical (PBP)   a human pharmaceutical produced in plants; also called plant-made pharmaceutical (PMP)

plant growth regulators   another name for plant hormones

plant hormones   the signaling molecules that, in certain concentrations, regulate growth and development, often by altering the expression of genes that trigger certain cell specialization and organ formation

plant tissue culture (PTC)   the process of growing small pieces of plants into small plantlets in or on sterile plant tissue culture media; plant tissue culture media have all of the required nutrients, chemicals, and hormones to promote cell division and specialization

plasma membrane   a specialized organelle of the cell that regulates the movement of materials into and out of the cell

plasmid   a tiny, circular piece of DNA, usually of bacterial origin; often used in recombinant DNA technologies

pluripotent   the stem cells that can specialize into any type of tissue

polar   the chemical characteristic of containing both a positive and negative charge on opposite sides of a molecule

pollination   the transfer of pollen (male gametes) to the pistil (the female part of the flower)

polyacrylamide   a polymer used as a gel material in vertical electrophoresis; used to separate smaller molecules, like proteins and very small pieces of DNA or RNA

polygenic   the traits that result from the expression of several different genes

polymer   a large molecule made up of many repeating subunits

polymerase chain reaction (PCR)   a technique that involves copying short pieces of DNA and then making millions of copies in a short time

polypeptide   a strand of amino acids connected to each other through peptide bonds

polysaccharide   a long polymer composed of many glucose (or variations of glucose) monomers

positive control   a group of data that will give predictable positive results

positive displacement micropipet   an instrument that is generally used to pipet small volumes of viscous (thick) fluids

potency assay   an experiment designed to determine the relative strength of a drug for the purpose of determining proper dosage

preparation   the process of extracting plasmids from cells

pressure-pumped columns   a column chromatography apparatus that uses pressure to force a sample through the resin bed

primary structure   the order and type of amino acids found in a polypeptide chain

primers   the small strands of DNA used as starting points for DNA synthesis or replication

primer annealing   the phase in PCR during which a primer binds to a template strand

primer design   a process by which a primer sequence is proposed and constructed

probes   the labeled DNA or RNA sequence (oligonucleotide) that is used for gene identification

prokaryotic/prokaryote   a cell that lacks membrane-bound organelles

promoter   the region at the beginning of a gene where RNA polymerase binds; the promoter "promotes" the recruitment of RNA polymerase and other factors required for transcription

proprietary rights   confidential knowledge or technology

proteases   proteins whose function is to break down other proteins

protein arrays   a fusion of technologies, where protein samples bound on the glass slide (protein chip) are assessed using antibodies or other recognition material

protein (x-ray) crystallography   a technique that uses x-ray wave diffraction patterns to visualize the positions of atoms in protein molecule to reveal its three-dimensional structure

proteins   one of the four classes of macromolecules; folded, functional polypeptides that conduct various functions within and around a cell (eg, adding structural support, catalyzing reactions, transporting molecules)

protein synthesis   the generation of new proteins from amino acid subunits; in the cell, it includes transcription and translation

proteome all of an organism's protein and protein-related material

proteomics the study of how, when, and where proteins are used in cells

protist an organism belonging to the Kingdom Protista, which includes protozoans, slime molds, and certain algae

protoplast a cell in which the cell wall has been degraded and is surrounded by only a membrane

Punnett Square a chart that shows the possible gene combinations that could result when crossing specific genotypes

pure science scientific research whose main purpose is to enrich the scientific knowledge base

purification the process of eliminating impurities from a sample; in protein purification, it is the separation of other proteins from the desired protein

purine a nitrogenous base composed of a double carbon ring; a component of DNA nucleotides

PVDF for polyvinylidene fluoride, a high molecular weight fluorocarbon with dielectric properties that make it suitable for Western blots because it is attractive to charged proteins

pyrimidine a nitrogenous base composed of a single carbon ring; a component of DNA nucleotides

Q

Quality Assurance (QA) a department that deals with quality objectives and how they are met and reported internally and externally

Quality Control (QC) a department in a company that monitors the quality of a product and all the instruments and reagents associated with it

quaternary structure the structure of a protein resulting from the association of two or more polypeptide chains

R

R plasmid a type of plasmid that contains a gene for antibiotic resistance

radicle an embryonic root-tip

reaction buffer a buffer in PCR that is used to maintain the pH of the synthesis reaction

reagent a chemical used in an experiment

real-time PCR use of fluorescent probe technology to measure PCR product as it is being produced, also called quantitative RT-PCR or qRT-PCR, for short

recessive how an allele for a gene is less strongly expressed than an alternate form (allele) of the gene; a gene must be homozygous recessive (ie, hh, rr) for an organism to demonstrate a recessive phenotype

recombinant DNA (rDNA) DNA created by combining DNA from two or more sources

recombinant DNA (rDNA) technology cutting and recombining DNA molecules

recovery the retrieval of a protein from broth, cells, or cell fragments

recovery period the period following transformation where cells are given nutrients and allowed to repair their membranes and express the "selection gene(s)"

regenerative medicine the field of medicine that focuses on replacement or restoration of damaged tissues or organs

research and development (R&D) the early stages in product development that include discovery of the structure and function of a potential product and initial small-scale production

respiration the breaking down of food molecules with the result of generating energy for the cell

restriction enzyme an enzyme that cuts DNA at a specific nucleotide sequence

restriction enzyme mapping determining the order of restriction sites of enzymes in relation to each other

restriction fragment length polymorphisms "RFLP" for short, restriction fragments of differing lengths due to differences in the genetic code for the same gene between two individuals

restriction fragments the pieces of DNA that result from a restriction enzyme digestion

reverse-transcription PCR use of reverse transcriptase to produce cDNA from mRNA for use in PCR, abbreviated "RT-PCR"

R group the chemical side-group on an amino acid; in nature, there are 20 different R groups that are found on amino acids

ribonucleic acid (RNA) the macromolecule that functions in the conversion of genetic instructions (DNA) into proteins

ribose the 5-carbon sugar found in RNA molecules

ribosome the organelle in a cell where proteins are made

RNAi an abbreviation for RNA interference, a type of double-stranded RNA that is chopped into small pieces when engulfed by the cell, and binds to and interferes with the cell's native RNA or DNA, blocking protein production

RNA polymerase an enzyme that catalyzes the synthesis of complementary RNA strands from a given DNA strand

RNA primase an enzyme that adds primers to template strands during in vivo DNA replication

RNase H an enzyme that functions to degrade RNA primers, during in vivo replication, that are bound to DNA template strands

runners the long, vine-like stems that grow along the soil surface

S

salting out   a technique for crystallizing proteins that involves precipitating a sample of pure protein using a stringent salt gradient of sodium chloride, ammonium sulfate, or some other salt

scale-up   the process of increasing the size or volume of the production of a particular product

screening   the assessment of hundreds, thousands, or even millions of molecules or samples

secondary structure   the structure of a protein (alpha helix and beta sheets) that results from hydrogen bonding

seed   the initial colony or a culture that is used as starter for a larger volume of culture

selection   the process of screening potential clones for the expression of a particular gene; for example, the expression of a resistance gene (such as resistance to ampicillin) in transformed cells

selective breeding   the parent selection and controlled breeding for a particular characteristic

semiconservative replication   a form of replication in which each original strand of DNA acts as a template, or model, for building a new side; in this model one of each new copy goes into a newly forming daughter cell during cell division

sexual reproduction   a process by which two parent cells give rise to offspring of the next generation by each contributing a set of chromosomes carried in gametes

shotgun cloning   a method of cloning commonly used during the sequencing of the human genome that involves digesting DNA into 500 bp pieces, generating libraries from those fragments, and eventually sequencing the libraries

silencer   a section of DNA that decreases the expression of a gene

silver stain   a stain used for visualizing proteins

siRNA   an abbreviation for short-interfering RNA, a type of single-stranded RNA oligo (fragment) that is created by scientists to target a gene for silencing

site-specific mutagenesis   a technique that involves changing the genetic code of an organism (mutagenesis) in certain sections (site-specific)

solute   the substance in a solution that is being dissolved

solution   a mixture of two or more substances where one (solute) completely dissolves in the other (solvent)

solvent   the substance that dissolves the solute

sonication   the use of high frequency sound waves to break open cells

Southern blotting   a process in which DNA fragments on a gel are transferred to a positively charged membrane (a blot) to be probed by labeled RNA or cDNA fragments

spectrophotometer   an instrument that measures the amount of light that passes through (is transmitted through) a sample

spinner flasks   a type of flask commonly used for scale-up in which there is a spinner apparatus (propeller blade) inside to keep cells suspended and aerated

stability assay   an experiment designed to determine the conditions that affect the shelf life of a drug

standard curve   a graph or curve generated from a series of samples of known concentration

standard deviation   a statistical measure of how much a dataset varies

starch   a polysaccharide that is composed of many glucose molecules

stationary phase   the latter period of a culture in which growth is limited due to the depletion of nutrients

steroids   a group of lipids whose functions include acting as hormones (testosterone and estrogen), venoms, and pigments

sticky ends   the restriction fragments in which one end of the double-stranded DNA is longer than the other; necessary for the formation of recombinant DNA

stock solution   a concentrated form of a reagent that is often diluted to form a "working solution"

substrate   the molecule that an enzyme acts on

sucrose   a disaccharide composed of glucose and fructose; also called table sugar

sugar   a simple carbohydrate molecule composed of hydrogen, carbon, and oxygen

supernatant   the (usually) clear liquid left behind after a precipitate has been spun down to the bottom of a vessel by centrifugation

synthetic biology   the application of biotechnologies to design and construct new biological systems, such as macromolecules, metabolic pathways, cells, tissues, or organisms

T

TAE buffer   a buffer that is often used for running DNA samples on agarose gels in horizontal gel boxes; contains TRIS, EDTA, and acetic acid

***Taq* polymerase**   a DNA synthesis enzyme that can withstand the high temperatures used in PCR

taxonomic relationships   how species are related to one another in terms of evolution

TE buffer   a buffer used for storing DNA; contains TRIS and EDTA

template   the strand of DNA from which a new complementary strand is synthesized

tertiary structure the structure of a protein that results from several interactions, the presence of charged or uncharged "R" groups, and hydrogen bonding

therapeutic an agent that is used to treat diseases or disorders

thermal cycler an instrument used to complete PCR reactions; automatically cycles through different temperatures

thin-layer chromatography a separation technique that involves the separation of small molecules as they move through a silica gel

Ti plasmid a plasmid found in Agrobacterium tumefaciens that is used to carry genes into plants, with the goal that the recipient plants will gain new phenotypes

tissue a group of cells that function together (eg, muscle tissue or nervous tissue)

tissue culture the process of growing plant or animal cells in or on a sterile medium containing all of the nutrients necessary for growth

topoisomerase an enzyme that acts to relieve tension in DNA strands as they unwind during in vivo DNA replication

toxicology assay an experiment designed to find what quantities of a drug are toxic to cells, tissues, and model organisms

t-PA short for tissue plasminogen activator; one of the first genetically engineered products to be sold; a naturally occurring enzyme that breaks down blood clots and clears blocked blood vessels

transcription the process of deciphering a DNA nucleotide code and converting it into an RNA nucleotide code; the RNA carries the genetic message to a ribosome for translation into a protein code

transcription factors molecules that regulate gene expression by binding onto enhancer or silencer regions of DNA and causing an increase or decrease in transcription of RNA

transduction the use of viruses to transform or genetically engineer cells

transfection the genetic engineering, or transformation, of mammalian cell lines

transformation the uptake and expression of foreign DNA by a cell

transformation efficiency a measure of how well cells are transformed to a new phenotype

transformed the cells that have taken up foreign DNA and started expressing the genes on the newly acquired DNA

transgenic the transfer of genes from one species to another, as in genetic engineering

transgenic plants the plants that contain genes from another species; also called genetically engineered or genetically modified plants

translation the process of reading a mRNA nucleotide code and converting it into a sequence of amino acids

transmittance the passing of light through a sample

triglycerides a group of lipids that includes animal fats and plant oils

TRIS a complex organic molecule used to maintain the pH of a solution

tRNA a type of ribonucleic acid (RNA) that shuttles amino acids into the ribosome for protein synthesis

tungsten lamp a lamp, used for VIS spectrophotometers, that produces white light (350–700 nm)

U

ultraviolet light (UV light) the high-energy light with wavelengths of about 100 to 350 nm; used to detect colorless molecules

unicellular composed of one cell

unit of measurement the form in which something is measured (g, mg, μg, L, mL, μL, km, cm, etc)

USDA abbreviation for United States Department of Agriculture; the federal agency that regulates the use and production of plants, plant products, plant pests, veterinary supplies and medications, and genetically modified plants and animals

V

vaccine an agent that stimulates the immune system to provide protection against a particular antigen or disease

variable anything that can vary in an experiment; the independent variable is tested in an experiment to see its effect on dependent variables

vector a piece of DNA that carries one or more genes into a cell; usually circular as in plasmid vectors

Vero cells African green monkey kidney epithelial cells

virus a particle containing a protein coat and genetic material (either DNA or RNA) that is not living and requires a host to replicate

visible light spectrum the range of wavelengths of light that humans can see, from approximately 350 to 700 nm; also called white light

VNTRs the abbreviation for variable number of tandem repeats, sections of repeated DNA sequences found at specific locations on certain chromosomes; the number of repeats in a particular VNTR can vary from person to person; used for DNA fingerprinting

volume   a measurement of the amount of space something occupies

W

weight   the force exerted on something by gravity; at sea level, it is considered equal to the mass of an object

Western blot   a process in which a gel with protein is transferred to a positively charged membrane (a blot) to be probed with antibodies

woody plants   the plants that add woody tissue; most woody plants have a long generation time of more than one year from seed to flower; most woody plants grow to be tall, thick, and hard

X

x-ray crystallography   a technique used to determine the three-dimensional structure of a protein

x-ray diffraction pattern   a pattern of light intensities that develops when an x-ray beam is passed through a mounted crystalline structure

Z

zygote   a cell that results from the fusion of a sperm nucleus and an egg nucleus

Index

Note: Figures are denoted with an *f*; tables are denoted with a *t*.

defined, 10
examples of agricultural products, 241–242, *347t*
examples of industrial products, 241
examples of pharmaceuticals, 241
first, 62
food crop yields and quality, 357–360, *357f, 358f*
insulin, 62–63
livestock, 345
plant-based pharmaceuticals, 350, *351f*
plants, *11f,* 61, 314–315, *315f,* 345
risk assessment and, 359
strawberries, *42f*
t-PA, 11, *12f*
using *A. thaliana,* 355, *356f*
using *A. tumefaciens,* 354, *355f*
genetic code, 60–61
genetic counselors, 386
genetic disorders
cystic fibrosis, 15
diagnosing, 385
examples, 385
inbreeding and, 344–345, *344f*
genetic engineering
antibodies produced through, 155–156
isolating genetic information, 126–129, 243–244, *243f*
overview, 240–245, *241f, 248f*
plants, *312f*
protocol, 240
using polymerase chain reaction, 244–245, *244f*
genetic engineers, 61
genetic nurse practitioners, 368
genetic panels, 368, 389
genetics, 24. *See also* DNA; RNA
genetic testing, 385–388, *386f,* 389
Genome Analyzer®, 427
genome annotation, 439
genome editing, 444
genomes
complete sequencing, 439
databases, 459
defined, 23
human genome, 121, *121f,* 439
Human Genome Project, 13, 23, 413, 423, 424, 426, 438, 439, *439f*
sizes, 114, *115t*
studying, 438–439
genomic DNA (gDNA)
defined, 347
isolation, *243f, 247t,* 347–348, *348f*
genomic (chromosomal) DNA, 222

genomics, 438–439
Genomyx, Inc., 14
Geno Technology Inc./G-Biosciences, 180
genotypes
overview, 320, *320f*
Punnett Square analysis and, 322–323, *322f, 322t, 323t*
Genzyme Corporation, 162
Geobacter sp, 457
German shepherds, 345
germination, 316–317, *317f*
gibberellin, 331
Gilead Sciences, Inc., 14, *371f*
Gladstone Institute at University of California, 7, *8f*
glassware (for containers), *83f*
GlaxoSmithKline, 241
glucocerebrosidase, 162, *162f*
glucose
described, 48, 54–55
production, 184
structural formula, *54f*
transport, 48
glue (from mussels), 460, *460f*
glycogen, 54
glycoprotein 120 and HIV, 153
glycoproteins, 153
glycosated, defined, 153
Goates, Blair, *359f*
Golden Rice, 357, *357f*
Golgi apparatus, *47f*
Good Manufacturing Practices (GMP) guidelines (FDA), 199, 293
goodness of fit, 326–327, *326t, 327t*
government research labs, 6–8
gowning up, *256f*
Goyal, Garima, *453f*
graduated cylinders
defined, 74
described, 76
reading, *77f*
gram (g), 81
grass plant clones, *329f*
gravity-flow chromatography, 286, 287, *287f*
gravity-flow columns, 278
green fluorescent protein (GFP), 253, *253f*
green molecules, 217
"green" plastics, 457
gRNA (guide RNA), 444, *444f*
GTC Biotherapeutics, Inc., 383
guide RNA (gRNA), 444, *444f*
GUS (beta-D-glucuronidase) genes, *328f,* 354

H

haploid, 319
harvest, 276
H-bonds, *115f,* 164
He, Molly, *171f*
heart disease, 373, *373f*
Heath, Colin M., 180, *180f*
HeLa cells, defined, 50
helicase, defined, 406
helicase molecules, 405–406
help desk technicians, *26f*
hemoglobin, 218, 219–220, *240f*
Heng, Meng, 274, *274f*
HEPA (high-efficiency particulate air) filters, use of, 256
herbaceous plants, 317
herbal remedies, 195, 196, *196f, 197t*
herbicide analysis, *229f*
herbs, *318f*
heroin testing, *229f*
herpes, *124f*
HER2 protein, 156, 389
heterozygous, 320
heterozygous crosses, 327
Hibiscus, 309f
high-fructose corn syrup, 357
High-Performance Integrated Virtual Environment (HIVE), *441f*
high-performance liquid chromatography (HPLC), *157f,* 212, *212f,* 227, *227f, 228f,* 289, *289f*
high-performance liquid chromatography (HPLC) instruments, 2, *2f*
high through-put screening, 135
*Hind*III, 134, 246, 250
histones, 123, *123f*
HIV/AIDS
antibody recruiting molecules, 384
blocker molecule treatment, 371, *371f*
ELISA and, 191, 257–258, *257f*
glycoprotein 120 and, 153
protease inhibitors, 375–376
HIVE Genomics Research Computer, *441f*
homologous pairs, 405, 406
homozygous, 320
homozygous dominant, 320
homozygous recessive, 320
Hood, Leroy, *426f*
horizontal gel boxes, 99, *132f*
hormones
described, 45
molecule size, 53
plant, 317, 331–332

ribonucleic acid (RNA). *See* RNA
ribose, 55, *55f*
ribosomes
    defined, 46
    mRNA code translated, 49
rice, 357, *357f*
Richardson, Sarah, *456f*
RIDASCREEN® FAST Peanut assay, 192
RIKEN Institute, 391
Rituxin, 64
RNA
    compared to DNA, 61–62
    contamination in DNA sample, 223, 224
    defined, 60
    in gel boxes, 130–131
    genomics and, 442–443, *443f*
    Northern blots, *410f*, 442
    screening, 412
RNA interference (RNAi), 442, *443f*
RNA polymerase, 119
RNA primase molecules, 406
RNase H, 406
Robinson, Brian, 40, *40f*
robots, advantages of use, 342
Romer Laboratories, 359
Roosevelt, Teddy, *318f*
rooting compounds, *311f*
roots, 312, 332, *332f, 346f*
Rosenbaum & Silvert, PC, *296f*
roses, 328, *328f*
Rotoli, Atticus, *294f*
rotovaps, *377f*
Roundup®, *42f*, 349–350
Roundup Ready® seeds, 149
Roundup Ready® soybeans, 341–342, 349–350
R plasmid, 117
rubber, 311
runners, 329
running buffers, *96f, 167f*

S

salicin, 181, 374, *374f*
salicylic acid, *374f*
saline buffers, 96
Salix alba, 374
Salmon, Ellie, 402, *402f*
salmon DNA in solution, *114f*
salting out, 446
San Francisco Chronicle, 8
Sanger Method, 423, 426, 427, 438
SARS (severe acute respiratory syndrome), *125f*
Saxe, Charles (Karl) L., III, *46f*

SB-Normal Stool Formula, 196
scale-up process, 253, *253f,* 255–258, *255f, 256f*
SciDATA, 464
Science (journal), 456
science technicians, 24
Scientific American Worldview, 464
scientific methodology, 7, 19–22, *20f, 21f, 22f*
Scios, Inc., 17
screening
    described, 371
    mass, 448, *448f*
SD (standard deviation), 325, *325f, 325t*
SDS (sodium dodecyl sulfate), 167, 276
sea squirts, 460
Seattle Genetics, Inc., 384
secondary structure of proteins, 151
seeds, 149, 259, 260, *260f,* 261, 306
selection, 251, 252, *261f*
selective breeding, 310
SelectUSA, 200
semiconservative replication, 116, *116f*
semi-dry transfers, 194
separation technology in biomanufacturing, *280f. See also* column chromatography; purification
    FPLC, *284f,* 287–288, *288f,* 289
Sephacryl resin beads, 289
Sequencing by Oligo Ligation Detection (SOLiD®), 427
severe acute respiratory syndrome (SARS), *125f*
sewage water treatment, 457, *457f*
sexual reproduction
    defined, 308
    genetic mixing, 344–345
    overview, 319
Shaman Pharmaceuticals, Inc., 195, 196
Shaman's Apprentice, The (Plotkin), 43
shoot (plant) development, 332, *332f*
short-interfering RNA (siRNA), 442, *443f*
shotgun cloning, 438
sickle cell disease, *61f, 154f,* 169, *169f,* 170, *240f,* 385
silencers, 122
silver stain, 168, *168f*
simple carbohydrates, 53. *See also* monosaccharides
simple sugars, 53. *See also* monosaccharides
single sugars, 53. *See also* monosaccharides

siRNA (siRNA), 442, *443f*
site-specific mutagenesis, 128, *128f*
6-carbon sugars, 54, *54f*
size-exclusion chromatography, 283–284, *284f*
skin, artificial, 456
sodium dodecyl sulfate (SDS), 167, 276
soil analysis, *229f*
Solazyme, Ic, *228f*
Solexa, 427
solutes
    calculating amount used in solution, 84
    defined, 81
    storing, *82f*
solutions
    buffered, 92–93
    defined, 81
    of differing % mass/volume concentration, 85–86
    of differing molar concentration, *87f,* 88–89
    dilutions of concentrated, 97–100, *98f*
    of DNA, *114f*
    of given mass/volume concentration, 84–85, *84f*
    making, 81–83, 88
    measuring pH, 90–91
    molar, 88–89
    stock, 97
solvents, 82, *90f, 90t*
somatostatin, 10
sonication, 276–277
Southern blots, 244, 410–411, *410f, 419f*
soybeans, 349–350
spacer DNA, 122
Species Survival Plan, *458f,* 461
specs. *See* spectrophotometers
spectrofluorometers, 226, *226f*
spectrometers
    function of, 224
    mass, 150, *150f, 331f*
    parts of, 215–216
spectrophotometers ("specs"). *See also* ultraviolet light (UV) spectrophotometers; visible light spectrum (VIS) spectrophotometers
    data analysis, 324, *324f*
    described, 214, *224f*
    functioning of, *215f, 216f*
    mass, 225–226, *225f, 226f*
    parts, 215–216
    using, to detect molecules, 212, *212f,* 214–218, *214f, 215f, 216f, 217f*